Neuropsychiatric Disorders and Epigenetics

Translational Epigenetics Series

Trygve O. Tollefsbol, Series Editor

Transgenerational Epigenetics
Edited by Trygve O. Tollefsbol, 2014

Personalized Epigenetics
Edited by Trygve O. Tollefsbol, 2015

Epigenetic Technological Applications
Edited by Y. George Zheng, 2015

Epigenetic Cancer Therapy
Edited by Steven G. Gray, 2015

DNA Methylation and Complex Human Disease
By Michel Neidhart, 2015

Epigenomics in Health and Disease
Edited by Mario F. Fraga and Agustin F. Fernández, 2015

Epigenetic Gene Expression and Regulation
Edited by Suming Huang, Michael Litt, and C. Ann Blakey, 2015

Epigenetic Biomarkers and Diagnostics
Edited by Jose Luis García-Giménez, 2015

Drug Discovery in Cancer Epigenetics
Edited by Gerda Egger and Paola Barbara Arimondo, 2015

Medical Epigenetics
Edited by Trygve O. Tollefsbol, 2016

Chromatin Signaling
Edited by Olivier Binda and Martin Fernandez-Zapico, 2016

Neuropsychiatric Disorders and Epigenetics

Edited by

Dag H. Yasui
University of California, Davis, CA, United States

Jacob Peedicayil
Department of Pharmacology and Clinical Pharmacology, Christian Medical College, Vellore Tamil Nadu, India

Dennis R. Grayson
Department of Psychiatry, College of Medicine University of Illinois, Chicago, IL, United States

AMSTERDAM • BOSTON • HEIDELBERG • LONDON
NEW YORK • OXFORD • PARIS • SAN DIEGO
SAN FRANCISCO • SINGAPORE • SYDNEY • TOKYO

Academic Press is an imprint of Elsevier

Academic Press is an imprint of Elsevier
125 London Wall, London EC2Y 5AS, United Kingdom
525 B Street, Suite 1800, San Diego, CA 92101-4495, United States
50 Hampshire Street, 5th Floor, Cambridge, MA 02139, United States
The Boulevard, Langford Lane, Kidlington, Oxford OX5 1GB, United Kingdom

Library of Congress Cataloging-in-Publication Data
A catalog record for this book is available from the Library of Congress

British Library Cataloguing-in-Publication Data
A catalogue record for this book is available from the British Library

ISBN: 978-0-12-800226-1

For information on all Academic Press publications
visit our website at https://www.elsevier.com/

Publisher: Mica Haley
Acquisitions Editor: Peter B. Linsley
Editorial Project Manager: Lisa Eppich
Production Project Manager: Chris Wortley
Designer: Mark Rogers

Typeset by Thomson Digital

Contents

List of Contributors

J.-M. Aubry
Department of Mental Health and Psychiatry, Service of Psychiatric Specialties, University Hospitals of Geneva; Department of Psychiatry, University of Geneva, Geneva, Switzerland

V. Buchholz
Department of Psychiatry, Social Psychiatry and Psychotherapy, Molecular Neuroscience Laboratory, Hannover Medical School (MHH), Hannover, Lower Saxony, Germany

G. Castelo-Branco
Laboratory of Molecular Neurobiology, Department of Medical Biochemistry and Biophysics, Karolinska Institutet, Stockholm, Sweden

K.-O. Cho
Department of Pharmacology, Catholic Neuroscience Institute, College of Medicine, The Catholic University of Korea, Seoul, Korea

F. Coppedè
Department of Translational Research and New Technologies in Medicine and Surgery, Section of Medical Genetics, University of Pisa, Pisa, Italy

A. Dayer
Department of Mental Health and Psychiatry, Service of Psychiatric Specialties, University Hospitals of Geneva; Department of Psychiatry, University of Geneva, Geneva, Switzerland

P.P. De Deyn
Department of Neurology and Alzheimer Research Center, University Medical Center Groningen, University of Groningen, Groningen, The Netherlands; Laboratory of Neurochemistry and Behavior, Institute Born-Bunge, University of Antwerp, Wilrijk, Belgium

A.D. Dekker
Department of Neurology and Alzheimer Research Center, University Medical Center Groningen, University of Groningen, Groningen, The Netherlands; Laboratory of Neurochemistry and Behavior, Institute Born-Bunge, University of Antwerp, Wilrijk, Belgium

J. Feng
Department of Biological Science, Neuroscience PhD Program, Florida State University, Tallahassee, FL, United States

H. Frieling
Department of Psychiatry, Social Psychiatry and Psychotherapy, Molecular Neuroscience Laboratory, Hannover Medical School (MHH), Hannover, Lower Saxony, Germany

S.H. Gan
Human Genome Centre, School of Medical Sciences, Universiti Sains Malaysia, Kubang Kerian, Kelantan, Malaysia

H. Gong
The Key Laboratory of Geriatrics, Beijing Institute of Geriatrics, Beijing Hospital, National Center of Gerontology, Beijing, China

D.R. Grayson
Department of Psychiatry, College of Medicine, University of Illinois, Chicago, IL, United States

M.M. Hefti
Department of Pathology, Icahn School of Medicine at Mount Sinai, New York, NY, United States

A. Hoffmann
Max Planck Institute of Psychiatry, Translational Research, Munich, Germany

T.-L. Huang
Department of Medical Research, Genomic and Proteomic Core Laboratory, Kaohsiung Chang Gung Memorial Hospital, Kaohsiung, Taiwan

M. Jagodic
Department of Clinical Neuroscience, Center for Molecular Medicine, Karolinska Institutet, Stockholm, Sweden

J. Jiménez-Conde
Department of Neurology, Hospital del Mar; Neurovascular Research Group, IMIM (Institut Hospital del Mar d'Investigacions Mèdiques); Department of Medicine, Autonomous University of Barcelona, Barcelona, Spain

T. Kubota
Department of Epigenetic Medicine, University of Yamanashi, Kofu-city, Yamanashi Prefecture, Japan

L. Kular
Department of Clinical Neuroscience, Center for Molecular Medicine, Karolinska Institutet, Stockholm, Sweden

C.-C. Lin
Department of Psychiatry, Kaohsiung Chang Gung Memorial Hospital and Chang Gung University College of Medicine, Kaohsiung, Taiwan

S.M. Nam
Department of Pharmacology, Catholic Neuroscience Institute, College of Medicine, The Catholic University of Korea, Seoul, Korea

G. Neri
Institute of Genomic Medicine, Catholic University, Rome, Italy

J. Peedicayil
Department of Pharmacology and Clinical Pharmacology, Christian Medical College, Vellore, Tamil Nadu, India

N. Perroud
Department of Mental Health and Psychiatry, Service of Psychiatric Specialties, University Hospitals of Geneva; Department of Psychiatry, University of Geneva, Geneva, Switzerland

J. Roquer
Department of Neurology, Hospital del Mar; Neurovascular Research Group, IMIM (Institut Hospital del Mar d'Investigacions Mèdiques); Department of Medicine, Autonomous University of Barcelona, Barcelona, Spain

M.G. Rots
Department of Pathology and Medical Biology, University Medical Center Groningen, University of Groningen, Groningen, The Netherlands

S. Sagarkar
Department of Biotechnology, Savitribai Phule Pune University, Ganeshkhind, Pune, Maharashtra, India

A. Sakharkar
Department of Biotechnology, Savitribai Phule Pune University, Ganeshkhind, Pune, Maharashtra, India

M.M. Shaik
Human Genome Centre, School of Medical Sciences, Universiti Sains Malaysia, Kubang Kerian, Kelantan, Malaysia

B.M. Shewchuk
Department of Biochemistry and Molecular Biology, Brody School of Medicine at East Carolina University, Greenville, NC, United States

C. Soriano-Tárraga
Department of Neurology, Hospital del Mar; Neurovascular Research Group, IMIM (Institut Hospital del Mar d'Investigacions Mèdiques); Department of Medicine, Autonomous University of Barcelona, Barcelona, Spain

D. Spengler
Max Planck Institute of Psychiatry, Translational Research, Munich, Germany

E. Tabolacci
Institute of Genomic Medicine, Catholic University, Rome, Italy

N. Tsankova
Department of Pathology; Department of Neuroscience; Friedman Brain Institute, Icahn School of Medicine at Mount Sinai, New York, NY, United States

S. Weibel
Department of Mental Health and Psychiatry, Service of Psychiatric Specialties, University Hospitals of Geneva, Geneva, Switzerland

X. Xu
Max Planck Institute for Biology of Ageing, Cologne, Germany; Department of Anesthesiology, Yale University School of Medicine, New Haven, CT, United States

D.H. Yasui
University of California, Davis, CA, United States

Preface

This book discusses the role of epigenetics in a wide range of neuropsychiatric disorders. All the disorders discussed in the book have the commonality that they all involve the brain. The disorders differ in that in some the clinical presentation is mainly psychiatric, while in others the clinical presentation is mainly neurological. However, all the discussed disorders have an epigenetic component in their pathogenesis, and a psychiatric component in their clinical presentation. Some of the chapters are specifically devoted to the translational aspects of the disorders, like the use of epigenetic drugs for the clinical management of the disorders. As is clear from many of the chapters of the book, we have a long way to go in the study of the role of epigenetics in neuropsychiatric disorders. We hope that this book will stimulate and catalyze the study of the basic and translational aspects of the epigenetics of neuropsychiatric disorders, and hasten the translation of such research into the clinical management of patients with these disorders.

The book is a member of the *Translational Epigenetics Series* being published by Elsevier, and overseen by the Series Editor Dr. Trygve O. Tollefsbol, University of Alabama at Birmingham, USA. Each of us thank Dr. Tollefsbol for giving us the opportunity to edit this book. We would also like to thank Ms. Lisa Eppich of Elsevier for her expert and most efficient assistance during the preparation of the book. We also thank each of the contributors of the chapters of the book for their fine contributions.

D.H. Yasui
Davis, CA, United States

J. Peedicayil
Vellore, India

D.R. Grayson
Chicago, IL, United States

SECTION

NEUROPSYCHIATRIC DISORDERS AND EPIGENETICS: GENERAL ASPECTS

I

CHAPTER

INTRODUCTION TO NEUROPSYCHIATRIC DISORDERS AND EPIGENETICS

1

J. Peedicayil*, D.R. Grayson, D.H. Yasui†**

**Department of Pharmacology and Clinical Pharmacology, Christian Medical College, Vellore, Tamil Nadu, India; **Department of Psychiatry, College of Medicine, University of Illinois, Chicago, IL, United States; †University of California, Davis, CA, United States*

CHAPTER OUTLINE

1.1 INTRODUCTION

Epigenetics, which literally means above or in addition to genetics, involves molecular mechanisms like DNA methylation, histone modification, and RNA-mediated regulation of gene expression. Epigenetics is attracting ever-increasing interest in virtually every branch of medicine [1–3], and indeed, has been referred to as the epicenter of modern medicine because it can help explain the relationships between an individual's genetic background, the environment, aging, and disease [4]. Epigenetics is predicted to greatly help in the prevention and treatment of disease by providing novel biomarkers, drug targets, and drugs [4,5]. It has also been suggested that epigenetics will help usher in the era of personalized medicine, the ability to treat patients on an individualized basis [6,7]. Like in other branches of medicine, the importance of epigenetics in the pathogenesis and management of psychiatric disorders is being increasingly appreciated, and epigenetics has become an active area of research in psychiatry since the first decade of this century [8]. This book discusses the role of epigenetics in the pathogenesis and management of neuropsychiatric disorders.

Neuropsychiatric Disorders and Epigenetics. http://dx.doi.org/10.1016/B978-0-12-800226-1.00001-0

1.2 NEUROPSYCHIATRY AND NEUROPSYCHIATRIC DISORDERS

Neuropsychiatry bridges conventional boundaries interposed between the mind and the brain [9]. It is an integrative and collaborative field that avoids speciality derived, reductionist categorizations that recognize and address only circumscribed features of a specific brain-based illness [9]. A major focus of neuropsychiatry is the assessment and treatment of the cognitive, behavioral, and affective (mood) symptoms of patients with disorders of the brain. Neuropsychiatry also encompasses the "gray area" between the subdisciplines of neurology and psychiatry [10]. Thus, there are very few neurological disorders without a psychiatric component, and very few psychiatric disorders without a neurological component. Neuropsychiatry also includes psychiatric disorders due to structural or pathological conditions outside the brain [11]. For example, several endocrine and metabolic disorders can result in psychiatric illness [12]. Neuropsychiatric disorders are sometimes referred to as neurobehavioral disorders [13,14].

1.3 HISTORICAL OUTLINE OF NEUROPSYCHIATRIC DISORDERS

The Greek physician Hippocrates (c. 460–375 BC) believed that the brain is the basis of the mind and that all psychopathology arose in the brain [15–17]. There were also opposing theories like the suggestion of Aristotle (384–322 BC) that the brain's main function was to cool the blood [15]. Moreover, through the middle ages aspects of our mental lives were often linked to organs other than the brain, such as the heart [15]. At the advent of modern medicine during the 17th century neurological and psychiatric disorders were not regarded as separate disciplines, but as one discipline, nervous diseases [18]. The concept of nervous diseases was consolidated and strengthened during the 18th century. It was during the 19th century that nervous diseases started diverging into two disciplines. Disorders with a structural abnormality in the brain tended to be called neurological disorders. Disorders of the "psychic apparatus" of the brain tended to be called psychiatric disorders. Psychiatry was thought to be neurology without clinical signs [15]. This separation of psychiatry from neurology extended into the 20th century. For much of the 20th century, neurology and psychiatry were separated by an artificial wall created by the divergence of the philosophical approaches to them, and also the research and therapeutic approaches toward them [19]. The major reasons for such a separation of these two disciplines was that on the one hand the triumphs of neuropathology and the clinico-pathologic method led to neurology as a structurally based discipline; and on the other hand, to the growth of psychodynamic psychiatry with a conceptual framework of a psychic (functional) apparatus either separated from, or obscurely linked to, the brain [18]. The advent of psychoanalysis in the United States of America during the 1930s sharpened and intensified the separation between psychiatry and neurology. Psychiatry and neurology became two separate disciplines. Neurological disorders were thought to be due to pathological lesions in the brain. Psychiatric disorders were thought to be due to abnormal functioning of the brain due to genetic and psychosocial factors. Psychiatric patients were isolated in mental hospitals and psychiatry was divorced from the rest of medicine.

Modern research highlights the difficulties caused by a predominantly organic, structurally based neurology and a predominantly "psychic" functionally based psychiatry, because in neurology there is growing evidence that structurally based disorders have a functional component. Thus, several neurological disorders like epilepsy, brain tumors, and cerebrovascular accidents are accompanied by behavioral, psychological, and cognitive problems [20–22]. In psychiatry the discovery of structural lesions by modern imaging techniques are proving very difficult to integrate with functional disorders [18]. For example, psychiatric disorders like schizophrenia [23] and bipolar disorder [24] are now known to be accompanied by structural lesions in the brain. Indeed, some workers regard schizophrenia [25] and bipolar disorder [26] as neuropsychiatric disorders. The neurological disorder *N*-methyl-D-aspartate (NMDA) encephalitis is clinically indistinguishable from acute schizophrenia, a psychiatric disorder [27,28]. The etiological role of psychosocial factors in disorders of the brain was thought to demarcate psychiatric disorders from neurological disorders, with psychosocial factors acting only in the former. However, now it is thought that psychosocial factors act via epigenetic mechanisms in the pathogenesis of psychiatric disorders, that is, by biological mechanisms [29–31]. In the light of the preceding data, there is a lot that unites psychiatry and neurology, and little that divides them. They are like two sides of the same coin. At the same time, there are differences between neurology and psychiatry. For example, psychotherapy is of major importance in the treatment of psychiatric disorders and only of minor importance in the treatment of neurological disorders. Hence, instead of merging psychiatry and neurology into one discipline, they can be considered to be two subdisciplines of neuropsychiatry [32,33], which in turn can be considered to come under neurosciences [19,34].

1.4 NEUROPSYCHIATRIC DISORDERS AND EPIGENETICS

As is well known, the brain is the most complex and complicated organ in the human body [35]. The human brain comprises about 100 billion neurons interconnected in neural circuits [36]. The human genome, which contains 20,000–24,000 genes [37], is unlikely to have the encoded information to specify this level of complexity. Hence, another layer of information, namely epigenetic regulation of gene expression, is made use of to encode the development and functioning of the human brain. Environmental factors are thought to affect gene expression by altering epigenetic mechanisms of gene expression, and RNA appears to be a major substrate for environment-epigenome interactions [38].

Epigenetics is known to play a major role in the development and functioning of the human brain. This is not surprising because epigenetic mechanisms are thought to have played a major role in the evolution of the human brain [39]. When an individual's brain develops, the three main types of cells in the brain, neurons and glial cells (astrocytes and oligodendrocytes), are formed from neural stem cells, which are cells that possess the ability to self-renew and differentiate into the three main types of cells in the brain [40]. Epigenetic mechanisms of gene expression play an important role in this process [40]. Epigenetic mechanisms of gene expression also probably play a major role in the whole-scale transformation over time of the midgestational human brain into the adult human brain. Indeed, since so many genes are translated differently, the fetal and adult human brains can almost be considered to be two different organs [41]. Even after birth, the human brain is an ever-changing organ encoding memories and directing behavior, and epigenetics is thought to play a major role in the changes involved [42]. Epigenetics is known to play a role in various functions and states of the brain like synaptic transmission [43], memory and neuronal plasticity [44], cognition and behavior [45], neuroendocrinology [46], neuroimmunology [47], and neuroinflammation [48]. Since epigenetic mechanisms of gene expression play a major role in the normal development and functioning of the brain, it is not surprising that abnormalities in epigenetic mechanisms of gene expression contribute to disorders of the brain. Indeed, there is growing evidence that epigenetics underlies a wide range of brain disorders.

This book discusses the role of epigenetics in a wide range of important neuropsychiatric disorders. Some of the disorders like pervasive developmental disorders, intellectual disability, attention-deficit hyperactivity disorder, and Down syndrome typically first manifest in infants and children. Others like cognitive disorders like Alzheimer's disease and Parkinson's disease, and cerebrovascular accidents typically first manifest in elderly individuals. Yet others like multiple sclerosis, migraine, drug addiction, and eating disorders typically first manifest between these age groups. An earlier book, *Epigenetics in Psychiatry* [49] discussed the role of epigenetics in disorders like schizophrenia, bipolar disorder, and major depressive disorder, which are not covered in detail in the current book. The earlier book also covered disorders like cognitive disorders, pervasive developmental disorders, intellectual disability, and drug addiction. Since the current book belongs to a *Translational Epigenetics Series*, translational aspects of the epigenetics of such disorders are discussed in the current book.

1.5 EPIGENETICS AND NEUROPSYCHIATRIC DISORDERS: TRANSLATIONAL ASPECTS

The ultimate objective of medical research is to improve the clinical management of patients, in terms of diagnosis, prevention, and treatment. This involves translational research, that is, translating what is found on the laboratory bench to the patient's bedside. Research on the epigenetics of neuropsychiatric disorders may well prove to be valuable and meaningful in this regard. Recently, it was suggested [50] that research on the epigenetics of neuropsychiatric disorders could prove to be useful in the diagnosis of these disorders by providing suitable biomarkers; in the prevention of these disorders since abnormal epigenetic patterns of gene expression are potentially reversible; and in the treatment of these disorders because drug therapy, regulation of diet, psychotherapy, electroconvulsive therapy, and physical exercise are known to correct abnormal epigenetic mechanisms of gene expression. All of these interventional modalities are used in the clinical management of patients with neuropsychiatric disorders.

ABBREVIATION

NMDA *N*-methyl-D-aspartate

GLOSSARY

Functional Without a physiological or anatomical cause

Psychic Of or relating to the mind

Psychoanalysis A systematic structure of theories concerning the relationship between conscious and unconscious psychological processes propounded by Sigmund Freud. A technical procedure for investigating unconscious mental processes and for treating neurotic disorders

Psychodynamic psychiatry An approach to the diagnosis and treatment of psychiatric patients based on the principles of psychoanalysis

ACKNOWLEDGMENT

The authors acknowledge the help of Dr. Abraham Verghese, Retired Professor of Clinical Psychiatry, Christian Medical College, Vellore, India, during the writing of this chapter.

REFERENCES

[1] Handel AE, Ebers GC, Ramagopalan SV. Epigenetics: molecular mechanisms and implications for disease. Trends Mol Med 2009;16:7–16.
[2] Portela A, Esteller M. Epigenetic modifications and human disease. Nat Biotechnol 2010;28:1057–68.
[3] Tollefsbol T, editor. Epigenetics in human disease. Waltham, MA: Elsevier; 2012.
[4] Feinberg AP. Epigenetics at the epicenter of modern medicine. JAMA 2008;299:1345–50.
[5] Feinberg AP, Fallin MD. Epigenetics at the crossroads of gene and the environment. JAMA 2015;314: 1129–30.
[6] Weber WW. The promise of epigenetics in personalized medicine. Mol Interv 2010;10:363–70.
[7] Peedicayil J. The epigenome in personalized medicine. Clin Pharmacol Ther 2013;93:149–50.
[8] Nestler EJ, Peńa CJ, Kundakovic M, Mitchell A, Akbarian S. Epigenetic basis of mental illness. Neuroscientist 2015 [Epub ahead of print].
[9] Yudofsky SC, Hales RE. Neuropsychiatry and the future of psychiatry and neurology. Am J Psychiatry 2002;159:1261–4.
[10] Carson AJ. Introducing a neuropsychiatry special issue: but what does that mean? J Neurol Neurosurg Psychiatry 2014;85:121–2.
[11] David AS. Basic concepts in neuropsychiatry. In: David AS, Fleminger S, Kopelman MD, Lovestone S, Mellers JDC, editors. Lishman's organic psychiatry. A textbook of neuropsychiatry. 4th ed. Chichester: Wiley-Blackwell; 2009. p. 3–25.
[12] Harrison NA, Kopelman MD. Endocrine diseases and metabolic disorders. In: David AS, Fleminger S, Kopelman MD, Lovestone S, Mellers JDC, editors. Lishman's organic psychiatry. A textbook of neuropsychiatry. 4th ed. Chichester: Wley-Blackwell; 2009. p. 617–88.
[13] Zasler ND, Martelli MF, Jacobs HE. Neurobehavioral disorders. Handb Clin Neurol 2013;110:377–88.
[14] Strub R. Neurobehavioral disorders (Organic brain syndromes). In: Weisberg LA, Garcia CA, Strub RL, editors. Essentials of clinical neurology. Maryland Heights, Missouri: Mosby; 1996. p. 408–28.
[15] Zeman A. Neurology is psychiatry—and vice versa. Pract Neurol 2014;14:136–44.
[16] Haas LF. Hippocrates 460–377 BC. J Neurol Neurosurg Psychiatry 1991;54:5.
[17] Costello EJ. Grand challenges in child and neurodevelopmental psychiatry. Front Psychiatry 2010;1:1–2.
[18] Reynolds EH. Structure and function in neurology and psychiatry. Br J Psychiatry 1990;157:481–90.
[19] Martin JB. The integration of neurology, psychiatry, and neuroscience in the 21st century. Am J Psychiatry 2002;159:695–704.
[20] Munger Clary HM. Anxiety and epilepsy: what neurologists and epileptologists should know. Curr Neurol Neurosci Rep 2014;14:445.
[21] Madhusoodanan S, Ting MB, Farah T, Ugur U. Psychiatric aspects of brain tumors: a review. World J Psychiatry 2015;5:273–85.
[22] Hackett ML, Pickles K. Frequency of depression after stroke: an updated systemic review and meta-analysis of observational studies. Int J Stroke 2014;9:1017–25.
[23] Bakhshi K, Chance SA. The neuropathology of schizophrenia: a selective review of past studies and emerging themes in brain structure and cytoarchitecture. Neuroscience 2015;303:82–102.
[24] Brooks JO, Vizueta N. Diagnostic and clinical implications of functional neuroimaging in bipolar disorder. J Psychiatr Res 2014;57:12–25.

[25] Dauverman MR, Whalley HC, Schmidt A, et al. Computational neuropsychiatry—schizophrenia as a cognitive brain network disorder. Front Psychiatry 2014;5:30.
[26] Ginsberg SD, Hemby SE, Smiley JF. Expression profiling in neuropsychiatric disorders: emphasis on glutamate receptors in bipolar disorder. Pharmacol Biochem Behav 2012;100:705–11.
[27] Barry H, Hardiman O, Healy DG, Keogan M, Moroney J, Molnar PP, et al. Anti-NMDA receptor encephalitis: an important differential diagnosis in psychosis. Br J Psychiatry 2011;199:508–9.
[28] Reilly TJ. The neurology-psychiatry divide: a thought experiment. BJPsych Bull 2015;39:134–5.
[29] Peedicayil J. Psychosocial factors may act via epigenetic mechanisms in the pathogenesis of mental disorders. Med Hypotheses 2008;70:700–1.
[30] Peedicayil J. Epigenetics as a link between psychosocial factors and mental disorders. Indian J Psychiatry 2015;57:218.
[31] Pidsley R, Mill J. Epigenetic studies of psychosis: current findings, methodological approaches, and implications for postmortem research. Biol Psychiatry 2011;69:146–56.
[32] Fitzgerald M. Do psychiatry and neurology need a close partnership or a merger? BJPsych Bull 2015;39:105–7.
[33] Sobanski JA, Dudek D. Psychiatry and neurology: from dualism to integration. Neurol Neurochir Pol 2013;47:577–83.
[34] Cowan WM, Kandel ER. Prospects for neurology and psychiatry. JAMA 2001;285:594–600.
[35] The Britannica Guide to the Brain. London: Constable & Robinson Ltd; 2008.
[36] Kandel ER, Hudspeth AJ. The brain and behavior. In: Kandel ER, Schwartz JH, Jessell TM, Siegelbaum SA, Hudspeth AJ, editors. Principles of neural science. 5th ed. New York: McGraw-Hill; 2013. p. 3–20.
[37] International Human Genome Sequencing Consortium. Finishing the euchromatic sequence of the human genome. Nature 2004;431:931–45.
[38] Mattick JS. The central role of RNA in human development and cognition. FEBS Lett 2011;585:1600–16.
[39] Peedicayil J. The importance of cultural inheritance. Med Hypotheses 2001;56:158–9.
[40] Sanosaka T, Namihira M, Nakashima K. Epigenetic mechanisms in sequential differentiation of neural stem cells. Epigenetics 2009;4:89–92.
[41] Nair P, Insel T. QnAs with Tom Insel. Proc Natl Acad Sci USA 2014;111:7884–5.
[42] Rubin TG, Gray JD, McEwen BS. Experience and the ever-changing brain: what the transcriptome can reveal. Bioessays 2014;36:1072–81.
[43] Nelson ED, Monteggia LM. Epigenetics in the mature mammalian brain: effects on behavior and synaptic transmission. Neurobiol Learn Mem 2011;96:53–60.
[44] Woldemichael BT, Bohacek J, Gapp K, Mansuy IM. Epigenetics of memory and plasticity. Prog Mol Biol Transl Sci 2014;122:305–40.
[45] Gräff J, Mansuy IM. Epigenetic codes in cognition and behaviour. Behav Brain Res 2008;192:70–87.
[46] Crews D. Epigenetics and its implications for behavioral neuroendocrinology. Front Neuroendocrinol 2008;29:344–57.
[47] Mathews HL, Janusek LW. Epigenetics and psychoneuroimmunology: mechanisms and models. Brain Behav Immunol 2011;25:25–39.
[48] Garden GA. Epigenetics and the modulation of neuroinflammation. Neurotherapeutics 2013;10:782–8.
[49] Peedicayil J, Grayson DR, Avramopoulos D, editors. Epigenetics in psychiatry. Waltham, MA: Elsevier; 2014.
[50] Peedicayil J. Epigenetics and the war on mental illness. Mol Psychiatry 2014;19:960.

CHAPTER

2

ENVIRONMENTAL FACTORS AND EPIGENETICS OF NEUROPSYCHIATRIC DISORDERS

A. Hoffmann, D. Spengler

Max Planck Institute of Psychiatry, Translational Research, Munich, Germany

CHAPTER OUTLINE

Neuropsychiatric Disorders and Epigenetics. http://dx.doi.org/10.1016/B978-0-12-800226-1.00002-2

2.1 INTRODUCTION: GENE X ENVIRONMENT INTERACTIONS REVISITED

The advent of molecular genetics in the beginning of last century and the identification of causative mutations involved in rare monogenic diseases has contributed to the popular belief that genetic variation may explain most, or even all, of our physical and mental conditions including our susceptibility to various diseases [1]. At the same time the complexity of the human genome has been simplistically reduced to an immutable master plan drafted with the inception of our lives that will inescapably determine our biologic destiny [1]. Still, an unexpected result from the completion of the Human Genome Project was the finding that the number of genes in the human genome is similar to that of *Caenorhabditis elegans* and that more than 95% of the human genome exhibits no protein-coding information [2,3]. These discoveries revived current interest in molecular epigenetic mechanisms in order to explain how the complex variation of human cells and tissues can be explained by the orchestrated expression of a limited number of genes [4]. Consequently, the epigenome is thought of as the entity of epigenetic marks that instructs each cell to correctly interpret the invariable DNA-based genetic blueprint. Importantly, epigenetic marks are responsive to the environment and can lead to long-term, but reversible, adjustments in gene-regulatory networks underlying the function of different physiological systems and whole organs. Among these, the human brain plays an eminent role and seems predestined to mediate between an ever changing environment and the preset genetic blueprint. In fact, the emerging field of neuroepigenomics suggests that our heritage undergoes steady transitions whereby genes influence our lives but also whereby our lives influence the actions of our genes [5].

In this chapter we will discuss the role of neuronal plasticity in light of new findings from epigenetics and briefly examine by which molecular mechanisms early life experiences are engraved into the epigenome. Following on, we will explore how psychological, social, and nutritional factors can affect physiological systems during sensitive windows of neurodevelopment to program long-term regulatory adjustments that can confer an increased risk for future psychiatric disease. We propose that early life adversity (ELA) and malnutrition share in common an activation of the stress system and that epigenetic mechanisms may sustain such activated states far beyond the initial trigger.

2.2 THE EPIGENETIC DIMENSION OF NEUROPLASTICITY

For a long time psychobiologists have been intrigued by the "nature versus nurture" dispute when considering individual differences in life course trajectories and the effects of early lifetime events on the prevalence, severity, and course of neuropsychiatric disorders [6]. While the genetic blueprint continuously instructs early development, the organism remains highly receptive to environmental cues. Hence, different phenotypes can arise from a single genotype by the processes of developmental plasticity [7]. As a case in point, the female honeybee can develop into either a worker or a queen dependent on how the early larva is fed. Such developmental plasticity enables an organism to cope with ever changing environments by modulation of its phenotypic development and life course. While commonly viewed as an adaptive reaction, environmentally induced adjustments are also thought to increase the risk for later disease when these changes do not match future needs (i.e., mismatch hypothesis) [7]. On the other hand, environmentally induced adjustments in regulatory set points and thresholds can also disrupt homeostatic control mechanisms under normative conditions and thus confer an increased risk for disease (i.e., risk hypothesis). While these hypotheses remain subject to ongoing debate, neuroplasticity is most prevalent when rapid changes in cell numbers, structure, and connectivity cooccur (pre- and postnatally, but also during puberty) and declines with increasing age [8].

The human brain develops through a highly organized process that initiates before birth and continues into adulthood [9]. At the end of the embryonic period rudimentary structures of the central nervous system are laid down, which evolve continuously through the end of gestation. This embryonic stage comprises the formation and fast growth of cortical and subcortical structures including the establishment of fiber pathways. Gross morphological changes of the prenatal neural system predetermine its later architectural organization and concur with the generation of neurons from human embryonic day 41 onward until major parts of neurogenesis are completed by midgestation. Rudimentary neural networks arise once neurons have migrated to their final destination and undergo further elaborations in response to various internal and environmental cues. Importantly, brain development subsists postnatally for an extended period of time, and structural changes frequently underlie changes in functional reorganization of processes controlling emotion, behavior, and cognition among other higher functions. Throughout the developing brain levels of connectivity surpass by far those seen in adults and are gradually pruned back through competitive processes that interconnect to experiences of an organism. These regulatory processes are essential for cellular plasticity and provide the capacity for adaptation—for better or for worse—with possibly lifelong consequences. The human brain preserves a lifelong capacity for structural neuroplasticity (i.e., turnover of synaptic connections, expansion, and contraction of dendritic trees and a limited amount of neurogenesis) although there are periods in life where the brain is more sensitive to the effects of experience. Such "sensitive periods" concern the formation of specific circuits that underlie specific abilities like vision and hearing, language acquisition, and higher cognitive functions.

A growing body of literature suggests that epigenetically mediated modulation of the expression of specific genes and pathways also plays an important role in developmental plasticity. While all cells of a multicellular organism are genetically identical, they are structurally and functional distinct due to the differential expression of their genes. Molecular epigenetic mechanisms controlling gene expression orchestrate various developmental processes including cell differentiation, X-chromosome inactivation in females, and genomic imprinting [10]. By extension, the same epigenetic mechanisms are thought to operate in neuroplasticity [5]. Since developmental plasticity involves dynamic changes in gene expression, epigenetic mechanisms coordinating gene regulation may be transient or persist across the life course. The honeybee referred earlier can exemplify how epigenetic processes contribute to developmental plasticity. Although the duration of access to royal jelly determines whether larvae develop into queens or worker, experimental silencing of the gene encoding DNA methyltransferase 3, an enzyme which adds new methylation marks to CpG dinucleotides (see section 2.3.1), causes most larvae to become queens [11]. In support of this finding, growing evidence indicates that epigenetic mechanisms can leave a lasting footprint at regulatory gene regions in response to the environment and thus contribute to the programming of risk phenotypes [12,13].

Taken together, neurodevelopment comprises critical windows of sensitivity during which environmental experiences can impact on neuronal substrates with lifelong consequences for the manifestation of distinct phenotypes. Although neuroplasticity has been traditionally ascribed to cellular and structural changes in response to environmental cues, molecular epigenetic mechanisms are increasingly recognized to modify the DNA and chromatin of neuronal cells in a stimulus-dependent manner and to trigger enduring changes in the expression of genes important for psychiatric phenotypes. In the next section, we will briefly consider principal aspects of molecular epigenetic mechanisms before turning our attention to the question of what kind of physiological systems can mediate the effects of early life experiences via epigenetic marking.

2.3 WHAT ARE MOLECULAR EPIGENETIC MECHANISMS?

The term epigenetics comes along with different historical flavors that need to be clarified before we can discuss newer insights into the underlying molecular mechanisms. In 1940, the biologist Conrad Waddington originally coined this neologism to conceptualize how genes might interact with the environment during development to give rise to different cellular and organismal phenotypes from the same genotype [14]. While Waddington did not postulate any mechanisms that could mediate these effects, Arthur Riggs thought of epigenetics as inheritable changes in gene expression regulating cell fate decisions and final phenotypes independently of changes in DNA sequence [15]. Since then, major advancements in the understanding of DNA methylation, posttranslational modifications of core histone, chromatin structure, nucleosome positioning, and more recently noncoding RNAs (ncRNA) among others have substantially advanced our perception of molecular epigenetics. Above all, DNA methylation and posttranslational histone modifications represent core concepts of molecular epigenetics that have been well studied over the last decades due to their role in the initiation and maintenance of long-lasting epigenetic states. In the context of this chapter, we will provide a snapshot of these two mechanisms to prepare the following sections and refer readers interested in a more comprehensive survey to a recent book [4].

2.3.1 EPIGENETIC TAGGING OF DNA

DNA methylation describes the addition of a methyl group to the fifth carbon of the nucleotide base cytosine (5mC) that takes place in somatic cells primarily in the context of palindromic CpG dinucleotides. The existence of DNA methylation was hypothesized as early as 1925 although its functional role in gene expression was only recognized by studies on X-chromosome inactivation and cancer [16,17]. In addition to canonical DNA methylation, non-CpG methylation (CpH; H = A, T, or C) with a major role in plants has been also detected in embryonic stem cells and the developing nervous system of mice and humans. However, further studies are still looked for to define the poorly understood function of non-CpG methylation and its possible relevance for gene regulation [18].

A family of highly conserved DNA methyltransferases (DNMT) consisting of DNMT1, DNMT3A, and DNMT3B catalyze DNA methylation at symmetric CpG residues in mammals [18]. DNMT1, jointly with its obligate partner UHRF1 (ubiquitin-like plant homeodomain and RING finger domain), recognizes preferentially hemimethylated DNA as a substrate during DNA replication and thus preserves the parental strand's methylation pattern. In contrast, de novo methylation of DNA is mainly catalyzed by the transfer of methyl groups through DNMT3A and DNMT3B in conjunction with DNMT3L. The latter encodes an enzymatically inactive homologue that due to its scaffolding function stimulates the catalytic activities of either DNMT3 (Fig. 2.1). Excepting for punctuated stretches of DNA with a high CpG content, so-called CpG islands (CGI), CpG sites are mostly depleted in the mammalian genome. CGIs localize to approximately 70% of all annotated promoters and commonly remain methylation free. On the contrary, CpGs outside of CGIs typically undergo DNA

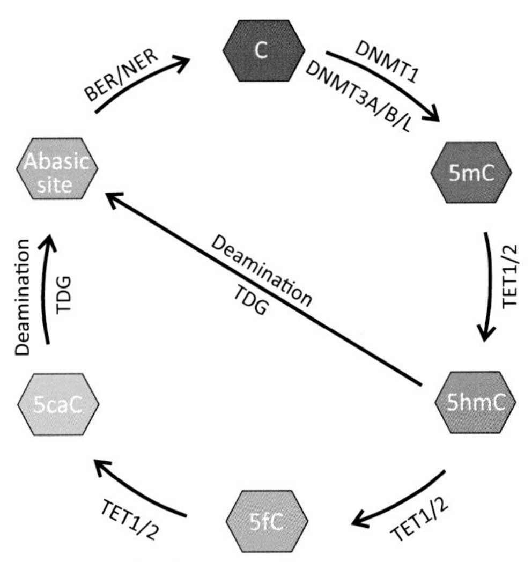

FIGURE 2.1 The Life Cycle of DNA Methylation in Mammalian Cells

Methylation of the nucleotide cytosine (C) occurs primarily in the context of CpG dinucleotides and is carried out either by the maintenance DNA methyltransferase DNMT1 during replication or by the de novo DNA methyltransferases DNMT3A or DNMT3B. Demethylation of 5-methylcytosine *(5mC)* occurs through iterative oxidation by ten-eleven translocation proteins *(TET1/2)* and generates first 5-hydroxymethylcytosine *(5-hmC)*, then 5-formylcytosine *(5-fc)*, and finally 5-carboxycytosine *(5-caC)*. The latter, oxidation product is then efficiently excised by the DNA repair machinery (base excision repair, *BER*; nucleotide excision repair *NER*). Alternatively, 5-hmC can be directly deaminated to thymine, which is then recognized by G/T mismatch-specific thymine DNA glycosylase *(TDG)*. Finally, the mismatched bases are replaced by the DNA repair machinery.

methylation [18]. Although supposed to stay methylation free, CGIs overlaying promoter regions frequently acquire DNA methylation and gene silencing during cancer development. Prompted by these findings, DNA methylation has been commonly defined as an all-purpose repressive mechanism in gene regulation.

This one-sided definition needs however to be revised in the light of recent findings from genome-wide DNA methylation studies showing that the effects of DNA methylation actually depend on genomic location, sequence composition, and transcriptional status and can contribute to either gene activation or repression [18,19]. In addition, the classical concept that DNA methylation is an irreversible, covalent bond laid down during development needs to be reconsidered in face of the transformative discovery of active demethylation by the family of ten-eleven translocation (Tet) proteins [20]. This group of enzymes catalyzes via iterative cycles of oxidation the conversion of 5mC into 5-hydroxymethylcytosine (5hmC), 5-formylcytosine (5fC), and ultimately 5-carboxycytosine (5aC) (Fig. 2.1). These oxidized derivatives are poorly recognized by maintenance DNMT complexes and once bound their catalytic activities are largely reduced. Consequently, Tet-mediated oxidation leads to replication-dependent dilution of the methylation mark on the parental strand during preimplantation and primordial development. Finally, excision of oxidized cytosines takes place by the DNA repair machinery [base excision repair (BER); nucleotide excision repair (NER)] as has been suggested for embryonic stem cells [20]. As an alternative route to active demethylation, 5hmC can be also directly deaminated to thymine which is further processed by DNA glycosylases like TDG (G/T mismatch-specific thymine DNA glycosylase) (Fig. 2.1). Notably, clusters of 5hmC seem to be enriched at regulatory sites throughout the genome suggestive of an instructive role in gene regulation. A plausible explanation for this distinct pattern is provided in the finding that oxidized cytosines can impair the binding of proteins that recognize 5mC residues and thus turn off their regulatory functions [19]. Taken together, the DNA methylome is more dynamic than originally thought and can undergo iterative cycles of methylation and demethylation at regulatory sites important for gene expression.

2.3.2 EPIGENETIC TAGGING OF HISTONES

Core histones represent highly basic proteins that tightly package DNA to fit the limited nuclear space. This compressed state, termed chromatin, also provides a platform on which gene regulation is carried out and modification of core histones represents an elaborate mechanism for epigenetic tagging of the genome. Histone modifications can take place in response to DNA methylation or by intracellular signaling pathways independent of DNA methylation [21].

In the nucleus, 146 base pairs of DNA are wound around a histone octamer composed of two copies of each core histone (i.e., H2A, H2B, H3, and H4) to generate a nucleosome, the building block of chromatin. Once nucleosomes are arrayed further into higher order structures in the presence of linker histones (i.e., H1) and other nonhistone proteins, they become mostly inaccessible to the transcriptional machinery. Although higher-order histone–DNA structures can be remodeled by different enzymatic complexes, in order to regain access to regulatory regions targeted by the transcriptional machinery [21]. Structural studies indicate that the N-terminal tails of histones protrude beyond the nucleosome surface and can form a signal integration platform, where posttranslational modifications provide a signature that ultimately coordinates the activity of numerous transcription factors, associated cofactors, and the transcription machinery in general [21]. Specific amino acid residues on the free amino-terminal tails, and also on the globular core domains, can serve as substrates for different kinds of posttranslational modifications comprising lysine acetylation, lysine and arginine methylation, serine phosphorylation, and covalent binding of ubiquitin among others [21]. Taken together, epigenetic tagging of chromatin plays an important role in DNA packaging, gene transcription, cross-talk between DNA methylation and chromatin configuration, and ultimately, integration of intrinsic and environmental signals into the epigenome.

2.4 ELA AS A MAJOR RISK FACTOR FOR NEUROPSYCHIATRIC DISORDERS

Over the last decades, a large body of epidemiological and experimental studies has examined the role of genetic and epidemiological factors in the onset, course, and treatment response of psychiatric disorders. Among these, stress-related disorders like major depressive disorder (MDD), posttraumatic stress disorder (PTSD), and anxiety disorders have been consistently shown to involve both genetic predisposition and environmental factors [22]. While the quality of early life (in utero, perinatal, and in infancy) has been well-recognized for determining later physical health, newer findings also support a role for mental health [7]. In this respect, ELA represents one of the strongest risk factors for the manifestation of several psychiatric diseases including MDD and PTSD, schizophrenia (SZ) and borderline personal disorder (BPD) [22–24]. ELA is defined by a number of conditions comprising parental maladjustment (mental illness, substance abuse, violence, and criminality), interpersonal loss (separation from parents or caregivers), life threatening childhood physical illness, severe childhood poverty, and maltreatment (sexual and physical abuse, neglect). For maltreatment 9.2 victims per 1,000 children were reported in 2012 amounting to approximately 668,000 victims of abuse and neglect in the USA [25]. It is important to emphasize that neglect during early childhood is the most common form of maltreatment (78.3% of all cases) although it receives much less public attention than physical (18.3%) or sexual (9.3%) abuse.

Similarly, perinatal adversity due to maternal MDD or PTSD frequently escapes notice although it manifests an estimated prevalence of 7–18% for depressive and 8.5% for generalized anxiety disorders [26,27]. Maternal mood disorders represent a major risk factor for the developing fetus and associate with an increased risk for premature delivery and restricted growth during the neonatal period of life. Moreover, offspring of depressed mothers are more likely to manifest impairments in mental, emotional, and motor development compared to control infants and to develop adolescent depression or anxiety [28,29].

While as early as the 1930s epidemiological data suggested the relevance of early life conditions for adult mortality, it was only in the 1970s that pioneering studies hypothesized the involvement of so-called perinatal programming factors to modulate the development of obesity, diabetes mellitus, and arteriosclerosis in adulthood [30]. Later, David Barker and colleagues demonstrated in their seminal investigation of historical human cohorts in England and Wales a strong inverse correlation between birth weight—a proxy for fetal nutrition and growth—and risk of mortality from cardiovascular disease in later life [31]. These findings have paved the way for the hypothesis of the developmental origins of health and diseases (DOHaD), which postulates a role for early life developmental factors in influencing an individual's response to the environment and susceptibility to disease in later life [7].

Maternal undernutrition during early gestation has been shown to associate with an increased risk for various metabolic pathologies in the offspring such as later life obesity, elevated lipid levels, and coronary artery disease. Since early undernutrition in humans frequently occurs in the context of severe socioeconomic distress or parental misbehavior, it has proved difficult to dissect the effects of undernutrition on general brain development from specific mental dysfunctions. Despite this limitation, the available data suggests that severe undernutrition can lead to lasting impairments in cognitive and social behaviors [32,33]. Today, maternal undernutrition has been largely overcome in Westernized societies, while the prevalence of obesity and excessive weight have reached epidemic proportions [34]. At this time, about one third of women in the USA are obese; a condition known to also program offspring for lifelong obesity and associated metabolic disorders and thus to initiate a vicious cycle of transmitting disease risk. In the context of this chapter it is important to note that obesity, metabolic syndrome, and insulin resistance are also increasingly recognized to affect offspring mental health function including cognition and heightened anxiety [35]. A growing body of animal studies shows that perinatal nutrition influences enduringly the likelihood of developing psychiatric disorders such as impaired social behavior and cognition [36–39], enhanced stress responsiveness, and altered reward-based behaviors [35,40–43]. Moreover, factors associated with maternal obesity in humans, such as hyperlipidemia, hyperglycemia, insulin resistance, and inflammation have been associated with an enhanced offspring risk for anxiety and depression [36,44–48], attention-deficit hyperactivity disorder [49], and autism spectrum disorders [50]. Taken together, ELA is a major risk factor for the development of different mood and cognitive disorders in offspring. At this time, neglect and abuse represent the most prevalent forms of ELA in Westernized socictics and frequently concur with maternal psychiatric illness. Moreover, maternal obesity and excessive weight are increasingly recognized as risk factors for child neurodevelopment and later mental function.

2.5 THE STRESS SYSTEM AS A MEDIATOR OF ELA

Epidemiological studies provide ample evidence for the eminent role of early risk factors such as neglect, parental maladjustment, and maternal obesity for the development of later cognitive and mood disorders. Why is that? Childhood maltreatment is wellknown to impose long-term effects on behavior by altering the activity and/or structure of neural circuits, which contribute to the regulation of the stress system [51]. Consistent with this view, individuals with a history of childhood maltreatment manifest altered stress reactivity characterized by an increased corticotropin-releasing hormone (CRH) secretion. Therefore, a useful starting point for this chapter is the hypothesis that early life risk factors are linked to lasting dysregulation of the stress system.

The term stress is commonly referred to the subjective state of sensing potential or actual threats in the environment [8]. Stress is conveyed by the interconnected norepinephrine and the hypothalamic pituitary adrenal system (HPA axis), which jointly act on specific neuronal populations that coordinate unique downstream responses such as an increase in heart rate, rise in blood pressure, and elevated levels of stress hormones [52]. This integrated stress response enables an organism to adapt to stressful stimuli through immediate as well as future modifications of behavior. Research in the field of stress biology has evidenced over the last decades a steadily growing number of neurotransmitters, neuropeptides, cytokines, enzyme systems, binding proteins, and transcription factors among others. As a whole, most of these factors share one common denominator—their contribution to the regulation of the HPA axis that plays a major part in the shared biology of early risk factors for psychiatric disorders. Central drivers of the HPA axis are the two neuropeptides CRH and arginine vasopressin (AVP), which are synthesized and released by the parvocellular neurons of the hypothalamic paraventricular nucleus (PVN). Once secreted into the hypothalamic hypophyseal portal system, they are transported to the anterior pituitary where they synergistically induce pro-opiomelanocortin (POMC) gene expression and secretion of its posttranslational product adrenocorticotropin (ACTH) [52]. Subsequently, ACTH enhances the secretion of the lipophilic hormones cortisol (in humans) and corticosterone (in humans, rats, and mice), which readily enter into the brain. There, they bind to nuclear glucocorticoid and mineralocorticoid receptors (GR and MR) encoded by *NR3C1* and *NR3C2*, respectively. The two receptor proteins are closely related ligand-gated transcription factors, which are highly coexpressed in different neurons of the limbic brain (Fig. 2.2).

The MR shows a high affinity for glucocorticoids and thus has been implicated in the initiation of the stress response, while the GR, requiring higher glucocorticoid concentrations, restrains, and ultimately shuts off the stress response. This negative feedback regulation of glucocorticoid secretion takes place at the level of the hippocampus, PVN, and anterior pituitary where GRs are highly expressed. Increased glucocorticoid secretion is the predominant response to stress resulting in energy mobilization, suppression of immune and reproductive functions, increased cardiac output, and behavioral changes. By these actions glucocorticoids promote adaptive responses although they may also lead to maladaptation when mechanisms thought to restrain HPA-axis activity ultimately fail. Such dysregulation

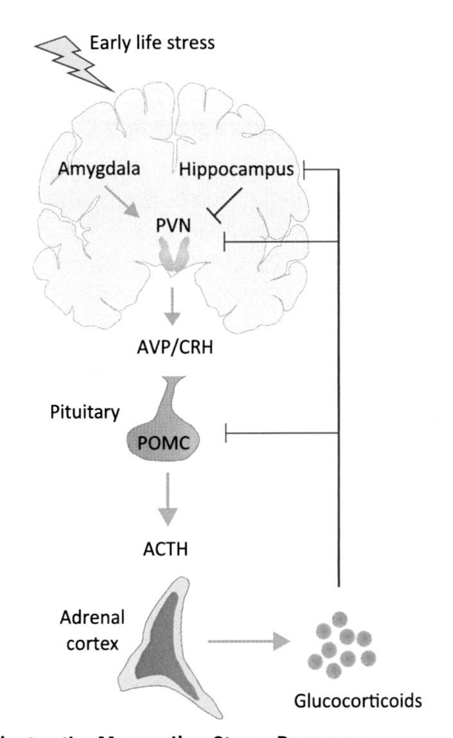

FIGURE 2.2 The HPA Axis Coordinates the Mammalian Stress Response

ELS is processed by higher brain centers such as the prefrontal cortex and hippocampus that in conjunction with other brain regions like the amygdala signal to the PVN. Neurons within this central node express the two neuropeptides CRH and AVP that are released into the hypothalamic-hypophyseal portal blood system. Once transported to the anterior pituitary, these neuropeptides bind to their respective G-protein coupled receptors and stimulate the expression of POMC and the secretion of its posttranslational product ACTH. Glucocorticoids are released from the adrenal gland in response to ACTH and readily enter the brain to bind to intracellular GR and MR receptors. These ligand-gated transcription factors are highly expressed in the hippocampus, PVN, and pituitary where they serve to initiate (i.e., MR) and terminate (i.e., GR) the stress response.

can evolve when the type, strength or duration of the stressor exceeds certain developmental or physiological thresholds. Sustained levels of glucocorticoids can enduringly (mal)program HPA-axis function via changes in developmental trajectories that coordinate the perception and subsequent response to various stressors. Together, these events can lead to persistent structural and regulatory changes predisposing to stress-related diseases later in life. For example, chronic stress can lead to structural changes such as shrinkage and less branching of dendrites concomitant with a reduced synaptic input [8]. Such cellular changes can concur or even interact with epigenetic marking, which instructs regulatory changes in gene expression. Here, we refer to the latter process as molecular plasticity and will discuss recent evidence for this hypothesis.

2.6 EPIGENETIC PROGRAMMING OF THE STRESS SYSTEM

In this section, we will examine the role of epigenetic programming of the HPA axis in response to ELA. In contrast to a growing number of animal studies that support long-lasting epigenetic effects of early stressful environments, studies on human postmortem brain are still few. This can be largely ascribed to the limited amount of material available from public depositories and a lack of well-documented clinical records. While improvements of this situation are urgently looked for, clinical scientists have turned to peripheral tissues such as blood or buccal swabs as a proxy for brain-related changes. Epigenetic changes in such tissues may provide useful biomarkers of past and/or present stress exposures; however, they appear less informative for deciphering how experience-dependent neuronal activity couples to the methylation machinery.

2.6.1 EPIGENETIC PROGRAMMING OF THE GR BY EARLY LIFE EXPERIENCES

Among the very first candidates implicated in stress-related epigenetic regulation was the GR due to its critical role in restraining HPA-axis activity. Human *NR3C1* comprises multiple first exons among which four exons are located in the distal promoter region (A_{1-3} and I) and ten (D, J, E, B, F, G, $C_{1-3,}$ and H) in the proximal composite promoter region [53]. The distal and proximal promoter regions locate 30 and 5 kb upstream of exon 2 that encodes the start codon. A CGI consisting of a high frequency of CpG residues straddles the proximal promoter and most alternative promoters are expressed in a tissue-specific and partly species-specific manner. The proximal *NR3C1* promoter is highly conserved between humans, rats, and mice and thus allows for comparing epigenetic marking in response to different environmental exposures across species [53].

As early as 2004, a seminal study in rats evidenced epigenetic programming of *Nr3c1* in response to differences in the quality of postnatal maternal care [54]. Pups reared by high-care taking mothers—defined by frequent pup licking and grooming—showed an enhanced turnover of serotonin in hippocampal tissues, in which GR expression was site-specifically increased. Once bound to G-protein coupled receptors, serotonin triggers intracellular cAMP production that in turn induces expression of the transcription factor EGR1 (early growth response 1, also called NGFI-A). EGR1 recognizes a DNA element at *Nr3c1* exon 1.7 (rat homologue to 1F in human) and facilitates the recruitment of general coactivators harboring histone acetyltransferase activity. This event promotes chromatin opening and *Nr3c1* transcription in response to high quality maternal care. Importantly, the same chain of events also initiates lasting demethylation at the exonic EGR1 DNA-binding site and thus leaves a molecular memory trace of the early life experience. This maternal-care dependent hypomethylation directs long-term increases in hippocampal GR expression and contributes to an improved negative feedback regulation and associated behavior [54]. However, epigenetic programming of *Nr3c1* can be reversed by timely pharmacological or behavioral interventions and thus illustrates the dynamic nature of experience-dependent methylation marks [55].

These findings from rats were subsequently translated to humans by investigating the *NR3C1* methylation status of adults with a history of early child abuse who died by suicide [56]. Hypermethylation of exon 1F was detected in the hippocampus of suicide completers with a history of child maltreatment and correlated with lower GR expression. In contrast, suicide completers without a history of childhood maltreatment showed neither a change in hippocampal *NR3C1* exon 1F methylation or in gene expression. Functionally, hypermethylation impaired EGR1 binding and subsequent transactivation of *NR3C1*. Together, these findings indicate that ELA can confer site-specific epigenetic programming of *NR3C1*and disrupt negative feedback regulation of the HPA axis in rodents and humans. Although these initial studies suggested that early life stress (ELS) targets preferentially exon 1F, this view has to be revised in light of later work that revealed epigenetic programming of multiple proximal promoter regions. In this context, expression of the noncoding exons 1B, 1C, and 1H, reported significant decrease in the hippocampus of suicide completers with a history of childhood abuse compared to nonabused suicides and controls [57]. This study also showed that *NR3C1* 1C methylation correlated inversely with GR expression in agreement with previous results on 1F in rats. What is the mechanism that coordinates lasting changes in multiple GR transcripts in response to ELS?

Part of the answer has been gained from a recent study in mice that used postnatal MS as a well-established model of ELS [58]. Separated pups develop sustained HPA-axis hyperactivity and behavioral impairments in conjunction with hypomethylation of hypothalamic *Avp* and pituitary *Pomc* (see subsequent section) [59,60]. Notably, GR expression was unaltered in mice hippocampus and pituitary, which represent major sites of negative feedback regulation. In contrast, multiple exon 1 transcripts were lastingly upregulated in the adult PVN of maternally separated mice and caused a net increase in total GR transcripts. This event translated into higher GR protein expression and an enhanced regulation of downstream glucocorticoid target genes. Interestingly, enhanced hypothalamic *Nr3c1* expression correlated with hypermethylation of the proximal CGI shore region. Functional analyses showed that the shore region encodes an insulator function thought to shield the proximal promoter from upstream regulatory influences. The shore region also contained a methylation-sensitive DNA-binding site for the transcriptional regulator YY1, which enforced insulator function upon binding. As a result, ELS dynamically regulated YY1 binding at the CGI shore region, associated insulator function, and transcription of multiple proximal GR transcripts (Fig. 2.3).

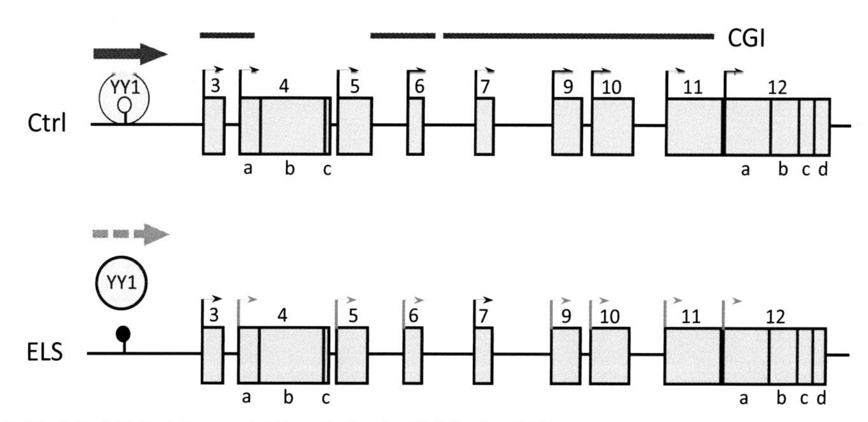

FIGURE 2.3 Model of *Nr3c1* Transcript Regulation by ELS in the PVN

The proximal *Nr3c1* promoter contains a DNA-binding site for the transcriptional repressor YY1 at the CGI-shore region. CGIs spanning the proximal *Nr3c1* promoter are depicted above the different nontranslated exons boxed in grey and numbered according to Bockmühl et al. [53]. In control mice (Ctrl), YY1 binding occurs in the absence of DNA methylation (open lollipop) and confers transcriptional repression of multiple GR-transcripts within the proximal promoter region. In contrast, increased methylation (black lollipop) at the YY1 DNA-binding site in response to ELS prevents YY1 binding and relieves transcriptional repression. As a result, multiple GR transcripts are upregulated and translated into GR protein.

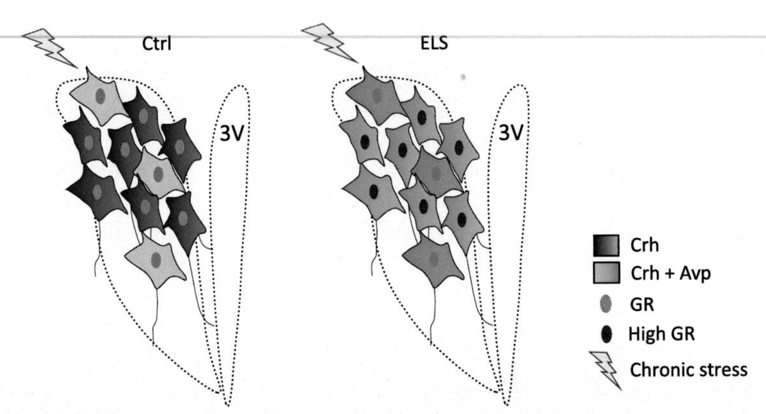

FIGURE 2.4 Cell Type-Specific, Epigenetic Programming of Hypothalamic *Nr3c1*

Mouse neurons in the PVN can express either Crh and Avp or solely Crh. Exposure to ELS leads to epigenetic programming and sustained upregulation of Avp when compared to controls (Co) (see also paragraph on Avp). Additionally, ELS induces epigenetic programming and upregulation of GR in Crh-expressing neurons. In adult mice, application of chronic stress results in upregulation of Crh in control mice, but not in mice with a history of ELS. As a result, ELS partly protects against chronic stress in adulthood as evidenced by a faster return to baseline glucocorticoid levels and less behavioral impairments.

Interestingly, ELS-induced *Nr3c1* upregulation was detected solely in parvocellular Crh-, but not in Avp-expressing neurons. Moreover, Crh expression was unaffected under resting conditions. In contrast, exposure to chronic mild stress led to a robust up-regulation of Crh in control mice but not in mice with a history of ELS (Fig. 2.4). Taken together, this work illustrates that the effects of cell type-specific epigenetic programming of *Nr3c1* can stay masked unless challenged by reexposure to stress. If so, ELS may actually protect against later stress. Prompted by these studies several groups have analyzed peripheral blood mononuclear cells (PBMC) from different populations of individuals who were exposed to varying forms of ELA. Although clinical phenotypes of the respondents, measures of ELA, and methylation analysis varied among these investigations, most of them agree on altered *NR3C1* methylation patterns in peripheral tissues raising the prospect of a potential biomarker for individuals exposed to ELA [61]. Overall, findings on epigenetic programming of *NR3C1* suggest that the effects depend on the quality of the initial stressor [maternal care vs. maternal separation (MS)], and are cell type-specific (hippocampal vs. PVN neurons), and context-dependent (naive vs. reexposed conditions).

2.6.2 EPIGENETIC PROGRAMMING OF FKBP5 BY GLUCOCORTICOIDS AND ELA

While GRs play an important role in the control of HPA-axis activity, their activity is in turn controlled by several cofactors including the FK506 binding protein 51 (FKBP51). This protein belongs to the family of immunophilins due to its ability to bind the immune suppressive drug FK506. FKBP51 was initially isolated in a complex with the progesterone receptor and serves among other functions as a cochaperone of the central chaperone HSP90 [62]. Biochemical studies showed that FKBP51 has peptidyl-propyl cis-trans isomerase activity and owns a tetratricopeptide repeat protein domain by which it can act as a cochaperone that regulates folding and activity of other proteins.

A potential role of FKBP51 in the regulation of the HPA axis was originally deduced from findings in new World primate species that showed increased plasma cortisol concentrations, decreased plasma cortisol binding globulin capacity and affinity, marked resistance of the HPA axis to suppression by dexamethasone, but no biologic signs of glucocorticoid excess [63]. Functionally, such glucocorticoid resistance has been ascribed to an increased expression of FKBP51. The inhibitory effect of FKBP51 on GR function can be assigned to at least two mechanisms: first, a reduced binding affinity of the GR-chaperone heterocomplex in the presence of FKBP51 and, second, a reduced interaction of the GR with the transport protein dynein. As a result, FKBP51 diminishes nuclear translocation and transcriptional activities of the GR [62]. It is important to note that *FKBP5* also represents a direct GR target gene due to the presence of several functional intronic glucocorticoid response elements. Glucocorticoid-dependent transactivation of *FKBP5* generates an ultrashort intracellular feedback loop that complements the systemic feedback regulation of the HPA axis in balancing the stress response.

As early as the 1990s, a direct, causal role of glucocorticoids in epigenetic programming had been suggested for the tyrosine aminotransferase gene, a well-known target gene of glucocorticoids. Exposure to glucocorticoids elicited a persistent change in DNA methylation, thus providing a memory of the first stimulation that facilitated subsequent glucocorticoid-induced transcription [64]. Prompted by these findings, Lee and coworkers investigated the potential effect and mechanism of chronic corticosterone treatment on *Fkbp5* expression and behavior in mice [65]. Sustained treatment (4 weeks) of adolescent mice triggered an enduring decline in DNA methylation in intron 5 (hypothalamus and hippocampus) and in intron 1 (blood) of the *Fkbp5* gene that persisted into adulthood. Moreover, lasting glucocorticoid-induced demethylation was recapitulated in a hippocampal mouse cell line indicating that results from in vivo studies are unlikely to derive from nonneuronal cells. Functionally, demethylation translated into increased *Fkbp5* expression in addition to glucocorticoid-induced changes in a variety of physiological parameters and heightened anxiety-like behavior. In a follow-up study it was further shown that the degree of *Fkbp5* demethylation in blood correlated well with the glucocorticoid burden (cortisol concentration x duration) and could serve as a peripheral biomarker of previous stress exposure [66].

Aging has been suggested to associate with HPA-axis disinhibition and increased glucocorticoids that due to their deleterious effects on brain function may set off a vicious cycle of hypercortisolism [8]. Consistent with this view, CpGs located within intron 5 of *Fkbp5* underwent demethylation in aging mice, a process that may explain the age-associated increase in FKBP51 protein. Epigenetic upregulation of *Fkbp5* with age also selectively impaired psychological stress-resiliency, but did not alter other glucocorticoid-mediated physiological processes [67]. In human brain, FKBP51 protein increased relative to age and Alzheimer's disease, corresponding with demethylation of intron 7 in *FKBP5*, whereby higher FKBP51 protein also associated with progression of Alzheimer's disease [68]. Together, these data suggest that enhanced glucocorticoid exposure, either due to pharmacological treatment or aging can drive epigenetic programming of *FKBP5* and impair HPA-axis function. Do these findings also apply to ELA?

In this respect a recent study suggests that the, cortisol reactivity of *FKBP5* can be programmed by ELS via a single nucleotide polymorphism (SNP) dependent and allele-specific demethylation in and around glucocorticoid response elements in intron 7 [69]. A single SNP close to a functional glucocorticoid response element in intron 2 of *FKBP5* facilitates (risk-allele) or impairs (protective-allele) glucocorticoid dependent transcription. These two alleles also determine the response to ELA: DNA methylation in and around intron 7 glucocorticoid response element is preserved for the protective allele, whereas demethylation of the risk allele promotes the interaction between intron 7 and the promoter. As a result, the risk allele confers *FKBP5* demethylation and enhanced transcription, reduced GR signaling, and a heightened risk for developing stress-related psychiatric disorders. Overall, this study suggests that genetic variation in GR signaling pathways is an important risk factor for psychiatric disorders. Such variation might also serve as a substrate for epigenetic programming by ELS and mediate the risk for later stress-related psychiatric disorders. Since these data were gained from PBMCs, further studies are needed to validate these findings in disease-relevant neural tissues and cells.

2.7 EPIGENETIC PROGRAMMING OF HYPOTHALAMIC NEUROPEPTIDES BY ELA

The two neuropeptide genes CRH and AVP are important drivers of the HPA axis and are both under the control of a negative feedback regulation by glucocorticoids. As will be discussed in this section, a growing body of studies suggests that either gene can serve as a substrate for epigenetic programming.

2.7.1 EPIGENETIC PROGRAMMING OF CRH

Severe maternal stress can overwhelm the protective function of the placenta and expose the developing fetus to heightened levels of glucocorticoids with an increased risk for psychiatric disorders later in life. In a mouse model of prenatal stress (embryonic days 1–-7), male but not female offspring showed anhedonia, reduced behavioral stress responsibility, and an enhanced response to application of a selective serotonin reuptake inhibitor [70]. This behavior associated with increased *Crh* expression in the amygdala concomitantly with decreased promoter methylation. Since the fetal brain would not have developed at the time when the stressor was applied, the authors of the study proposed that these effects may be explained through effects on the placenta such as lower Dnmt1 expression in males as compared to females. In another study, rats with a history of MS were analyzed for HPA-axis responsiveness to acute stress [71]. While females showed higher resting and acute stress-induced corticosterone levels, increased corticosterone levels were detected in males only after acute stress. These hormonal profiles were matched by corresponding increases in hypothalamic, but not amygdala, Crh expression that associated with *Crh* promoter hypomethylation. Hence, two ELS responsive CpG residues mapped next to and at the dyad symmetry axis of a cyclic AMP response element and reduced DNA binding of the activated form of the CRE-binding protein. In addition to its major role in neuroendocrine regulation, CRH is also known to affect learning and memory via extrahypothalamic sites of expression in a time-and dose-dependent manner. Brief, physiological increases in CRH favor memory formation within the range of seconds to minutes, whereas prolonged increases elicit spine retraction and synapse loss leading to impairments in learning and premature cognitive decline [72]. In the developing hippocampus several cell populations including Cajal–Retzius cells start to express CRH, whereas in adults expression is largely confined to interneurons residing in the pyramidal cell layers of the CA1 and CA3 and is released from axon terminals during stress [72].

Adult rats exposed to MS manifest reduced hippocampal glutamatergic synaptic plasticity and memory formation that can be efficiently blocked by a Crh receptor 1 antagonist [73]. Consistent with these findings, Crh mRNA and protein expression was enhanced in hippocampal CA1 concomitantly with a decline in *Crh* promoter methylation. As a result, binding of the methyl-CpG-binding protein 2 (Mecp2, a versatile transcriptional regulator discussed later) decreased, whereas neuronal activity-dependent phosphorylation of Mecp2 at serine 421 (Mecp2-S421), favoring DNA dissociation, increased. In support of Mecp2's role as suppressor, sequential chromatin immunoprecipitation analysis (ChIP) evidenced an interaction with Hdac2(histone deacetylase 2) or Dnmt1. Also, environmental enrichment attenuated the effects of ELS on adult behavior, protected from enhanced Crh expression and Mecp2 phosphorylation, and restored Mecp2 binding to the Crh promoter in maternal separated rats [74]. This study conclusively indicates a role for ELS in hippocampal memory function via epigenetic programming of *Crh* and the potential reversibility of these effects by timely interventions.

While ELS is a major risk factor for psychiatric disorders, chronic stress in adulthood can also trigger psychiatric disease in susceptible individuals. In support of this view, application of social defeat imposes a strong stressor in male mice leading to social avoidance and anhedonia [74]. This behavioral phenotype associated with increased Crh expression and demethylation of the proximal promoter, whereby decreases in Hdac2 and Dnmt3b expression preceded *Crh* demethylation. At the same time, expression of the growth arrest- and DNA damage-inducible gene *Gadd45* was increased and pointed to a possible role of the DNA repair machinery in *Crh* demethylation. Furthermore, 3 weeks treatment with the widely-prescribed antidepressant imipramine normalized social avoidance behavior and restored *Crh* promoter methylation. Taken together, this study supports the idea that epigenetic programming extends to the adolescent brain and remains responsive to timely pharmacological interventions like in earlier stages of life. Overall, findings from this section ascribe to epigenetic programming of Crh an important role for effects of ELS on later neuroendocrine and behavioral phenotypes. Hence, different stressors can program enduringly, albeit reversibly, Crh expression in a tissue- and cell- type-specific manner.

2.7.2 EPIGENETIC PROGRAMMING OF AVP BY ELA

While CRH plays an increasingly important role for the regulation of the HPA axis as the organism matures, AVP is thought to be more relevant for an appropriate adrenocortical response to stress during perinatal life when the endocrine response to stress is still dampened [52,75]. This hypothesis prompts the question whether *AVP* could be epigenetically programmed by ELS as well.

MS is a well-known model to induce ELS in mice and causes a lasting hyperactivity of the HPA axis manifested by corticosterone hypersecretion under basal conditions and hyperresponsiveness to an acute stressor later in life [59]. Also, adult mice with a history of ELS showed additionally an enhanced immobility in the forced swim test and memory deficits in an inhibitory avoidance task. At the same time, MS elicited an acute and long-lasting upregulation of Avp, but not of Crh, in the hypothalamus. On the other hand, Avp expression in the nucleus supraopticus with a major role in fluid homeostasis remained unaltered [59]. Interestingly, enhanced Avp expression concurred with a reduced DNA methylation at multiple CpG dinucleotides throughout the downstream enhancer region, which underpins tissue-specific expression and maps to a CpG island of intermediate CpG density. This ELS-induced hypomethylation was most distinct in late adolescence and young adulthood and declined one year after the triggering event. What mechanism could underlie epigenetic programming of *Avp*?

Notably, those CpG residues with the greatest decline in DNA methylation fulfilled the criteria of a high affinity MECP2 DNA-bindings site [76]. In accord with this prediction, a ChIP analysis evidenced that Mecp2, but not other methyl-CpG-binding proteins, was selectively bound to the ELS responsive enhancer region. While enhancer hypomethylation correlated with reduced Mecp2 binding in adult mice this was also the case following termination of MS when DNA methylation was still maintained. This unexpected result indicated that mechanisms independent of DNA methylation contribute to Mecp2 DNA binding. Consistent with this possibility, depolarization-dependent Ca^{2+} influx and activation of calmodulin kinases has been reported to elicit Mecp2 phosphorylation. This modification facilitates in turn Mecp2's dissociation from the DNA and derepression of target genes [77]. In support of this hypothesis, enhanced Mecp2 phosphorylation was detected in hypothalamic Avp-expressing neurons during and after application of MS [59]. This result indicates that Mecp2 occupancy at the *Avp* enhancer is first controlled by neuronal activity-dependent phosphorylation and secondly by DNA methylation. While MECP2 has been commonly assigned a role as repressor due to its interaction with HDACs and DNMTs, this canonical view had to be revised in light of recent studies that suggested an additional role in gene activation both at the local and genome-wide scale [77]. Does Mecp2 then actually repress hypothalamic *Avp*? A further study showed that Mecp2 preferentially associated with repressive histone marks at the *Avp* enhancer as evidenced by sequential ChIP analysis and once bound, Mecp2 recruited de novo Dnmts (Dnmt3a and Dnmt3b), but not Hdacs (Hdac1 or Hdac2) [78]. Taken together, these studies indicate that *Avp* can serve as a substrate for epigenetic programming in response to ELS and underpin lasting neuroendocrine and behavioral changes related to stress- and trauma-induced psychiatric disorders. Hence, Mecp2 plays a critical role in the mediation of early life experiences. First, neuronal activity-dependent depolarization triggers Mecp2 phosphorylation, dissociation from the *Avp* enhancer, and increased expression. Second, loss of Dnmts recruited by Mecp2 is proposed to pave enhancer hypomethylation and thus leave a lasting memory trace of the initial event. In this context, it is important to remember that the neuronal methylome undergoes extensive reconfiguration during perinatal life and neuronal activity-driven changes in the methylation machinery might offer a common route to integrate early life experiences into the epigenome [79].

2.7.3 EPIGENETIC PROGRAMMING OF PITUITARY POMC BY ELA

CRH and AVP bind to G-protein coupled receptors at the anterior pituitary and elicit an increase in intracellular levels of cAMP and Ca^{2+}. These second messengers stimulate in turn the expression of pituitary POMC and the secretion of its posttranslational product ACTH. This hierarchical order of the HPA axis prompts the question whether epigenetic programming of hypothalamic *Crh* and *Avp* by ELS does also extend to pituitary *Pomc*. In support of this hypothesis, maternally separated mice showed higher Pomc expression under resting conditions and an increased secretion of ACTH following application of the AVP/CRH challenge test [60].

This endocrine phenotype associated with lasting hypomethylation of multiple CpG residues in the proximal *Pomc* promoter region. Notably, the CpGs with the greatest decline in methylation affected those located in regulatory regions conferring transactivation by the upstream secretagogues Crh and Avp. Functionally, in vitro methylation of this region strongly reduced the activity of corresponding *Pomc* promoter reporter constructs in transfection assays. Also, additional experiments evidenced that the ELS responsive CpG residues conformed to the criteria of a high affinity MECP2 DNA-binding site. In accord with this prediction, in vivo ChIP experiments evidenced that Mecp2 occupied the proximal *Pomc* promoter region in anterior pituitary cells and associated hereby with repressive chromatin marks, Dnmt1, and Hdac2. Importantly, Mecp2 binding was reduced in ELS-treated mice at early stages and in adulthood when compared to controls and indicated that Mecp2 can mediate epigenetic programming of *Pomc* as well. In contrast to the PVN, however, Mecp2-S421 phospho-immunoreactivity was not increased after exposure to the early life stressor indicating that unrelated phosphorylation sites may control Mecp2 DNA binding in corticotroph cells [80].

2.8 EPIGENETIC PROGRAMMING OF THE HPA AXIS BY EARLY NUTRITION

While ELS has been increasingly recognized as a risk factor for the development of different psychiatric disorders, the role of nutrition in early life in psychiatric disorders remains poorly understood. Here, we propose that nutritional imbalances and ELS share in common a deregulation of the HPA axis and that nutrition may also through epigenetic programming increase the risk for adult psychopathology.

2.8.1 EPIGENETIC PROGRAMMING OF THE GR BY EARLY UNDERNUTRITION

Periconceptional undernutrition in sheep accelerates maturation of the fetal HPA axis in late gestation and manifests elevated fetal baseline cortisol concentrations [81]. Notably, in fetuses from ewes undernourished from -60 to +30 days around conception, hypothalamic *Nr3c1* expression was increased in conjunction with proximal promoter hypomethylation. In contrast, *Nr3c1* expression and DNA methylation were unaltered in the anterior pituitary or hippocampus [82]. Moreover, the promoter region of *Pomc* from late-gestation fetal hypothalami underwent hypomethylation, although expression was unaffected. Together, these findings suggest that epigenetic changes may serve as a mechanism to predispose hypothalamic feeding centers to deregulation in later life. In support of this prediction, these epigenetic changes were maintained up to 5 years after the maternal insult [83]. Maternal undernutrition-induced hypomethylation concurred with increased active chromatin marks at the proximal *Nr3c1* promoter and underpinned enhanced mRNA and protein expression in either sex. As a result, hypothalamic Pomc expression, which inhibits food intake (see subsequent section), was decreased in 5-year-old adult males and associated with enhanced obesity. A possible explanation for the absence of obesity in females is that estrogens could attenuate the effects from programming in the hypothalamus and provide compensatory pathways to regulate food intake and energy expenditure. A limitation of these studies is the fact that differences in cortisol levels between undernutrition-exposed and control offspring were little and that behavior was not assessed [82]. Hence, further studies in complementary model systems appear desirable.

2.8.2 EPIGENETIC PROGRAMMING OF HYPOTHALAMIC POMC BY EARLY OVERNUTRITION

Obesity has reached epidemic proportions in the last few decades and affects about one third of the adult population and almost one fifth of the youth population in western countries like United States of

America [84]. Sustained obesity has profound consequences for an individual's life ranging from psychological symptoms to serious comorbidities (diabetes, cancer, cardiovascular, and psychiatric disease among others) that markedly reduce both the quality and length of life.

A broad range of genetic and environmental factors contributes to an imbalance between energy intake and energy expenditure. Signals controlling energy homeostasis involve immediate signals to a meal and long-term satiety signals. One of the most important regions in the brain that coordinates such long-term signals is represented by the arcuate nucleus (ARC). The ARC resides in the mediobasal hypothalamus and comprises appetite suppressor (anorexigenic) and appetite stimulator (orexiogenic) populations of neurons that respond to circulating peripheral hormones such as leptin, insulin, and ghrelin [85]. Reduced levels of leptin and insulin stimulate expression of orexiogenic neuropeptides (e.g., neuropeptide Y, NPY; agouti-related protein, AgRp), which are counterbalanced by anorexigenic neuropeptides (e.g., cocaine- and amphetamine-regulated transcript, CART, and alpha-melanocyte-stimulating hormone, α-MSH). The neuropeptide α-MSH is generated from a posttranslational cleavage product of POMC through tissue-specific processing of the prohormone by convertases 1 and 2. As a result, a variety of bioactive peptides are generated including α-MSH, β-endorphin, β-lipotropin, and ACTH.

Energy balance in mammals is defined as a negative ground state that needs to be counteracted by signals derived from elevated fat stores. Otherwise, the brain continuously senses an energy deficit promoting feeding and reduction of energy expenditure by default [86]. For example, an impaired response of the brain to leptin underlies a sustained dysregulation of food intake and energy homeostasis. This situation is frequently met with in obesity and has been termed leptin resistance. In this chapter we will therefore explore the possibility whether epigenetic programming of *POMC* may contribute to this phenotype. In support of this hypothesis, a number of recent studies evidenced a role for epigenetic programming of hypothalamic *POMC* in obesity. Both male and female postweaning rats fed a high-fat diet (HFD) manifest an increased body weight concomitantly with high leptin and insulin levels. While the orexiogenic system seems to be intact (lower Npy and AgRp mRNA levels), both sexes manifest a deficit in the anorexigenic system with no change (males) or decreased (females) Pomc mRNA levels [87]. Notably, analysis of the *Pomc* promoter detected hypermethylation at several DNA-binding sites for the transcription factor Sp1(specificity protein 1) in HFD-treated rats when compared to controls and correlated with reduced Sp1 DNA binding as evidenced by ChIP experiments. Expression of Sp1 is stimulated by insulin and since Sp1 binding to the *Pomc* promoter mediates leptin's anti-obesity effects, these results indicate that HFD epigenetically programs *Pomc* expression by disrupting the anti-obesity action of leptin [88,89]. This conclusion is further supported by a recent study in neonatal mice in which a maternal diet rich in conjugated linoleic acids likewise elicited hypermethylation of the Sp1 DNA-binding sites in the proximal *Pomc* promoter and thus predisposed to metabolic disease in adulthood [90]. Similarly, perinatal exposure to a HFD in weanling rat offspring showed a significant hypermethylation across the entire *Pomc* promoter [91]. While there were no group differences in Pomc expression between HFD and control offspring under resting conditions, the Pomc/leptin ratio was impaired in case of *Pomc* hypermethylation indicating an impaired signaling.

Chronic low-grade inflammation is frequently associated with obesity, while acute inflammation reduces food intake and leads to a negative energy balance. Although both types of inflammation stimulate NF-κB (nuclear factor kappa-light-chain-enhancer of activated B cells) signaling, only acute inflammation led to NF-κB activation and increased Pomc transcription [92]. In contrast, HFD-induced chronic inflammation increased *Pomc* promoter methylation at a region which inhibited RELA (the most abundant subunit of NF-κB) binding and transactivation. RELA further bound to STAT3 (signal transducer and activator of transcription 3) and inhibited STAT3-mediated leptin activation of the *Pomc* promoter. In conclusion, this study provides a mechanism for the involvement of RELA in the divergent regulation of energy homeostasis in acute and chronic inflammation and the role of obesity associated low-grade inflammation in epigenetic programming of hypothalamic *Pomc*. Epigenetic programming of hypothalamic *Pomc* by nutritional factors has been corroborated by a number of additional studies including high-fat, high sucrose diet [93], high folate diet [94], leptin treatment [95], and over- or undernutrition [82,96]. At the same time, the importance of perinatal diet consumption for later behavioral and psychiatric disorders is increasingly realized although there is still a lack of studies in which epigenetic programming of metabolism is investigated for its role as risk factor for the development of behavioral, cognitive, and mood disorders later in life [35]. Future studies are therefore needed to address possible relationships between these two phenotypes and to develop therapeutic interventions acting on either or both.

2.9 OTHER EPIGENETIC ASPECTS OF NUTRITIONAL EFFECTS ON NEUROPSYCHIATRIC DISORDERS

As suggested earlier, an individual's nutrition during the perinatal period can influence his or her developing a neuropsychiatric disorder later in life. In addition, an individual's nutrition during prenatal life and during childhood and adulthood can influence his or her developing a neuropsychiatric disorder. Just as in the perinatal period, this can happen due to changes in DNA methylation, histone modifications, and RNA-mediated regulation of gene expression. For more details of epigenetic effects of nutrition on neuropsychiatric disorders the reader is referred elsewhere [97].

2.10 THERAPEUTIC PROSPECTS

How genes and environment conspire together is of greatest interest for psychiatric disorders. Pioneering research over the last decade strongly supports the notion that genetic factors and environmental influences do not operate independently of each other and that past experiences can provide critical instructions in determining the functional output of the information, that is, stored in the genome [12,13]. A large body of human epidemiological studies has firmly established that prenatal and early postnatal environmental factors influence the adult risk for developing various common psychiatric disorders. Hence, molecular epigenetic mechanisms can inscribe early life experiences at the level of the DNA and chromatin of neural cells and thus act at the interface of genes and the environment. Such epigenetic marks can form a kind of molecular memory and provide a plausible route to couple key experiences to alterations in gene expression leading to distinct behavioral and psychiatric phenotypes. In this regard, epigenetic marking of PBMCs has been suggested to provide a useful proxy for neural cells. Although this hypothesis raises many issues, such signatures may still serve as a valuable biomarker for monitoring past stress exposures.

While epigenetic marks such as DNA methylation were originally thought of as irreversible, recent insights into the process of active demethylation point to a more dynamic role of DNA methylation in gene regulation. Consistent with this view, timely behavioral or pharmacological interventions can reverse stress-induced DNA methylation marks and restore behavioral and neuroendocrine phenotypes in rodents. This is indeed good news and suggests that psychiatric diseases resultant from epigenetic marking in response to ELS or trauma are more amenable to therapeutic interventions then those hard-coded by genetic variations. In this respect, combined psychotherapeutic and pharmacological treatments may be more effective than either alone; future translational studies are needed to address this important question at the molecular scale.

ABBREVIATIONS

ACTH	Adrenocorticotropin
AgRp	Agouti-related protein
AVP	Arginine vasopressin
ARC	Arcuate nucleus
CGI	CpG island
ChIP	Chromatin immunoprecipitation
BPD	Borderline personal disorder
CRH	Corticotropin-releasing hormone
DNA	Deoxyribonucleic acid
DNMT	DNA methyltransferase
EGR1	Early growth response 1
ELA	Early life adversity
ELS	Early life stress
GR	Glucocorticoid receptor
HDAC	Histone deacetylase
HFD	High-fat diet
5hmC	5-Hydroymethylcytosine
HPA axis	Hypothalamic pituitary adrenal axis
5mC	5-Methylcytosine
Mecp2	Methyl-CpG-binding protein 2
Mecp2-S421	Serine residue 421 of Mecp2
MDD	Major depressive disorder
MR	Mineralocorticoid receptor
MS	Maternal separation
NPY	Neuropeptide Y
NR3C1	Numan glucocorticoid receptor gene
PBMC	Peripheral blood mononuclear cells
POMC	Pro-opiomelanocortin
PTSD	Posttraumatic stress disorder
PVN	Hypothalamic paraventricular nucleus
SNP	Single nucleotide polymorphism
SP1	Specificity protein 1
SZ	Schizophrenia
Tet	Ten-eleven translocation proteins

GLOSSARY

Glucocorticoids Steroid hormones synthesized and released by the adrenal cortex and which regulate carbohydrate metabolism

Hypothalamic neuropeptides A group of peptides released from hypothalamic neurons which reach the anterior pituitary through the hypothalamic-adenohypophyseal portal system and which regulate the secretion of anterior pituitary hormones

Mineralocorticoids Steroid hormones synthesized and released by the adrenal cortex and which regulate the electrolyte balance of the body

Neuronal plasticity The ability of neurons to stabilize or alter synapses

ACKNOWLEDGMENT

We are thankful to our colleagues for stimulating discussions and advice on the studies dealt with in this chapter.

REFERENCES

[1] Jablonka E, Lamb MJ, Zeligowski A. Evolution in four dimensions. Cambridge, MA: MIT Press; 2005. Verfügbar unter: http://www.gbv.de/dms/mpib-toc/392894742.pdf.

[2] Lander ES, Linton LM, Birren B, Nusbaum C, Zody MC, Baldwin J, et al. Initial sequencing and analysis of the human genome. Nature 2001;409(6822):860–921.

[3] Venter JC, Adams MD, Myers EW, Li PW, Mural RJ, Sutton GG, et al. The sequence of the human genome. Science 2001;291(5507):1304–51.

[4] Allis CD, Caparros M-L, Jenuwein T, Reinberg D. Epigenetics. 2nd Edition Cold Spring Harbor NY: Cold Spring Harbor Laboratory Press; 2015.

[5] Murgatroyd C, Wu Y, Bockmühl Y, Spengler D. Genes learn from stress: how infantile trauma programs us for depression. Epigenetics 2010;5(3):194–9.

[6] Sameroff A. A unified theory of development: a dialectic integration of nature and nurture. Child Dev 2010;81(1):6–22.

[7] Bateson P, Gluckman P, Hanson M. The biology of developmental plasticity and the predictive adaptive response hypothesis. J Physiol 2014;592(11):2357–68.

[8] Lupien SJ, McEwen BS, Gunnar MR, Heim C. Effects of stress throughout the lifespan on the brain, behaviour and cognition. Nat Rev Neurosci 2009;10(6):434–45.

[9] Stiles J, Jernigan TL. The basics of brain development. Neuropsychol Rev 2010;20(4):327–48.

[10] Jaenisch R, Bird A. Epigenetic regulation of gene expression: how the genome integrates intrinsic and environmental signals. Nat Genet 2003;33 Suppl:245–54.

[11] Kucharski R, Maleszka J, Foret S, Maleszka R. Nutritional control of reproductive status in honeybees via DNA methylation. Science 2008;319(5871):1827–30.

[12] Hoffmann A, Spengler D. The lasting legacy of social stress on the epigenome of the hypothalamic-pituitary-adrenal axis. Epigenomics 2012;4(4):431–44.

[13] Hoffmann A, Spengler D. DNA memories of early social life. Neuroscience 2014;264:64–75.

[14] Waddington CH. The strategy of the genes: a discussion of some aspects of theoretical biology. New York: Macmillan; 1957.

[15] Russo VEA, Martienssen RA, Riggs AD. Epigenetic mechanisms of gene regulation. Plainview N.Y: Cold Spring Harbor Laboratory Press; 1996.

[16] Holliday R. A new theory of carcinogenesis. Br J Cancer 1979;40(4):513–22.
[17] Riggs AD. X inactivation, differentiation, and DNA methylation. Cytogenet Cell Genet 1975;14(1):9–25.
[18] Schübeler D. Function and information content of DNA methylation. Nature 2015;517(7534):321–6.
[19] Lister R, Mukamel EA, Nery JR, Urich M, Puddifoot CA, Johnson ND, et al. Global epigenomic reconfiguration during mammalian brain development. Science 2013;341(6146):1237905.
[20] Wu H, Zhang Y. Reversing DNA methylation: mechanisms, genomics, and biological functions. Cell 2014;156(1–2):45–68.
[21] Rothbart SB, Strahl BD. Interpreting the language of histone and DNA modifications. Biochim Biophys Acta 2014;1839(8):627–43.
[22] Heim C, Binder EB. Current research trends in early life stress and depression: review of human studies on sensitive periods, gene-environment interactions, and epigenetics. Exp Neurol 2012;233(1):102–11.
[23] Green JG, McLaughlin KA, Berglund PA, Gruber MJ, Sampson NA, Zaslavsky AM, et al. Childhood adversities and adult psychiatric disorders in the national comorbidity survey replication I: associations with first onset of DSM-IV disorders. Arch Gen Psychiatry 2010;67(2):113–23.
[24] McLaughlin KA, Green JG, Gruber MJ, Sampson NA, Zaslavsky AM, Kessler RC. Childhood adversities and adult psychiatric disorders in the national comorbidity survey replication II: associations with persistence of DSM-IV disorders. Arch Gen Psychiatry 2010;67(2):124–32.
[25] NCANDS. Child Maltreatment 2012 370 L'Enfant Promenade, S.W. Washington, DC. 20447. U.S. Department of Health and Human Services; Report No.: 23 year of reporting. http://www.acf.hhs.gov/sites/default/files/cb/cm2012.pdf
[26] Bennett HA, Einarson A, Taddio A, Koren G, Einarson TR. Prevalence of depression during pregnancy: systematic review. Obstet Gynecol 2004;103(4):698–709.
[27] Ross LE, McLean LM. Anxiety disorders during pregnancy and the postpartum period: a systematic review. J Clin Psychiatry 2006;67(8):1285–98.
[28] Gerardin P, Wendland J, Bodeau N, Galin A, Bialobos S, Tordjman S, et al. Depression during pregnancy: is the developmental impact earlier in boys? A prospective case-control study. J Clin Psychiatry 2011;72(3):378–87.
[29] Silberg JL, Maes H, Eaves LJ. Genetic and environmental influences on the transmission of parental depression to children's depression and conduct disturbance: an extended children of twins study. JChild Psychol Psychiatry 2010;51(6):734–44.
[30] Plagemann A. Perinatal programming and functional teratogenesis: impact on body weight regulation and obesity. Physiol Behav 2005;86(5):661–8.
[31] Barker DJ, Osmond C, Golding J, Kuh D, Wadsworth ME. Growth in utero, blood pressure in childhood and adult life, and mortality from cardiovascular disease. BMJ 1989;298(6673):564–7.
[32] Duncan GJ, Ziol-Guest KM, Kalil A. Early-childhood poverty and adult attainment, behavior, and health. Child Dev 2010;81(1):306–25.
[33] Fox NA, Almas AN, Degnan KA, Nelson CA, Zeanah CH. The effects of severe psychosocial deprivation and foster care intervention on cognitive development at 8 years of age: findings from the Bucharest Early Intervention Project. J Child Psychol Psychiatry 2011;52(9):919–28.
[34] Alwan A. Global Status Report on Noncommunicable Diseases 2014. World Health Organization. http://apps.who.int/iris/bitstream/10665/148114/1/9789241564854_eng.pdf
[35] Patchev AV, Rodrigues AJ, Sousa N, Spengler D, Almeida OFX. The future is now: early life events preset adult behaviour. Acta Physiol 2014;210(1):46–57.
[36] Bilbo SD, Tsang V. Enduring consequences of maternal obesity for brain inflammation and behavior of offspring. FASEB J 2010;24(6):2104–15.
[37] Raygada M, Cho E, Hilakivi-Clarke L. High maternal intake of polyunsaturated fatty acids during pregnancy in mice alters offsprings' aggressive behavior, immobility in the swim test, locomotor activity and brain protein kinase C activity. J Nutr 1998;128(12):2505–11.

[38] Tozuka Y, Kumon M, Wada E, Onodera M, Mochizuki H, Wada K. Maternal obesity impairs hippocampal BDNF production and spatial learning performance in young mouse offspring. Neurochem Int 2010;57(3):235–47.

[39] Yu H, Bi Y, Ma W, He L, Yuan L, Feng J, et al. Long-term effects of high lipid and high energy diet on serum lipid, brain fatty acid composition, and memory and learning ability in mice. Int J Dev Neurosci 2010;28(3):271–6.

[40] Augustyniak RA, Singh K, Zeldes D, Singh M, Rossi NF. Maternal protein restriction leads to hyperresponsiveness to stress and salt-sensitive hypertension in male offspring. Am J Physiol Regul Integr Comp Physiol 2010;298(5):R1375–1382.

[41] D'Asti E, Long H, Tremblay-Mercier J, Grajzer M, Cunnane SC, Di Marzo V, et al. Maternal dietary fat determines metabolic profile and the magnitude of endocannabinoid inhibition of the stress response in neonatal rat offspring. Endocrinology 2010;151(4):1685–94.

[42] Naef L, Moquin L, Dal Bo G, Giros B, Gratton A, Walker C-D. Maternal high-fat intake alters presynaptic regulation of dopamine in the nucleus accumbens and increases motivation for fat rewards in the offspring. Neuroscience 2011;176:225–36.

[43] Naef L, Srivastava L, Gratton A, Hendrickson H, Owens SM, Walker C-D. Maternal high fat diet during the perinatal period alters mesocorticolimbic dopamine in the adult rat offspring: reduction in the behavioral responses to repeated amphetamine administration. Psychopharmacology 2008;197(1):83–94.

[44] Scott KM, McGee MA, Wells JE, Oakley Browne MA. Obesity and mental disorders in the adult general population. J Psychosom Res 2008;64(1):97–105.

[45] Strine TW, Mokdad AH, Dube SR, Balluz LS, Gonzalez O, Berry JT, et al. The association of depression and anxiety with obesity and unhealthy behaviors among community-dwelling US adults. Gen Hosp Psychiatry 2008;30(2):127–37.

[46] Sullivan EL, Nousen EK, Chamlou KA. Maternal high fat diet consumption during the perinatal period programs offspring behavior. Physiol Behav. 2014;123:236–42.

[47] Sullivan EL, Grayson B, Takahashi D, Robertson N, Maier A, Bethea CL, et al. Chronic consumption of a high-fat diet during pregnancy causes perturbations in the serotonergic system and increased anxiety-like behavior in nonhuman primate offspring. J Neurosci 2010;30(10):3826–30.

[48] Vickers MH, Breier BH, Cutfield WS, Hofman PL, Gluckman PD. Fetal origins of hyperphagia, obesity, and hypertension and postnatal amplification by hypercaloric nutrition. Am J Physiol Endocrinol Metab 2000;279(1):E83–7.

[49] Ray GT, Croen LA, Habel LA. Mothers of children diagnosed with attention-deficit/hyperactivity disorder: health conditions and medical care utilization in periods before and after birth of the child. Med Care 2009;47(1):105–14.

[50] Ashwood P, Kwong C, Hansen R, Hertz-Picciotto I, Croen L, Krakowiak P, et al. Brief report: plasma leptin levels are elevated in autism: association with early onset phenotype? J Autism Dev Disord 2008;38(1):169–75.

[51] Heim C, Newport DJ, Mletzko T, Miller AH, Nemeroff CB. The link between childhood trauma and depression: insights from HPA axis studies in humans. Psychoneuroendocrinology 2008;33(6):693–710.

[52] De Kloet ER, Joëls M, Holsboer F. Stress and the brain: from adaptation to disease. Nat Rev Neurosci 2005;6(6):463–75.

[53] Bockmühl Y, Murgatroyd CA, Kuczynska A, Adcock IM, Almeida OFX, Spengler D. Differential regulation and function of 5'-untranslated GR-exon 1 transcripts. Mol Endocrinol 2011;25(7):1100–10.

[54] Weaver IC, Cervoni N, Champagne FA, D'Alessio AC, Sharma S, Seckl JR, et al. Epigenetic programming by maternal behavior. Nat Neurosci 2004;7(8):847–54.

[55] Weaver ICG, Champagne FA, Brown SE, Dymov S, Sharma S, Meaney MJ, et al. Reversal of maternal programming of stress responses in adult offspring through methyl supplementation: altering epigenetic marking later in life. J Neurosci 2005;25(47):11045–54.

[56] McGowan PO, Sasaki A, D'Alessio AC, Dymov S, Labont B, Szyf M, et al. Epigenetic regulation of the glucocorticoid receptor in human brain associates with childhood abuse. Nat Neurosci 2009;12(3):342–8.

[57] Labonte B, Yerko V, Gross J, Mechawar N, Meaney MJ, Szyf M, et al. Differential glucocorticoid receptor exon 1(B), 1(C), and 1(H) expression and methylation in suicide completers with a history of childhood abuse. Biol Psychiatry 2012;72(1):41–8.
[58] Bockmühl Y, Patchev AV, Madejska A, Hoffmann A, Sousa JC, Sousa N, et al. Methylation at the CpG island shore region upregulates Nr3c1 promoter activity after early-life stress. Epigenetics 2015;10(3):247–57.
[59] Murgatroyd C, Patchev AV, Wu Y, Micale V, Bockmühl Y, Fischer D, et al. Dynamic DNA methylation programs persistent adverse effects of early-life stress. Nat Neurosci 2009;12(12):1559–66.
[60] Wu Y, Patchev AV, Daniel G, Almeida OFX, Spengler D. Early-life stress reduces DNA methylation of the Pomc gene in male mice. Endocrinology 2014;155(5):1751–62.
[61] Turecki G, Meaney MJ. Effects of the social environment and stress on glucocorticoid receptor gene methylation: a systematic review. Biol Psychiatry 2016;79(2):87–96.
[62] Storer CL, Dickey CA, Galigniana MD, Rein T, Cox MB. FKBP51 and FKBP52 in signaling and disease. Trends Endocrinol Metab 2011;22(12):481–90.
[63] Chrousos GP, Loriaux DL, Tomita M, Brandon DD, Renquist D, Albertson B, et al. The new world primates as animal models of glucocorticoid resistance. Adv Exp Med Biol 1986;196:129–44.
[64] Thomassin H, Flavin M, Espinás ML, Grange T. Glucocorticoid-induced DNA demethylation and gene memory during development. EMBO J 2001;20(8):1974–83.
[65] Lee RS, Tamashiro KLK, Yang X, Purcell RH, Harvey A, Willour VL, et al. Chronic corticosterone exposure increases expression and decreases deoxyribonucleic acid methylation of Fkbp5 in mice. Endocrinology 2010;151(9):4332–43.
[66] Lee RS, Tamashiro KLK, Yang X, Purcell RH, Huo Y, Rongione M, et al. A measure of glucocorticoid load provided by DNA methylation of Fkbp5 in mice. Psychopharmacology 2011;218(1):303–12.
[67] Sabbagh JJ, O'Leary JC, Blair LJ, Klengel T, Nordhues BA, Fontaine SN, et al. Age-associated epigenetic upregulation of the FKBP5 gene selectively impairs stress resiliency. PLoS One 2014;9(9):e107241.
[68] Blair LJ, Nordhues BA, Hill SE, Scaglione KM, O'Leary JC, Fontaine SN, et al. Accelerated neurodegeneration through chaperone-mediated oligomerization of tau. J Clin Invest 2013;123(10):4158–69.
[69] Klengel T, Mehta D, Anacker C, Rex-Haffner M, Pruessner JC, Pariante CM, et al. Allele-specific FKBP5 DNA demethylation mediates gene-childhood trauma interactions. Nat Neurosci 2013;16(1):33–41.
[70] Mueller BR, Bale TL. Sex-specific programming of offspring emotionality after stress early in pregnancy. J Neurosci 2008;28(36):9055–65.
[71] Chen J, Evans AN, Liu Y, Honda M, Saavedra JM, Aguilera G. Maternal deprivation in rats is associated with corticotrophin-releasing hormone (CRH) promoter hypomethylation and enhances CRH transcriptional responses to stress in adulthood. J Neuroendocrinol 2012;24(7):1055–64.
[72] Maras PM, Baram TZ. Sculpting the hippocampus from within: stress, spines, and CRH. Trends Neurosci 2012;35(5):315–24.
[73] Wang A, Nie W, Li H, Hou Y, Yu Z, Fan Q, et al. Epigenetic upregulation of corticotrophin-releasing hormone mediates postnatal maternal separation-induced memory deficiency. PloS One 2014;9(4):e94394.
[74] Elliott E, Ezra-Nevo G, Regev L, Neufeld-Cohen A, Chen A. Resilience to social stress coincides with functional DNA methylation of the Crf gene in adult mice. Nat Neurosci 2010;13(11):1351–3.
[75] Goncharova ND. Stress responsiveness of the hypothalamic-pituitary-adrenal axis: age-related features of the vasopressinergic regulation. Front Endocrinol 2013;4:26.
[76] Klose RJ, Sarraf SA, Schmiedeberg L, McDermott SM, Stancheva I, Bird AP. DNA binding selectivity of MeCP2 due to a requirement for A/T sequences adjacent to methyl-CpG. Mol cell 2005;19(5):667–78.
[77] Zimmermann CA, Hoffmann A, Raabe F, Spengler D. Role of mecp2 in experience-dependent epigenetic programming. Genes 2015;6(1):60–86.
[78] Murgatroyd C, Spengler D. Polycomb binding precedes early-life stress responsive DNA methylation at the Avp enhancer. PloS One 2014;9(3):e90277.
[79] Lister R, Pelizzola M, Dowen RH, Hawkins RD, Hon G, Tonti-Filippini J, et al. Human DNA methylomes at base resolution show widespread epigenomic differences. Nature 2009;462(7271):315–22.

[80] Li H, Chang Q. Regulation and function of stimulus-induced phosphorylation of MeCP2. Front Biol 2014;9(5):367–75.
[81] Bloomfield FH, Oliver MH, Hawkins P, Holloway AC, Campbell M, Gluckman PD, et al. Periconceptional undernutrition in sheep accelerates maturation of the fetal hypothalamic-pituitary-adrenal axis in late gestation. Endocrinology 2004;145(9):4278–85.
[82] Stevens A, Begum G, Cook A, Connor K, Rumball C, Oliver M, et al. Epigenetic changes in the hypothalamic proopiomelanocortin and glucocorticoid receptor genes in the ovine fetus after periconceptional undernutrition. Endocrinology 2010;151(8):3652–64.
[83] Begum G, Davies A, Stevens A, Oliver M, Jaquiery A, Challis J, et al. Maternal undernutrition programs tissue-specific epigenetic changes in the glucocorticoid receptor in adult offspring. Endocrinology 2013;154(12):4560–9.
[84] Ogden CL, Carroll MD, Kit BK, Flegal KM. Prevalence of childhood and adult obesity in the United States, 2011–2012. JAMA 2014;311(8):806–14.
[85] Bouret SG, Simerly RB. Developmental programming of hypothalamic feeding circuits. Clin Genet 2006;70(4):295–301.
[86] Myers MG, Leibel RL, Seeley RJ, Schwartz MW. Obesity and leptin resistance: distinguishing cause from effect. Trends Endocrinol Metab 2010;21(11):643–51.
[87] Marco A, Kisliouk T, Weller A, Meiri N. High fat diet induces hypermethylation of the hypothalamic Pomc promoter and obesity in post-weaning rats. Psychoneuroendocrinology 2013;38(12):2844–53.
[88] Pan X, Solomon SS, Borromeo DM, Martinez-Hernandez A, Raghow R. Insulin deprivation leads to deficiency of Sp1 transcription factor in H-411E hepatoma cells and in streptozotocin-induced diabetic ketoacidosis in the rat. Endocrinology 2001;142(4):1635–42.
[89] Yang G, Lim C-Y, Li C, Xiao X, Radda GK, Li C, et al. FoxO1 inhibits leptin regulation of pro-opiomelanocortin promoter activity by blocking STAT3 interaction with specificity protein 1. J Biol Chem 2009;284(6):3719–27.
[90] Zhang X, Yang R, Jia Y, Cai D, Zhou B, Qu X, et al. Hypermethylation of Sp1 binding site suppresses hypothalamic POMC in neonates and may contribute to metabolic disorders in adults: impact of maternal dietary CLAs. Diabetes 2014;63(5):1475–87.
[91] Marco A, Kisliouk T, Tabachnik T, Meiri N, Weller A. Overweight and CpG methylation of the Pomc promoter in offspring of high-fat-diet-fed dams are not "reprogrammed" by regular chow diet in rats. FASEB J 2014;28(9):4148–57.
[92] Shi X, Wang X, Li Q, Su M, Chew E, Wong ET, et al. Nuclear factor κB (NF-κB) suppresses food intake and energy expenditure in mice by directly activating the Pomc promoter. Diabetologia 2013;56(4):925–36.
[93] Zheng J, Xiao X, Zhang Q, Yu M, Xu J, Wang Z, et al. Maternal and post-weaning high-fat, high-sucrose diet modulates glucose homeostasis and hypothalamic POMC promoter methylation in mouse offspring. Metab Brain Dis 2015;30(5):1129–37.
[94] Cho CE, Sánchez-Hernández D, Reza-López SA, Huot PSP, Kim Y-I, Anderson GH. High folate gestational and post-weaning diets alter hypothalamic feeding pathways by DNA methylation in Wistar rat offspring. Epigenetics 2013;8(7):710–9.
[95] Palou M, Picó C, McKay JA, Sánchez J, Priego T, Mathers JC, et al. Protective effects of leptin during the suckling period against later obesity may be associated with changes in promoter methylation of the hypothalamic pro-opiomelanocortin gene. Br J Nutr 2011;106(5):769–78.
[96] Plagemann A, Harder T, Brunn M, Harder A, Roepke K, Wittrock-Staar M, et al. Hypothalamic proopiomelanocortin promoter methylation becomes altered by early overfeeding: an epigenetic model of obesity and the metabolic syndrome. J Physiol 2009;587(Pt 20):4963–76.
[97] Peedicayil J. Nutritional effects on epigenetics in psychiatry. In: Peedicayil J, Grayson DR, Avramopoulos D, editors. Epigenetics in psychiatry. Waltham, MA: Elsevier; 2014. p. 563–75.

CHAPTER

EPIGENETIC BIOMARKERS IN NEUROPSYCHIATRIC DISORDERS

3

C.-C. Lin*, T.-L. Huang**

**Department of Psychiatry, Kaohsiung Chang Gung Memorial Hospital and Chang Gung University College of Medicine, Kaohsiung, Taiwan; **Department of Medical Research, Genomic and Proteomic Core Laboratory, Kaohsiung Chang Gung Memorial Hospital, Kaohsiung, Taiwan*

CHAPTER OUTLINE

3.1 INTRODUCTION

Neuropsychiatric disorders are complex, heterogeneous conditions resulting from the interaction of factors including genetic, neurobiological, and cultural factors, and life experiences. Increasing evidence has suggested that epigenetic modifications are involved in the pathophysiology of major neuropsychiatric disorders, namely, schizophrenia (SZ), bipolar disorder (BD), and major depressive disorder (MDD). A modern definition for epigenetics is the structural adaptation of chromosomal regions, which may enhance or impair DNA transcription [1]. There are multiple layers and players in

Neuropsychiatric Disorders and Epigenetics. http://dx.doi.org/10.1016/B978-0-12-800226-1.00003-4

epigenetics which are associated with human neurodevelopmental diseases [2]. Some of the most studied epigenetic mechanisms are DNA methylations, histone modifications, and microRNAs (miRNAs).

DNA methylation is probably the best studied epigenetic mechanism, and it correlates with gene repression of the methylated region [3]. In DNA methylation, a methyl group is added to cytosine, mostly associated with CpG dinucleotides, by DNA methyltransferases (DNMT). Mammals have five known DNMT members: DNMT1, DNMT2, DNMT3A, DNMT3B, and DNMT3L, but only DNMT1, DNMT3A, and DNMT3B have methyltransferase activity [4]. The ten-eleven translocase (TET) protein family converts 5-methylcytosine (5mC) to 5-hydroxymethylcytosine (5hmC) which initiates demethylation [5,6].

Histone modifications include histone acetylation and methylation. Histone acetylation is regulated by histone acetyltransferases (HATs) and histone deacetylases (HDACs), which add or remove an acetyl group to the histones, respectively. Histone acetylation makes DNA more accessible for transcription, and is usually enriched at enhancer elements and gene promoters to facilitate access of transcription factors [7]. Histones could be methylated, by histone methyltransferases (HMTs). Methylation of histone H3 at lysine 9 (H3K9) and 27 (H3K27) are usually repressive. Methylation of histones at other sites such as H3K4 may be activating.

Noncoding RNAs are transcribed from DNA but not translated into proteins. The most well studied noncoding RNAs are miRNAs, which are 21–23 nucleotides long and can inhibit the translation of messenger RNA, thus causing posttranscriptional gene silencing [8].

While there are animal models for neuropsychiatric disorders such as SZ and MDD, certain symptoms of such disorders can only be observed in human subjects, which will be the focus of this chapter. While psychiatric disorders are considered brain diseases, brain tissues are difficult to access. Some studies have indicated that peripheral blood cells can be used as a proxy for brain tissue [9,10]. In this chapter, we will review the recent findings of aforementioned epigenetic mechanisms in SZ, BD, and MDD. Under each disease category, the most commonly investigated candidate molecules will be reviewed, followed by a section on genome-wide studies.

3.2 SCHIZOPHRENIA

SZ is a heterogeneous disease characterized by positive symptoms (such as delusions and hallucinations), negative symptoms (such as flat affect and poverty of speech), and cognitive dysfunction (such as impaired working memory or executive function). Even in monozygotic twins, the concordance rate is merely 50%, indicating that the environment plays an important role in the development of SZ. It is worth mentioning that in studies utilizing postmortem brains, patients with SZ and BD are often grouped together, as patients with "major psychosis."

3.2.1 DNA METHYLATION

DNA methylation and demethylation require DNMTs and TETs, respectively, and there is evidence that levels of these enzymes are altered in the patients with SZ [11]. DNMTs and TETs are abnormally increased in the brains of patients with SZ, and these abnormalities also influence the expression of several candidate genes of SZ, such as brain-derived neurotrophic factor (BDNF), glucocorticoid receptor (GR), glutamic acid decarboxylase 67 (GAD67), and reelin (RELN). The expression of DNA methylating/demethylating enzymes and SZ candidate genes such as BDNF and GR are altered in the same direction in both brain tissue and blood lymphocytes [12]. The TET gene family and activation-induced deaminase/apolipoprotein B mRNA-editing enzymes (AID/APOBEC) are critical in the DNA demethylation pathway. In the inferior parietal lobule of patients with SZ and BD, TET1, but not TET2 and TET3, mRNA and protein expression, as well as the level of 5hmC were increased compared with controls. APOBEC3A mRNA was significantly decreased in the SZ, BD, and MDD groups, while APOBEC3C was decreased only in SZ and BD groups [13]. Increases of DNMT1 and TET1 in peripheral blood lymphocytes of patients with SZ are comparable to those reported in the brain of patients with SZ. Decreased levels of BDNF and GR mRNA were also reported in patients with SZ [14]. Using postmortem brains from 24 patients with SZ and 24 unaffected controls, genome-wide DNA methylation analysis revealed differentially methylated regions (DMR) in the gene encoding DNMT1, among other genes [15].

In the peripheral blood of a south Indian population, DNMT1 rs2114724 and rs2228611 were found to be significantly associated with SZ at genotypic and allelic levels. DNMT3B rs2424932 was associated with the risk of developing SZ in males only. DNMT3B rs1569686 was associated with early onset of SZ and also with family history. DNMT3L rs2070565 was associated with an increased risk of developing SZ at an early age in individuals with a family history [16]. These studies suggest that in patients with SZ, enzymes associated with DNA methylation, such as DNMT and TET, have increased levels compared to controls, both in peripheral blood and in the brain. The changes in the DNA methylation enzymes also influence the expression of genes implicated in the pathophysiology of SZ, such as GAD67, RELN, and BDNF.

3.2.1.1 Gamma-Aminobutyric Acid

Glutamic acid decarboxylase 67 (GAD67) is encoded by the GAD1 gene, crucial for gamma-aminobutyric acid (GABA) neurotransmission in the brain. Decreased expression of GAD1 has been widely replicated in patients with SZ [17–19]. In the PFC of patients with SZ, a significant, on average eightfold, deficit in repressive chromatin-associated DNA methylation (H3K27me3) at the GAD1 promoter was found, suggesting that chromatin remodeling mechanisms are involved in dysregulated GABAergic gene expression in SZ [20]. In the anterior lateral temporal lobe, GAD1 methylation level was found to increase with age, but was not differentially methylated in patients with SZ [21]. In the postmortem inferior parietal lobe and cerebellum, higher 5hmC levels were also detected at the GAD1 promoter region in SZ and BD groups [13]. In the hippocampus of patients with SZ and BD, methylation levels of 1308 GAD1 regulatory network-associated CpG sites were assessed. Fifty four differentially methylated regions were identified in single-group comparisons. Methylation changes were enriched in MSX1, CCND2, and DAXX at specific loci [22]. In a Han Chinese sample of patients with SZ, rs3219151 ($C > T$, GABA A receptor, alpha 6 (GABRA6)) showed significant decreased risk for SZ [23]. In a miRNA-derived network analysis of differentially methylated genes in SZ, GABA receptor B1, although no direct cause and effect have been shown, was found to have a central importance [24].

3.2.1.2 Reelin

The reelin (RELN) gene encodes a protein necessary for neuronal migration, axonal branching, synaptogenesis, and cell signaling. Many studies have shown that RELN expression is reduced in the brains of patients with SZ [18,19]. In the PFC from ten male patients with SZ, RELN promoter hypermethylation was observed, and the level of DNA methylation had an inverse relationship with the expression of the RELN gene [25]. In another study, increased methylation of the RELN promotor was also found in the occipital lobes of patients with SZ [26]. In a Japanese study, in the PFC of controls, significant correlation was found between levels of RELN promoter methylation and age, but no such correlation was detected in the brains of patients with SZ and BD [27]. In another Japanese study, no detectable DNA methylation (<5%) was found in the PFC of both controls and patients with SZ [28]. In peripheral blood, within-pair-difference of methylation of RELN promoter was lower in monozygotic twins than in dizygotic twins [29]. A methylome-wide association study using blood samples of 759 SZ cases and 738 controls found associations to FAM63B, a part of networks regulated by miRNAs that can be linked to neuronal differentiation and dopaminergic gene expression and RELN [30].

3.2.1.3 Brain-Derived Neurotrophic Factor

The BDNF gene consists of 11 exons and 9 promoters [31]. Decreased expression of BDNF has been observed in the PFC of patients with SZ [32]. A SNP of BDNF gene, Val66Met (rs6265), was related to the risk of SZ [33].

In a genome-wide epigenomic analysis of the frontal cortex of patients with SZ and BD, no significant methylation difference was detected in BDNF gene (including upstream of exons I and VI, immediately downstream of exon III, and exon IX) from controls. Val homozygotes of Val66Met SNP in exon IX had significantly higher DNA methylation in the four nearby CpG sites than Val/Met or Met/Met carriers [34]. In the PFC of 15 patients with SZ and 15 controls, analysis of a large genomic region surrounding BDNF promoter IV revealed that a single CpG site at -93 was hypomethylated in patients with SZ [35].

Using DNA from blood samples of patients with SZ and healthy controls, frequency of the BDNF gene methylation was highlighted as a statistically significant relationship between patients and controls regarding decreased risk of disease in comparison to unmethylated patterns [36]. In a Japanese study, BDNF promoters I and IV methylation was analyzed using peripheral blood cells. The patients with SZ had a higher level of methylation at BDNF promoter I compared to controls, although the difference was small. However, no significant difference was found in methylation levels of BDNF promoter IV [37].

3.2.1.4 Sex-Determining Region Y-Box Containing Gene 10

Sex-determining region Y-box containing gene 10 (SOX10) is an oligodendrocyte-specific transcription factor. Downregulation of oligodendrocyte-related genes has been observed in prefrontal lobes of patients with SZ [38]. In brains of patients with SZ, the CpG sites of SOX10 were highly methylated, and the degree of methylation was positively correlated with reduced expression of SOX10 [39]. Using postmortem brains from 24 patients with SZ and 24 controls, a genome-wide DNA methylation analysis revealed DMR in several genes, including SOX10, which had a lower methylation level in patients than in controls [15].

In the peripheral blood of discordant twins, a relative hypermethylation of the SOX10 promoter was found [29].

3.2.1.5 Catechol-O-Methyltransferase

Catechol-O-methyltransferase (COMT) is responsible for degrading catecholamines, such as dopamine, which is disrupted in SZ. While earlier work did not show strong evidence of COMT or its SNP Val158Met (rs4680) directly contributing to the risk of SZ [40,41], rs4680 may be related to certain phenotypes, such as response to antipsychotic treatment [42]. In the frontal lobes of SZ and BD patients, membrane-bound COMT (MB-COMT) promoter DNA is frequently hypomethylated compared with controls, especially in the left frontal lobes [43]. In gene-specific methylation analyses using DNA from leukocytes, the gene encoding soluble COMT was hypermethylated in patients with SZ [44]. In the peripheral blood of 85 patients with SZ, two methylation sites of the soluble COMT promoter region and COMT Val158Met polymorphism were analyzed. In Val/Val homozygous patients, COMT promoter region methylation was negatively correlated to physical activity. In Met/Met homozygous patients, COMT promoter region methylation was positively correlated with physical activity as well as metabolic syndrome [45]. In a study utilizing functional MRI and cryoconserved blood samples, a positive association between MB-COMT promoter methylation and neural activity in the left dorsolateral prefrontal cortex was observed [46].

3.2.1.6 Genome-Wide Analysis

With the advancement of better technology, genome-wide scans of methylation differences between patients with SZ and controls are becoming more affordable. Genome-wide analysis could confirm the significance of known candidate genes, as well as provide insights of associations of new pathways. A genome-wide epigenomic analysis of frontal cortex of patients with SZ and BD revealed methylation differences in loci involved in glutamatergic and GABAergic neurotransmission, brain development, and other processes functionally linked to disease etiology. The same study also detected a strong correlation between DNA methylation in the mitogen-activated protein kinase kinase 1 (MEK1) gene promoter region and lifetime antipsychotic use in patients with SZ [34]. Another postmortem genome-wide DNA methylation study of SZ and BD found that in frontal cortex of both disease groups, widespread hypomethylation as compared to normal controls was found, but in the anterior cingulate gyrus of both disease groups, extensive methylation was found instead. Functional enrichment analysis indicated that important psychiatric disorder-related biological processes such as neuron development, differentiation and projection may be altered by epigenetic changes located in the intronic regions [47]. Using postmortem brains from 24 patients with SZ and 24 unaffected controls, another genome-wide DNA methylation analysis revealed DMR in the genes encoding nitric oxide synthase 1 (NOS1), RAC-alpha serine/threonine-protein kinase (AKT1), dystrobrevin-binding protein 1 (DTNBP1), serine/threonine-protein phosphatase 2B catalytic subunit gamma isoform (PPP3CC), as well as DNMT1 and SOX10 [15]. In the DLPFC of pateints with SZ, a genome-wide DNA methylation profiling revealed 107 differentially methylated CpG sites, and 79 sites (73.8%) were hypermethylated [48]. While most studies have focused on PFC, some studies investigated cerebellum. In the cerebellum of patients with SZ and BD, genome-wide methylation and expression analyzes were compared with controls. Four CpGs were identified to significantly correlate with the differential expression and methylation of genes encoding phosphoinositide-3-kinase, regulatory subunit 1 (PIK3R1), butyrophilin, subfamily 3, member A3 (BTN3A3), nescient helix-loop-helix 1 (NHLH1), and solute carrier family 16, member 7 (SLC16A7) in patients with SZ and BD [49].

Using peripheral blood-derived DNA, a discordant monozygotic twin study analyzed the genome-wide DNA methylation patterns of patients with SZ and BD, and found that the top psychosis-associated, DMR, which was significantly hypomethylated in affected twins, was located in the promoter of ST6GALNAC1, a gene involved in stress-activated kinase signaling [50]. Subsequent pathway analysis showed that the top scoring functional network in the SZ gene list was associated with "nervous system development and function [50]. Global DNA methylation was decreased in SZ twins, but the reduced methylation was significant in males only [29]. Using DNA from leukocytes, global methylation analysis revealed highly significant hypomethylation in patients with SZ. Haloperidol treatment was associated with higher methylation, while early disease onset was associated with lower methylation in patients with SZ [44]. In the peripheral blood- derived DNA of patients with SZ and BD who are Hispanics along the US–Mexico border, the number of hypomethylated DMR is greater than that of hypermethylated DMRs, indicating a global hypomethylation and local hypermethylation in major psychosis [51]. Taken together, global hypomethylation in patients with SZ was observed in most genome-wide studies, and treatment with antipsychotics reversed that phenomenon.

3.2.1.7 Miscellaneous

This section is dedicated to studies that could not fit well into the above categories. Given that many proposed pathways in the pathophysiology of psychiatric disorders interact at many levels, this section is as important as the others. In the postmortem brain tissues of patients with SZ, hypermethylation was observed in serotonin transporter (5HTT) promoter region, especially in drug-free patients. A trend for DNA methylation in the 5HTT promoter region was observed in brain tissues of patients with BD, but was not statistically significant [52]. Using leukocyte DNA, patients with SZ and BD had increased methylation of the promoter region of the serotonin receptor 5HTR1A [53]. From saliva DNA of patients with SZ and BD as well as their first degree relatives, the cytosine of the T102C polymorphic site of 5HTR2A was hypomethylated [54]. In Chinese patients with paranoid and undifferentiated SZ, significant association between average methylation of six CpG sites in the dopamine receptor D4 (DRD4) gene was found, but no association was found between DRD4 methylation status and subtypes of SZ. Further analysis found that the significant association between DRD4 methylation and SZ was found in males, but not females [55]. In another Chinese study, DRD3 was investigated in patients with paranoid and undifferentiated SZ. CpG2 was significantly associated with SZ. Other CpG sites were associated with different subtypes in males or females [56].

In the postmortem brains of patients with SZ and BD, human leukocyte antigen (HLA) complex group 9 gene showed significant differences in both CpG and CpH modifications from controls [57]. In the postmortem DLPFC, hippocampus, anterior cingulate cortex and cerebellum of older patients (average age 77.6 ± 10.1 years) with SZ, analysis of methylation patterns of 19 genes of major neurotransmitter systems (including BDNF, RELN, DRD, COMT, serotonin transporter SLC6A4, GAD1, etc.) found no significant difference between patients and controls [58]. Using three datasets of genome-wide methylation profilings of PFC of patients with SZ, a combined analysis identified seven regions that are consistently differentially methylated in SZ. The regions were near CERS3, DPPA5, PRDM9, DDX43, REC8, LY6G5C, and a region on chromosome 10. PRDM9 encodes a histone methyltransferase [59].

From whole blood DNA of 99 individuals, reduced methylation of leucine-rich repeat transmembrane protein (LRRTM1) promoter showed a significant association with SZ. Sibling pairs concordant for SZ had more similar methylation levels at the LRRTM1 promoter than diagnostically discordant pairs. The study suggests that hypomethylation at the LRRTM1 promoter, particularly of the paternally inherited allele, was a risk factor for the development of SZ [60]. In leukocytes of patients with SZ, rs2661319 in the regulator of G protein signaling 4 (RGS4) gene was significantly associated with suicide attempt. The total methylation of analyzed 119 CpG SNPs was not associated with suicide attempt [61]. In an Iranian study, hypermethylation of glutathione S-transferase T1 (GSTT1) and glutathione S-transferase P1 (GSTP1) promoter regions were associated with SZ [62]. In the peripheral blood of patients with first-episode psychosis, GTP cyclohydrolase 1 (GCH1) promoter region was more methylated than in controls [63].

DNA methylation appeared altered in many systems in patients with SZ. While global DNA hypomethylation was observed, hypermethylation was frequently reported in the promoter regions of candidate genes implicated in SZ. Early results have been promising and warrant further investigations in this area.

3.2.2 HISTONE MODIFICATIONS

Levels of enzymes responsible for histone modifications are also found to be altered in patients with SZ. A study measuring mRNA expression of three HMTs (GLP, SETDB1, and G9a) found increased GLP, SETDB1 mRNA expression in both parietal cortex and lymphocyte samples from patients with SZ. Increased G9a mRNA expression was found in lymphocytes as well. Lymphocyte levels of G9a mRNA is positively correlated with the positive and negative syndrome scale (PANSS) negative subscale total, while GLP mRNA is positively correlated with the PANSS general subscale total. Increased lymphocyte SETDB1 mRNA levels are associated with longer durations of illness and a family history of SZ [64].

There are more studies on HDACs. In the postmortem prefrontal cortices, HDAC1 expression levels were significantly higher in SZ versus controls. Higher HDAC1 expression levels were observed compared to controls, though not significantly. The GAD67 mRNA level was negatively correlated with the HDAC1, HDAC3, and HDAC4 mRNA expression levels [65]. Among 601 tag SNPs in HDAC genes, 10 markers were in significant association with SZ in the exploratory sample, but in the whole sample, only one marker rs14251 in HDAC3 gene was replicated and remained significant [66]. GWAS data from over 60,000 participants from the Psychiatric Genomics Consortium were analyzed for enriched pathways in 3 adult psychiatric disorders: SZ, BD, and MDD, and the histone H3–K4 methylation pathway was one of the top 5 pathways in SZ. Integrative pathway analysis showed that histone methylation showed strongest association across these three disorders [67]. In a Chinese sample of patients with SZ, genotype frequencies of four tag SNPs on HDAC2 (rs10499080, rs6568819, rs2499618, and rs13204445) and two tag SNPs on HDAC3 (rs11741808, rs2530223) genes were not significantly different from controls. Further haplotype-based haplotype relative risk test and the transmission disequilibrium test data found no association between those SNPs and SZ [68]. In another Chinese study, among the 19 SNPs of the HDAC2 gene, 3 SNPs, rs13212283, rs6568819, and rs9488289, were nominally associated with SZ in the initial test group. However, these results could not be verified in the validation group consisting of 896 cases and 1815 controls. Haplotype analysis also failed to confirm the associations [69]. While some studies showed that altered levels of HMT and HDAC as well as their SNPs may be associated with SZ, the two Chinese studies found no positive association thus far. Ethnic difference could affect the association between different subtypes of enzymes and SZ.

A study investigating the phosphorylation, acetylation, and methylation of 6 lysine, serine, and arginine residues of histones H3 and H4, and 16 metabolic gene transcripts of prefrontal cortex of 41 subjects with SZ and 41 matched controls found no significant alterations in histone profiles or gene expression between the two groups. A subgroup of eight patients with SZ with levels of H3-(methyl)arginine 17, H3meR17, exceeding control values by 30% was associated with decreased expression of four metabolic transcripts [70]. In human prefrontal cortex of patients with SZ, decreased GAD1 expression and H3K4-trimethylation (H3K4me3), predominantly in females and in conjunction with a risk haplotype at the 5′ end of GAD1, were observed. Mixed-lineage leukemia 1 (Mll1) is a histone methyltransferase expressed in GABAergic and other cortical neurons. In the same study, heterozygosity for a truncated, lacZ knock-in allele of Mll1 resulted in decreased H3K4 methylation at GABAergic gene promoters. In contrast, after clozapine treatment, GAD1 H3K4me3 and Mll1 occupancy was increased in the cerebral cortex of mice. These effects were not mimicked by haloperidol or genetic ablation of dopamine D2 and D3 receptors [71]. In the PFC of patients with SZ, a significantly lower methylation level of eight CpG sites of the GAD1 gene was found in repressive chromatin (H3K27me3) compared to controls. Methylation levels of GAD1 did not differ significantly between patients and controls with open chromatin (H3K4me3). These findings suggest that histone modification also affect DNA methylation [20]. In the postmortem frontal cortices of patients with SZ and BD, promoter-associated acetylated H3K9K14, two epigenetic marks associated with transcriptionally active chromatin, are correlated with gene expression levels of GAD1, 5-hydroxytryptamine receptor 2C (HTR2C), translocase of outer mitochondrial membrane 70 homolog A (TOMM70A) and protein phosphatase 1E (PPM1E) [72]. Another study found increased H3K9me2 levels in both postmortem parietal cortex and lymphocyte samples of patients with SZ [64]. In peripheral blood, patients with BD had significantly higher baseline levels of H3K9 and K14 acetylation compared to patients with SZ [73]. In lymphocytes, patients with SZ had significantly higher mean baseline levels of H3K9me2, a repressive chromatin mark, than healthy controls. The study also found a negative correlation between age at onset of illness and levels of H3K9me2 [74]. Taken together, histone modifications in SZ appeared to be altered, but more data are required to have a better understanding of the roles of different modifications on different histones at different genes.

3.2.3 MicroRNAs

Just as in the case of DNA methylation and histone modifications, there is also evidence that proteins associated with miRNA processing may be altered in SZ. In Chinese patients with SZ, two SNPs in the DGCR8 and DICER, both miRNA processing genes, were significantly associated with altered risk of SZ. The genotype or allele frequency of rs3742330 in DICER was significantly different in patients [75].

There are a large number of miRNAs and so the studies about them are also abundant. In a study comparing the expression levels of 264 miRNAs from postmortem prefrontal cortex tissue of patients with SZ or schizoaffective disorder with controls, 16 miRNAs were differentially expressed, and 15 were downregulated in SZ, though the maximum increase in change was only 1.77 fold [76]. Using postmortem cortical gray matter from the superior temporal gyrus (STG), an analysis of global miRNA expression found significant upregulation of miR-181b expression in schizophrenia [77]. miRNA-346 is encoded by a gene located in intron 2 of the glutamate receptor ionotropic delta 1 (GRID1) gene. In a study using DLPFC samples of patients with SZ, BD, and healthy controls, the mean miR-346 expression levels were significantly lower in patients with SZ compared to controls. There was a similar, though not statistically significant, trend for patients with BD. The mean GRID1 expression levels in patients with SZ and BD were lower compared to controls, though not significantly. The expression levels of miR-346 and GRID1 is less correlated in patients with SZ than in patients with BD or in controls [78]. In another miRNA expression profiling of STG and the DLPFC, miRNA expression profiles and a global expression increase were observed in similar patterns in both regions (21% of expressed miRNAs in the STG and 9.5% of expressed miRNAs in the DLPFC) [79]. Four miRNAs of the Beveridge et al. studies (miR-24, miR-26b, miR-29c, and miR-7) overlapped with the set of significantly changed miRNAs from the Perkins et al. study. However, in the Perkins et al. study, all four were downregulated, while in the Beveridge et al. studies all were upregulated [80]. Using DLPFC of patients with SZ and BD, 441 miRNAs were analyzed. Twenty two differentially expressed miRNAs from controls were identified. Seven dysregulated miRNAs were identified in patients with SZ, and they were miR-7, miR-34a, miR-132, miR-132*, miR-154*, miR-212, and miR-544. A negative correlation was found between miR-132 and miR-212 expression and their targets (tyrosine hydroxylase and phosphogluconate dehydrogenase) expression [81]. In the postmortem DLPFC of patients with SZ and schizoaffective disorder, 28 differentially expressed miRNAs were identified, and 89% of them were increased compared to controls [82]. Four thirty five miRNAs and 18 small nucleolar RNAs were analyzed in postmortem brain tissue samples from patients with SZ or BD, and 19% were differentially expressed in SZ or BD. Both diagnoses were associated with reduced miRNA expression levels, though a more pronounced effect was observed for BD [83]. In the PFC, miR-30b expression was significantly reduced in the cerebral cortex of female but not male patients with SZ. In addition, disease-related changes in miR-30b expression in females were further modulated by estrogen receptor alpha (Esr1) SNP rs2234693 genotype [84]. In the PFC of patients with SZ, miR-132 was downregulated, among 854 miRNAs investigated [85]. Comparing exosomal miRNAs from frozen postmortem prefrontal cortices of patients with SZ and controls, miR-497 had increased expression in the SZ group [86]. In the dentate gyrus (DG) granule cells in postmortem human hippocampus, miR-182 target gene expression did not differ significantly between carriers of rs76481776 T-allele and noncarriers in patients with SZ and MDD [87]. The above studies of human brains reveal associations of a number of miRNAs with SZ, though the results are not always consistent. Interestingly, several miRNAs were affected by antipsychotic treatment, so the miRNAs have the potential to predict treatment response.

In a study investigating eighteen known SNPs within or near brain-expressed miRNAs from DNA extracted from whole blood samples of Danish, Swedish, and Norwegian patients with SZ and controls, two SNPs rs17578796 and rs1700 in mir-206 and mir-198 showed nominal significant allelic association with SZ in the Danish and Norwegian samples respectively, but only rs17578796 was significant in the joint sample [88]. Analysis of 59 miRNA genes on the X-chromosome from males with and without SZ spectrum disorders revealed that eight ultrarare variants in the precursor or mature miRNA were identified in eight distinct miRNA genes in 4% of analyzed males with SZ, while only one ultrarare variant was identified in a control sample (with a history of depression) [89]. In a Chinese sample, ss178077483 located in the pre-mir-30e was found to be strongly associated with SZ. ss178077483, combined with mir-30e rs7556088 and miR-24-MAPK14 rs3804452, showed a weak gene–gene interaction for SZ risk. Patients with SZ also had higher levels of mature mir-30e in peripheral leukocytes than controls [90]. In the mononuclear leukocytes of Taiwanese patients with SZ, seven miRNAs (hsa-miR-34a, miR-449a, miR-564, miR-432, miR-548d, miR-572, and miR-652) were found to be differentially expressed in a genome-wide comparison with controls. Those miRNAs were then validated in another sample of patients, but only miR-34a was differentially expressed [91]. In a genome-wide association study of SZ, five new loci and two previously implicated were found. The strongest new finding was with rs1625579 within an intron of a putative primary transcript for miRNA-137. Four other SZ loci were predicted targets of miR-137. In a joint analysis with a BD sample, three loci reached genome-wide significance: calcium channel CACNA1C (rs4765905), ankyrin-3 (ANK3, rs10994359), and the inter-alpha-trypsin inhibitor heavy chain H3 (ITIH3)-ITIH4 region (rs2239547) [92]. In serum samples of patients with SZ, miR-92a, miR-181b, miR-219-2-3p, miR-1308, and let-7g levels were increased, while miR-17 and miR-195 levels were decreased, as compared to controls. When compared with familial SZ patients, miR-219-2-3p, miR-92a, miR-346, let-7g, and miR-17 were significantly higher in sporadic SZ. On the contrary, miR-181b, miR-195, and miR-1308 were downregulated in sporadic SZ compared with SZ patients with a positive family history. miR-346 level was increased in the risperidone treatment group compared to healthy controls and other drug treatment group (clozapine, olanzapine, and aripiprazole) [93]. In a sample of Chinese patients with SZ, hsa-pre-mir-146a rs2910164 and hsa-mir-499 rs3746444 were not associated with SZ, though certain genotypes were associated with clinical symptoms such as hallucinations and avolition [94]. In a Han Chinese sample of patients with SZ, rs3219151 [C > T, GABA A receptor, alpha 6 (GABRA6)] showed significantly decreased risk for SZ. miR-124 showed significantly repressed 3' UTR binding to regulator of G protein signaling 4 (RGS4) mRNA from the rs10759-C allele [23]. Several miRNA studies focused on Japanese patients. rs112439044, in the miRNA-30e gene, was investigated in Japanese patients with SZ and healthy controls. The genotype distributions of rs112439044 were not significantly different between patients and controls, but there was a significant association between the T allele and SZ [95]. The mutated T allele frequency of rs139365823 in miR-138-2 gene was significantly lower in Japanese patients with SZ than in controls [96]. Resequencing of the miR-185 gene in patients with SZ identified two rare patient-specific novel variants flanking the miR-185 gene, but follow-up genotyping provided no further evidence of their involvement in SZ [97]. In the plasma samples of 40 patients with first-episode SZ treated with risperidone for a year and who had achieved remission, miR-365 and miR-520c-3p levels were significantly decreased after 1 year of risperidone treatment. Those two miRNA levels were not significantly correlated with clinical symptoms, however [98]. In 20 drug-free patients with SZ, miRNA-181b, miRNA-30e, miRNA-34a, and miRNA-7 levels were higher, as compared to controls. After 6 weeks of antipsychotic (olanzapine, quetiapine, ziprasidone, and risperidone) treatment, miR-81b was significantly downregulated. Furthermore, the change of miRNA-181b expression was positively correlated with the improvement of some of the items on the symptom scales [99]. Plasma levels of 10 miRNAs of patients with SZ were compared with healthy controls. A panel of miR-30e, miR-181b, miR-34a, miR-346, and miR-7 had significantly increased levels with significant combined diagnostic value for SZ. After 6 weeks of antipsychotic treatment, miR-132, miR-181b, miR-432, and miR-30e levels decreased significantly. The improvement of clinical symptomatology was correlated with the changes of miR-132, miR-181b, miR-212, and miR-30e levels. Decreases of levels of miR-132 and miR-432 were greater in patients with better treatment response [100]. miR-132, miR-134, miR-1271, miR-664*, miR-200c, and miR-432 levels were significantly decreased in PBMCs of patients with SZ compared with healthy controls. After antipsychotic treatment, increase in miR-132, miR-664*, and miR-1271 levels in patients with SZ compared with the baseline levels [101].

Similar to the findings of miRNA studies of postmortem brains, a large number of miRNAs may be associated with SZ in the peripheral blood, and many of them change according to antipsychotic treatment or status of treatment response. More robust evidence is needed to confirm their usefulness as biomarkers, because many of those findings were not well replicated. In contrast, miRNA-137 and its SNP rs1625579 have been studied more extensively.

Since a genome-wide association study found rs1625579 of the miR-137 gene was the strongest new finding [92], many studies sought to verify the importance of rs1625579 and miR-137 since then. In patients with SZ, "G" allele on the miR-137 SNP rs1625579 predicted deficiency of cognitive functions in combination with higher severity of negative symptoms [102]. In a study genotyping rs1625579 of the miR-137 gene of patients with SZ, schizoaffective disorder, and bipolar I disorder, carriers of "T" allele had more cognitive deficits involving episodic memory and attention control [103].

In a study utilizing functional magnetic resonance imaging and genotyping of SNP rs1625579 of miR-137, no difference was found in rs1625579 allele frequency distribution between patients with SZ and controls. Patients with SZ showed significantly higher left DLPFC retrieval activation on working memory load 3, lower working memory performance, and longer response times compared with controls, though the genotype had no effect on working memory performance or response times. Individuals with the rs1625579 TT genotype had higher left DLPFC activation than those with the GG/GT genotypes [104]. In another study to investigate rs1625579 genotype and brain region volumes with structural magnetic resonance imaging, no significant relationship was found between rs1635579 genotype and brain volume in patients with SZ or controls [105]. In patients with SZ, the rs1625579 TT genotype was associated with attenuated reduction of midposterior corpus callosum volume [106].

In a Han Chinese sample of patients with SZ, among 33 SNPs of the miR137 gene and the target gene CACNA1C of miR-13, rs1625579 in the miR-137 gene and rs1006737, and rs4765905 in the CACNA1C gene, had significantly different allele and genotype frequencies from controls. Haplotype rs1006737-rs4765905-rs882194 in CACNA1C showed significant association with SZ, and two haplotypes (ACC and ACT) in the block were significantly increased in the patients [107]. SNP rs1625579 of miR-137 gene was genotyped in Han Chinese patients with SZ, and genotype and allele distributions of the rs1625579 were significantly different between patients and controls [108]. In another sample of Han Chinese patients with SZ, genotype and allele frequencies of allelic variants of rs66642155, a variable number tandem repeat polymorphism, and the SNP rs1625579 were not significantly different between patient and control populations. Age at onset was much later in wild type rs66642155 carriers than in mutation carriers among patients [109]. In a Han Chinese population, no significant difference was detected in either allele or genotype frequency in rs1625579 between patients with SZ and controls [110]. In a Japanese study, resequencing of the miR-137 gene in patients with SZ revealed four sequence variations in the 5′ and 3′ flanking regions, but those were not significantly associated with SZ [111]. The studies of miR-137 and rs1625579 were promising, although two Chinese studies and one Japanese study found no association with SZ. Ethnic difference may be in play here, but further research will be needed to confirm that.

3.3 BIPOLAR DISORDER

Bipolar I disorder is a unique psychiatric disease which features three distinct mood states: manic, depressive, and euthymic states. In a severe manic state, patients could experience severe agitation and delusions, clinically indistinguishable from SZ (although a thorough history taking could distinguish the two). As previously mentioned, many postmortem studies grouped patients with SZ and BD together, as "major psychosis." In a depressive state, the clinical symptoms are similar to MDD. In bipolar II disorder, patients can be in a less disturbing, hypomanic state. These changing mood states have to be taken into consideration when interpreting studies of BD.

3.3.1 DNA METHYLATION

mRNA expression levels of four DNMT isoforms were analyzed in the peripheral white blood cells of patients with BD and MDD in euthymic or depressive states. DNMT1 mRNA levels were significantly decreased in the depressive but not in the euthymic state of MDD and BD compared to DNMT3B mRNA levels [112].

3.3.1.1 BDNF

Blood BDNF protein levels and their relationship with various mood states of BD have been well studied [113,114]. The manic state of BD is usually managed by both mood stabilizers and antipsychotics, and earlier studies found an increase in BDNF levels following treatment for acute mania [113].

A genome-wide epigenomic analysis of the frontal cortex of patients with SZ and BD found no significant methylation difference in the tested regions of the BDNF gene (upstream of exons I and VI, immediately downstream of exon III, and exon IX) between patients with SZ and BD and controls. Val homozygotes of Val66Met SNP in exon IX had significantly higher DNA methylation in the four nearby CpG sites [34]. Another study using postmortem frontal cortex of BD patients found hypermethylation of CpG islands in BDNF promoter regions. The study also found global hypermethylation in BD patients, as well as significantly decreased BDNF mRNA expression [115].

In leukocyte-derived DNA, global methylation levels did not differ between euthymic patients with BD and control subjects [116]. Using peripheral blood mononuclear cells (PBMC) of patients with bipolar I and II disorder and controls, the DNA methylation levels of promoter region of BDNF exon I were analyzed. Hypermethylation of the promoter region and downregulation of BDNF gene expression were observed in patients with bipolar II disorder, but not in those with bipolar I disorder. DNA methylation levels of different mood states were evaluated in this study. When looking at bipolar I and bipolar II disorder separately, no statistically significant methylation difference was detected between different mood states, although the manic state in the bipolar II group appeared to have a lower methylation level. When patients with bipolar I and II disorder were pooled together, DNA methylation level was reduced in patients in mania/hypomania/mixed state, as compared to patients in euthymic state or those in depressive state. Higher levels of DNA methylation were observed in patients treated with both mood stabilizers and antidepressants than patients treated with mood stabilizers only. Treatment with lithium and sodium valproate was also associated with DNA hypomethylation compared with treatment with other drugs [117]. Using venous blood, methylation fractions differed between patients with bipolar I disorder (higher methylation) and controls for 11 out of 36 CpG sites investigated in BDNF promoters III and V. Four CpG sites in promoter V and one in promoter III remained significant after false discovery rate correction [118]. Using DNA extracted from PBMC, 207 patients with MDD, 59 patients with BD, and 278 controls were included for BDNF exon I promoter methylation. Significantly increased methylation was found in patients with MDD compared to patients with BD and controls. The increased methylation in patients with MDD was associated with antidepressant treatment. None of the 12 investigated SNPs, including Val66Met, showed significant genotype-methylation interactions [119]. In a follow-up study of the previous investigation [117], using PBMC DNA from 43 patients with MDD, 61 patients with bipolar I disorder, and 50 patients with bipolar II disorder, as well as 44 age-matched controls, 17 CpG sites in the BDNF exon I promoter were evaluated for methylation levels. Increased methylation levels were found in patients with MDD and bipolar II disorder, but not bipolar I disorder. When looking at different mood states, methylation levels of patients in depressive state were significantly higher, compared to the levels of patients in manic/mixed states. Patients treated with mood stabilizers lithium and sodium valproate showed overall lower levels [120].

In summary, BDNF exon I was well studied, but different studies might not always cover the exact same region [121]. Methylation changes in bipolar I disorder appear less obvious than changes in bipolar II disorder or MDD. Treatment with mood stabilizers lithium or valproate was associated with lower methylation levels of the studied regions, while antidepressant treatment was associated with higher methylation levels.

3.3.1.2 Serotonin (5HT)

The serotonin (5HT) system is the target of many antidepressants, and is implicated in mood disorders. The relationship between the insertion/deletion polymorphism 5HTTLPR of 5HT transporter gene SLC6A4 has not been consistent [122], but decreased 5HT transporter has been observed in positron emission tomography and single photon emission computed tomography [123,124].

In a promoter-wide DNA methylation analysis of lymphoblastoid cell lines derived from two pairs of monozygotic twins discordant for BD, hypermethylation of the SLC6A4 promoter was observed. That finding was confirmed in postmortem brains of patients with BD [125]. In the postmortem brain tissues of patients with SZ, hypermethylation was observed in the SLC6A4 promoter region, especially in drug-free patients. A trend for DNA methylation in SLC6A4 promoter region was observed in brain tissues of patients with BD, but was not statistically significant [52]. The two studies using peripheral tissues for analysis have been mentioned in the SZ section before. Using leukocyte DNA, patients with SZ and BD had increased methylation of the promoter region of the serotonin receptor 5HTR1A [53]. From saliva DNA of patients with SZ or BD as well as their first degree relatives, the cytosine of the T102C polymorphic site of HTR2A was hypomethylated [54].

3.3.1.3 GABA

GABA dysregulation has been studied in both SZ and BD [126]. The following postmortem studies have been mentioned previously. A genome-wide epigenomic analysis revealed methylation differences in loci involved in GABAergic neurotransmission [34]. Higher 5hmC levels were also detected at the GAD1 promoter region in SZ and BD groups [13]. Fifty four differentially methylated regions were identified in GAD1 regulatory network-associated CpG loci in both SZ and BD [22].

3.3.1.4 Human Leukocyte Antigen Complex Group 9 Gene

Increased methylation of CpG sites upstream of HLA complex group 9 gene (HCG9) was found in the postmortem PFC of female patients with SZ and BD as compared with controls (20% vs. 15%) [34]. In a follow-up study investigating DNA methylation differences of HCG9 in BD using DNA samples from postmortem brains, peripheral white blood cells, and sperm, lower DNA methylation was found in the analyzed region in the samples from patients with BD. It is interesting that the three distinctively different tissues showed similar methylation patterns [127]. The HCG gene in the postmortem brains of patients with SZ and BD showed significant differences in both CpG and CpH modifications from controls [57].

3.3.1.5 Genome-Wide Analysis

Several of the studies below had been mentioned in the section on SZ earlier. A postmortem genome-wide DNA methylation study of SZ and BD found that in frontal cortex of both disease groups, widespread hypomethylation as compared to normal controls was found, but in the anterior cingulate gyrus of both disease groups, extensive methylation was found instead [47]. In the cerebellum of patients with SZ and BD, genome-wide methylation and expression analyses found four CpGs that were differentially methylated [49]. A discordant monozygotic twin study that used genome-wide analysis of DNA methylation on peripheral blood DNA samples of SZ and BD found that ST6GALNAC1 was significantly hypomethylated in affected twins. Subsequent pathway analysis showed that the top scoring functional network in the BD gene list, in the top network was "a developmental, genetic and neurological disorder [50]. In the peripheral blood-derived DNA of patients with SZ and BD who are Hispanics along the US–Mexico border, a global hypomethylation and local hypermethylation in major psychosis were observed [51]. In leukocyte DNA from whole blood of patients with bipolar I disorder, global methylation was significantly influenced by insulin resistance, second-generation antipsychotic use, and smoking [128].

3.3.1.6 *Miscellaneous*

Again, some of the studies mentioned here had been reported in the SZ section before. In the frontal lobes of SZ and BD patients, MB-COMT promoter DNA was frequently hypomethylated compared with controls [43]. Similar to the findings in brain, in DNA from the saliva of patients with SZ and BD, MB-COMT promoter was hypomethylated compared to controls [129]. In peripheral blood- derived DNA, patients with bipolar I disorder in euthymic state had increased methylation at the FK506-binding protein 5 (FKBP5) gene, as compared with controls [130]. DNA methylation studies in BD appear less abundant than SZ, although many candidate molecules appeared promising, such as BDNF. However, more work is needed to better understand the role of DNA methylation in the pathophysiology of BD.

3.3.2 HISTONE MODIFICATIONS

In the postmortem prefrontal cortices, higher HDAC1 expression levels were observed in patients with BD, but not significantly, compared to controls [65]. GWAS data from over 60,000 participants from the Psychiatric Genomics Consortium were analyzed for enriched pathways in 3 adult psychiatric disorders: SZ, BD, and MDD. The major pathway for BD was histone H3–K4 methylation. Integrative pathway analysis showed that histone methylation showed strongest association across the three psychiatric disorders [67].

In the postmortem frontal cortices of patients with SZ and BD, promoter-associated acetylated histone H3 at lysines 9/14 (ac-H3K9K14), two epigenetic marks associated with transcriptionally active chromatin, are correlated with gene expression levels of GAD1, HTR2C, TOMM70A, and PPM1E [72]. In postmortem PFC of patients with BD and MDD, splice variants of the synapsin family and its relationship with H3K4me3 promoter enrichment were investigated. H3K4me3 enrichment in the SYN2 promoter correlated with the expression upregulation of SYN2a in BD [131]. The data on histone modifications in BD patients seem scarce, and most were analyzed alongside with SZ. Further research in this field will be needed. Clinical studies that investigate changes in different mood states will be beneficial to the understanding of BD.

3.3.3 MicroRNAs

miR-346 is encoded by a gene located in intron 2 of the glutamate receptor ionotropic delta 1 (GRID1) gene. In a study using DLPFC samples of patients with SZ, BD, and healthy controls, the mean miR-346 expression levels were lower, but not statistically significant, in patients with BD compared to controls. The expression levels of miR-346 and GRID1 is less correlated in patients with SZ than in patients with BD or in controls [78]. Using DLPFC cortices of patients with SZ and BD, 441 miRNAs were analyzed. Of the 15 dysregulated miRNAs found in BD samples, 7 were over-expressed and 8 were underexpressed [81]. Four thirty five miRNAs and 18 small nucleolar RNAs were analyzed in postmortem brain tissue samples from patients with SZ or BD, and 19% were differentially expressed in SZ or BD. Both diagnoses were associated with reduced miRNA expression levels, though a more pronounced effect was observed for BD [83]. Comparing exosomal miRNAs from frozen postmortem prefrontal cortices of patients with BD and controls, miR-29c had increased expression in the bipolar group [86]. In the DLPFC of patients with SZ and BD, miR-137 expression in the DLPFC did not differ between diagnoses. rs1625579 genotyping only found that significantly lower miR-137 expression levels were observed in the homozygous TT subjects compared to TG and GG subjects in the control group, but not in the disease groups or overall samples. Reduced miR-137 levels in TT subjects corresponded to increased levels of the miR-137 target gene TCF4, but did not correspond to levels of other target genes (ZNF804A and CACNA1C) [132]. In the DG granule cells in postmortem human hippocampus, carriers of rs76481776 T-allele in patients with BD and controls had significantly different miR-182 target gene expression levels from noncarriers [87].

In the peripheral blood, a significant association was found between the T allele of the rs76481776 polymorphism in the pre-miR-182 and late insomnia in patients with major depressive episode (217 with MDD and 142 individuals with bipolar depression) [133]. In the plasma of 21 manic patients with bipolar I disorder, miR-134 levels in drug-free, 2-week medicated, and 4-week medicated patients were significantly decreased when compared with controls, and the level increased following medication. Decreased miR-134 levels in drug-free and medicated patients were negatively correlated with the severity [134]. Carriers of the "T" allele of rs1625579 of the miR-137 gene had more cognitive deficits involving episodic memory and attention control among patients with SZ, schizoaffective disorder, and bipolar I disorder [103]. In the blood of bipolar I patients treated with lithium or sodium valproate, miR-206 rs16882131 and BDNF rs6265 were genotyped. Individually, neither SNP was associated with risk of bipolar I disorder and treatment response. However, a significant gene to gene interaction between the miR-206 rs16882131 and BDNF rs6265 polymorphisms was found to contribute to bipolar I disorder susceptibility and treatment response. Patients with miR-206 TT + TC and BDNF AA genotypes had a significantly lower mean treatment score than those with miR-206 CC and BDNF AA + AG as well as those with miR-206 CC and BDNF GG genotypes [135]. In a genome-wide association data set of 9,747 patients with BD and 14,278 controls, 609 miRNAs were scanned, and 9 miRNAs showed a significant association with BD after correction for multiple testing. The most promising were miR-499, miR-708, and miR-1908. Pathway analyses revealed pathways associated with brain development and neuron projection [136]. Compared with SZ, there are fewer studies in BD regarding epigenetics. BDNF appeared to be the most frequently studied molecule, but even then, a firm conclusion can not be drawn. Further research will be needed to better understand the role of epigenetic modifications in bipolar disorder, especially in different mood states.

3.4 MAJOR DEPRESSIVE DISORDER

MDD is a disabling disease characterized by depressed mood, loss of interest, impaired vegetative functions, worthlessness, and most dangerous of all, suicidal ideation and behavior. Less severe depression can exist as dysthymic disorder or adjustment disorder.

3.4.1 DNA METHYLATION

DNMT-1 and DNMT-3b expression were altered in several brain regions of brains of suicide completers [137]. mRNA expression levels of four DNMT isoforms were analyzed in the peripheral white blood cells of patients with BD and MDD in euthymic or depressive states. DNMT1 mRNA levels were significantly decreased in the depressive but not in the euthymic state of MDD and BD. DNMT3B mRNA levels were significantly increased in the depressive but not in the euthymic state in MDD only [112]. mRNA level of activation-induced deaminase/apolipoprotein B mRNA-editing enzymes (AID/APOBEC), important in the DNA demethylation pathway, was decreased in the SZ, BD, and MDD groups [13].

3.4.1.1 BDNF

BDNF has been chosen as a candidate molecule for the development of depression [138]. A lower serum level of BDNF was found in patients in MDD [139]. The BDNF SNP Val66Met/rs6265 has been well studied in MDD [140]. Comparing the methylation levels in the Wernicke area of brains of suicide completers and controls, increased methylation was found in BDNF exon IV promoters in the brains of suicide completers, and the higher methylation level corresponded with lower BDNF mRNA levels [141]. However, not all suicide completers are always patients having MDD.

Genomic DNA from peripheral blood of 20 Japanese patients with MDD and 18 controls were analyzed. Twenty nine CpG sites out of 35 CpG sites in BDNF exon I were differentially methylated between the two groups. No significant methylation difference was detected in CpG sites of BDNF exon IV between the two groups [142]. In a study analyzing DNA from peripheral blood of 108 patients with MDD, increased methylation of BDNF promoter IV was significantly associated with a previous suicidal attempt history, suicidal ideation during treatment, and suicidal ideation at last evaluation and with higher Beck Scale for suicide ideation scores and poor treatment outcomes for suicidal ideation [143]. In the PBMC of MDD patients, increased methylation of BDNF exon I promoter and decreased BDNF gene expression were observed. No significant difference was observed between depressed and euthymic mood states. Patients treated with antidepressants alone had increased methylation compared to patients treated with both mood stabilizers and antidepressants. When both mood states and medication were analyzed together, patients in depressed state treated with only antidepressants had increased methylation levels than depressed patients treated with both mood stabilizers and antidepressants, a phenomenon not observed among patients in euthymic states [144]. In leukocytes of patients with MDD, methylation levels of 12 CpG sites of BDNF exon IV promoter were analyzed. Lack of methylation of CpG site −87 was associated with treatment nonresponse [145]. Using DNA extracted from PBMC, 207 patients with MDD, 59 patients with BD, and 278 controls were included for the study of BDNF exon I promoter methylation. Significantly increased methylation was found in patients with MDD compared to patients with BD and controls. The increased methylation in patients with MDD was associated with antidepressant treatment. None of the 12 investigated SNPs, including Val66Met, showed significant genotype–methylation interactions [119]. Following up previous studies [117,144], PBMC DNA from 43 patients with MDD, 61 patients with bipolar I disorder, and 50 patients with bipolar II disorder, as well as 44 age-matched controls were analyzed for methylation levels of 17 CpG sites in the BDNF exon I promoter. Increased methylation levels were found in patients with MDD and bipolar II disorder, but not bipolar I disorder. When looking at different mood states, methylation levels of patients in depressive state were significantly higher, compared to the levels of patients in manic/mixed states. Patients treated with mood stabilizers (lithium or sodium valproate) showed overall lower levels [120]. Seven thirty two Koreans aged ≥65 years were followed up for 2 years, with 521 of them having no depression at baseline. Higher BDNF methylation was independently associated with the prevalence and incidence of depression and severe depressive symptoms. No significant interaction was found between BDNF methylation levels and Val66Met genotype [146]. The majority of the data focused on BDNF exon I and IV, and the results on exon I appeared more promising. Increased BDNF methylation was more consistently observed in MDD and depressive mood state. In summary, similar to BDNF studies of BD, exon I was the most frequently investigated, partly because several studies investigated both diagnoses simultaneously. More work is needed to confirm these findings.

3.4.1.2 Monoamines

As mentioned previously, serotonin (5HT) is a monoamine and is the target of many antidepressants. Polymorphisms in the promoter region (5HTTLPR) of the 5HT transporter SLC6A4 have been widely studied in MDD [147]. Monoamine oxidase A (MAO-A) is involved in monoamine metabolism, and is also the target of antidepressants (MAO inhibitors). A variable-number-tandem-repeat (VNTR) polymorphism in the promoter region of the MAO-A gene had been studied in MDD and treatment response [148]. In the buccal cells of 25 depressed and 125 nondepressed adolescents, depressive symptoms were more common among those with increased SLC6A4 methylation who carried the 5HTTLPR short-allele [149]. In 108 patients with MDD undergoing 12-week treatment with antidepressants, hypermethylation of SLC6A4 promoter was associated with childhood adversities and worse clinical presentations, but not with treatment outcomes [150]. Using DNA from peripheral blood leukocytes of 84 monozygotic twin pairs, intrapair difference in SLC6A4 promoter methylation variation at 10 of the 20 investigated CpG sites correlated with intrapair differences in depression scores [151]. In a study using blood DNA from patients with MDD or panic disorder, methylation levels of the SLC6A2 gene promoter were not significantly different from healthy controls, and no significant change was observed after antidepressant treatment [152]. In peripheral blood of MDD patients, methylation rates for several CpG sites of SLC6A4 differed significantly after treatment, and the methylation rate of CpG 3 in patients with better treatment responses was significantly higher than that in patients with poorer responses. However, the methylation profiles of SLC6A4 were not significantly different between MDD patients and controls, or between unmedicated patients and medicated patients. The 5HTTLPR allele was also not associated with the methylation profiles [153]. In whole blood DNA, lower average methylation of nine CpG sites in the SLC6A4 transcriptional control region upstream of exon 1A was found to be associated with impaired escitalopram response after 6 weeks. 5HTTLPR haplotype was neither associated with SLC6A4 DNA methylation nor treatment response [154].

The same group examined the 43 CpG sites in MAO-A regulatory and exon1/intron1 regions, and also genotyped the MAO-A VNTR, in patients with MDD treated with escitalopram for 6 weeks. No significant association was found between methylation status and treatment response. No association was detected between MAO-A VNTR genotypes and MAO-A methylation status [155]. In saliva from females with MDD, mixed anxiety depression or dysthymia, early parental death was associated with hypermethylation of one CpG site close to an NGFI-A binding site of the NR3C1 gene covering 1F. Regression analysis showed that this association could be mediated by the MAO-A long allele [156]. In a follow-up replication study, 10 CpG sites of MAO-A gene from saliva were analyzed. In females, linear regression generated a model that could explain 32% of the variance in average MAO-A methylation, and in this model, subjects with a history of MDD, mixed anxiety depression or dysthymia were hypomethylated compared to controls. In males, the linear regression did not generate a significant model [157]. In summary, increased methylation was frequently observed in the promoter region of the SLC6A4 gene. Data on 5HTTLPR appeared less consistent. More studies are needed on MAO-A to draw a conclusion.

3.4.1.3 Hypothalamic-Pituitary-Adrenal Axis

Hyperactivity of the hypothalamic-pituitary-adrenal (HPA) axis is one of the most consistent findings in MDD [158]. The GR is the most important regulator of the HPA axis, and GR sensitivity has been shown to be reduced in MDD [159].

In a study of postmortem brains of patients with MDD, low methylation levels of GR promoters were observed in both the patient and control groups [160]. In blood samples of 101 patients with borderline personality disorder with a high rate of childhood maltreatment and 99 patients with MDD with a low rate of childhood maltreatment as well as 15 MDD patients with comorbid posttraumatic stress disorder (PTSD), increased methylation of 8 CpG sites of the GR gene (NR3C1), including a portion of exon 1F, correlated with the severity, the number, and the type of maltreatments of childhood sexual abuse [161]. In saliva from females with MDD, mixed anxiety depression or dysthymia, early parental death was associated with hypermethylation of one CpG site close to an NGFI-A binding site of the NR3C1 gene covering 1F [156]. Patients with MDD had significantly lower methylation than controls at 2 CpG sites in the NR3C1 promoter region using peripheral blood DNA. In MDD, methylation had a positive correlation with the bilateral cornu ammonis (CA) 2–3 and CA4-DG subfields, but in controls, methylation had a positive correlation with the subiculum and presubiculum. No differences in total and subfield volumes of the hippocampus were found between patients with MDD and controls, but patients with MDD had a significantly thinner cortex in the left rostromiddle frontal, right lateral orbitofrontal, and right pars triangularis areas, as compared to controls [162]. A significant number of epigenetic studies of GR focused on suicide completers, childhood abuse, and lifetime adversities. However, here we have only included studies with depression.

3.4.1.4 Genome-Wide Analysis

One of the first genome-wide DNA methylation scans focused on MDD compared 39 postmortem frontal cortices of patients with MDD with those of 26 controls. Of the 3.5 million CpGs covered, 224 candidate regions showed methylation differences of greater than 10%. The greatest methylation difference was in the roline-rich membrane anchor 1 (PRIMA1) gene, which encodes a protein anchor for acetylcholinesterase in neuronal membranes [163]. In 18 pairs of monozygotic adolescent twins in which one member scored consistently higher on self-rated depression than the other, genome-wide DNA methylation analysis using buccal cell DNA identified a list of differentially methylated probes, which were then tested in postmortem cerebellum from patients with MDD. Two probes were identified in the process. One was hypermethylated in MDD cerebellum, and was located within STK32C, which encodes a serine/threonine kinase. The other was hypomethylated, mapping to the first intron of DEP domain containing the protein DEPDC7. No differences in overall mean genome-wide DNA methylation was found between twins with depression and unaffected twins. However, the difference in variance was significant [164]. In ventral PFC of 53 suicide completers with MDD, an eightfold greater number of methylated CpG sites relative to controls and greater DNA methylation changes than the increased methylation observed in normal aging were identified [165].

Genome-wide methylation profile analysis for over 14,000 genes using DNA extracted from whole blood revealed that individuals who reported a lifetime history of depression had increased methylation in regions involved in brain development and tryptophan metabolism and decreased methylation in regions involved in lipoprotein metabolism, compared to individuals without a history of depression [166]. In a genome-wide scan of CpG sites using WBC-derived DNA between 12 pairs of monozygotic twins discordant for MDD (cases) and 12 pairs of monozygotic twins concordant for no MDD (controls), no overall difference in mean global methylation between cases and controls was found, but the difference in variance across all probes between cases and controls was highly significant. Female cases had significantly increased methylation throughout the genome when compared with controls, suggesting a gender difference in methylation patterns [167]. In 17 pairs of monozygotic twins, those with at least one lifetime diagnosis of anxious or depressive disorder were identified, and the twins were categorized into 7 healthy pairs, 6 discordant pairs, and 4 concordant pairs. Using their peripheral blood DNA to identify the differentially methylated probes, one of the most relevant outcomes was the association between hypomethylation of cg01122889 in WD repeat-containing protein 26 (WDR26) gene and a lifetime diagnosis of depression [168]. In the whole blood DNA of 50 monozygotic twins discordant for MDD, hypermethylation within the coding region of zinc finger and BTB domain containing 20 (ZBTB20) gene, an important factor in the development of the hippocampus, was hypermethylated in the twins with MDD, who also showed increased global variation in methylation [169]. Genome-wide DNA methylation profiling of peripheral leukocytes found differential methylation of 363 CpG sites, and all of them were hypomethylated in patients with MDD as compared to controls. 85.7% of those CpG sites were located in the gene promoter regions. Those markers were validated in another group of MDD patients [170].

3.4.1.5 Miscellaneous

In the $GABA_A$ receptor alpha1 subunit promoter region of the brains of suicide completers, 3 sites of 16 analyzed CpG sites were hypermethylated compared to controls [137]. In peripheral blood DNA of 17 pairs of monozygotic twins, intrapair DNA methylation differences in an intron of DEPDC7 were associated with intrapair differences in current depressive symptoms, evaluated with brief symptom inventory (BSI). Hypomethylation difference of the particular site correlated with differences between BSI scores between the twins [171]. Methylation in regions proximal to the transcription start site (TSS) of long-chain polyunsaturated fatty acid biosynthetic genes, fatty acid desaturases 1 (Fads1) and 2 (Fads2), and elongation of very long-chain fatty acids protein 5 (Elovl5) were studied in a group of MDD patients and controls. MDD patients had lower CpG methylation within the Fads2 upstream but higher methylation at the Elovl5 upstream proximal regions from the TSS as compared to controls. Among the MDD patients, suicide attempters had significantly lower CpG methylation levels within the downstream Elovl5 TSS region, but higher methylation in the upstream Elovl5 region compared to suicide nonattempters [172]. Overall, a significant amount of work found methylation abnormalities in candidate genes or genome-wide studies. However, not all results were consistent, so more work is still needed to better understand the pathophysiological mechanisms underlying MDD.

3.4.2 HISTONE MODIFICATIONS

GWAS data from over 60,000 participants from the Psychiatric Genomics Consortium were analyzed for enriched pathways in 3 adult psychiatric disorders: SZ, BD, and MDD. Integrative pathway analysis showed that histone methylation showed strongest association across the three psychiatric disorders [67]. In postmortem prefrontal lobes, MDD patients with antidepressant treatment history had significantly lower H3K27 methylation levels in BDNF exon IV than MDD patients without antidepressant treatment history and controls [173]. In postmortem PFC of patients with BD and MDD, H3K4me3 enrichment in the SYN2 promoter correlated with the expression upregulation of SYN2b in MDD [131].

In peripheral blood of drug-naive MDD patients, a significant decrease in H3K27me3 levels at BDNF promoter IV was detected after 8 weeks of citalopram treatment. That decrease was mainly contributed by changes among treatment responders, and there was no significant change among nonresponders. Between responders and nonresponders, the H3K27me3 levels were the same before treatment, but responders had lower levels after treatment. Negative correlations were also found between changes in depression severity and changes in H3K27me3 expression, as well as BNDF mRNA levels and H3K27me3 levels [174]. There has not been much work regarding histone modifications in MDD, especially lacking is genome-wide analysis. We will need more research to address this isuse [175].

3.4.3 MicroRNAs

Abnormalities of miRNA processing genes such as DGR8 and DICER have been observed in patients with depression. In patients with MDD, SNPs of miRNA processing genes were different from healthy controls. A variant allele of DGCR8 rs3757 was associated with increased risk of suicidal tendency and improved response to antidepressant treatment. A variant allele of AGO1 rs636832 showed decreased risk of suicidal tendency, suicidal behavior, and recurrence. No significant differences were found in GEMIN4 rs7813 between patients and healthy controls, however [176]. In the blood of patients with PTSD with comorbid depression, DICER1 expression was significantly reduced as compared to controls. A DICER1 SNP, rs10144436, was significantly associated with DICER1 expression and with the depression-associated PTSD. Genome-wide analysis of miRNAs showed significant downregulation in depressed PTSD patients [177].

In the prefrontal cortex of depressed, drug-naive suicide completers, 21 miRNAs were significantly decreased as compared to controls [178]. In the ventrolateral PFC of depressed patients, miR-1202 expression was decreased. In another sample, baseline blood levels of miR-1202 were significantly decreased in remitters after 8 weeks of citalopram treatment, compared to nonresponders and controls. After treatment, the blood levels of miR-1202 significantly increased, and the depression severity score was negatively correlated with change in miR-1202 expression [179]. In the postmortem prefrontal areas of suicide completers with MDD history, four miRNAs levels (those of miR-34c-5p, miR-139-5p, miR-195, and miR-320c) were increased, among the ten miRNAs predicted to target the 3'UTR of polyamine genes SAT1 and SMOX. Those miRNA levels and the expression levels of SAT1 and SMOX were also correlated [180]. In the DG granule cells in postmortem human hippocampus, miR-182 target gene expression did not differ significantly between carriers of rs76481776 T-allele and noncarriers in patients with SZ and MDD [87]. The CSF analysis of 179 miRNAs of MDD patients identified that 11 miRNAs showed increased levels and 5 miRNAs showed lower levels when compared to controls. Consequent serum analysis revealed that three miRNAs (miR-221-3p, miR-34a-5p, and let-7d-3p) had increased levels and 1 miRNA (miR-451a) had decreased levels. Those results were validated in another group of MDD patients [181]. In drug-free MDD patients, CSF miR-16 was significantly lower than that in controls, and was negatively correlated with depression scores and positively associated with CSF serotonin. Blood miR-16 was not significantly different between patients and controls and was not statistically correlated with CSF miR-16 [182].

A significant association was found between the T allele of the rs76481776 polymorphism in the pre-miR-182 and late insomnia in patients with major depressive episode (217 with MDD and 142 individuals with bipolar depression) [133]. In the PBMC of MDD patients, two miRNAs (miR-941 and miR-589) were differentially expressed at baseline and at 8 weeks, and maintained stable over-expression during an 8-week period [183]. In the blood of 10 depressed subjects after 12 weeks of treatment with escitalopram, 28 miRNAs were increased, and 2 miRNAs were decreased. Pathway analysis revealed association with neuronal brain function [184]. In serum, patients with depression had lower serum BDNF levels and higher serum miR-132 and miR-182 levels, as compared to controls. A positive correlation was found between serum miR-132 level and self-rating depression scale score. A negative correlation was found between serum BDNF levels and miR-132/miR-182 levels in depression [185]. In PBMC of MDD patients, five miRNA (miRNA-26b, miRNA-1972, miRNA-4485, miRNA-4498, and miRNA-4743) levels were increased in MDD patients, in the initial scan of 723 human miRNAs. Those miRNAs were enriched in pathways related to nervous system and brain function [186]. A GWAS found that SNP rs41305272, a predicted miR-330-3p target site, in mitogen-activated protein kinase kinase 5 (MAP2K5) mRNA was associated with MDD in African-American samples, but not in European-American samples. rs41305272*T carrier frequency was also correlated with the number of anxiety and depressive disorders diagnosed per subject [187]. In the peripheral blood of 169 patients with depression, anxiety or stress, and adjustment disorder, plasma miR-144-5p levels at baseline were significantly lower compared with the healthy controls, and increased significantly after an 8-week follow-up. The plasma miR-144-5p expression level was also inversely correlated with the self-rated depression score [188]. Two SNPs, rs10877887 CC genotype and rs13293512 CC genotype, in the promoters of let-7 family were associated with an increased risk of MDD. rs13293512 TC and CC genotypes had an increased risk of recurrence after treatment [189]. In drug-free MDD patients, CSF miR-16 was significantly lower than that in controls, and was negatively correlated with depression scores and positively associated with CSF serotonin. Blood miR-16 was not significantly different between patients and controls and was not statistically correlated with CSF miR-16 [182]. A number of miRNAs were associated with the diagnosis, phenotype, or treatment response in depressive disorders. However, more work is still needed to confirm their usefulness as biomarkers.

3.5 CONCLUSIONS

Epigenetics in neuropsychiatric disorders is a new frontier to help us better understand these brain disorders. Neuropsychiatric disorders are heterogenous disorders, often involve interactions of multiple known and possibly unknown pathways. Epigenetic research of known candidate genes gives us a better understanding of the known pathways, and genome-wide analysis could confirm those findings as well as provide insights into new pathways or new interactions between pathways.

A trend in recent epigenetic researches is the increasing utilization of in silico analysis. Pathway analysis may be used before the experiments, to help identify the targets for study. For example, in silico analysis helped narrow down the number of miRNAs to measure in experiments. Pathway analysis could also be used to interpret a large amount of data. Genome-wide analysis often generates a large set of data, and pathway analysis could help simplify and rank such data. Another trend is the increasing complexity of experimental designs. Most recent epigenetic studies have utilized a 2-step or 3-step experimental design. For example, use of in silico analysis to identify a set of desired targets, then measuring those targets in the brain or peripheral tissues of patients, and then verifying the findings in another sample of patients, other tissues, cell cultures, or animal models. In human studies, a majority of the studies have focused on PFC (or a region of the PFC), with a few focusing on the cerebellum. In animal models, a circuitry of epigenetic abnormalities has been established [190], but in humans, most postmortem studies were limited to the PFC, with a few studies with cerebellum [175]. Studies investigating a wider area of brains, or scanning for more components of several pathways, will help us understand the interactions between brain regions and pathways better.

Human epigenetic studies are in their early stages. Several better studied candidate genes, such as RELN and GAD1 in SZ and BDNF in BD and MDD, seem promising as future biomarkers. Given the heterogeneity of psychiatric disorders, a panel of molecules will be needed to establish a satisfying validity. In that aspect, more clinical research will be needed in the future, given psychiatric disorders' varying phenotypes (such as the varying mood states of BD). Certain molecules might not be associated with the diagnosis, but they could be relevant in critical phenotypes such as suicide and treatment response. Such data require thorough history taking and clinical observations. In conclusion, epigenetics in neuropsychiatric disorders still requires more work, but this research presents a window to have a glimpse into the complexity of the human brain.

ABBREVIATIONS

BD Bipolar disorder
BDNF Brain-derived neurotrophic factor
COMT Catechol-O-methyl transferase
CSF Cerebrospinal fluid
DLPFC Dorsolateral prefrontal cortex
DMR Differentially methylated regions
GAD1 Glutamic acid decarboxylase 1
GR Glucocorticoid receptor
GWAS Genome-wide association study
MAO Monoamine oxidase
MDD Major depressive disorder
miRNA MicroRNA
mRNA Messenger RNA
PBMC Peripheral blood mononuclear cells
PFC Prefrontal cortex
PTSD Posttraumatic stress disorder
SNP Single nucleotide polymorphism
SZ Schizophrenia
VNTR Variable number of tandem repeats

GLOSSARY

Biomarker A naturally occurring molecule or characteristic using which a particular pathological or physiological process, disease, etc., can be identified

ACKNOWLEDGMENTS

The research data mentioned in the chapter are supported by clinical research grants (CMRPG8B0761, CMRPG8C0831, and CMRPG8D1471) from Kaohsiung Chang Gung Memorial Hospital and clinical research grants (NSC 99-2628-B-182-002-MY2 and MOST 103-2314-B-182A-012) from the Ministry of Science and Technology (previously known as National Science Council) in Taiwan.

REFERENCES

[1] Bird A. Perceptions of epigenetics. Nature 2007;447(7143):396–8.
[2] LaSalle JM, Powell WT, Yasui DH. Epigenetic layers and players underlying neurodevelopment. Trends Neurosci 2013;36(8):460–70.
[3] Bergman Y, Cedar H. DNA methylation dynamics in health and disease. Nat Struct Mol Biol 2013;20(3): 274–81.
[4] Portela A, Esteller M. Epigenetic modifications and human disease. Nat Biotechnol 2010;28(10):1057–68.
[5] Gavin DP, Chase KA, Sharma RP. Active DNA demethylation in post-mitotic neurons: a reason for optimism. Neuropharmacology 2013;75:233–45.
[6] Guo JU, Su Y, Zhong C, Ming GL, Song H. Hydroxylation of 5-methylcytosine by TET1 promotes active DNA demethylation in the adult brain. Cell 2011;145(3):423–34.
[7] Wang Z, Zang C, Rosenfeld JA, Schones DE, Barski A, Cuddapah S, et al. Combinatorial patterns of histone acetylations and methylations in the human genome. Nat Genet 2008;40(7):897–903.
[8] Peedicayil J. Epigenetic approaches for bipolar disorder drug discovery. Expert Opin Drug Discov 2014;9(8):917–30.
[9] Boulle F, van den Hove DL, Jakob SB, Rutten BP, Hamon M, van Os J, et al. Epigenetic regulation of the BDNF gene: implications for psychiatric disorders. Mol Psychiatry 2012;17(6):584–96.
[10] Davies MN, Volta M, Pidsley R, Lunnon K, Dixit A, Lovestone S, et al. Functional annotation of the human brain methylome identifies tissue-specific epigenetic variation across brain and blood. Genome Biol 2012;13(6):R43.
[11] Grayson DR, Guidotti A. The dynamics of DNA methylation in schizophrenia and related psychiatric disorders. Neuropsychopharmacology 2013;38(1):138–66.
[12] Guidotti A, Auta J, Davis JM, Dong E, Gavin DP, Grayson DR, et al. Toward the identification of peripheral epigenetic biomarkers of schizophrenia. J Neurogenet 2014;28(1–2):41–52.
[13] Dong E, Gavin DP, Chen Y, Davis J. Upregulation of TET1 and downregulation of APOBEC3A and APOBEC3C in the parietal cortex of psychotic patients. Transl Psychiatry 2012;2:e159.
[14] Auta J, Smith RC, Dong E, Tueting P, Sershen H, Boules S, et al. DNA-methylation gene network dysregulation in peripheral blood lymphocytes of schizophrenia patients. Schizophr Res 2013;150(1):312–8.
[15] Wockner LF, Noble EP, Lawford BR, Young RMcD, Morris CP, Whitehall VLJ, et al. Genome-wide DNA methylation analysis of human brain tissue from schizophrenia patients. Transl Psychiatry 2014;4:e339.
[16] Saradalekshmi KR, Neetha NV, Sathyan S, Nair IV, Nair CM, Banerjee M. DNA methyl transferase (DNMT) gene polymorphisms could be a primary event in epigenetic susceptibility to schizophrenia. PloS One 2014;9(5):e98182.
[17] Akbarian S, Kim JJ, Potkin SG, Hagman JO, Tafazzoli A, Bunney WE Jr, et al. Gene expression for glutamic acid decarboxylase is reduced without loss of neurons in prefrontal cortex of schizophrenics. Arch Gen Psychiatry 1995;52(4):258–66.
[18] Guidotti A, Auta J, Davis JM, Di-Giorgi-Gerevini V, Dwivedi Y, Grayson DR, et al. Decrease in reelin and glutamic acid decarboxylase67 (GAD67) expression in schizophrenia and bipolar disorder: a postmortem brain study. Arch Gen Psychiatry 2000;57(11):1061–9.

[19] Torrey EF, Barci BM, Webster MJ, Bartko JJ, Meador-Woodruff JH, Knable MB. Neurochemical markers for schizophrenia, bipolar disorder, and major depression in postmortem brains. Biol Psychiatry 2005;57(3):252–60.
[20] Huang HS, Akbarian S. GAD1 mRNA expression and DNA methylation in prefrontal cortex of subjects with schizophrenia. PloS One 2007;2(8):e809.
[21] Siegmund KD, Connor CM, Campan M, Long TI, Weisenberger DJ, Biniszkiewics D, et al. DNA methylation in the human cerebral cortex is dynamically regulated throughout the life span and involves differentiated neurons. PloS One 2007;2(9):e895.
[22] Ruzicka WB, Subburaju S, Benes FM. Circuit- and diagnosis-specific DNA methylation changes at gamma-aminobutyric acid-related genes in postmortem human hippocampus in schizophrenia and bipolar disorder. JAMA Psychiatry 2015;72(6):541–51.
[23] Gong Y, Wu CN, Xu J, Feng G, Xing QH, Fu W, et al. Polymorphisms in microRNA target sites influence susceptibility to schizophrenia by altering the binding of miRNAs to their targets. Eur Neuropsychopharmacol 2013;23(10):1182–9.
[24] Gumerov V, Hegyi H. MicroRNA-derived network analysis of differentially methylated genes in schizophrenia, implicating GABA receptor B1 [GABBR1] and protein kinase B [AKT1]. Biol Direct 2015;10:59.
[25] Abdolmaleky HM, Cheng KH, Russo A, Smith CL, Faraone SV, Wilcox M, et al. Hypermethylation of the reelin (RELN) promoter in the brain of schizophrenic patients: a preliminary report. Am J Med Genet 2005;134B(1):60–6.
[26] Grayson DR, Jia X, Chen Y, Sharma RP, Mitchell CP, Guidotti A, et al. Reelin promoter hypermethylation in schizophrenia. Proc Natl Acad Sci USA 2005;102(26):9341–6.
[27] Tamura Y, Kunugi H, Ohashi J, Hohjoh H. Epigenetic aberration of the human REELIN gene in psychiatric disorders. Mol Psychiatry 2007;12(6):519. 593–600.
[28] Tochigi M, Iwamoto K, Bundo M, Komori A, Sasaki T, Kato N, et al. Methylation status of the reelin promoter region in the brain of schizophrenic patients. Biol Psychiatry 2008;63(5):530–3.
[29] Bonsch D, Wunschel M, Lenz B, Janssen G, Weisbrod M, Sauer H. Methylation matters? Decreased methylation status of genomic DNA in the blood of schizophrenic twins. Psychiatry Res 2012;198(3):533–7.
[30] Aberg KA, McClay JL, Nerella S, Clark S, Kumar G, Chen W, et al. Methylome-wide association study of schizophrenia: identifying blood biomarker signatures of environmental insults. JAMA Psychiatry 2014;71(3):255–64.
[31] Pruunsild P, Kazantseva A, Aid T, Palm K, Timmusk T. Dissecting the human BDNF locus: bidirectional transcription, complex splicing, and multiple promoters. Genomics 2007;90(3):397–406.
[32] Hashimoto T, Bergen SE, Nguyen QL, Xu B, Monteggia LM, Pierri JN, et al. Relationship of brain-derived neurotrophic factor and its receptor TrkB to altered inhibitory prefrontal circuitry in schizophrenia. J Neurosci 2005;25(2):372–83.
[33] Gratacos M, Gonzalez JR, Mercader JM, de Cid R, Urretavizcaya M, Estivill X. Brain-derived neurotrophic factor Val66Met and psychiatric disorders: meta-analysis of case-control studies confirm association to substance-related disorders, eating disorders, and schizophrenia. Biol Psychiatry 2007;61(7):911–22.
[34] Mill J, Tang T, Kaminsky Z, Khare T, Yazdanpanah S, Bouchard L, et al. Epigenomic profiling reveals DNA-methylation changes associated with major psychosis. Am J Hum Genet 2008;82(3):696–711.
[35] Keller S, Errico F, Zarrilli F, Florio E, Punzo D, Mansueto S, et al. DNA methylation state of BDNF gene is not altered in prefrontal cortex and striatum of schizophrenia subjects. Psychiatry Res 2014;220(3):1147–50.
[36] Kordi-Tamandani DM, Sahranavard R, Torkamanzehi A. DNA methylation and expression profiles of the brain-derived neurotrophic factor (BDNF) and dopamine transporter (DAT1) genes in patients with schizophrenia. Mol Biol Rep 2012;39(12):10889–93.
[37] Ikegame T, Bundo M, Murata Y, Kasai K, Kato T, Iwamoto K. DNA methylation of the BDNF gene and its relevance to psychiatric disorders. J Hum Genet 2013;58(7):434–8.

[38] Sugai T, Kawamura M, Iritani S, Araki K, Makifuchi T, Imai C, et al. Prefrontal abnormality of schizophrenia revealed by DNA microarray: impact on glial and neurotrophic gene expression. Ann NY Acad Sci 2004;1025:84–91.
[39] Iwamoto K, Bundo M, Yamada K, Takao H, Iwayama-Shigeno Y, Yoshikawa T, et al. DNA methylation status of SOX10 correlates with its downregulation and oligodendrocyte dysfunction in schizophrenia. J Neurosci 2005;25(22):5376–81.
[40] Hosak L. Role of the COMT gene Val158Met polymorphism in mental disorders: a review. Eur Psychiatry 2007;22(5):276–81.
[41] Okochi T, Ikeda M, Kishi T, Kawashima K, Kinoshita Y, Kitajima T, et al. Meta-analysis of association between genetic variants in COMT and schizophrenia: an update. Schizophr Res 2009;110(1–3):140–8.
[42] Huang E, Zai CC, Lisoway A, Maciukiewicz M, Felsky D, Tiwari AK, et al. Catechol-O-methyltransferase Val158Met polymorphism and clinical response to antipsychotic treatment in schizophrenia and schizoaffective disorder patients: a meta-analysis. Int J Neuropsychopharmacol 2016;19(5):1–12.
[43] Abdolmaleky HM, Cheng KH, Faraone SV, Wilcox M, Glatt SJ, Gao F, et al. Hypomethylation of MB-COMT promoter is a major risk factor for schizophrenia and bipolar disorder. Hum Mol Genet 2006;15(21): 3132–45.
[44] Melas PA, Rogdaki M, Osby U, Schalling M, Lavebratt C, Ekstrom TJ. Epigenetic aberrations in leukocytes of patients with schizophrenia: association of global DNA methylation with antipsychotic drug treatment and disease onset. FASEB J 2012;26(6):2712–8.
[45] Lott SA, Burghardt PR, Burghardt KJ, Bly MJ, Grove TB, Ellingrod VL. The influence of metabolic syndrome, physical activity and genotype on catechol-O-methyl transferase promoter-region methylation in schizophrenia. Pharmacogenomics J 2013;13(3):264–71.
[46] Walton E, Liu J, Hass J, White T, Scholz M, Roessner V, et al. MB-COMT promoter DNA methylation is associated with working-memory processing in schizophrenia patients and healthy controls. Epigenetics 2014;9(8):1101–7.
[47] Xiao Y, Camarillo C, Ping Y, Arana TB, Zhao H, Thompson PM, et al. The DNA methylome and transcriptome of different brain regions in schizophrenia and bipolar disorder. PloS One 2014;9(4):e95875.
[48] Numata S, Ye T, Herman M, Lipska BK. DNA methylation changes in the postmortem dorsolateral prefrontal cortex of patients with schizophrenia. Front Genet 2014;5:280.
[49] Chen C, Zhang C, Cheng L, Reilly JL, Bishop JR, Sweeney JA, et al. Correlation between DNA methylation and gene expression in the brains of patients with bipolar disorder and schizophrenia. Bipolar Disord 2014;16(8):790–9.
[50] Dempster EL, Pidsley R, Schalkwyk LC, Owens S, Georgiades A, Kane F, et al. Disease-associated epigenetic changes in monozygotic twins discordant for schizophrenia and bipolar disorder. Hum Mol Genet 2011;20(24):4786–96.
[51] Li Y, Camarillo C, Xu J, Arana TB, Xiao Y, Zhao Z, et al. Genome-wide methylome analyses reveal novel epigenetic regulation patterns in schizophrenia and bipolar disorder. Biomed Res Int 2015;2015:201587.
[52] Abdolmaleky HM, Nohesara S, Ghadirivasfi M, Lambert AW, Ahmadkhaniha H, Ozturk S, et al. DNA hypermethylation of serotonin transporter gene promoter in drug naive patients with schizophrenia. Schizophr Res 2014;152(2–3):373–80.
[53] Carrard A, Salzmann A, Malafosse A, Karege F. Increased DNA methylation status of the serotonin receptor 5HTR1A gene promoter in schizophrenia and bipolar disorder. J Affect Disord 2011;132(3):450–3.
[54] Ghadirivasfi M, Nohesara S, Ahmadkhaniha HR, Eskandari MR, Mostafavi S, Thiagalingam S, et al. Hypomethylation of the serotonin receptor type-2A Gene (HTR2A) at T102C polymorphic site in DNA derived from the saliva of patients with schizophrenia and bipolar disorder. Am J Med Genet 2011;156B(5): 536–45.
[55] Cheng J, Wang Y, Zhou K, Wang L, Li J, Zhuang Q, et al. Male-specific association between dopamine receptor D4 gene methylation and schizophrenia. PloS One 2014;9(2):e89128.

[56] Dai D, Cheng J, Zhou K, Lv Y, Zhuang Q, Zheng R, et al. Significant association between DRD3 gene body methylation and schizophrenia. Psychiatry Res 2014;220(3):772–7.
[57] Pal M, Ebrahimi S, Oh G, Khare T, Zhang A, Kaminsky ZA, et al. High precision DNA Modification Analysis of HCG9 in major psychosis. Schizophr Bull 2016;42(1):170–7.
[58] Alelu-Paz R, Gonzalez-Corpas A, Ashour N, Escanilla A, Monje A, Guerrero Marquez C, et al. DNA methylation pattern of gene promoters of major neurotransmitter systems in older patients with schizophrenia with severe and mild cognitive impairment. Int J Geriatr Psychiatry 2015;30(6):558–65.
[59] Wockner LF, Morris CP, Noble EP, Lawford BR, Whitehall VL, Young RM, et al. Brain-specific epigenetic markers of schizophrenia. Transl Psychiatry 2015;5:e680.
[60] Brucato N, DeLisi LE, Fisher SE, Francks C. Hypomethylation of the paternally inherited LRRTM1 promoter linked to schizophrenia. Am J Med Genet 2014;165B(7):555–63.
[61] Bani-Fatemi A, Goncalves VF, Zai C, de Souza R, Le Foll B, Kennedy JL, et al. Analysis of CpG SNPs in 34 genes: association test with suicide attempt in schizophrenia. Schizophr Res 2013;147(2–3):262–8.
[62] Kordi-Tamandani DM, Mojahed A, Sahranavard R, Najafi M. Association of glutathione s-transferase gene methylation with risk of schizophrenia in an Iranian population. Pharmacology 2014;94(3–4):179–82.
[63] Ota VK, Noto C, Gadelha A, Santoro ML, Spindola LM, Gouvea ES, et al. Changes in gene expression and methylation in the blood of patients with first-episode psychosis. Schizophr Res 2014;159(2–3):358–64.
[64] Chase KA, Gavin DP, Guidotti A, Sharma RP. Histone methylation at H3K9: evidence for a restrictive epigenome in schizophrenia. Schizophr Res 2013;149(1–3):15–20.
[65] Sharma RP, Grayson DR, Gavin DP. Histone deactylase 1 expression is increased in the prefrontal cortex of schizophrenia subjects: analysis of the National Brain Databank microarray collection. Schizophr Res 2008;98(1–3):111–7.
[66] Kebir O, Chaumette B, Fatjo-Vilas M, Ambalavanan A, Ramoz N, Xiong L, et al. Family-based association study of common variants, rare mutation study and epistatic interaction detection in HDAC genes in schizophrenia. Schizophr Res 2014;160(1–3):97–103.
[67] Network, Pathway Analysis Subgroup of Psychiatric Genomics Consortium. Psychiatric genome-wide association study analyses implicate neuronal, immune and histone pathways. Nat Neurosci 2015;18(2):199–209.
[68] Han H, Yu Y, Shi J, Yao Y, Li W, Kong N, et al. Associations of histone deacetylase-2 and histone deacetylase-3 genes with schizophrenia in a Chinese population. Asia Pac Psychiatry 2013;5(1):11–6.
[69] Chen G, Guan F, Lin H, Li L, Fu D. Genetic analysis of common variants in the HDAC2 gene with schizophrenia susceptibility in Han Chinese. J Hum Genet 2015;60(9):479–84.
[70] Akbarian S, Ruehl MG, Bliven E, Luiz LA, Peranelli AC, Baker SP, et al. Chromatin alterations associated with downregulated metabolic gene expression in the prefrontal cortex of subjects with schizophrenia. Arch Gen Psychiatry 2005;62(8):829–40.
[71] Huang HS, Matevossian A, Whittle C, Kim SY, Schumacher A, Baker SP, et al. Prefrontal dysfunction in schizophrenia involves mixed-lineage leukemia 1-regulated histone methylation at GABAergic gene promoters. J Neurosci 2007;27(42):11254–62.
[72] Tang B, Dean B, Thomas EA. Disease- and age-related changes in histone acetylation at gene promoters in psychiatric disorders. Trans Psychiatry 2011;1:e64.
[73] Gavin DP, Kartan S, Chase K, Jayaraman S, Sharma RP. Histone deacetylase inhibitors and candidate gene expression: an in vivo and in vitro approach to studying chromatin remodeling in a clinical population. J Psychiatric Res 2009;43(9):870–6.
[74] Gavin DP, Rosen C, Chase K, Grayson DR, Tun N, Sharma RP. Dimethylated lysine 9 of histone 3 is elevated in schizophrenia and exhibits a divergent response to histone deacetylase inhibitors in lymphocyte cultures. J Psychiatry Neurosci 2009;34(3):232–7.
[75] Zhou Y, Wang J, Lu X, Song X, Ye Y, Zhou J, et al. Evaluation of six SNPs of MicroRNA machinery genes and risk of schizophrenia. J Mol Neurosci 2013;49(3):594–9.

[76] Perkins DO, Jeffries CD, Jarskog LF, Thomson JM, Woods K, Newman MA, et al. microRNA expression in the prefrontal cortex of individuals with schizophrenia and schizoaffective disorder. Genome Biol 2007;8(2):R27.
[77] Beveridge NJ, Tooney PA, Carroll AP, Gardiner E, Bowden N, Scott RJ, et al. Dysregulation of miRNA 181b in the temporal cortex in schizophrenia. Hum Mol Genet 2008;17(8):1156–68.
[78] Zhu Y, Kalbfleisch T, Brennan MD, Li Y. A microRNA gene is hosted in an intron of a schizophrenia-susceptibility gene. Schizophr Res 2009;109(1–3):86–9.
[79] Beveridge NJ, Gardiner E, Carroll AP, Tooney PA, Cairns MJ. Schizophrenia is associated with an increase in cortical microRNA biogenesis. Mol Psychiatry 2010;15(12):1176–89.
[80] Miller BH, Wahlestedt C. MicroRNA dysregulation in psychiatric disease. Brain Res 2010;1338:89–99.
[81] Kim AH, Reimers M, Maher B, Williamson V, McMichael O, McClay JL, et al. MicroRNA expression profiling in the prefrontal cortex of individuals affected with schizophrenia and bipolar disorders. Schizophr Res 2010;124(1–3):183–91.
[82] Santarelli DM, Beveridge NJ, Tooney PA, Cairns MJ. Upregulation of dicer and microRNA expression in the dorsolateral prefrontal cortex Brodmann area 46 in schizophrenia. Biol Psychiatry 2011;69(2):180–7.
[83] Moreau MP, Bruse SE, David-Rus R, Buyske S, Brzustowicz LM. Altered microRNA expression profiles in postmortem brain samples from individuals with schizophrenia and bipolar disorder. Biol Psychiatry 2011;69(2):188–93.
[84] Mellios N, Galdzicka M, Ginns E, Baker SP, Rogaev E, Xu J, et al. Gender-specific reduction of estrogen-sensitive small RNA, miR-30b, in subjects with schizophrenia. Schizophr Bull 2012;38(3):433–43.
[85] Miller BH, Zeier Z, Xi L, Lanz TA, Deng S, Strathmann J, et al. MicroRNA-132 dysregulation in schizophrenia has implications for both neurodevelopment and adult brain function. Proc Natl Acad Sci USA 2012;109(8):3125–30.
[86] Banigan MG, Kao PF, Kozubek JA, Winslow AR, Medina J, Costa J, et al. Differential expression of exosomal microRNAs in prefrontal cortices of schizophrenia and bipolar disorder patients. PloS One 2013;8(1):e48814.
[87] Kohen R, Dobra A, Tracy JH, Haugen E. Transcriptome profiling of human hippocampus dentate gyrus granule cells in mental illness. Transl Psychiatry 2014;4:e366.
[88] Hansen T, Olsen L, Lindow M, Jakobsen KD, Ullum H, Jonsson E, et al. Brain expressed microRNAs implicated in schizophrenia etiology. PloS One 2007;2(9):e873.
[89] Feng J, Sun G, Yan J, Noltner K, Li W, Buzin CH, et al. Evidence for X-chromosomal schizophrenia associated with microRNA alterations. PloS One 2009;4(7):e6121.
[90] Xu Y, Li F, Zhang B, Zhang F, Huang X, Sun N, et al. MicroRNAs and target site screening reveals a pre-microRNA-30e variant associated with schizophrenia. Schizophr Res 2010;119(1–3):219–27.
[91] Lai CY, Yu SL, Hsieh MH, Chen CH, Chen HY, Wen CC, et al. MicroRNA expression aberration as potential peripheral blood biomarkers for schizophrenia. PloS One 2011;6(6):e21635.
[92] Schizophrenia Psychiatric Genome-Wide Association Study Consortium. Genome-wide association study identifies five new schizophrenia loci. Nat Genet 2011;43(10):969–76.
[93] Shi W, Du J, Qi Y, Liang G, Wang T, Li S, et al. Aberrant expression of serum miRNAs in schizophrenia. J Psychiatric Res 2012;46(2):198–204.
[94] Zou M, Li D, Lv R, Zhou Y, Wang T, Liu J, et al. Association between two single nucleotide polymorphisms at corresponding microRNA and schizophrenia in a Chinese population. Mol Biol Rep 2012;39(4):3385–91.
[95] Watanabe Y, Iijima Y, Egawa J, Nonokawa A, Kaneko N, Arinami T, et al. Replication in a Japanese population that a MIR30E gene variation is associated with schizophrenia. Schizophr Res 2013;150(2–3):596–7.
[96] Watanabe Y, Shibuya M, Nunokawa A, Kaneko N, Igeta H, Egawa J, et al. A rare MIR138-2 gene variation is associated with schizophrenia in a Japanese population. Psychiatry Res 2014;215(3):801–2.

[97] Forstner AJ, Basmanav FB, Mattheisen M, Bohmer AC, Hollegaard MV, Janson E, et al. Investigation of the involvement of MIR185 and its target genes in the development of schizophrenia. J Psychiatry Neurosci 2014;39(6):386–96.

[98] Liu S, Yuan YB, Guan LL, Wei H, Cheng Z, Han X, et al. MiRNA-365 and miRNA-520c-3p respond to risperidone treatment in first-episode schizophrenia after a 1 year remission. Chin Med J (Engl) 2013;126(14):2676–80.

[99] Song HT, Sun XY, Zhang L, Zhao L, Guo ZM, Fan HM, et al. A preliminary analysis of association between the downregulation of microRNA-181b expression and symptomatology improvement in schizophrenia patients before and after antipsychotic treatment. J Psychiatric Res 2014;54:134–40.

[100] Sun XY, Zhang J, Niu W, Guo W, Song HT, Li HY, et al. A preliminary analysis of microRNA as potential clinical biomarker for schizophrenia. Am J Med Genet B Neuropsychiatr Genet 2015;168B(3):170–8.

[101] Yu HC, Wu J, Zhang HX, Zhang GL, Sui J, Tong WW, et al. Alterations of miR-132 are novel diagnostic biomarkers in peripheral blood of schizophrenia patients. Prog Neuropsychopharmacol Biol Psychiatry 2015;63:23–9.

[102] Green MJ, Cairns MJ, Wu J, Dragovic M, Jablensky A, Tooney PA, et al. Genome-wide supported variant MIR137 and severe negative symptoms predict membership of an impaired cognitive subtype of schizophrenia. Mol Psychiatry 2013;18(7):774–80.

[103] Cummings E, Donohoe G, Hargreaves A, Moore S, Fahey C, Dinan TG, et al. Mood congruent psychotic symptoms and specific cognitive deficits in carriers of the novel schizophrenia risk variant at MIR-137. Neurosci Lett 2013;532:33–8.

[104] van Erp TG, Guella I, Vawter MP, Turner J, Brown GG, McCarthy G, et al. Schizophrenia miR-137 locus risk genotype is associated with dorsolateral prefrontal cortex hyperactivation. Biol Psychiatry 2014;75(5):398–405.

[105] Rose EJ, Morris DW, Fahey C, Cannon D, mcDonald C, Scanlon C, et al. The miR-137 schizophrenia susceptibility variant rs1625579 does not predict variability in brain volume in a sample of schizophrenic patients and healthy individuals. Am J Med Genet 2014;165B(6):467–71.

[106] Patel VS, Kelly S, Wright C, Gupta CN, Arias-Vasquez A, Perrone-Bizzozero N, et al. MIR137HG risk variant rs1625579 genotype is related to corpus callosum volume in schizophrenia. Neuroscience Lett 2015;602:44–9.

[107] Guan F, Zhang B, Yan T, Li L, Liu F, Li T, et al. MIR137 gene and target gene CACNA1C of miR-137 contribute to schizophrenia susceptibility in Han Chinese. Schizophr Res 2014;152(1):97–104.

[108] Ma G, Yin J, Fu J, Luo X, Zhou H, Tao H, et al. Association of a miRNA-137 polymorphism with schizophrenia in a Southern Chinese Han population. Biomed Res Int 2014;2014:751267.

[109] Wang S, Li W, Zhang H, Wang X, Yang G, Zhao J, et al. Association of microRNA137 gene polymorphisms with age at onset and positive symptoms of schizophrenia in a Han Chinese population. Int J Psychiatry Med 2014;47(2):153–68.

[110] Yuan J, Cheng Z, Zhang F, Zhou Z, Yu S, Jin C. Lack of association between microRNA-137 SNP rs1625579 and schizophrenia in a replication study of Han Chinese. Mol Genet Genom 2015;290(1):297–301.

[111] Egawa J, Nunokawa A, Shibuya M, Wantanabe Y, Kaneko N, Igeta H, et al. Resequencing and association analysis of MIR137 with schizophrenia in a Japanese population. Psychiatry Clin Neurosci 2013;67(4):277–9.

[112] Higuchi F, Uchida S, Yamagata H, Otsuki K, Hobara T, Abe N, et al. State-dependent changes in the expression of DNA methyltransferases in mood disorder patients. J Psychiatr Res 2011;45(10):1295–300.

[113] Fernandes BS, Gama CS, Ceresér KM, Yatham LN, Fries GR, Colpo G, et al. Brain-derived neurotropic factor as a state-marker of mood episodes in bipolar disorders: a systematic review and meta-regression analysis. J Psychiatr Res 2011;45:995–1004.

[114] Polyakova M, Stuke K, Schuemberg K, Mueller K, Schoenknecht P, Schroeter ML. BDNF as a biomarker for successful treatment of mood disorders: a systemic & quantitative meta-analysis. J Affect Disord 2015;174:432–40.

[115] Rao JS, Keleshian VL, Klein S, Rapoport SI. Epigenetic modifications in frontal cortex from Alzheimer's disease and bipolar disorder patients. Transl Psychiatry 2012;2:e132.
[116] Bromberg A, Bersudsky Y, Levine J, Agam G. Global leukocyte DNA methylation is not altered in euthymic bipolar patients. J Affect Disord 2009;118(1–3):234–9.
[117] D'Addario C, Dell'Osso B, Palazzo MC, Benatti B, Lietti L, Cattaneo E, et al. Selective DNA methylation of BDNF promoter in bipolar disorder: differences among patients with BDI and BDII. Neuropsychopharmacology 2012;37(7):1647–55.
[118] Strauss JS, Khare T, De Luca V, Jeremian R, Kennedy JL, Vincent JB, et al. Quantitative leukocyte BDNF promoter methylation analysis in bipolar disorder. Int J Bipolar Disord 2013;1:28.
[119] Carlberg L, Scheibelreiter J, Hassler MR, Schloegelhofer M, Schmoeger M, Ludwig B, et al. Brain-derived neurotrophic factor (BDNF)-epigenetic regulation in unipolar and bipolar affective disorder. J Affect Disord 2014;168:399–406.
[120] Dell'Osso B, D'Addario C, Carlotta Palazzo M, Benatti B, Camuri G, Galimberti D, et al. Epigenetic modulation of BDNF gene: differences in DNA methylation between unipolar and bipolar patients. J Affect Disord 2014;166:330–3.
[121] Mitchelmore C, Gede L. Brain derived neurotrophic factor: epigenetic regulation in psychiatric disorders. Brain Res 2014;1586:162–72.
[122] Mansour HA, Talkowski ME, Wood J, Pless L, Bamne M, Chowdari KV, et al. Serotonin gene polymorphisms and bipolar I disorder: focus on the serotonin transporter. Ann Med 2005;37(8):590–602.
[123] Oquendo MA, Hastings RS, Huang YY, Simpson N, Ogden RT, Hu XZ, et al. Brain serotonin transporter binding in depressed patients with bipolar disorder using positron emission tomography. Arch Gen Psychiatry 2007;64(2):201–8.
[124] Chou YH, Wang SJ, Lin CL, Mao WC, Lee SM, Liao MH. Decreased brain serotonin transporter binding in the euthymic state of bipolar I but not bipolar II disorder: a SPECT study. Bipolar Disord 2010;12(3):312–8.
[125] Sugawara H, Iwamoto K, Bundo M, Ueda J, Miyauchi T, Komori A, et al. Hypermethylation of serotonin transporter gene in bipolar disorder detected by epigenome analysis of discordant monozygotic twins. Transl Psychiatry 2011;1:e24.
[126] Benes FM, Berretta S. GABAergic interneurons: implications for understanding schizophrenia and bipolar disorder. Neuropsychopharmacology 2001;25(1):1–27.
[127] Kaminsky Z, Tochigi M, Jia P, Pal M, Mill J, Kwan A, et al. A multi-tissue analysis identifies HLA complex group 9 gene methylation differences in bipolar disorder. Mol Psychiatry 2012;17(7):728–40.
[128] Burghardt KJ, Goodrich JM, Dolinoy DC, Ellingrod VL. DNA methylation, insulin resistance and second-generation antipsychotics in bipolar disorder. Epigenomics 2015;7(3):343–52.
[129] Nohesara S, Ghadirivasfi M, Mostafavi S, Eskandari MR, Ahmadkhaniha H, Thiagalingam S, et al. DNA hypomethylation of MB-COMT promoter in the DNA derived from saliva in schizophrenia and bipolar disorder. J Psychiatric Res 2011;45(11):1432–8.
[130] Fries GR, Vasconcelos-Moreno MP, Gubert C, dos Santos BT, Sartori J, Eisele B, et al. Hypothalamic-pituitary-adrenal axis dysfunction and illness progression in bipolar disorder. Int J Neuropsychopharmacol 2015;18(1):1–10.
[131] Cruceanu C, Alda M, Nagy C, Freemantle E, Rouleau GA, Turecki G. H3K4 tri-methylation in synapsin genes leads to different expression patterns in bipolar disorder and major depression. Int J Neuropsychopharmacol 2013;16(2):289–99.
[132] Guella I, Sequeira A, Rollins B, Morgan L, Torri F, van Erp TG, et al. Analysis of miR-137 expression and rs1625579 in dorsolateral prefrontal cortex. J Psychiatric Res 2013;47(9):1215–21.
[133] Saus E, Soria V, Escaramis G, Vivarelli F, Crespo JM, Kagerbauer B, et al. Genetic variants and abnormal processing of pre-miR-182, a circadian clock modulator, in major depression patients with late insomnia. Hum Mol Genet 2010;19(20):4017–25.

[134] Rong H, Liu TB, Yang KJ, Yang HC, Wu DH, Liao CP, et al. MicroRNA-134 plasma levels before and after treatment for bipolar mania. J Psychiatric Res 2011;45(1):92–5.

[135] Wang Z, Zhang C, Huang J, Yuan C, Hong W, Chen J, et al. MiRNA-206 and BDNF genes interacted in bipolar I disorder. J Affect Disord 2014;162:116–9.

[136] Forstner AJ, Hofmann A, Maaser A, Sumer S, Khudayberdiev S, Muhleisen TW, et al. Genome-wide analysis implicates microRNAs and their target genes in the development of bipolar disorder. Transl Psychiatry 2015;5:e678.

[137] Poulter MO, Du L, Weaver IC, Palkovits M, Faludi G, Merali Z, et al. GABAA receptor promoter hypermethylation in suicide brain: implications for the involvement of epigenetic processes. Biol Psychiatry 2008;64(8):645–52.

[138] Dwivedi Y. Brain-derived neurotrophic factor: role in depression and suicide. Neuropsychiatr Dis Treat 2009;5:433–49.

[139] Karege F, Bondolfi G, Gervasoni N, Schwald M, Aubry JM, Bertschy G. Low brain-derived neurotrophic factor (BDNF) levels in serum of depressed patients probably results from lowered platelet BDNF release unrelated to platelet reactivity. Biol Psychiatry 2005;57(9):1068–72.

[140] Tsai SJ, Cheng CY, Yu YW, Chen TJ, Hong CJ. Association study of a brain-derived neurotrophic-factor genetic polymorphism and major depressive disorders, symptomatology, and antidepressant response. Am J Med Genet 2003;123B(1):19–22.

[141] Keller S, Sarchiapone M, Zarrilli F, Videtic A, Ferraro A, Carli V, et al. Increased BDNF promoter methylation in the Wernicke area of suicide subjects. Arch Gen Psychiatry 2010;67(3):258–67.

[142] Fuchikami M, Morinobu S, Segawa M, Okamoto Y, Yamawaki S, Ozaki N, et al. DNA methylation profiles of the brain-derived neurotrophic factor (BDNF) gene as a potent diagnostic biomarker in major depression. PloS One 2011;6(8):e23881.

[143] Kang HJ, Kim JM, Lee JY, Kim SY, Bae KY, Kim SW, et al. BDNF promoter methylation and suicidal behavior in depressive patients. J Affect Disord 2013;151(2):679–85.

[144] D'Addario C, Dell'Osso B, Galimberti D, Palazzo MC, Benatti B, Di Francesco A, et al. Epigenetic modulation of BDNF gene in patients with major depressive disorder. Biol Psychiatry 2013;73(2):e6–7.

[145] Tadic A, Muller-Engling L, Schlicht KF, Kotsiari A, Dreimuller N, Kleinmann A, et al. Methylation of the promoter of brain-derived neurotrophic factor exon IV and antidepressant response in major depression. Mol Psychiatry 2014;19(3):281–3.

[146] Kang HJ, Kim JM, Bae KY, Kim SW, Shin IS, Kim HR, et al. Longitudinal associations between BDNF promoter methylation and late-life depression. Neurobiol Aging 2015;36(4):1764. e1-7.

[147] Lotrich FE, Pollock BG. Meta-analysis of serotonin transporter polymorphisms and affective disorders. Psychiatr Genet 2004;14(3):121–9.

[148] Yu YW, Tsai SJ, Hong CJ, Chen TJ, Chen MC, Yang CW. Association study of a monoamine oxidase a gene promoter polymorphism with major depressive disorder and antidepressant response. Neuropsychopharmacology 2005;30(9):1719–23.

[149] Olsson CA, Foley DL, Parkinson-Bates M, Byrnes G, McKenzie M, Patton GC, et al. Prospects for epigenetic research within cohort studies of psychological disorder: a pilot investigation of a peripheral cell marker of epigenetic risk for depression. Biol Psychol 2010;83(2):159–65.

[150] Kang HJ, Kim JM, Stewart R, Kim SY, Bae KY, Kim SW, et al. Association of SLC6A4 methylation with early adversity, characteristics and outcomes in depression. Prog Neuropsychopharmacol Biol Psychiatry 2013;44:23–8.

[151] Zhao J, Goldberg J, Bremner JD, Vaccarino V. Association between promoter methylation of serotonin transporter gene and depressive symptoms: a monozygotic twin study. Psychosom Med 2013;75(6):523–9.

[152] Bayles R, Baker EK, Jowett JB, Barton D, Esler M, El-Osta A, et al. Methylation of the SLC6a2 gene promoter in major depression and panic disorder. PloS One 2013;8(12):e83223.

[153] Okada S, Morinobu S, Fuchikami M, Segawa M, Yokomaku K, Kataoka T, et al. The potential of SLC6A4 gene methylation analysis for the diagnosis and treatment of major depression. J Psychiatric Res 2014;53:47–53.
[154] Domschke K, Tidow N, Schwarte K, Deckert J, Lesch KP, Arolt V, et al. Serotonin transporter gene hypomethylation predicts impaired antidepressant treatment response. Int J Neuropsychopharmacol 2014;17(8):1167–76.
[155] Domschke K, Tidow N, Schwarte K, Ziegler C, Lesch KP, Deckert J, et al. Pharmacoepigenetics of depression: no major influence of MAO-A DNA methylation on treatment response. J Neural Transm (Vienna) 2015;122(1):99–108.
[156] Melas PA, Wei Y, Wong CC, Sjoholm LK, Aberg E, Mill J, et al. Genetic and epigenetic associations of MAOA and NR3C1 with depression and childhood adversities. Int J Neuropsychopharmacol 2013;16(7):1513–28.
[157] Melas PA, Forsell Y. Hypomethylation of MAOA's first exon region in depression: a replication study. Psychiatry Res 2015;226(1):389–91.
[158] Pariante CM, Lightman SL. The HPA axis in major depression: classical theories and new developments. Trends Neurosci 2008;31(9):464–8.
[159] Claes S. Glucocorticoid receptor polymorphisms in major depression. Ann NY Acad Sci 2009;1179:216–28.
[160] Alt SR, Turner JD, Klok MD, Meijer OC, Lakke EA, Derijk RH, et al. Differential expression of glucocorticoid receptor transcripts in major depressive disorder is not epigenetically programmed. Psychoneuroendocrinology 2010;35(4):544–56.
[161] Perroud N, Paoloni-Giacobino A, Prada P, Olie E, Salzmann A, Nicastro R, et al. Increased methylation of glucocorticoid receptor gene (NR3C1) in adults with a history of childhood maltreatment: a link with the severity and type of trauma. Transl Psychiatry 2011;1:e59.
[162] Na KS, Chang HS, Won E, Han KM, Choi S, Tae WS, et al. Association between glucocorticoid receptor methylation and hippocampal subfields in major depressive disorder. PloS One 2014;9(1):e85425.
[163] Sabunciyan S, Aryee MJ, Irizarry RA, Rongione M, Webster MJ, Kaufman WE, et al. Genome-wide DNA methylation scan in major depressive disorder. PloS One 2012;7(4):e34451.
[164] Dempster EL, Wong CC, Lester KJ, Burrage J, Gregory Am, Mill J, et al. Genome-wide methylomic analysis of monozygotic twins discordant for adolescent depression. Biol Psychiatry 2014;76(12):977–83.
[165] Haghighi F, Xin Y, Chanrion B, O'Donnell AH, Ge Y, Dwork AJ, et al. Increased DNA methylation in the suicide brain. Dialogues Clin Neurosci 2014;16(3):430–8.
[166] Uddin M, Koenen KC, Aiello AE, Wildman DE, de los Santos R, Galea S. Epigenetic and inflammatory marker profiles associated with depression in a community-based epidemiologic sample. Psychol Med 2011;41(5):997–1007.
[167] Byrne EM, Carrillo-Roa T, Henders AK, Bowdler L, mcRae AF, Health AC, et al. Monozygotic twins affected with major depressive disorder have greater variance in methylation than their unaffected co-twin. Transl Psychiatry 2013;3:e269.
[168] Cordova-Palomera A, Fatjo-Vilas M, Gasto C, Navarro V, Krebs MO, Fananas L. Genome-wide methylation study on depression: differential methylation and variable methylation in monozygotic twins. Transl Psychiatry 2015;5:e557.
[169] Davies MN, Krause L, Bell JT, Gao F, Ward KJ, Wu H, et al. Hypermethylation in the ZBTB20 gene is associated with major depressive disorder. Genome Biol 2014;15(4):R56.
[170] Numata S, Ishii K, Tajima A, iga J, Kinoshita M, Wantanabe S, et al. Blood diagnostic biomarkers for major depressive disorder using multiplex DNA methylation profiles: discovery and validation. Epigenetics 2015;10(2):135–41.
[171] Cordova-Palomera A, Fatjo-Vilas M, Palma-Gudiel H, Blasco-Fontecilla H, Kebir O, Fananas L. Further evidence of DEPDC7 DNA hypomethylation in depression: a study in adult twins. Eur Psychiatry 2015;30(6):715–8.

[172] Haghighi F, Galfalvy H, Chen S, Huang YY, Cooper TB, Burke AK, et al. DNA methylation perturbations in genes involved in polyunsaturated fatty acid biosynthesis associated with depression and suicide risk. Front Neurol 2015;6:92.

[173] Chen ES, Ernst C, Turecki G. The epigenetic effects of antidepressant treatment on human prefrontal cortex BDNF expression. Int J Neuropsychopharmacol 2011;14(3):427–9.

[174] Lopez JP, Mamdani F, Labonte B, Beaulieu MM, Yang JP, Berlim MT, et al. Epigenetic regulation of BDNF expression according to antidepressant response. Mol Psychiatry 2013;18(4):398–9.

[175] Nestler EJ, Pena CJ, Kundakovic M, Mitchell A, Akbarian S. Epigenetic basis of mental illness. Neuroscientist 2015;1–16.

[176] He Y, Zhou Y, Xi Q, Cui H, Luo T, Song H, et al. Genetic variations in microRNA processing genes are associated with susceptibility in depression. DNA Cell Biol 2012;31(9):1499–506.

[177] Wingo AP, Almli LM, Stevens JJ, Klengel T, Uddin M, Li Y, et al. DICER1 and microRNA regulation in post-traumatic stress disorder with comorbid depression. Nat Commun 2015;6:10106.

[178] Smalheiser NR, Lugli G, Rizavi HS, Torvik VI, Turecki G, Dwivedi Y. MicroRNA expression is downregulated and reorganized in prefrontal cortex of depressed suicide subjects. PloS One 2012;7(3):e33201.

[179] Lopez JP, Lim R, Cruceanu C, Crapper L, Fasano C, Labonte B, et al. miR-1202 is a primate-specific and brain-enriched microRNA involved in major depression and antidepressant treatment. Nat Med 2014;20(7):764–8.

[180] Lopez JP, Fiori LM, Gross JA, Labonte B, Yerko V, Mechawar N, et al. Regulatory role of miRNAs in polyamine gene expression in the prefrontal cortex of depressed suicide completers. Int J Neuropsychopharmacol 2014;17(1):23–32.

[181] Wan Y, Liu Y, Wang X, Wu J, Liu K, Zhou J, et al. Identification of differential microRNAs in cerebrospinal fluid and serum of patients with major depressive disorder. PloS One 2015;10(3):e0121975.

[182] Song MF, Dong JZ, Wang YW, He J, Ju X, Zhang L, et al. CSF miR-16 is decreased in major depression patients and its neutralization in rats induces depression-like behaviors via a serotonin transmitter system. J Affect Disord 2015;178:25–31.

[183] Belzeaux R, Bergon A, Jeanjean V, Loriod B, Formisano-Treziny C, Verrier L, et al. Responder and non-responder patients exhibit different peripheral transcriptional signatures during major depressive episode. Transl Psychiatry 2012;2:e185.

[184] Bocchio-Chiavetto L, Maffioletti E, Bettinsoli P, Giovannini C, Bignotti S, Tardito D, et al. Blood microRNA changes in depressed patients during antidepressant treatment. Euro Neuropsychopharmacol 2013;23(7):602–11.

[185] Li YJ, Xu M, Gao ZH, Wang YQ, Yue Z, Zhang YX, et al. Alterations of serum levels of BDNF-related miRNAs in patients with depression. PloS One 2013;8(5):e63648.

[186] Fan HM, Sun XY, Guo W, Zhong AF, Niu W, Zhao L, et al. Differential expression of microRNA in peripheral blood mononuclear cells as specific biomarker for major depressive disorder patients. J Psychiatric Res 2014;59:45–52.

[187] Jensen KP, Kranzler HR, Stein MB, Gelernter J. The effects of a MAP2K5 microRNA target site SNP on risk for anxiety and depressive disorders. Am J Med Genet 2014;165B(2):175–83.

[188] Wang X, Sundquist K, Hedelius A, Palmer K, Memon AA, Sundquist J. Circulating microRNA-144-5p is associated with depressive disorders. Clin Epigenet 2015;7(1):69.

[189] Liang Y, Zhao G, Sun R, Mao Y, Li G, Chen X, et al. Genetic variants in the promoters of let-7 family are associated with an increased risk of major depressive disorder. J Affect Disord 2015;183:295–9.

[190] Pena CJ, Bagot RC, Labonte B, Nestler EJ. Epigenetic signaling in psychiatric disorders. J Mol Biol 2014;426(20):3389–412.

SECTION

EPIGENETICS OF NEUROPSYCHIATRIC DISORDERS II

CHAPTER

4

EPIGENETICS AND COGNITIVE DISORDERS—TRANSLATIONAL ASPECTS

F. Coppedè

Department of Translational Research and New Technologies in Medicine and Surgery, Section of Medical Genetics, University of Pisa, Pisa, Italy

CHAPTER OUTLINE

4.1 INTRODUCTION: EPIGENETICS AND COGNITIVE DISORDERS

Epigenetic processes, including DNA methylation and posttranslational modifications of histone tails, influence the chromatin structure and gene expression levels without involving changes of the primary DNA sequence [1]. Those mechanisms are physiologically required in embryonic development, cell

Neuropsychiatric Disorders and Epigenetics. http://dx.doi.org/10.1016/B978-0-12-800226-1.00004-6

differentiation, X-chromosome inactivation, genomic imprinting, repression of repetitive elements, and maintenance of cellular identity [1]. Furthermore, DNA methylation and histone tail modifications are dynamically regulated in the adult central nervous system (CNS) in response to experiential stimuli, acting as mediators of neuronal plasticity and playing a role in the formation, consolidation, and storage of memory, as well as in behavior [2–5]. There is some evidence that impairments of epigenetic processes could contribute to human conditions characterized by cognitive decline [6]. Indeed, epigenetic changes have been observed in cell culture models, animal models, and human tissues of three major neurodegenerative diseases, namely Alzheimer's disease (AD), Parkinson's disease (PD), and Huntington's disease (HD) [6]. A brief summary of the main findings of those studies is provided in this chapter, which is however focused on the description of the translational aspects of that evidence, that is, the possible implications of epigenetic tools in diagnosis and treatment of neurodegenerative disorders.

4.2 DNA METHYLATION AND HISTONE TAIL MODIFICATIONS

DNA methylation is an epigenetic mechanism consisting of the addition of a methyl group to the DNA, mediated by enzymes called DNA methyltransferases (DNMTs) (Fig. 4.1). The best-characterized DNA methylation process is the addition of a methyl group to cytosine in a CpG dinucleotide context, forming 5-methylcytosine (5-mC). Sites of CpG clusters are called CpG islands (regions of at least 200-bp, with a GC percentage greater than 50%, and with an observed-to-expected CpG ratio greater than 60%) or CpG island shores (regions of lower CpG density that flank both sides of CpG islands). When a CpG island in the promoter region of a gene is methylated, the expression of that gene is repressed because methyl-CpG-binding domain (MBD) proteins recognize and bind to the methylated

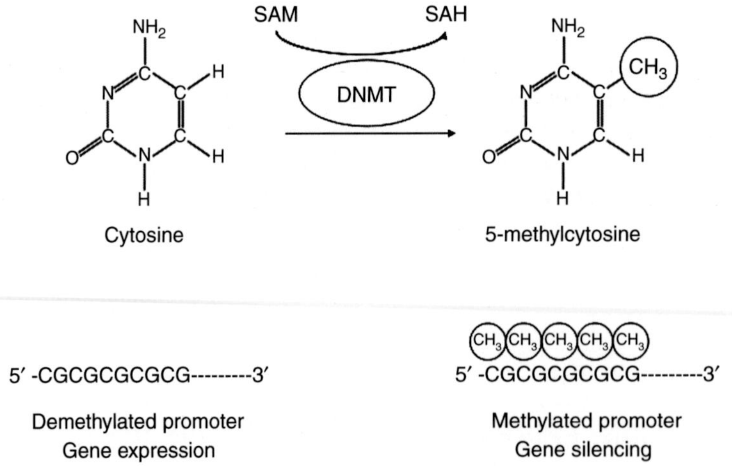

FIGURE 4.1 DNA Methylation

DNA methyltransferases *(DNMTs)* transfer the methyl group from S-adenosylmethionine *(SAM)* to cytosines in the DNA, leading to the formation of 5-methylcytosine and S-adenosylhomocysteine *(SAH)*.

DNA, and in turn recruit other epigenetic factors to enhance chromatin remodeling and transcriptional repression [7–9].

DNA methylation is largely dependent on the cellular availability of dietary folates and other B-group vitamins, all required for the production of S-adenosylmethionine (SAM), the intracellular donor compound of methyl groups [10]. Multiple DNMT families have been described in mammals. Among them, those with enzymatic activity include DNMT1 which is required for the maintenance of established patterns of DNA methylation during cell division, and the de novo DNMT3a and DNMT3b enzymes that establish DNA methylation patterns during early development [8]. 5-Hydroxymethylcytosine (5-hmC) is another modification of cytosine resulting from the oxidation of 5-mC mediated by members of the ten-eleven translocation (TET) protein family [11]. The specific distribution of 5-hmC in mammalian brain regions and its lower affinity for MBD proteins than 5-mC, reveal that 5-hmC is another important epigenetic mark with suggested roles in neurodevelopmental and neurodegenerative disorders [12].

Several posttranslational modifications occur on the histone-tails of nucleosomes and are associated with either open or condensed chromatin structure (Fig. 4.2). Those modifications include acetylation, methylation, phosphorylation, ubiquitylation, sumoylation, and other posttranslational modifications that exert their effects via directly influencing the overall structure of chromatin or by regulating the binding of effector molecules. Collectively, those modifications are involved in the regulation of gene expression, as well as in DNA repair, replication and recombination processes [13]. Among them, acetylation and methylation on histone tail residues represent the two best-characterized epigenetic marks regulating the chromatin structure [14]. Acetylation of lysine residues on histone tails neutralizes their

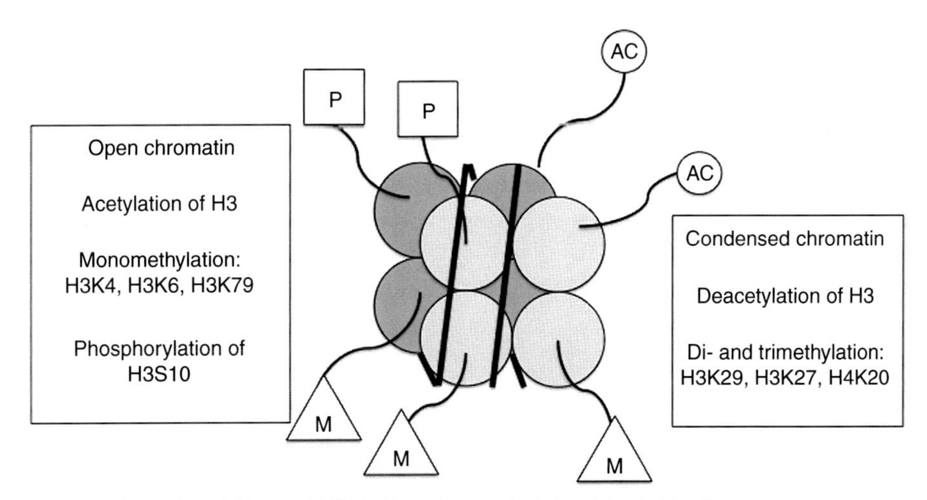

FIGURE 4.2 Some Examples of Histone Tail Modifications and Their Biologic Significance

Acetylation *(Ac)* of lysine residues is mediated by histone acetyltransferases *(HATs)*, and this mark is associated with an open chromatin structure. Removal of histone acetylation by histone deacetylases *(HDACs)* is considered a repressive mark linked to condensed chromatin. Methylation *(M)* of histone tails, and other modifications such as for example phosphorylation (*P*), can be associated with either opened or condensed chromatin structures, some examples are provided in the figure.

positive charge and decreases their affinity for the DNA, resulting in an open chromatin structure that allows transcription. On the contrary, histone deacetylation results in increased affinity for the DNA, leading to a condensed chromatin structure linked to transcriptional repression. Proteins called histone acetyltransferases (HATs) and histone deacetylases (HDACs) catalyze histone tail acetylation, and deacetylation, respectively [14]. Methylation can occur on either lysine or arginine residues of the tails of core histones H3 and H4, and can be associated with either chromatin condensation or relaxation, due to the fact that several sites for methylation are present on each tail. For example, methylation of H3K4, H3K36, and H3K79 and histone acetylation promote transcription, whereas histone hypoacetylation and increased di- and trimethylation at H3K9, and H3K27, are considered repressive marks [13,14]. Histone lysine methyltransferases (HMTs) and protein arginine methyltransferases (PRMTs) are required for the methylation of histone tail residues, and histone demethylases (HDMTs) are the "erasers" of that mark [15]. Additional histone-modifying enzymes are involved in the establishment of complex patterns of posttranslational modifications of histone tails (the so-called histone code) that, in concert with DNA methylation, tightly regulate gene expression levels [13–15].

4.3 EPIGENETIC CHANGES IN ALZHEIMER'S DISEASE

It is estimated that 46.8 million people worldwide suffer from dementia, with projections suggesting that this number will increase to 131.5 million by 2050, and AD is actually the most common neurodegenerative disorder and the primary form of dementia in the elderly. The term "dementia" describes a condition of loss of cognitive functioning, such as thinking, remembering, and reasoning, coupled with mood changes severe enough to reduce a person's ability to perform daily activities [16]. Aging is the major AD risk factor, and most of the cases (almost 95%) are regarded as late-onset AD (LOAD) forms, with disease onset after 65 years of age. The severity of dementia ranges with time in affected individuals following the progression of the neurodegenerative process that takes place in selected brain regions, including the temporal and parietal lobes and restricted regions within the frontal cortex and the cingulate gyrus. After a predementia stage, often defined as amnestic mild cognitive impairment (MCI), dementia usually starts as a mild condition that, with time, progresses up to the most severe stage, when patients are completely dependent on caregivers for the performance of daily activities. Unfortunately, AD is actually regarded as a disorder that cannot be cured or even slowed, and the available treatments are only symptomatic as they can temporarily slow the worsening of some symptoms [16,17].

Affected brain regions in AD are characterized by the accumulation of extracellular aggregates of the amyloid β (Aβ) peptide, denoted as senile plaques or amyloid plaques, and by intraneuronal aggregates of hyperphosphorylated tau protein called neurofibrillary tangles. In a minority of the cases AD is inherited as an early-onset (<65 years) autosomal dominant disease caused by mutations in one of three causative genes, namely *APP, PSEN1,* and *PSEN2*, coding for the amyloid precursor protein and presenilin proteins, respectively, all involved in the production of the Aβ peptide. By contrast, sporadic late-onset forms are likely the result of complex lifelong interactions among genetic, environmental, stochastic and epigenetic factors, superimposed on the age related accumulation of neuronal damage [18]. Among genetic factors, the *APOE* gene coding for apolipoprotein E, represents the major susceptibility factor for LOAD. Indeed, the *APOE ε4* allele is associated with increased LOAD risk, while the *APOE ε2* allele is protective [18]. Additional susceptibility LOAD loci have been identified in recent years by means of genome-wide association studies (GWAS) [17–19]. Among environmental factors, serious head injuries have been linked to increased LOAD risk. Furthermore, growing evidence links heart and brain health, and factors such as type 2 diabetes, high blood pressure, midlife obesity, hyperhomocysteinemia, and stroke, are suspected to contribute to cognitive decline. By contrast, a healthy dietary pattern with high consumption of fruits and vegetables, low consumption of red and processed meats, and favoring mono- and polyunsaturated fats over saturated fats, coupled with physical activity, brain stimulation and social engagement, might reduce the risk of cognitive impairment, AD, and related forms of dementia [17].

4.3.1 DNA METHYLATION IN ALZHEIMER'S DISEASE

Regarding the analysis of epigenetic marks in postmortem AD brains, results are still conflicting and often limited by the small sample size of case-control cohorts, so that it is currently impossible to draw any definitive conclusions regarding the contribution of DNA methylation changes to AD pathology [6,19,20]. In this regard, the only study performed in a pair of monozygotic twins discordant for AD revealed decreased global 5-mC content in the AD brain, pointing to a possible epigenetic contribution to disease pathogenesis [21]. Further studies addressing global DNA methylation levels in postmortem brains reported either a slight decrease in methylation, no significant difference between AD and control groups, or even hypermethylation in the investigated regions, which include the frontal and temporal cortices and the hippocampus [22–26]. Differences between studies are likely the result of small sample size and age-related interindividual variability in global DNA methylation levels, coupled with different technical approaches for handling postmortem materials as well as for the assessment of the global methylation status [6]. Similar conflicting results were obtained from the assessment of 5-hmC content in postmortem tissues [20].

The search for gene-specific methylation changes in postmortem AD brain samples was initially driven by the results of either in vitro or animal studies, and focused on genes required for the production of the Aβ peptide (*APP*, *PSEN1*, *PSEN2*, *BACE1*) or coding for protein tau (*MAPT*) involved in neurofibrillary tangle formation [6]. Despite that some of those genes are epigenetically deregulated in animal or cell culture models of AD following environmental stimulation [6], results in postmortem AD brains are still largely inconclusive, and none of the major AD genes (*APP*, *PSEN1*, *PSEN2*) has emerged as a clear disease methylation biomarker in recent large-cohort studies [27–30]. Changes in the methylation pattern of other loci, including *ANK1*, *SORBS3*, *SORL1*, *ABCA7*, *BIN1*, *TMEM59* and others, have been reported in postmortem brain analysis of AD and control samples [6,27–31], but the clinical significance of those findings is still uncertain as it is still unclear whether the observed changes are causative of AD or rather a consequence of the degenerative process taking place in the brain [19,20]. Furthermore, epigenetic changes are tissue or even cell type specific, so that the translational application of methylation marks detected postmortem in selected brain regions is still uncertain.

Unfortunately, also the studies of DNA methylation in blood cells of living AD patients are still in their beginning with overall inconclusive results [6]. Global changes in DNA methylation levels have been reported by some, but not by all authors [32–34]. Methylation studies of major AD genes, including *APP*, *PSEN1*, and *PSEN2*, gave overall inconclusive results and none of them has emerged as a clear peripheral methylation biomarker of AD [35–37]. Several other potential loci, such as *BDNF*, *SIRT1*, *FAAH*, *5-LOX*, and others have been suggested as peripheral methylation biomarkers, but results are still largely conflicting between studies or lack replication [6,19,20,36,37], so that the clinical utility of those biomarkers is still uncertain. Similarly, studies in blood DNA of early disease stages, such as MCI, are scarce and yet to be confirmed [6,19].

4.3.2 HISTONE TAIL MODIFICATIONS IN ALZHEIMER'S DISEASE

For what is concerning histone tail modifications in AD, studies on human samples are still scarce, but those performed with animal models revealed a contribution of those epigenetic modifications to learning and memory processes, and accumulating evidence suggests that the manipulation of histone acetylation with HDAC inhibitors (HDACi) can ameliorate the cognitive deficit and improve memory recovery in those animals [38]. Those studies will be detailed in the next sections of this chapter. Concerning postmortem specimens of AD individuals, decreased levels of H3 K18/K23 acetylation were reported in the temporal lobe with respect to controls [39]. Furthermore, increased levels of HDAC6 and HDAC2 have been reported in AD brain regions, the first linked to increased tau phosphorylation and the latter to the repression of memory-related genes [40,41].

4.4 EPIGENETIC CHANGES IN PARKINSON'S DISEASE

Parkinson's disease has a prevalence of 1–2% in individuals aged more than 65 years and is the second most common neurodegenerative disorder after AD, with an estimate that 4–5 million affected individuals are in the world's most populous nations, and projections suggest that this number will double by 2030 [42]. The disease is the result of a progressive loss of neuromelanin-producing dopaminergic neurons in the *substantia nigra pars compacta* (SN), with the presence of intracytoplasmic inclusions termed Lewy bodies (LBs: containing aggregates of α-synuclein, as well as other proteins such as ubiquitin, neurofilament protein, and others), and Lewy neurites in surviving neurons. Clinically, PD is characterized by classic motor-symptoms that include resting tremor, rigidity, bradykinesia, and postural instability. Nonmotor symptoms are also frequent, including autonomic insufficiency, sleep disorders, olfactory dysfunction, psychiatric disorders (apathy, depression), and cognitive impairment. Also for PD, current treatments based on levodopa and dopaminergic therapy offer only some improvements of the symptoms, but are not able to halt disease progression [42,43].

Families with inherited PD are rare (10–15% of all PD patients), and in most of the cases the disease is sporadic, arising from complex gene–environment interactions likely involving epigenetic mechanisms [43]. Among the several genes linked to familial PD, five of them have been confirmed to cause inherited forms of typical PD, including a-synuclein (*SNCA*) and leucine-rich repeat kinase 2 (*LRRK2*) genes that cause autosomal dominant PD, and parkin (*PARK2*), phosphatase and tensin homologue-induced putative kinase 1 (*PINK1*), and DJ-1 (*PARK7*) genes, all linked to autosomal recessive PD [43]. In addition, common polymorphisms (SNPs) of *SNCA* and *LRRK2* genes are among the most replicated susceptibility factors for sporadic PD, together with SNPs in glucocerebrosidase (*GBA*) and *MAPT* genes, overall suggesting that accumulation and degradation of misfolded proteins, the autophagy–lysosomal pathway, and mitochondrial dysfunctions are central pathways in PD pathogenesis [43]. A role for mitochondrial dysfunction and mitochondrial DNA damage in PD pathogenesis is further suggested by environmental factors such as the mitochondrial toxins 1-methyl-4-phenyl-1,2,3,6-tetrahydropyridine (MPTP), rotenone, paraquat and maneb, all linked to human PD [44]. Indeed living in rural areas and several other environmental factors have been linked to PD risk, and a recent metaanalysis revealed that among them there is evidence of a protective effect for physical activity and smoking, while head injuries showed association with increased disease risk [45].

4.4.1 DNA METHYLATION IN PARKINSON'S DISEASE

Epigenetic studies in PD mainly focus on promoter methylation levels of disease-related genes [43]. Particularly, there is accumulating evidence of decreased *SNCA* promoter and intron 1 methylation in several brain regions of sporadic PD patients including the SN, likely contributing to an increased production of the α-synuclein protein [46–49]. Very interestingly, decreased *SNCA* methylation has been also reported in blood DNA of PD patients, suggesting that methylation patterns in brain are mirrored in the blood, and that the analysis of *SNCA* methylation in blood DNA warrants further investigation as a potential diagnostic biomarker [49–52]. Moreover, it was shown that α-synuclein sequesters DNMT1 from the nucleus into the cytoplasm, leading to global DNA hypomethylation in PD and dementia with Lewy bodies [53]. A large dataset of postmortem brain samples revealed changes in methylation and gene expression associated with PD risk variants in *PARK16*, transmembrane glycoprotein NMB (*GPNMB*), and syntaxin 1B (*STX1B*) genes [54]. Moreover, a recent study revealed a loss of mitochondrial 5-mC levels in the *SN* of PD patients [54]. Overall, those studies argue in favor of altered DNA methylation in PD.

4.4.2 HISTONE TAIL MODIFICATIONS IN PARKINSON'S DISEASE

Most of the data suggesting a possible involvement of histone tail modifications in PD have been obtained in cell culture models or animal models of the disease [43]. Those studies have revealed that α-synuclein binds to histones and inhibits histone acetylation [55]. There is also evidence that HDAC6 promotes the formation of α-synuclein inclusions and might represent a cytoprotective response to sequester toxic proteins [56]. Furthermore, PINK1 interacts with proteins involved in histone methylation (EED/WAIT1) leading to reduced trimethylation of H3K27 [57].

4.5 EPIGENETIC CHANGES IN HUNTINGTON'S DISEASE

Huntington's disease (HD) is the most common polyglutamine (polyQ) disorder with a prevalence of 5–10 cases per 100,000 individuals worldwide. The disease is inherited as an autosomal dominant trait resulting from a CAG repeats expansion (36 repeats or more) within exon 1 of the gene encoding huntingtin (*HTT*). Disease penetrance is complete in individuals with 40 or more CAG repeats, but age at onset of the symptoms is inversely correlated with the number of repeats. However, most HD individuals worldwide possess between 40 and 50 CAG repeats and age at onset of the symptoms occurs in midlife. For those individuals with a CAG repeat expansion between 36 and 40 repeats penetrance is incomplete, meaning that they may or may not develop the disease. There is no cure for HD and death of the patients usually occurs between 15 and 20 years after the onset of the symptoms [58]. A loss of function of normal HTT and/or a toxic gain of function of the mutant protein are believed to contribute to the disruption of multiple intracellular pathways, ultimately leading to a progressive neurodegeneration resulting in choreiform movements and cognitive impairment; psychiatric and behavioral disturbances are also frequently observed in the patients [58]. In this regard, accumulating evidence suggests interactions between HTT and proteins regulating the histone code, pointing to an involvement of altered epigenetic pathways in disease pathogenesis and opening the way to potential therapies with epigenetic drugs [59].

4.5.1 DNA METHYLATION IN HUNTINGTON'S DISEASE

The suggestion that changes in DNA methylation might be involved in HD came from a study investigating murine striatal derived cell lines expressing either *wild-type* or mutant HTT [60]. That study revealed that a large fraction of the genes that change in expression in the presence of mutant HTT demonstrate significant changes in DNA methylation [60]. In addition, a genome-wide distribution analysis of 5-hmC revealed decreased 5-hmC levels in both striatum and cortex of HD transgenic mice [61]. However, gene-specific methylation analysis and studies in humans are still scarce [59]. In this regard, a recent study in postmortem HD brains has revealed that epigenetic deregulation of the hairy and enhancer of split 4 (*HES4*) gene, including DNA methylation changes in the promoter, could play a critical role in modifying disease pathogenesis and severity [62].

4.5.2 HISTONE TAIL MODIFICATIONS IN HUNTINGTON'S DISEASE

Concerning histone tail modifications, there is substantial evidence that mutant HTT directly interacts with proteins regulating the histone code [59]. Particularly, studies in cell cultures and animal models of the disease revealed that mutant HTT interacts with the HAT protein CBP (CREB-binding protein), leading to depletion of soluble CBP or sequestration into HTT aggregates, and inducing histone hypoacetylation and reduced expression of genes related to memory functions [63,64]. In addition to changes in histone tail acetylation, animal and cell culture models of HD revealed changes in histone methylation and ubiquitylation [59]. Studies in postmortem HD brains confirmed most of the changes observed in disease models, including histone hypoacetylation in caudate nucleus and Purkinje cells, increase of HDAC4 and HDAC5 in the cingulate cortex, and gene-specific changes in H3K4 trimethylation [59,62,65]. In this regard, a recent study revealed a specific perturbation of neuronal H3K4 trimethylation in the prefrontal cortex of HD cases, affecting 136 loci [62]. Particularly, the promoter of the *HES4* gene showed an association between reduction of H3K4 trimethylation and DNA hypermethylation that was related to altered expression of *HES4* and its target genes, all involved in striatal development [62].

4.6 EPIGENETIC DRUGS

Epigenetic drugs fall into two main categories: DNMT inhibitors (DNMTi) and HDAC inhibitors (HDACi). Those compounds have been largely used in different tissues and model systems to study the role of DNA methylation and histone tail acetylation, including animal and cell culture models of neurodegenerative diseases [38].

4.6.1 DNMT INHIBITORS

DNMTi include nucleoside analog inhibitors and nonnucleoside inhibitors. 5-Azacytidine and 5-aza-2′-deoxycytidine (decitabine) are analogs of cytidine with a nitrogen atom in place of carbon in the fifth position of the ring. The use of those compounds for cancer therapy is valuable for hematological malignancies, and both decitabine and 5-azacytidine received US Food and Drug Administration (FDA) approval for myelodysplastic syndromes [66]. A more recent DNMTi, zebularine, is more stable and less toxic than the two previous ones [67]. Despite that those drugs require to be incorporated into the replicating DNA to exert their function, increasing evidence suggests that nucleoside analogs can induce DNA demethylation also in postmitotic neurons, but their mechanisms of action are still unclear [67]. To date, the only FDA approved DNMTi belong to the family of nucleoside analogs, but all display relevant side effects, including cellular and clinical toxicity as well as chemical instability [66]. Therefore, there is a keen interest to develop novel and more specific drugs able to directly target DNMTs such as RG108, a nonnucleoside small molecule capable of direct enzyme inhibition [67].

4.6.2 HDAC INHIBITORS

Eighteen HDAC enzymes are known in mammals, and are grouped into four classes (I–IV). Table 4.1 provides a summary of their function and subcellular localization. Several compounds have the potential to inhibit HDACs, they are collectively known as HDACi and belong to four main structural classes:

Table 4.1 Mammalian Histone Deacetylases (HDAC) and Some of Their Inhibitors (HDACi)

Proteins	Function	Inhibitors
Class I HDACs (Zinc dependent) HDAC 1 HDAC 2 HDAC 3 HDAC 8	Regulation of gene-specific transcription through the formation of stable transcriptional complexes	Pan-inhibitors: trichostatin A, vorinostat, butyrate, phenylbutyrate, valproate Selective inhibitors: MS-275 (HDAC1), 4b (HDAC1 and 3)
Class IIa HDACs (zinc dependent) HDAC 4 HDAC 5 HDAC 7 HDAC 9	Shuttle proteins between nucleus and cytoplasm to interact with both nuclear and cytoplasmic proteins, have histone deacetylase activity by interacting with HDAC3	Pan-inhibitors: trichostatin A, vorinostat, butyrate, phenylbutyrate, valproate
Class IIb HDACs (zinc dependent) HDAC 6 HDAC 10	HDAC6: takes part in the microtubule network by acting on α-tubulin and tau proteins HDAC10: cytoplasmic deacetylase	Pan-inhibitors: trichostatin A, vorinostat
Class III HDACs: Sirtuins (nicotinamide dependent) SIRT 1 SIRT 2 SIRT 3 SIRT 4 SIRT 5 SIRT 6 SIRT 7	SIRT 1-3: deacetylases SIRT4: ADP-ribosyltransferase SIRT 5: deacetylase and other activities SIRT 6: ADP-ribosyltransferase and deacetylase SIRT 7: deacetylase	Pan-inhibitor: nicotinamide Selective inhibitors: AK-1, AGK-2, AK-7 (sirtuin 2)
Class IV (zinc dependent) HDAC 11	Member of the survival of motor neuron complex, has a functional role in mRNA splicing	LAQ824

short chain fatty acids, hydroxamic acids, benzamides, and cyclic tetrapeptides [68]. Short chain fatty acids, such as sodium butyrate and valproate are mostly nonspecific (pan-inhibitors) and inhibit the majority of Class I and Class IIa HDACs (Table 4.1). Another group of pan HDACi are hydroxamates, such as trichostatin A (TSA) and vorinostat (suberoylanilide hydroxamic acid: SAHA) that inhibit most of Class I and Class II enzymes (Table 4.1). Among these drugs only vorinostat and romidepsin, this latter belonging to the cyclic tetrapeptides, have both received FDA approval for the treatment of human diseases, and particularly for the treatment of T cell cutaneous lymphoma. Several efforts are ongoing to develop more selective HDACi. For example, MS-275 (entinostat) is a benzamide that only blocks Class I HDACs (HDAC1-3), and preferentially HDAC1 [68]. Table 4.1 summarizes most of the HDACi that have been used in animal and cell culture models of neurodegenerative diseases.

4.7 NATURAL COMPOUNDS EXERTING EPIGENETIC PROPERTIES

Most of the previously discussed epigenetic drugs are toxic and nonspecific compounds, leading the scientific community in a continuous effort to develop more specific and less toxic ones. Increasing evidence suggests that several natural compounds could help to achieve this goal (Table 4.2). Several flavonoids, antraquinones, polyphenols, and other natural compounds are able to inhibit DNMT expression and/or activity. For example, epigallocatechin-3-gallate (EGCG) found in green tea is a potent DNMT1 inhibitor [69,70]. Dietary folates and related B-group vitamins (B2, B6, and B12), which are found in many foods, work as either methyl donors or cofactors in the production of SAM [10]. Dietary restriction of B-group vitamins could impair the intracellular methylation potential, defined as the ratio

Table 4.2 Some Examples of Natural Compounds With Epigenetic Properties

Compound	Dietary Source	Function
Curcumin	Turmeric	HAT inhibitor (p300/CBP) HDAC inhibitor DNMT inhibitor
Epigallocatechin-3-gallate (EGCG)	Green tea	HAT inhibitor DNMT inhibitor
Folate(Vitamin B9)	Dark green leafy vegetables, legumes, citrus fruits, grapes, berries, nuts, grains, eggs, meat and poultry	Methyl donor
Genistein	Soybeans and soy product	HDAC inhibitor
Quercetin	Citrus fruits, apples, onions, tea, berries	HAT activator HDAC inhibitor DNMT inhibitor
Resveratrol	Red grapes, wine, eucalyptus	HDAC inhibitor
Sulforaphane	Cruciferous vegetables such as broccoli and cabbage	HDAC inhibitor
Vitamin B12	Meat, fish, poultry, eggs, and dairy products	Cofactor for the synthesis of the methyl donor compound SAM

between S-adenosylmethionine and S-adenosylhomocysteine, thus inducing global changes in DNA or protein methylation [10]. In addition, compounds such as curcumin, genistein, quercetin, resveratrol, and many others are able to interfere with histone tail modifications regulating or inhibiting HAT and HDAC activities [70,71]. Some examples of major natural compounds exerting epigenetic properties are provided in Table 4.2.

4.8 MANIPULATION OF EPIGENETIC MECHANISMS FOR THE TREATMENT OF ALZHEIMER'S DISEASE

The growing body of evidence suggesting that epigenetic modifications could play a role in AD pathogenesis has been followed by several attempts to investigate the potential of either natural or synthetic epigenetic compounds to counteract disease symptoms [38]. In this regard, human neuroblastoma cells and transgenic AD mice maintained under conditions of B-group vitamin deficiency showed *PSEN1* promoter demethylation, followed by an increased expression of proteins required for Aβ peptide production, such as presenilin 1 and beta secretase 1 (BACE1), and Aβ peptide deposition in the animal brains. By contrast, SAM supplementation in the same experimental models reversed the negative effects of vitamins B deprivation [72,73]. Additional studies in AD models confirmed that treatment with folic acid or related methyl donor compounds was linked to changes in DNMT activities and in gene-specific promoter methylation, overall resulting in reduced Aβ peptide production and accumulation [74,75]. Studies in humans are still scarce, but there is some evidence that circulating folate levels in AD patients might be linked to promoter methylation levels of genes relevant for one carbon metabolism, DNA methylation, and apoptosis [36,76]. For what is concerning the efficacy of vitamins B supplementation in slowing the rate of cognitive, behavioral, functional and global decline in individuals with MCI or AD results are still uncertain [77]. There is some indication from elderly human subjects suggesting that B-group vitamin supplementation can slow the atrophy of specific brain regions that are associated with cognitive decline [77], but a recent metaanalysis of available clinical trials revealed only a weak evidence of benefits for the domains of memory in patients with MCI, but no adequate evidence of an effect of vitamins B on general cognitive function, executive function and attention in people with MCI [78]. Similarly, folic acid alone or vitamins B in combination were unable to stabilize or slow the decline in cognition, function, behavior, and global change of AD patients [78]. Another recent clinical trial that included 2919 elderly participants (65 years and older) revealed that a 2-year folic acid and vitamin B12 supplementation did not beneficially affect performance on cognitive domains in individuals with elevated Hcy levels, and only slightly slowed the rate of decline of global cognition [79]. Recent results of clinical trials with a nutritional formulation (folate, vitamin B12, SAM, alpha-tocopherol, *N*-acetyl cysteine, acetyl-L-carnitine) for cognition and mood in AD and MCI are a little bit more encouraging, suggesting that the nutraceutical formulation maintained or improved cognitive performance and mood/behavior [80,81]. Similarly, clinical studies in AD cohorts treated with curcumin, that possesses epigenetic, antioxidant, and antiinflammatory properties, have revealed limited effects to date, but there is some optimism for a potential beneficial effect in certain predementia phases [82].

Another line of active research in AD involves the manipulation of histone acetylation in animal models of the disease [38]. Early studies in the field revealed that intraperitoneal sodium butyrate administration for 4 weeks was able to improve learning and memory in transgenic AD mice

(CK-p25) [83]. It was subsequently observed that a 5-week intraperitoneal sodium butyrate injection in a transgenic model of AD (Tg2576 mice) led to decreased tau phosphorylation and restoration of dendritic spine density in hippocampal neurons [84]. Similarly, chronic intraperitoneal injection of sodium butyrate improved associative memory in a transgenic mouse model for amyloid deposition (APPPS1-21 mice) [85]. Francis et al. [86] observed that fear conditioning in APP/PS1 mice resulted in a significant reduction of hippocampal histone 4 (H4) acetylation. However, an acute treatment with TSA prior to fear conditioning rescued both acetylated H4 levels and contextual freezing performances in the animals [86]. Many similar examples are available in the literature linking HDACi administration with improved cognition and memory functions in AD animal models, some of which are listed in Table 4.3. For example, a 10-day oral administration of entinostat, a selective inhibitor of HDAC1, improved behavior and reduced neuroinflammation and amyloid plaque deposition in APP/PS1 mice [87]. More recently, intraperitoneal injection of TSA for 2 months resulted in decreased HDAC activity and increased the levels of the antiamyloidogenic protein gelsolin in the hippocampus and cortex of APPswe/PS1(δE9) mice [88].

Neprilysin is a major Aβ peptide-degrading enzyme, and it was observed that repeated oral gamma-hydroxybutyrate administration in APPSWE mice induced neprilysin overexpression, reduced cerebral Aβ contents, and prevented cognitive deficits [89]. Gamma-hydroxybutyrate is an endogenous compound involved in the regulation of GABAergic brain activities, and a drug that potentiate the GABAergic system that shows HDACi activity at pharmacologic doses [89]. Interestingly, also intragastric administration of the natural compound EGCG for 60 days reduced Aβ peptide accumulation in vitro and rescued cognitive deterioration in senescence-accelerated mice P8 (SAMP8) by upregulating neprilysin expression [90].

Table 4.3 Some Examples of the Effects of Histone Deacetylase Inhibitors (HDACi) in Models of Alzheimer's Disease

Compound	Experimental Model	Results[a]
Sodium butyrate	Transgenic AD mice (CK-p25)	Improved learning and memory
Sodium butyrate	Transgenic AD mice (Tg2576)	Decreased tau phosphorylation and restoration of dendritic spine density in hippocampal neurons
Sodium butyrate	Transgenic AD mice (APP/PS1)	Improved associative memory
γ-Hydroxybutyrate	Transgenic AD mice (APPSWE)	Induction of neprylisin overexpression and prevention of cognitive deficit
MS-275 (entinostat)	Transgenic AD mice (APP/PS1)	Amelioration of neuroinflammation and cerebral amyloidosis, and improved behavior
Trichostatin A	Transgenic AD mice (APP/PS1)	Acute treatment prior to fear conditioning training rescued hippocampal H4 acetylation levels and contextual freezing performances
Trichostatin A	Transgenic AD mice [APPswe/PS1(δE9)]	Decreased HDAC activity and increased expression of gelsolin in cortex and hippocampus

[a] *See the text for references.*

Collectively, those studies suggest that targeting histone modifications with HDACi reduce AD-like features in animal models of the disease (Table 4.3). However, the results of studies in individuals suffering from AD treated with natural compounds exerting epigenetic (HDACi) properties did not hold the promise of animal model investigations. For example, treatment with soy isoflavones was associated with improved nonverbal memory, construction abilities, verbal fluency, and speeded dexterity compared to treatment with placebo in cognitively healthy older adults [91], but a similar treatment did not benefit cognition in older men and women with AD [92]. Similarly, despite that resveratrol is a potent antioxidant, epigenetic, and antiinflammatory compound that facilitates the nonamyloidogenic cleavage of APP and promotes the clearance of the neurotoxic Aβ peptide, it is unlikely to be effective as monotherapy in AD individuals due to its poor bioavailability, biotransformation, and requisite synergism with other dietary factors [93]. Taken collectively, those studies suggest that targeting the epigenome with compounds able to interfere with DNA methylation and histone tail modifications, likely using combined therapies, might represent a promising approach for AD prevention, but treating individuals with already diagnosed dementia might be too late to achieve a desired effect [79–93].

4.9 MANIPULATION OF EPIGENETIC MECHANISMS FOR THE TREATMENT OF PARKINSON'S DISEASE

Most of our current knowledge of the effects of targeting the epigenome in PD comes from studies in animal models of the disease obtained by overexpression of mutant human α-synuclein, or by exposure to mitochondrial toxins [38]. The role of DNMTi in the treatment of PD is yet to be determined [94], but many studies in animal and cell culture models of the disease support a beneficial effect for HDACi treatment [38,94]. For example, nigral neurons of mice exposed to the herbicide paraquat, and fly models of the disease (*Drosophila* models) overexpressing α-synuclein, revealed that α-synuclein translocates into the nucleus, binds to histones, and inhibits histone acetylation [55]. The toxicity of α-synuclein was rescued by the administration of sodium butyrate or vorinostat [55] or by selective inhibition of sirtuin 2 in several models of PD overexpressing the mutant protein [55,95]. Similarly, an increasing number of studies suggest protective effects of HDACi in animal and cell culture models of the disease obtained by exposure to neurotoxic agents [94]. For example, TSA selectively rescued mitochondrial fragmentation and cell death induced by MPP+ treatment in human neuroblastoma cells [96], and sodium butyrate improved locomotor functions in a rotenone-induced *Drosophila* model of PD [97] and upregulated DJ-1 protein expression to protect against MPP+ toxicity and mitochondrial damage [98]. Several other investigators observed that valproate, vorinostat, TSA, and sodium butyrate upregulate glial cell line-derived neurotrophic factor (GDNF) and brain-derived neurotrophic factor (BDNF) in astrocytes and protect dopaminergic neurons through HDAC inhibition and consequential H3 acetylation [99,100]. Collectively, those studies revealed that the neuroprotective effects of various HDACi are mainly accomplished through increased histone acetylation, through an increased transcription of genes encoding neurotropic factors or involved in the protection against mitochondrial DNA damage, or by counteracting the negative effects of α-synuclein accumulation [38,100]. Interestingly, there is accumulating evidence that also natural compounds exerting antioxidant and epigenetic properties, such as EGCG, curcumin, and resveratrol, are neuroprotective in animal models of the disease, and could play an important role in delaying the onset or halting the progression of PD [101–104]. However, despite significant data on their potential neuroprotective effects, clinical studies are in most

Table 4.4 Some Examples of the Effects of Histone Deacetylase Inhibitors (HDACi) in Models of Parkinson's Disease

Compound	Experimental Model	Results[a]
AK-1 or AGK-2 (selective inhibithors of SIRT2)	*Drosophila* model of PD overexpressing α-synuclein	Reduction of α-synuclein mediated toxicity
Sodium butyrate or vorinostat	*Drosophila* model of PD overexpressing α-synuclein	Reduction of α-synuclein mediated toxicity
Sodium butyrate	MPTP-induced PD mouse model	Upregulation of DJ-1 expression and reduced neurotoxicity
Sodium butyrate	Rotenone-induced PD fly model	Improved locomotor impairment and early mortality
Trichostatin A	Neuroblastoma cells treated with MPP+	Rescued mitochondrial fragmentation and cell death induced by MPP+ treatment
Valproate, vorinostat, butyrate, trichostatin A	Cultures of neuronal and glial cells	Upregulation of glial GDNF and BDNF in astrocytes and protection of dopaminergic neurons

[a]*See the text for references.*

of the cases very limited [104]. Various clinical trials have also been initiated to investigate the possible therapeutic potential of HDACi in individuals suffering from PD, but results are still pending [100] (Table 4.4).

4.10 MANIPULATION OF EPIGENETIC MECHANISMS FOR THE TREATMENT OF HUNTINGTON'S DISEASE

To date, no DNMTi has been tested in a HD model [59]. However, there is substantial evidence supporting the importance of targeting HDACs in HD models. Studies in a *Caenorhabditis elegans* model expressing a human HTT fragment with an expanded polyglutamine tract revealed that targeting HDACs resulted in a reduction of the associated neurodegeneration in worm neurons [105]. Similarly, HDACi (butyrate or SAHA) arrested the ongoing progressive neuronal degeneration induced by polyglutamine repeat expansion in two *Drosophila* models of polyglutamine disease [106], and reduction of specific HDACs and sirtuins suppressed pathogenesis in a *Drosophila* model of HD [107]. Following the results obtained in fly models [107], several investigators have shown that targeting HDACs can ameliorate Huntington motor deficits in mouse models of the disease [108–110], and some examples are listed in Table 4.5. For example, a subcutaneous injection of SAHA improved motor impairment in the R6/2 mouse model of HD, and led to the degradation of HDAC4 in cortex and brain stem and to decreased HDAC2 levels [110]. Also more recent studies using selective HDACi revealed a protective role. For example, a 4 weeks intraperitoneal injection of AK-7, a selective inhibitor of sirtuin 2, improved motor functions, extended survival and reduced mutant HTT aggregation in two mouse models of HD [111]. Similarly, the histone deacetylase inhibitor 4b, which preferentially targets HDAC1 and HDAC3, was

Table 4.5 Some Examples of the Effects of Histone Deacetylase Inhibitors (HDACi) in Models of Huntington's Disease

Compound	Experimental Model	Results[a]
Sodium butyrate or SAHA	*Drosophila* models of polyglutamine disease	Halt of the neurodegenerative process
Sodium butyrate	R6/2 mouse model of HD	Improved body weight and motor impairment
SAHA	R6/2 mouse model of HD	Increased histone acetylation in the brain and improved motor impairment
SAHA	R6/2 mouse model of HD	Improved motor impairment, and reduction of HDAC2 and HDAC4 levels
AK-7 (selective inhibitor of sirtuin 2)	R6/2 mouse model and140CAG knock-in Htt mouse model	Improved motor functions, extended survival, and reduction of mutant huntingtin aggregation
4b (targets HDAC1 and HDAC3)	N171-82Q transgenic mice	Improved motor functions, ameliorated cognitive decline

[a]*See the text for references.*

found to ameliorate HD-related phenotypes in different HD model systems [112]. For example, 10–12 weeks injection prevented body weight loss, improved several parameters of motor function and ameliorated Htt-elicited cognitive decline in N171-82Q transgenic mice [110]. More recently, it was found that the HDACi 4b can elicit transgenerational effects, via cross talk between different epigenetic mechanisms, to have an impact on disease phenotypes in a beneficial manner [113]. Those studies opened the way to clinical trials in human HD individuals, which have been recently reviewed [114]. Those trials revealed that sodium phenylbutyrate has a maximum tolerance dose of 15 g/day, and other trials investigated the effects of targeting sirtuin 1, but additional data are required to clarify their beneficial effects in humans [114]. Another line of research is addressing the potential of dietary manipulation in order to induce HDAC inhibition while avoiding the toxic effects of drugs. In this regard, a recent study in the YAC128 mouse model of HD (that contains the complete human HTT) has revealed that dietary restriction was able to lower HTT levels and corrected many effects of the transgene including increased body weight, decreased blood glucose, and impaired motor function [115].

4.11 CONCLUSIONS

Epigenetic processes, including DNA methylation and histone tail modifications, play a fundamental role in the neuronal plasticity required for learning and memory processes and their impairment is likely to contribute to cognitive decline. Indeed, many authors observed either global or gene-specific epigenetic changes in cell culture models, animal models or postmortem human neurons of individuals suffering from Alzheimer's disease, Parkinson's disease, and Huntington's disease. However, despite that there is accumulating evidence of those changes in affected brain regions, their diagnostic value is still largely debated. This is likely resulting from the fact that the available studies are often conflicting and limited by small sample size. Furthermore, changes observed in postmortem brains are not always mirrored by changes in blood DNA of living individuals, and studies in blood cells of early disease stages are still scarce. Therefore, researchers are still questioning whether the observed epigenetic changes are causative of neurodegeration or rather a consequence of the neurodegenerative processes taking place in selected brain regions. However, there is no doubt that those changes occur in the brain of affected individuals, and the treatment of disease animal models with epigenetic compounds, including inhibitors of histone deacetylases and methyl donor compounds, were neuroprotective and often ameliorated cognitive decline or motor symptoms, opening the way for a potential application in humans. Unfortunately, many epigenetic drugs are toxic and nonspecific, posing the problem of safer and more useful compounds. In this regard, there is accumulating evidence that several natural compounds exerting epigenetic properties, such as dietary B-vitamins, curcumin, resveratrol, epigallocatechin-3-gallate, and many others, could hold promise of being very useful compounds to counteract the age-related cognitive decline. One of the questions which is still open is timing, as many approaches in patients showed limited beneficial effects, suggesting that intervention might work better at earlier stages of the disease or even as a preventative strategy in the elderly. However, targeting the epigenome with either natural or synthetic compounds remains one of the most promising strategies to counteract cognitive decline and neurodegeneration in the elderly.

ABBREVIATIONS

5-aza-dC 5-Aza-2′-Deoxycytidine (decitabine)
Aβ Amyloid-beta peptide
ABCA7 ATP-binding cassette, sub-family A, member 7
AD Alzheimer's disease
ANK1 Ankirin 1
APP Amyloid precursor protein
APOE Apolipoprotein E
BACE1 β-Site APP cleaving enzyme 1 (beta secretase)
BDNF Brain-derived neurotrophic factor
BIN1 Bridging integrator 1
CBP CREB-binding protein
CNS Central nervous system
DNMTi DNA methyltransferase inhibitors
DNMTs DNA methyltransferases
EGCG Epigallocatechin-3-gallate
FAAH Fatty acid amide hydrolase
FDA Food and drug administration
GDNF Glial cell line-derived neurotrophic factor
GBA Glucocerebrosidase
GPNMB Transmembrane glycoprotein NMB
GWAS Genome-wide association study
HATs Histone acetyltransferases
HES Hairy and enhancer of split 4
HCY Homocysteine

HD	Huntington's disease
HDACi	Inhibitors of histone deacetylases
HDACs	Histone deacetylases
HDMTs	Histone demethylases
5-hmC	5-Hydroxymethylcytosine
HMTs	Histone lysine methyltransferases
HTT	Huntingtin
LBs	Lewy bodies
LOAD	Late-onset Alzheimer's disease
5-LOX	5-Lipoxygenase
LRRK2	Luecine-rich repeat kinase 2
MAPT	Microtubule associated protein tau
5-mC	5-Methylcytosine
MBDs	Methyl-CpG-binding domain proteins
MCI	Mild cognitive impairment
MPP+	1-Methyl-4-phenylpyridinium
MPTP	1-Methyl-4-phenyl-1,2,3,6-tetrahydropyridine
MS-275	Entinostat
PARK2	Parkin
PARK7	DJ-1
PD	Parkinson's disease
PINK1	PTEN-induced putative kinase 1
PRMTs	Protein arginine methyltransferases
PSEN1	Presenilin 1
PSEN2	Presenilin 2
SAH	*S*-adenosylhomocysteine
SAHA	Suberoylanilide hydroxamic acid (vorinostat)
SAM	S-adenosylmethionine
SIRT1	Sirtuin 1
SN	Substantia nigra
SNCA	Alpha-synuclein
SORBS3	Sorbin and SH3 domain containing 3
SORL1	Sortilin-related receptor 1
STX1B	Syntaxin 1B
TET	Ten-eleven translocation
TMEM59	Transmembrane protein 59
TSA	Trichostatin A

GLOSSARY

Alzheimer's disease The most common neurodegenerative disorder and the primary form of dementia in the elderly. Neurodegeneration in Alzheimer's disease leads to memory loss accompanied by changes of behavior and personality severe enough to affect daily life

DNA methylation The addition of a methyl group (CH_3) to cytosine forming 5-methylcytosine (5-mC), usually in a CpG dinucleotide context. DNA methylation in the promoter region is usually associated with gene silencing

DNA methyltransferase inhibitors (DNMTi) Compounds able to inhibit DNA methylation processes

Histone deacetylase inhibitors (HDACi) Compounds able to inhibit the activity of histone deacetylases

Histone tail modifications Covalent posttranslational modifications of amino acidic residues on the histone tails, such as acetylation, methylation, phosphorylation, ubiquitylation, and sumoylation. The combination of those modifications, and their interplay with DNA methylation and chromatin remodeling proteins, regulates the chromatin structure in a dynamic fashion

Huntington's disease Progressive neurodegenerative disorder resulting in cognitive impairment, choreiform movements, psychiatric and behavioral disturbances

Parkinson's disease The second most common neurodegenerative disorder after Alzheimer's disease, clinically characterized by resting tremor, rigidity, bradykinesia, and postural instability as well as nonmotor symptoms such as autonomic insufficiency, cognitive impairment, and sleep disorders

REFERENCES

[1] Martín-Subero JI. How epigenomics brings phenotype into being. Pediatr Endocrinol Rev 2011;9 (Suppl. 1):506–10.

[2] Guo JU, Ma DK, Mo H, Ball MP, Jang MH, Bonaguidi MA, et al. Neuronal activity modifies the DNA methylation landscape in the adult brain. Nat Neurosci 2011;14:1345–51.

[3] Sultan FA, Day JJ. Epigenetic mechanisms in memory and synaptic function. Epigenomics 2011;3:157–81.

[4] Lattal KW, Wood MA. Epigenetics and persistent memory: implications for reconsolidation and silent extinction beyond the zero. Nat Neurosci 2013;16:124–9.

[5] Puckett RE, Lubin FD. Epigenetic mechanisms in experience-driven memory formation and behaviour. Epigenomics 2011;3:649–64.

[6] Coppedè F. Epigenetics and cognitive disorders. In: Peedicayil J, Avramopoulos D, Grayson DR, editors. Epigenetics in psychiatry. San Diego: Academic Press; 2014. p. 343–67.

[7] Jones PA. Functions of DNA methylation: islands, start sites, gene bodies and beyond. Nat Rev Genet 2012;13:484–92.

[8] Goll MG, Bestor TH. Eukaryotic cytosine methyltransferases. Annu Rev Biochem 2005;74:481–514.

[9] Fournier A, Sasai N, Nakao M, Defossez PA. The role of methyl-binding proteins in chromatin organization and epigenome maintenance. Brief Funct Genom 2012;11:251–64.

[10] Coppedè F. One-carbon metabolism and Alzheimer's disease: focus on epigenetics. Curr Genom 2010;11:246–60.

[11] Guo JU, Su Y, Zhong C, Ming GL, Song H. Hydroxylation of 5-methylcytosine by TET1 promotes active DNA demethylation in the adult brain. Cell 2011;145:423–34.

[12] Cheng Y, Bernstein A, Chen D, Jin P. 5-Hydroxymethylcytosine: a new player in brain disorders? Exp Neurol 2015;268:3–9.

[13] Bannister AJ, Kouzarides T. Regulation of chromatin by histone modifications. Cell Res 2011;21:381–95.

[14] Berger SL. The complex language of chromatin regulation during transcription. Nature 2007;447:407–12.

[15] Martin C, Zhang Y. The diverse functions of histone lysine methylation. Nat Rev Mol Cell Biol 2005;6: 838–49.

[16] World Alzheimer Report 2015. London: Alzheimer's Disease International; 2015.

[17] Reitz C, Mayeux R. Alzheimer disease: epidemiology, diagnostic criteria, risk factors and biomarkers. Biochem Pharmacol 2014;88:640–51.

[18] Migliore L, Coppedè F. Genetics, environmental factors and the emerging role of epigenetics in neurodegenerative diseases. Mutat Res 2009;667:82–97.

[19] Bennet DA, Yu L, Yang J, Srivastava GP, Aubin C, De Jager P. Epigenomics of Alzheimer's disease. Transl Res 2015;165:200–20.

[20] Sanchez-Mut JV, Gräff J. Epigenetic alterations in Alzheimer's disease. Front Behav Neurosci 2015;9:347.

[21] Mastroeni D, McKee A, Grover A, Rogers J, Coleman PD. Epigenetic differences in cortical neurons from a pair of monozygotic twins discordant for Alzheimer's disease. PLoS ONE 2009;4:e6617.
[22] Mastroeni D, Grover A, Delvaux E, Whiteside C, Coleman PD, Rogers J. Epigenetic changes in Alzheimer's disease: decrements in DNA methylation. Neurobiol Aging 2010;31:2025–37.
[23] Bakulski KM, Dolinoy DC, Sartor MA, et al. Genome-wide DNA methylation differences between late-onset Alzheimer's disease and cognitively normal controls in human frontal cortex. J Alzheimers Dis 2012;29:571–88.
[24] Coppieters N, Dieriks BV, Lill C, Faull RL, Curtis MA, Dragunow M. Global changes in DNA methylation and hydroxymethylation in Alzheimer's disease human brain. Neurobiol Aging 2014;35:1334–44.
[25] Chouliaras L, Mastroeni D, Delvaux E, et al. Consistent decrease in global DNA methylation and hydroxymethylation in the hippocampus of Alzheimer's disease patients. Neurobiol Aging 2013;34:2091–9.
[26] Lashley T, Gami P, Valizadeh N, Li A, Revesz T, Balazs R. Alterations in global DNA methylation and hydroxymethylation are not detected in Alzheimer's disease. Neuropathol Appl Neurobiol 2015;41:497–506.
[27] Barrachina M, Ferrer I. DNA methylation of Alzheimer disease and tauopathy-related genes in postmortem brain. J Neuropathol Exp Neurol 2009;68:880–91.
[28] Lunnon K, Smith R, Hannon E, De Jager PL, Srivastava G, Volta M, et al. Methylomic profiling implicates cortical deregulation of ANK1 in Alzheimer's disease. Nat Neurosci 2014;17:1164–70.
[29] De Jager PL, Srivastava G, Lunnon K, Burgess J, Schalkwyk LC, Yu L, et al. Alzheimer's disease: early alterations in brain DNA methylation at ANK1, BIN1, RHBDF2 and other loci. Nat Neurosci 2014;17:1156–63.
[30] Yu L, Chibnik LB, Srivastava GP, Pochet N, Yang J, Xu J, et al. Association of Brain DNA methylation in SORL1, ABCA7, HLA-DRB5, SLC24A4, and BIN1 with pathological diagnosis of Alzheimer disease. JAMA Neurol 2015;72:15–24.
[31] Sanchez-Mut JV, Aso E, Panayotis N, Lott I, Dierssen M, Rabano A, et al. DNA methylation map of mouse and human brain identifies target genes in Alzheimer's disease. Brain 2013;136:3018–27.
[32] Bollati V, Galimberti D, Pergoli L, Dalla Valle E, Barretta F, Cortini F, et al. DNA methylation in repetitive elements and Alzheimer disease. Brain Behav Immun 2011;25:1078–83.
[33] Hernández HG, Mahecha MF, Mejía A, Arboleda H, Forero DA. Global long interspersed nuclear element 1 DNA methylation in a Colombian sample of patients with late-onset Alzheimer's disease. Am J Alzheimers Dis Other Demen 2014;29:50–3.
[34] Di Francesco A, Arosio B, Falconi A, Micioni Di Bonaventura MV, Karimi M, et al. Global changes in DNA methylation in Alzheimer's disease peripheral blood mononuclear cells. Brain Behav Immun 2015;45:139–44.
[35] Piaceri I, Raspanti B, Tedde A, Bagnoli S, Sorbi S, Nacmias B. Epigenetic modifications in Alzheimer's disease: cause or effect? J Alzheimers Dis 2015;43:1169–73.
[36] Tannorella P, Stoccoro A, Tognoni G, Petrozzi L, Salluzzo MG, Ragalmuto A, et al. Methylation analysis of multiple genes in blood DNA of Alzheimer's disease and healthy individuals. Neurosci Lett 2015;600:143–7.
[37] Carboni L, Lattanzio F, Candeletti S, Porcellini E, Raschi E, Licastro F, et al. Peripheral leukocyte expression of the potential biomarker proteins Bdnf, Sirt1, and Psen1 is not regulated by promoter methylation in Alzheimer's disease patients. Neurosci Lett 2015;605:44–8.
[38] Coppedè F. The potential of epigenetic therapies in neurodegenerative diseases. Front Genet 2014;5:220.
[39] Zhang K, Schrag M, Crofton A, Trivedi R, Vinters H, Kirsch W. Targeted proteomics for quantification of histone acetylation in Alzheimer's disease. Proteomics 2012;12:1261–8.
[40] Gräff J, Rei D, Guan JS, Wang WY, Seo J, Hennig KM, et al. An epigenetic blockade of cognitive functions in the neurodegenerating brain. Nature 2012;483:222–6.
[41] Ding H, Dolan PJ, Johnson GV. Histone deacetylase 6 interacts with the microtubule-associated protein tau. J Neurochem 2008;106:2119–30.

[42] Dorsey ER, Constantinescu R, Thompson JP, Biglan KM, Holloway RG, Kieburtz K, et al. Projected number of people with Parkinson disease in the most populous nations, 2005 through 2030. Neurology 2007;68:384–6.
[43] Coppedè F. Genetics and epigenetics of Parkinson's disease. Sci World J 2012;2012:489830.
[44] Subramaniam SR, Chesselet MF. Mitochondrial dysfunction and oxidative stress in Parkinson's disease. Prog Neurobiol 2013;106–107:17–32.
[45] Bellou V, Belbasis L, Tzoulaki I, Evangelou E, Ioannidis JP. Environmental risk factors and Parkinson's disease: an umbrella review of meta-analyses. Parkinsonism Relat Disord 2016;23:1–9.
[46] Jowaed A, Schmitt I, Kaut O, Wüllner U. Methylation regulates alpha-synuclein expression and is decreased in Parkinson's disease patients' brains. J Neurosci 2010;30:6355–9.
[47] Matsumoto L, Takuma H, Tamaoka A, Kurisaki H, Date H, Tsuji S, et al. CpG demethylation enhances alphasynuclein expression and affects the pathogenesis of Parkinson's disease. PLoS ONE 2010;5:e15522.
[48] de Boni L, Riedel L, Schmitt I, Kraus TF, Kaut O, Piston D, et al. DNA methylation levels of α-synuclein intron 1 in the aging brain. Neurobiol Aging 2015;36:3334e7–3334e11.
[49] Pihlstrøm L, Berge V, Rengmark A, Toft M. Parkinson's disease correlates with promoter methylation in the α-synuclein gene. Mov Disord 2015;30:577–80.
[50] Tan YY, Wu L, Zhao ZB, Wang Y, Xiao Q, Liu J, et al. Methylation of α-synuclein and leucine-rich repeat kinase 2 in leukocyte DNA of Parkinson's disease patients. Parkinsonism Relat Disord 2014;20:308–13.
[51] Ai SX, Xu Q, Hu YC, Song CY, Guo JF, Shen L, et al. Hypomethylation of SNCA in blood of patients with sporadic Parkinson's disease. J Neurol Sci 2014;337:123–8.
[52] Schmitt I, Kaut O, Khazneh H, deBoni L, Ahmad A, Berg D, et al. L-dopa increases α-synuclein DNA methylation in Parkinson's disease patients in vivo and in vitro. Mov Disord 2015;30:1794–801.
[53] Desplats P, Spencer B, Coffee E, Patel P, Michael S, Patrick C, et al. Alpha-synuclein sequesters Dnmt1 from the nucleus: a novel mechanism for epigenetic alterations in Lewy body diseases. J Biol Chem 2011;286:9031–7.
[54] Blanch M, Mosquera JL, Ansoleaga B, Ferrer I, Barrachina M. Altered mitochondrial DNA methylation pattern in Alzheimer disease-related pathology and in Parkinson disease. Am J Pathol 2016;186:385–97.
[55] Kontopoulos E, Parvin JD, Feany MB. Alpha-synuclein acts in the nucleus to inhibit histone acetylation and promote neurotoxicity. Hum Mol Genet 2006;15:3012–23.
[56] Richter-Landsberg C, Leyk J. Inclusion body formation, macroautophagy, and the role of HDAC6 in neurodegeneration. Acta Neuropathol 2013;126:793–807.
[57] Berthier A, Jiménez-Sáinz J, Pulido R. PINK1 regulates histone H3 trimethylation and gene expression by interaction with the polycomb protein EED/WAIT1. Proc Natl Acad Sci USA 2013;110:14729–34.
[58] Reiner A, Dragatsis I, Dietrich P. Genetics and neuropathology of Huntington's disease. Int Rev Neurobiol 2011;98:325–72.
[59] Valor LM, Guiretti D. What's wrong with epigenetics in Huntington's disease? Neuropharmacology 2014;80:103–14.
[60] Ng CW, Yildirim F, Yap YS, Dalin S, Matthews BJ, Velez PJ, et al. Extensive changes in DNA methylation are associated with expression of mutant huntingtin. Proc Natl Acad Sci USA 2013;110:2354–9.
[61] Wang F, Yang Y, Lin X, Wang JQ, Wu YS, Xie W, et al. Genome-wide loss of 5-hmC is a novel epigenetic feature of Huntington's disease. Hum Mol Genet 2013;22:3641–53.
[62] Bai G, Cheung I, Shulha HP, Coelho JE, Li P, Dong X, et al. Epigenetic dysregulation of hairy and enhancer of split 4 (HES4) is associated with striatal degeneration in postmortem Huntington brains. Hum Mol Genet 2015;24:1441–56.
[63] Jiang H, Poirier MA, Liang Y, Pei Z, Weiskittel CE, Smith WW, et al. Depletion of CBP is directly linked with cellular toxicity caused by mutant huntingtin. Neurobiol Dis 2006;23:543–51.
[64] Giralt A, Puigdellívol M, Carretón O, Paoletti P, Valero J, Parra-Damas A, et al. Long-term memory deficits in Huntington's disease are associated with reduced CBP histone acetylase activity. Hum Mol Genet 2012;21:1203–16.

[65] Yeh HH, Young D, Gelovani JG, Robinson A, Davidson Y, Herholz K, et al. Histone deacetylase class II and acetylated core histone immunohistochemistry in human brains with Huntington's disease. Brain Res 2013;1504:16–24.

[66] Tsai HC, Li H, Van Neste L, Cai Y, Robert C, Rassool FV, et al. Transient low doses of DNA-demethylating agents exert durable antitumor effects on hematological and epithelial tumor cells. Cancer Cell 2012;21(3):430–46.

[67] Kundakovic M. DNA methyltransferase inhibitors and psychiatric disorders. In: Peedicayil S, Avramopoulos D, Grayson DR, editors. Epigenetics in psychiatry. San Diego: Academic Press; 2014. p. 497–514.

[68] Chakravarty S, Bhat UA, Reddy GR, Gupta P, Kumar A. Histone deacetylase inhibitors and psychiatric disorders. In: Peedicayil J, Avramopoulos D, Grayson DR, editors. Epigenetics in psychiatry. San Diego: Academic Press; 2014. p. 515–44.

[69] Zwergel C, Valente S, Mai A. DNA methyltransferase inhibitors from natural sources. Curr Top Med Chem 2016;16:680–96.

[70] Aggarwal R, Jha M, Shrivastava A, Jha AK. Natural compounds: role in reversal of epigenetic changes. Biochemistry 2015;80:972–89.

[71] Vahid F, Zand H, Nosrat-Mirshekarlou E, Najafi R, Hekmatdoost A. The role dietary of bioactive compounds on the regulation of histone acetylases and deacetylases: a review. Gene 2015;562:8–15.

[72] Fuso A, Seminara L, Cavallaro RA, D'Anselmi F, Scarpa S. S-adenosylmethionine/homocysteine cycle alterations modify DNA methylation status with consequent deregulation of PS1 and BACE and beta-amyloid production. Mol Cell Neurosci 2005;28:195–204.

[73] Fuso A, Nicolia V, Ricceri L, Cavallaro RA, Isopi E, Mangia F, et al. S-adenosylmethionine reduces the progress of the Alzheimer-like features induced by B-vitamin deficiency in mice. Neurobiol Aging 2012;33: e1–16.

[74] Li W, Jiang M, Zhao S, Liu H, Zhang X, Wilson JX, et al. Folic acid inhibits amyloid β-peptide production through modulating DNA methyltransferase activity in N2a-APP cells. Int J Mol Sci 2015;16:25002–13.

[75] Li W, Liu H, Yu M, Zhang X, Zhang M, Wilson JX, et al. Folic acid administration inhibits amyloid β-peptide accumulation in APP/PS1 transgenic mice. J Nutr Biochem 2015;26:883–91.

[76] Wang Y, Xu S, Cao Y, Xie Z, Lai C, Ji X, et al. Folate deficiency exacerbates apoptosis by inducing hypomethylation and resultant overexpression of DR4 together with altering DNMTs in Alzheimer's disease. Int J Clin Exp Med 2014;7:1945–57.

[77] Douaud G, Refsum H, de Jager CA, Jacoby R, Nichols TE, Smith SM, et al. Preventing Alzheimer's disease-related gray matter atrophy by B-vitamin treatment. Proc Natl Acad Sci USA 2013;110:9523–8.

[78] Li MM, Yu JT, Wang HF, Jiang T, Wang J, Meng XF, et al. Efficacy of vitamins B supplementation on mild cognitive impairment and Alzheimer's disease: a systematic review and meta-analysis. Curr Alzheimer Res 2014;11:844–52.

[79] van der Zwaluw NL, Dhonukshe-Rutten RA, van Wijngaarden JP, Brouwer-Brolsma EM, van de Rest O, In 't Veld PH, et al. Results of 2-year vitamin B treatment on cognitive performance: secondary data from an RCT. Neurology 2014;83:2158–66.

[80] Remington R, Bechtel C, Larsen D, Samar A, Doshanjh L, Fishman P, et al. A phase II randomized clinical trial of a nutritional formulation for cognition and mood in Alzheimer's disease. J Alzheimers Dis 2015;45:395–405.

[81] Remington R, Lortie JJ, Hoffmann H, Page R, Morrell C, Shea TB. A nutritional formulation for cognitive performance in mild cognitive impairment: a placebo-controlled trial with an open-label extension. J Alzheimers Dis 2015;48:591–5.

[82] Goozee KG, Shah TM, Sohrabi HR, Rainey-Smith SR, Brown B, Verdile G, et al. Examining the potential clinical value of curcumin in the prevention and diagnosis of Alzheimer's disease. Br J Nutr 2016;115:449–65.

[83] Fischer A, Sananbenesi F, Wang X, Dobbin M, Tsai LH. Recovery of learning and memory is associated with chromatin remodelling. Nature 2007;447:178–82.

[84] Ricobaraza A, Cuadrado-Tejedor M, Marco S, Pérez-Otaño I, García-Osta A. Phenylbutyrate rescues dendritic spine loss associated with memory deficits in a mouse model of Alzheimer disease. Hippocampus 2012;22:1040–50.
[85] Govindarajan N, Agis-Balboa RC, Walter J, Sananbenesi F, Fischer A. Sodium butyrate improves memory function in an Alzheimer's disease mouse model when administered at an advanced stage of disease progression. J Alzheimers Dis 2011;26:187–97.
[86] Francis YI, Fà M, Ashraf H, Zhang H, Staniszewski A, Latchman DS, et al. Dysregulation of histone acetylation in the APP/PS1 mouse model of Alzheimer's disease. J Alzheimers Dis 2009;18:131–9.
[87] Zhang ZY, Schluesener HJ. Oral administration of histone deacetylase inhibitor MS-275 ameliorates neuroinflammation and cerebral amyloidosis and improves behavior in a mouse model. J Neuropathol Exp Neurol 2013;72:178–85.
[88] Yang W, Chauhan A, Wegiel J, Kuchna I, Gu F, Chauhan V. Effect of trichostatin A on gelsolin levels, proteolysis of amyloid precursor protein, and amyloid beta-protein load in the brain of transgenic mouse model of Alzheimer's disease. Curr Alzheimer Res 2014;11:1002–11.
[89] Klein C, Mathis C, Leva G, Patte-Mensah C, Cassel JC, Maitre M, et al. γ-Hydroxybutyrate (Xyrem) ameliorates clinical symptoms and neuropathology in a mouse model of Alzheimer's disease. Neurobiol Aging 2015;36:832–44.
[90] Chang X, Rong C, Chen Y, Yang C, Hu Q, Mo Y, et al. (−)-Epigallocatechin-3-gallate attenuates cognitive deterioration in Alzheimer's disease model mice by upregulating neprilysin expression. Exp Cell Res 2015;334:136–45.
[91] Gleason CE, Carlsson CM, Barnet JH, Meade SA, Setchell KD, Atwood CS, et al. A preliminary study of the safety, feasibility and cognitive efficacy of soy isoflavone supplements in older men and women. Age Ageing 2009;38:86–93.
[92] Gleason CE, Fischer BL, Dowling NM, Setchell KD, Atwood CS, Carlsson CM, et al. Cognitive effects of soy isoflavones in patients with Alzheimer's disease. J Alzheimers Dis 2015;47:1009–19.
[93] Braidy N, Jugder BE, Poljak A, Jayasena T, Mansour H, Nabavi SM, et al. Resveratrol as a potential therapeutic candidate for the treatment and management of Alzheimer's disease. Curr Top Med Chem 2016;16:1951–60.
[94] Feng Y, Jankovic J, Wu YC. Epigenetic mechanisms in Parkinson's disease. J Neurol Sci 2015;349:3–9.
[95] Outeiro TF, Kontopoulos E, Altmann SM, Kufareva I, Strathearn KE, Amore AM, et al. Sirtuin 2 inhibitors rescue alpha-synuclein-mediated toxicity in models of Parkinson's disease. Science 2007;317:516–9.
[96] Zhu M, Li WW, Lu CZ. Histone decacetylase inhibitors prevent mitochondrial fragmentation and elicit early neuroprotection against MPP+. CNS Neurosci Ther 2014;20:308–16.
[97] St Laurent R, O'Brien LM, Ahmad ST. Sodium butyrate improves locomotor impairment and early mortality in a rotenone-induced Drosophila model of Parkinson's disease. Neuroscience 2013;246:382–90.
[98] Zhou W, Bercury K, Cummiskey J, Luong N, Lebin J, Freed CR. Phenylbutyrate up-regulates the DJ-1 protein and protects neurons in cell culture and in animal models of Parkinson disease. J Biol Chem 2011;286:14941–51.
[99] Wu X, Chen PS, Dallas S, Wilson B, Block ML, Wang CC, et al. Histone deacetylase inhibitors up-regulate astrocyte GDNF and BDNF gene transcription and protect dopaminergic neurons. Int J Neuropsychopharmacol 2008;11:1123–34.
[100] Sharma S, Taliyan R. Targeting histone deacetylases: a novel approach in Parkinson's disease. Parkinsons Dis 2015;2015:303294.
[101] Peng K, Tao Y, Zhang J, Wang J, Ye F, Dan G, et al. Resveratrol regulates mitochondrial biogenesis and fission/fusion to attenuate rotenone-induced neurotoxicity. Oxid Med Cell Longev 2016;2016:6705621.
[102] Fu W, Zhuang W, Zhou S, Wang X. Plant-derived neuroprotective agents in Parkinson's disease. Am J Transl Res 2015;7:1189–11202.

[103] Spinelli KJ, Osterberg VR, Meshul CK, Soumyanath A, Unni VK. Curcumin treatment improves motor behavior in α-synuclein transgenic mice. PLoS One 2015;10(6):e0128510.
[104] Caruana M, Vassallo N. Tea polyphenols in Parkinson's disease. Adv Exp Med Biol 2015;863:117–37.
[105] Bates EA, Victor M, Jones AK, Shi Y, Hart AC. Differential contributions of *Caenorhabditis elegans* histone deacetylases to huntingtin polyglutamine toxicity. J Neurosci 2006;26:2830–8.
[106] Steffan JS, Bodai L, Pallos J, Poelman M, McCampbell A, Apostol BL, et al. Histone deacetylase inhibitors arrest polyglutamine-dependent neurodegeneration in Drosophila. Nature 2001;413:739–43.
[107] Pallos J, Bodai L, Lukacsovich T, Purcell JM, Steffan JS, Thompson LM, et al. Inhibition of specific HDACs and sirtuins suppresses pathogenesis in a Drosophila model of Huntington's disease. Hum Mol Genet 2008;17:3767–75.
[108] Hockly E, Richon VM, Woodman B, Smith DL, Zhou X, Rosa E, et al. Suberoylanilide hydroxamic acid, a histone deacetylase inhibitor, ameliorates motor deficits in a mouse model of Huntington's disease. Proc Natl Acad Sci USA 2003;100:2041–6.
[109] Ferrante RJ, Kubilus JK, Lee J, Ryu H, Beesen A, Zucker B, et al. Histone deacetylase inhibition by sodium butyrate chemotherapy ameliorates the neurodegenerative phenotype in Huntington's disease mice. J Neurosci 2003;23:9418–27.
[110] Mielcarek M, Benn CL, Franklin SA, Smith DL, Woodman B, Marks PA, et al. SAHA decreases HDAC 2 and 4 levels in vivo and improves molecular phenotypes in the R6/2 mouse model of Huntington's disease. PLoS One 2011;6:e27746.
[111] Chopra V, Quinti L, Kim J, Vollor L, Narayanan KL, Edgerly C, et al. The sirtuin 2 inhibitor AK-7 is neuroprotective in Huntington's disease mouse models. Cell Rep 2012;2:1492–7.
[112] Jia H, Kast RJ, Steffan JS, Thomas EA. Selective histone deacetylase (HDAC) inhibition imparts beneficial effects in Huntington's disease mice: implications for the ubiquitin-proteasomal and autophagy systems. Hum Mol Genet 2012;21:5280–93.
[113] Jia H, Morris CD, Williams RM, Loring JF, Thomas EA. HDAC inhibition imparts beneficial transgenerational effects in Huntington's disease mice via altered DNA and histone methylation. Proc Natl Acad Sci USA 2015;112:E56–64.
[114] Valor LM. Epigenetic-based therapies in the preclinical and clinical treatment of Huntington's disease. Int J Biochem Cell Biol 2015;67:45–8.
[115] Moreno CL, Ehrlich ME, Mobbs CV. Protection by dietary restriction in the YAC128 mouse model of Huntington's disease: relation to genes regulating histone acetylation and HTT. Neurobiol Dis 2016;85:25–34.

CHAPTER

EPIGENETICS IN PERVASIVE DEVELOPMENTAL DISORDERS: TRANSLATIONAL ASPECTS

5

T. Kubota

Department of Epigenetic Medicine, University of Yamanashi, Kofu-city, Yamanashi Prefecture, Japan

CHAPTER OUTLINE

5.1 INTRODUCTION

Pervasive developmental disorders (PDDs) are neurodevelopmental disorders that were proposed initially in the Diagnostic and Statistical Manual of Mental Disorders 4 (DSM-4), and consist of five subtypes: autistic disorder, Asperger disorder, Rett disorder (Rett syndrome: RTT), childhood disintegrative disorder, and PDD not otherwise specified (PDD-NOS) [1]. In DSM-5, autism spectrum disorder (ASD) was proposed, which includes all subtypes of PDD except RTT, since the causative gene of RTT has been identified [2]. This finding indicates the importance of epigenetics in PDDs because the gene for RTT encodes a protein involved in epigenetic gene regulation [3]. In this chapter, we discuss the current epidemiological and genetic understanding of PDDs, epigenetic abnormalities initially discovered in "congenital" PDDs including RTT, environmentally-induced epigenetic changes associated with "acquired" PDDs, epigenetics-based empirical medicine for congenital Prader–Willi syndrome (PWS), and its future application for acquired PDDs.

Neuropsychiatric Disorders and Epigenetics. http://dx.doi.org/10.1016/B978-0-12-800226-1.00005-8

5.2 EPIDEMIOLOGICAL UNDERSTANDING OF PDD

ASD is characterized by impairments in social interactions and communication, as well as repetitive behaviors and restricted interests [2]. A number of social and environmental factors are known to be involved in ASD. For example, inappropriate child rearing (e.g., child abuse, malnutrition) by parents with psychiatric problems are suggested to be involved in ASD [4–7]. Viral infection (e.g., rubella, cytomegalovirus), environmental chemicals, and drugs are also known to be associated with ASD [8–10]. In fact, inflammation via microglia, potentially induced by various environmental factors including infection, has been demonstrated in the postmortem brains and by neuroimaging in ASD patients [11–13].

5.3 GENETIC UNDERSTANDING OF PDD

The combination of the determination of the human genome sequence and the advent of DNA sequencing technology (e.g., next-generation sequencing) has led to an acceleration of the discovery of disease-causing genes. As a result, mutations in genes encoding neuronal molecules have been identified in ASD patients, which facilitate the deterioration in function of various types of neurons, such as cortical interneurons, pyramidal neurons, and the medium spiny neurons in the striatum [14,15]. For example, mutations in *SHANK*, which encodes a synaptic scaffold protein, in a subset of ASD patients, are thought to contribute to the pathogenesis of ASD, and a *Shank* mutant mouse model exhibits neurological abnormalities [16,17]. Furthermore, mutations in genes encoding molecules associated with common pathways of synaptic development are also associated with ASD, suggesting that ASD can be recognized as a "synaptic disorder" [18,19]. Recently, mutations in genes encoding chromatin-remodeling factors, for example, histone modification enzymes and chromodomain helicase DNA-binding domain 8, have also been identified in ASD patients [20–22]. These findings indicate that ASD can also be recognized as a "chromatin disorder" [20].

Chromatin is a structure of DNA surrounded by histone proteins, which is modified by "epigenetic" factors (e.g., DNA methyltransferase, histone deacetylase) that can alter gene expression. Therefore, abnormalities in epigenetic factors lead to aberrant gene expression, which is associated with various diseases. Thus, aberrant expression caused by epigenetic changes in neuronal cells can be the cause of ASD [23,24]. Indeed, the first examples of human diseases caused by epigenetic abnormalities were PDDs, such as RTT and PWS [3,25].

5.4 EPIGENETIC ABNORMALITIES IN CONGENITAL PDD

5.4.1 ABNORMALITIES IN GENOMIC IMPRINTING

Genomic imprinting is an epigenetic phenomenon that was discovered initially in mammals, which determines parental-specific monoallelic expression of genes that is established during oogenesis and spermatogenesis. Genomic imprinting is imparted in the germ line, where inherited maternal and paternal imprints are erased and new imprinting patterns are established according to the individual's sex [26]. For example, a differential methylated region 2 (DMR2) of the mouse *Igf2* gene is hypomethylated in both male and female primordial gene cells, and then maternal allele-specific DNA methylation

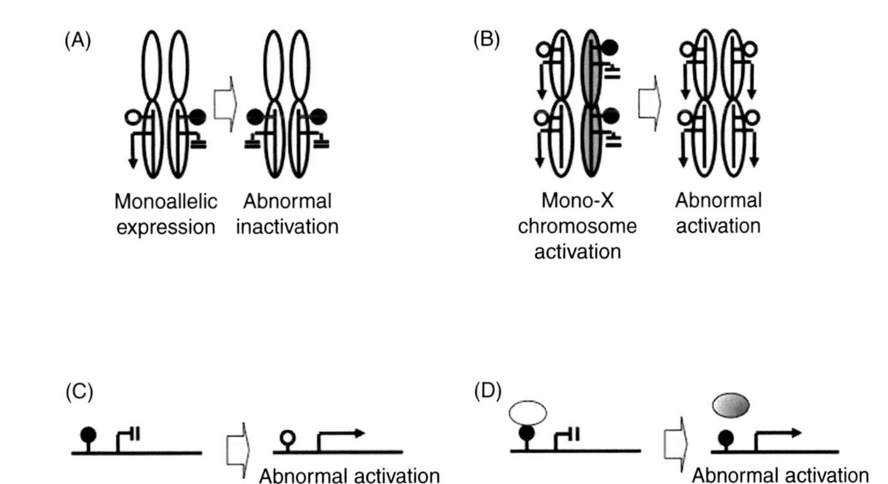

FIGURE 5.1 Patterns of Epigenetic Abnormalities Involved in Congenital Pervasive Developmental Disorders

(A) Genomic imprinting. (B) X chromosome inactivation. (C) DNA-methylating enzyme. (D) Methylated DNA-binding protein.

that contributes to its monoallelic expression is established in DMR2 [27]. This indicates that the erasure and reestablishment of imprinting occurs early in primordial germ cell (PGC) development [27]. Although DNA methylation is the best established epigenetic mark that is critical for allele-specific expression of imprinted genes, many aspects of the regulation of DNA methylation, including how methylation complexes determine the target region and the molecular mechanisms underlying DNA demethylation, remains largely unknown [28]. In this context, recent analysis of the mechanism underlying genome-wide DNA demethylation to reset the epigenome for totipotency revealed that erasure of DNA methylation in PGCs occurs via its conversion to 5-hydroxymethylcytosine (5hmC), driven by ten-eleven translocation methylcytosine dioxygenase 1 (TET1) and TET2, and that global conversion to 5hmC is initiated asynchronously in PGCs for imprint erasure [29]. Once a parent-of-origin DNA methylation pattern is established in imprinted genes, the pattern will not change over a lifetime, thereby preserving monoallelic expression patterns. However, if there is a defect in the expressed allele of an imprinted gene, this leads to genomic imprinting disorders, which are PDDs including Prader–Willi, Angelman and Beckwith–Wiedeman syndromes [25,30] (Fig. 5.1A).

5.4.2 ABNORMALITIES IN X-CHROMOSOME INACTIVATION

The X chromosome is much larger than the Y chromosome; therefore it carries substantially more active genes than the Y chromosome. Consequently, females with an XX karyotype could have higher gene expression from their two X chromosomes than males with an XY karyotype that have a single X. However, this potential imbalance between females and males is prevented by the epigenetic inactivation of one of the X chromosomes in females. If X-chromosome inactivation (XCI) does not occur properly, it can cause lethality in the affected female embryo; this effect is evident in embryonic clones produced by somatic nuclear transfer in which the majority of clones abort due to the failure of XCI in mice [31] (Fig. 5.1B).

In humans, if one of the X chromosomes in a female is very small (e.g., a ring X chromosome generated by a chromosomal rearrangement that deletes the XCI locus (XIST gene) at Xq13, the female has a normal and a small X chromosome that are both active. Although this does not result in embryonic lethality, affected females generally show extremely severe neurodevelopmental delay [32]. These results indicate that the proper inactivation of genes is essential for normal birth and development.

5.4.3 ABNORMALITIES IN ENZYMES ASSOCIATED WITH EPIGENETIC GENE REGULATION

A variety of enzymes chemically modify DNA and histone proteins to determine the epigenetic pattern of the genome (epigenome), which establishes cell identity. Therefore, abnormalities in these enzymes lead to aberrant gene expression, which affects cell function, potentially resulting in human diseases.

For example, methyltransferases (DNMTs) mediate the addition of a methyl group to CpG dinucleotides, and mutations in *DNMT3B*, which encodes a DNMT, cause ICF syndrome, which is characterized by *i*mmunodeficiency, *c*entromere instability, and *f*acial anomalies with mild mental retardation [33,34]. In this syndrome, dysregulation of genes caused by aberrant DNA methylation due to DNMT mutations is expected (Fig. 5.1C). A recent study using induced pluripotent stem cells (iPSCs) derived from ICF patients demonstrates an ICF-specific DNA methylation pattern by genome-scale bisulfite sequencing [35]. However, the key genes whose DNA methylation patterns are abnormal, which are associated with development clinical features, such as immunodeficiency, are still unidentified.

Similarly, it was reported recently that mutations in enzymes responsible for histone modifications are a new cause of PDDs. For example, mutations in the gene encoding euchromatin histone methyltransferase 1 (*EHMT1*), which dimethylates histone 3 at lysine 9, causes Kleefstra syndrome, which is characterized by developmental delay, intellectual disability, hypotonia, and distinct facial features [36]. Furthermore, a heterozygous *Ehmt1* knockout mouse shows hippocampal dysfunction with deficits in fear extinction learning and in novel and spatial object recognition, which could be associated with the phenotype of PDDs [37].

5.5 UPDATED UNDERSTANDING OF RETT SYNDROME

RTT is a representative neurodevelopmental disease. Patients show apparently normal psychomotor development during the first year of life, then enter a short period of developmental stagnation that progresses rapidly with characteristic features, repetitive and stereotypic hand movements, seizures, gait ataxia, and autism [38].

The incidence of RTT is 1 in 10,000–15,000 female births [39,40]. RTT patients are girls (boys are presumed to be embryonic lethal, although there are exceptions [41]), indicating that the disease is inherited in an X-linked dominant manner and that the causative gene should be located on the X chromosome. Linkage analysis identified the candidate locus of the gene between Xq27 and Xqter [42], and subsequent mutation analysis revealed that the causative gene is *methyl-CpG-binding protein 2 (MECP2),* which does not encode an expected molecule associated with synaptic function, but encodes an unexpected molecule associated with epigenetic regulation [3]. This finding introduced a new hypothesis to researchers that epigenetics is an important paradigm for understanding PDD.

MeCP2 functions as a transcriptional repressor (Fig. 5.1D). The central portion of this protein (methyl-binding domain: MBD) binds to methylated DNA and represses transcription [43,44], and the downstream portion (transcription repression domain: TRD) interacts with the Sin3A/HDAC complex or the chromatin remodeling system in mammalian and Xenopus cells [45–47]. MeCP2 is also known to form a bridge between DNA methylation and repressive histone modifications (e.g., H3K9) at its target genomic regions [48].

MECP2 mutations in RTT patients include single point mutations (57–60%), insertions/deletions (15–18%), and large deletions (7–10%) [3,49]. Almost all RTT patients reported to date have sporadic mutations in MECP2, with limited familial exceptions. Among these, the T158M and R168X mutations are the most common. However, neither of these reduce the ability of the MBD to bind to the methylated CpG of target genomic regions; T158M results in only a mild decrease in affinity for methylated DNA. These findings indicate that MeCP2 has functions other than methyl-CpG binding [50,51]. Furthermore, a T158A knock-in mouse model (*Mecp2*$^{T158A/y}$) recapitulates nearly all of the neurological features, including progressive motor, and cognitive impairments and premature lethality, seen in a knock-out mouse model (*Mecp2*$^{-/y}$), suggesting that a single missense mutation is sufficient to cause various features of RTT [51].

Although the deleterious effects of MeCP2 mutations are variable depending on the portion of MeCP2 affected (e.g., MBD, TRD) and type of mutation (e.g., missense, nonsense), all mutations are expected to induce the aberrant expression of MeCP2 target genes (Fig. 5.1D), of which the gene encoding brain-derived neurotrophic factor (*BDNF*) was the first identified [52,53]; in these two studies, Greenberg's group and Sun's group searched for regulatory proteins of *BDNF* because BDNF is decreased in patients with depression [54,55]. As a result, MeCP2 was coincidentally discovered as a regulatory protein of *BDNF*. Thereafter, a number of MeCP2 target genes have been identified, including those encoding proteins associated with synaptic function, such as *DLX5, ID, CRH, IGFBP3, CDKL1, PCDHB1,* and *PCDH7* [56–60], suggesting that RTT is a synaptic disorder, which is consistent with the physiological finding that MeCP2 controls excitatory synaptic strength by regulating the number of glutamatergic synapses [61].

An initial investigation into the spatial and temporal distributions of MeCP2 during development show that it is highly expressed in the brain, lung, and spleen, is slightly expressed in the heart and kidney, and is barely detectable in the liver, stomach, and small intestine. However, there was no obvious correlation between mRNA and protein levels, suggesting that MeCP2 is posttranscriptionally regulated by tissue-specific factors [62]. That study demonstrated that MeCP2 appears in Cajal–Retzius cells in the neocortex and hippocampus, is expressed in the neurons of the deeper and superficial cortical layers [62].

MeCP2 is also expressed in nonneuronal (glial) cells during embryogenesis [63]. The loss of MeCP2 expression was found in glial cells of RTT brains as well as in neuronal cells. Mutant glial cells derived from a RTT mouse model deteriorate the dendritic development of wild-type hippocampal neurons [64]. These findings indicate that astrocytes carrying MeCP2 mutations have a noncell autonomous effect on neuronal development. Furthermore, MeCP2 deficiency in astrocytes causes significant abnormalities in BDNF regulation, cytokine production, and neuronal dendritic induction, suggesting that astrocytes, as well as neurons, contribute to abnormal brain development in RTT [65]. The importance of *Mecp2* expression in astrocytes is also evidenced by experiments showing that reexpression of *Mecp2* preferentially in astrocytes significantly improves locomotion and anxiety levels, restores respiratory abnormalities to a normal pattern, and greatly prolongs the lifespan in globally *Mecp2*-deficient mice, and that restoration of *Mecp2* in astrocytes in the mutant mice exert a noncell-autonomous positive effect on neurons, restoring normal dendritic morphology and increasing the levels of the excitatory glutamate transporter VGLUT1 [66]. These findings suggest that astrocytes are a target tissue for therapy to improve RTT symptoms.

Several lines of evidence also suggest that *Mecp2* is expressed in microglial cells. *Mecp2* (+/−) female mice exhibited significant improvements as a result of wild-type microglial engraftment [67]. The finding indicates that microglia is an alternative major player in the pathophysiology of RTT and that bone marrow transplantation could be a feasible therapeutic approach for RTT patients [68].

In order to capture the developmental stage of RTT, we generated neural cells from RTT patient-derived human-induced pluripotent stem cells (hiPSC) from a rare case of RTT in monozygotic twins [69]. We took advantage of the nonrandom pattern of X-chromosome inactivation in female hiPSC, and the shared genetic background of the RTT monozygotic twins. Isogenic pairs of wild type (expressing normal MeCP2) hiPSC lines and mutant (expressing mutant MeCP2) hiPSC lines from the female RTT patients were generated. A comparative study of these two hiPSC lines demonstrated that a subset of astrocyte-specific genes (e.g., *GFAP*) is aberrantly expressed in neural cells differentiated from the mutant hiPSC lines because of the lack of MeCP2, which regulates *GFAP* expression [69]. This finding indicates that abnormal neurons with astrocyte properties may be associated with the neurological features of RTT. Furthermore, these isogenic iPSC lines provide a unique resource for the further study of disease pathology and for screening novel drugs [70].

5.6 ACQUIRED EPIGENETIC CHANGES ASSOCIATED WITH PDD

In 2004, Meaney's group found that a short-term mental stress changed DNA methylation status in the brain [71]. More precisely, maternal separation during the first week of life induced hypermethylation within the promoter region of the glucocorticoid receptor (*Gr*) gene (gene symbol: *Nr3c1*), which encodes a hormone associated with resilience, in rat brains, and this epigenetic change and suppression of *Gr* is prolonged, resulting in abnormal behavior [71]. At that time, it was believed that epigenetic status established during development is stable and it should not be changed postnataly. The only exception to this was in cancer cells in which remarkable epigenomic changes caused by long-term environmental stress are observed. Therefore, the finding of Meaney's group stimulated researchers to investigate short-term environmental stress–induced epigenetic changes introduced by various environmental factors. This observation also suggests that short-term epigenetic changes may be a possible etiological mechanism of PDDs associated with childhood neglect during early periods of life. Indeed, the ability of stress to induce long-term epigenetic changes has been confirmed in postmortem brains of suicide victims with a history of childhood abuse. In these individuals, hypermethylation of the neuron-specific promoter of *NR3C1*, as well as its reduced expression, were demonstrated in the hippocampal region [72]. These results indicate that mental stress in early life can induce long-lasting epigenetic changes that have a lifelong effect on personality [73].

Epidemiological studies suggest that offspring born to mothers exposed to famine have a lower birth weight than offspring born to the mothers not exposed to famine [74], and that these offspring have increased risks of metabolic disorders (e.g., obesity, diabetes mellitus) and mental disorders [75,76]. In Japan, birth weight is decreased due to intentional dieting in young women since the 1970s, which can lead to fetal malnutrition [77]. Rat experiments demonstrate that malnutrition during the fetal period decreases DNA methylation and increases the expression of the peroxisome proliferator-activated receptor alpha (*PPARa*) gene in the liver, which is associated with metabolic disorders [78,79]. Altered DNA methylation status was identified in the whole blood of individuals who suffered malnutrition during a period of famine in the Netherlands, in which the methylation of imprinted loci (e.g., *IGF2*, *INSIGF*) was lower among individuals who were exposed to famine periconceptionally [80,81].

Environmental chemicals are known to have long-lasting effects on development, metabolism, and health, and epigenetics is thought to be the underlying mechanisms based on insights emerging from experimental model systems and human epidemiological studies. For example, a comprehensive epigenomic approach using BeadChip technology revealed that tobacco smoke, which contains more than 4000 chemicals, reduces DNA methylation at the CpG site in *F2RL3*, which potentially increases the risk of cardiovascular diseases [82] and in *AHRR* that is involved in detoxification metabolism [83]. Epidemiological studies on smokers during pregnancy demonstrate that DNA methylation status is altered at a number of CpG loci in a fetus-derived umbilical cord blood [84] (Fig. 5.2). Furthermore, persistent alterations in DNA methylation are observed at multiple CpG loci, including *AHRR* and *MYO1G* (which encodes a regulator of B lymphocytes), in offspring at the age of 17 years who were born to mothers with a history of smoking during pregnancy [85]. Epigenetic changes are observed in *MYO1G* potentially associated with the immune system and that may have a role in bronchial asthma, a feature of offspring born to mothers who smoked during pregnancy [86].

Such environmental chemically induced DNA methylation changes are not only preserved in individuals over their lifetime but can also be inherited by the next generation: the transgenerational effects of endocrine disruptors, such as vinclozolin, methoxychlor, and bisphenol A have been shown in

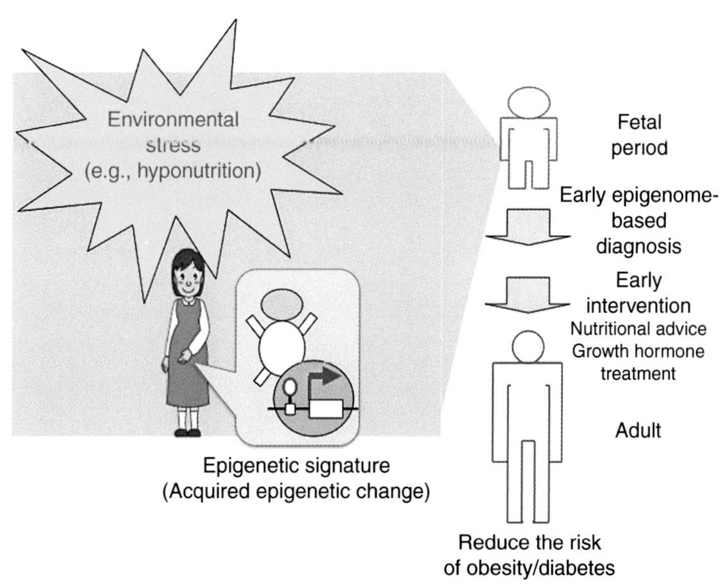

FIGURE 5.2 Schema of Reversible Characteristics in Epigenetics

Several environmental factors are known to change epigenetic status (e.g., DNA methylation), leading to aberrant gene expression. However, aberrant epigenetic status can be reversed by providing appropriate conditions, which subsequently recovers gene expression.

rats through the F3 generation with increased risk of male infertility and pubertal abnormalities with altered DNA methylation patterns in the sperm [87,88]. Furthermore, short-term mental stress due to maternal-neonate separation immediately after birth not only alters the epigenetic status in the brain of the neonate, which is associated with the subsequent persistent abnormal behavior [71], but also alters the epigenetic status (CpG methylation in the *Mecp2*, cannabinoid receptor-1, corticotropin releasing factor receptor 2 and estrogen receptor-alpha1b genes) of sperm with abnormal behavior in subsequent generations [89]. These findings provide biological evidence suggesting that environmental factors, such as traumatic experiences in early life, are risk factors for the development of behavioral and emotional disorders not only in one generation but also in successive generations.

5.7 CONCLUSIONS

"Preemptive medicine" is a type of personalized medicine that is based on the individual and is thus different from population-based "preventive medicine." In preemptive medicine, a practical approach is to detect high-risk individuals by epigenome-based blood biomarkers, and to follow this up with early intervention at the preclinical stage to prevent serious events, such as type 2 diabetes, coronary heart disease, and psychiatric disorders [90].

An epigenome-based preemptive approach has been established for the congenital PDD, PWS. High-risk individuals (i.e., PWS patients) are identified by epigenomic-based blood testing in infancy [25], and a variety of physical and drug treatments are provided to prevent future symptoms (e.g., obesity, type 2 diabetes) by offering a well-balanced low-calorie diet and physical therapy at the ages of 2–4 years of age [91,92]. Furthermore, growth hormone (GH) treatment improves body composition, cognition in childhood, and ability to adapt to society in adulthood [93–98] (Fig. 5.3).

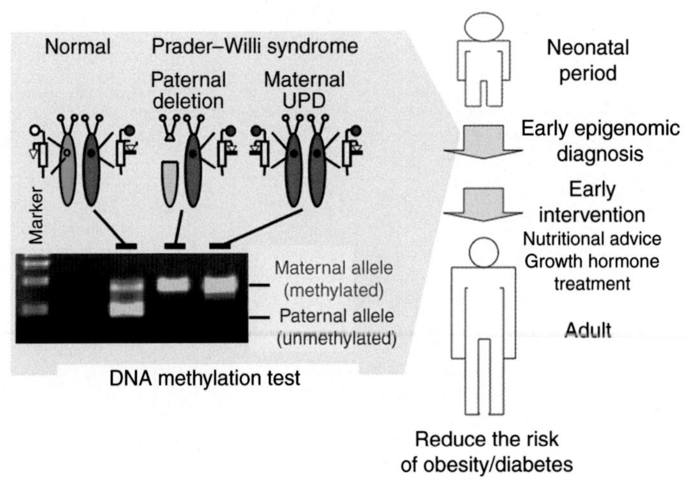

FIGURE 5.3 Epigenome-Based Preemptive Medicine for Prader–Willi Syndrome (PWS)

Both methylated and unmethylated alleles of the imprinted *SNRPN* gene are detected in normal individuals, whereas only methylated allele is detected in PWS patients. The DNA-methylation based diagnostic assay is helpful for early detection of PWS and subsequent early intervention to prevent from subsequent complications.

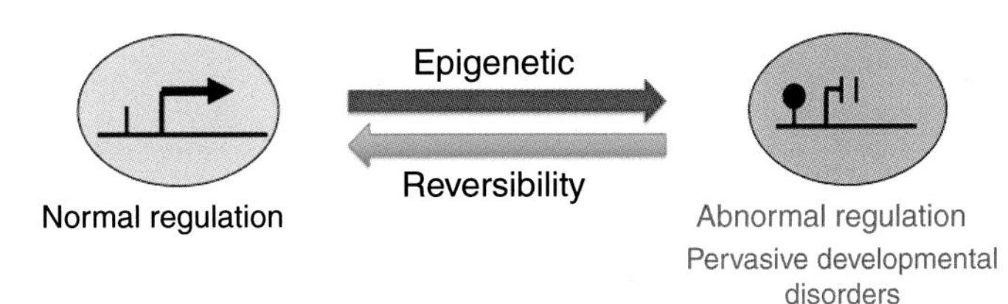

FIGURE 5.4 Epigenome-Based Preemptive Medicine for "Acquired" Pervasive Developmental Disorders

To detect epigenetic changes induced by environmental factors (e.g., maternal smoking), it is necessary to know whether individuals have been exposed to such factors in the past. The early detection of an "epigenetic signature" will enable the initiation of early interventions to prevent from subsequent complications.

In order to establish an intervention program for acquired PDDs similar to that for PWS, it will be necessary to identify "environmentally induced epigenomic signatures" that should be detectable in peripheral tissues (e.g., lymphocytes) (Fig. 5.4). At the same time, it should be feasible to take advantage of epigenomic reversibility to develop effective treatments (e.g., appropriate nutrients, drugs to restore environmentally induced epigenomic changes). In fact, folic acid and histone deacetylase (HDAC) inhibitors are known to restore environmentally induced epigenomic changes [99–101]. In this context, drugs are under development by many pharmacological companies for such therapeutic purposes. The early prediction of latent diseases allows early intervention; such preemptive treatments are the ultimate future goals of medicine [90].

GLOSSARY

Angelman syndrome A congenital syndrome caused by the opposite parent-of origin specific imprinting error to PWS, which is suppression of maternally expressed gene(s) on chromosome 15q12. Patients show mental retardation much more severe than that of PWS and intractable epilepsy

Beckwith–Wiedemann syndrome A congenital syndrome caused by complex abnormal epigenetic patterns on chromosome 11p15 that is characterized by overgrowth in the fetal period, resulting in macrosomia (giant baby), macroglossia, omphalocele, visceromegaly, and embryonal tumors (e.g., Wilms tumor), hepatoblastoma, neuroblastoma, and rhabdomyosarcoma) with specific ear creases/pits

Epigenome An epigenetic pattern of the genome (whole chromosome region), which determines the expression patterns of genes. The epigenome is different between different types of cells (e.g., neurons, astrocytes, skin fibroblasts) and between the immature state and a mature state (e.g., neural stem cells, mature neurons)

Prader–Willi syndrome (PWS) A congenital syndrome caused by suppression of paternally expressed gene(s) on chromosome 15q12 due to abnormal DNA methylation patterns of the gene(s). The patients are weak during infancy, but start to over-eat due to endocrinological abnormalities and subsequently develop obesity and type 2 diabetes

X-chromosome inactivation (XCI) A female-specific genetic mechanism, in which only one of the X chromosomes is active to compensate for gene dosage differences between females and males, who carry only one X chromosome. The chromosome is inactivated by the expression of X-inactivation-specific transcript, which is located near the center of the X-chromosome (Xq13)

REFERENCES

[1] American Psychiatric Association. Diagnostic and Statistical Manual of Mental Disorders 4. Washington, DC: American Psychiatric Association; 1994.

[2] American Psychiatric Association. Proposed revision: a 05 autism spectrum disorder. DSM-5 development. Available from: http://web-beta.archive.org/web/20121115164727/http://www.dsm5.org/ProposedRevision/Pages/proposedrevision.aspx?rid=94

[3] Amir RE, Van den Veyver IB, Wan M, Tran CQ, Francke U, Zoghbi HY. Rett syndrome is caused by mutations in X-linked MECP2, encoding methyl-CpG-binding protein 2. Nat Genet 1999;23(2):185–8.

[4] Larsson HJ, Eaton WW, Madsen KM, Vestergaard M, Olesen AV, Agerbo E, et al. Risk factors for autism: perinatal factors, parental psychiatric history, and socioeconomic status. Am J Epidemiol 2005;161(10):916–25.

[5] Jokiranta E, Brown AS, Heinimaa M, Cheslack-Postava K, Suominen A, Sourander A. Parental psychiatric disorders and autism spectrum disorders. Psychiatry Res 2013;207(3):203–11.

[6] Roberts AL, Lyall K, Rich-Edwards JW, Ascherio A, Weisskopf MG. Association of maternal exposure to childhood abuse with elevated risk for autism in offspring. JAMA Psychiatry 2013;70(5):508–15.

[7] Belluscio LM, Berardino BG, Ferroni NM, Ceruti JM, Cánepa ET. Early protein malnutrition negatively impacts physical growth and neurological reflexes and evokes anxiety and depressive-like behaviors. Physiol Behav 2014;129:237–54.

[8] Berger BE, Navar-Boggan AM, Omer SB. Congenital rubella syndrome and autism spectrum disorder prevented by rubella vaccination–United States, 2001–2010. BMC Public Health 2011;11:340.

[9] Sakamoto A, Moriuchi H, Matsuzaki J, Motoyama K, Moriuchi M. Retrospective diagnosis of congenital cytomegalovirus infection in children with autism spectrum disorder but no other major neurologic deficit. Brain Dev 2015;37(2):200–5.

[10] Heilbrun LP, Palmer RF, Jaen CR, Svoboda MD, Miller CS, Perkins J. Maternal chemical and drug intolerances: potential risk factors for autism and attention deficit hyperactivity disorder (ADHD). J Am Board Fam Med 2015;28(4):461–70.

[11] Rodriguez JI, Kern JK. Evidence of microglial activation in autism and its possible role in brain underconnectivity. Neuron Glia Biol 2011;7(2–4):205–13.

[12] Theoharides TC, Asadi S, Patel AB. Focal brain inflammation and autism. J Neuroinflammation 2013;10:46.

[13] McDougle CJ. Toward an immune-mediated subtype of autism spectrum disorder. Brain Res 2015;1617: 72–92.

[14] Ronemus M, Iossifov I, Levy D, Wigler M. The role of de novo mutations in the genetics of autism spectrum disorders. Nat Rev Genet 2014;15(2):133–41.

[15] Chang J, Gilman SR, Chiang AH, Sanders SJ, Vitkup D. Genotype to phenotype relationships in autism spectrum disorders. Nat Neurosci 2015;18(2):191–8.

[16] Sala C, Vicidomini C, Bigi I, Mossa A, Verpelli C. Shank synaptic scaffold proteins: keys to understanding the pathogenesis of autism and other synaptic disorders. J Neurochem 2015;135(5):849–58.

[17] Schmeisser MJ. Translational neurobiology in Shank mutant mice–model systems for neuropsychiatric disorders. Ann Anat 2015;200:115–7.

[18] Zoghbi HY. Postnatal neurodevelopmental disorders: meeting at the synapse? Science 2003;302(5646):826–30.
[19] Ebrahimi-Fakhari D, Sahin M. Autism and the synapse: emerging mechanisms and mechanism-based therapies. Curr Opin Neurol 2015;28(2):91–102.
[20] De Rubeis S, He X, Goldberg AP, Poultney CS, Samocha K, Cicek AE, et al. Synaptic, transcriptional and chromatin genes disrupted in autism. Nature 2014;515(7526):209–15.
[21] Balan S, Iwayama Y, Maekawa M, Toyota T, Ohnishi T, Toyoshima M, et al. Exon resequencing of H3K9 methyltransferase complex genes, EHMT1, EHTM2 and WIZ, in Japanese autism subjects. Mol Autism 2014;5(1):49.
[22] Barnard RA, Pomaville MB, O'Roak BJ. Mutations and modeling of the chromatin remodeler CHD8 define an emerging autism etiology. Front Neurosci 2015;9:477.
[23] Schanen NC. Epigenetics of autism spectrum disorders. Hum Mol Genet 2006;15(2):R138–50.
[24] Grafodatskaya D1, Chung B, Szatmari P, Weksberg R. Autism spectrum disorders and epigenetics. J Am Acad Child Adolesc Psychiatry 2010;49(8):794–809.
[25] Kubota T, Das S, Christian SL, Baylin SB, Herman JG, Ledbetter DH. Methylation-specific PCR simplifies imprinting analysis. Nat Genet 1997;16(1):16–7.
[26] Szabó PE, Mann JR. Biallelic expression of imprinted genes in the mouse germ line: implications for erasure, establishment, and mechanisms of genomic imprinting. Genes Dev 1995;9(15):1857–68.
[27] Sato S, Yoshimizu T, Sato E, Matsui Y. Erasure of methylation imprinting of Igf2r during mouse primordial germ-cell development. Mol Reprod Dev 2003;65(1):41–50.
[28] Weaver JR, Susiarjo M, Bartolomei MS. Imprinting and epigenetic changes in the early embryo. Mamm Genome 2009;20(9–10):532–43.
[29] Hackett JA, Sengupta R, Zylicz JJ, Murakami K, Lee C, Down TA, et al. Germline DNA demethylation dynamics and imprint erasure through 5-hydroxymethylcytosine. Science 2013;339(6118):448–52.
[30] Kubota T, Saitoh S, Matsumoto T, Narahara K, Fukushima Y, Jinno Y, Niikawa N. Excess functional copy of allele at chromosomal region 11p15 may cause Wiedemann–Beckwith (EMG) syndrome. Am J Med Genet 1994;49(4):378–83.
[31] Nolen LD, Gao S, Han Z, Mann MR, Gie Chung Y, Otte AP, et al. X chromosome reactivation and regulation in cloned embryos. Dev Biol 2005;279(2):525–40.
[32] Kubota T, Wakui K, Nakamura T, Ohashi H, Watanabe Y, Yoshino M, et al. Proportion of the cells with functional X disomy is associated with the severity of mental retardation in mosaic ring X Turner syndrome females. Cytogenet Genome Res 2002;99(1–4):276–84.
[33] Shirohzu H, Kubota T, Kumazawa A, Sado T, Chijiwa T, Inagaki K, Suetake I, Tajima S, Wakui K, Miki Y, Hayashi M, Fukushima Y, Sasaki H. Three novel DNMT3B mutations in Japanese patients with ICF syndrome. Am J Med Genet 2002;112(1):31–7.
[34] Kubota T, Furuumi H, Kamoda T, Iwasaki N, Tobita N, Fujiwara N, et al. ICF syndrome in a girl with DNA hypomethylation but without detectable DNMT3B mutation. Am J Med Genet A 2004;129A(3):290–3.
[35] Huang K, Wu Z, Liu Z, Hu G, Yu J, Chang KH, et al. Selective demethylation and altered gene expression are associated with ICF syndrome in human-induced pluripotent stem cells and mesenchymal stem cells. Hum Mol Genet 2014;23(24):6448–57.
[36] Kleefstra T, Kramer JM, Neveling K, Willemsen MH, Koemans TS, Vissers LE, et al. Disruption of an EHMT1-associated chromatin-modification module causes intellectual disability. Am J Hum Genet 2012;91(1):73–82.
[37] Balemans MC, Kasri NN, Kopanitsa MV, Afinowi NO, Ramakers G, Peters TA, et al. Hippocampal dysfunction in the Euchromatin histone methyltransferase 1 heterozygous knockout mouse model for Kleefstra syndrome. Hum Mol Genet 2013;22(5):852–66.
[38] Christodoulou J. GeneReview-MeCP2 related disorders. Available from: http://www.ncbi.nlm.nih.gov/books/NBK1497/

[39] Hagberg B, Aicardi J, Dias K, Ramos O. A progressive syndrome of autism, dementia, ataxia, and loss of purposeful hand use in girls: Rett's syndrome: report of 35 cases. Ann Neurol 1983;14(4):471–9.
[40] Bienvenu T, Philippe C, De Roux N, Raynaud M, Bonnefond JP, Pasquier L, et al. The incidence of Rett syndrome in France. Pediatr Neurol 2006;34(5):372–5.
[41] Reichow B, George-Puskar A, Lutz T, Smith IC, Volkmar FR. Brief report: systematic review of Rett syndrome in males. J Autism Dev Disord 2015;45(10):3377–83.
[42] Webb T, Clarke A, Hanefeld F, Pereira JL, Rosenbloom L, Woods CG. Linkage analysis in Rett syndrome families suggests that there may be a critical region at Xq28. J Med Genet 1998;35:997–1003.
[43] Nan X, Campoy FJ, Bird A. MeCP2 is a transcriptional repressor with abundant binding sites in genomic chromatin. Cell 1997;88(4):471–81.
[44] Kudo S. Methyl-CpG-binding protein MeCP2 represses Sp1-activated transcription of the human leukosialin gene when the promoter is methylated. Mol Cell Biol 1998;18(9):5492–9.
[45] Wolffe AP. Histone deacetylase: a regulator of transcription. Science 1996;272(5260):371–2.
[46] Jones PL, Veenstra GJ, Wade PA, Vermaak D, Kass SU, Landsberger N, et al. Methylated DNA and MeCP2 recruit histone deacetylase to repress transcription. Nat Genet 1998;19(2):187–91.
[47] Nan X, Bird A. The biological functions of the methyl-CpG-binding protein MeCP2 and its implication in Rett syndrome. Brain Dev 2001;23(Suppl. 1):S32–7.
[48] Fuks F, Hurd PJ, Wolf D, Nan X, Bird AP, Kouzarides T. The methyl-CpG-binding protein MeCP2 links DNA methylation to histone methylation. J Biol Chem 2003;278(6):4035–40.
[49] Ballestar E, Yusufzai TM, Wolffe AP. Effects of Rett syndrome mutations of the methyl-CpG binding domain of the transcriptional repressor MeCP2 on selectivity for association with methylated DNA. Biochemistry 2000;39(24):7100–6.
[50] Adkins NL, Georgel PT. MeCP2: structure and function. Biochem Cell Biol 2011;89(1):1–11.
[51] Goffin D, Allen M, Zhang L, Amorim M, Wang IT, Reyes AR, et al. Rett syndrome mutation MeCP2 T158A disrupts DNA binding, protein stability and ERP responses. Nat Neurosci 2011;15(2):274–83.
[52] Chen WG, Chang Q, Lin Y, Meissner A, West AE, Griffith EC, et al. Derepression of BDNF transcription involves calcium-dependent phosphorylation of MeCP2. Science 2003;302(5646):885–9.
[53] Martinowich K, Hattori D, Wu H, Fouse S, He F, Hu Y, et al. DNA methylation-related chromatin remodeling in activity-dependent BDNF gene regulation. Science 2003;302(5646):890–3.
[54] Karege F, Perret G, Bondolfi G, Schwald M, Bertschy G, Aubry JM. Decreased serum brain-derived neurotrophic factor levels in major depressed patients. Psychiatry Res 2002;109(2):143–8.
[55] Thompson Ray M, Weickert CS, Wyatt E, Webster MJ. Decreased BDNF, trkB-TK+ and GAD67 mRNA expression in the hippocampus of individuals with schizophrenia and mood disorders. J Psychiatry Neurosci 2011;36(3):195–203.
[56] Horike S, Cai S, Miyano M, Cheng JF, Kohwi-Shigematsu T. Loss of silent-chromatin looping and impaired imprinting of DLX5 in Rett syndrome. Nat Genet 2005;37:31–40.
[57] Peddada S, Yasui DH, LaSalle JM. Inhibitors of differentiation (ID1, ID2, ID3 and ID4) genes are neuronal targets of MeCP2 that are elevated in Rett syndrome. Hum Mol Genet 2006;15:2003–14.
[58] Itoh M, Ide S, Takashima S, Kudo S, Nomura Y, Segawa M, et al. Methyl CpG-binding protein 2 (a mutation of which causes Rett syndrome) directly regulates insulin-like growth factor binding protein 3 in mouse and human brains. J Neuropathol Exp Neurol 2007;66(2):117–23.
[59] Carouge D, Host L, Aunis D, Zwiller J, Anglard P. CDKL5 is a brain MeCP2 target gene regulated by DNA methylation. Neurobiol Dis 2010;38(3):414–24.
[60] Miyake K, Hirasawa T, Soutome M, et al. The protocadherins, PCDHB1 and PCDH7, are regulated by MeCP2 in neuronal cells and brain tissues: implication for pathogenesis of Rett syndrome. BMC Neurosci 2011;12:81.
[61] Chao HT, Zoghbi HY, Rosenmund C. MeCP2 controls excitatory synaptic strength by regulating glutamatergic synapse number. Neuron 2007;56(1):58–65.

[62] Shahbazian MD, Antalffy B, Armstrong DL, Zoghbi HY. Insight into Rett syndrome: MeCP2 levels display tissue- and cell-specific differences and correlate with neuronal maturation. Hum Mol Genet 2002;11(2):115–24.
[63] Nagai K, Miyake K, Kubota T. A transcriptional repressor MeCP2 causing Rett syndrome is expressed in embryonic non-neuronal cells and controls their growth. Brain Res Dev Brain Res 2005;157(1):103–6.
[64] Ballas N, Lioy DT, Grunseich C, Mandel G. Non-cell autonomous influence of MeCP2-deficient glia on neuronal dendritic morphology. Nat Neurosci 2009;12(3):311–7.
[65] Maezawa I, Swanberg S, Harvey D, LaSalle JM, Jin LW. Rett syndrome astrocytes are abnormal and spread MeCP2 deficiency through gap junctions. J Neurosci 2009;29(16):5051–61.
[66] Lioy DT, Garg SK, Monaghan CE, Raber J, Foust KD, Kaspar BK, et al. A role for glia in the progression of Rett's syndrome. Nature 2011;475(7357):497–500.
[67] Derecki NC, Cronk JC, Lu Z, Xu E, Abbott SB, Guyenet PG, et al. Wild-type microglia arrest pathology in a mouse model of Rett syndrome. Nature 2012;484(7392):105–9.
[68] Derecki NC, Cronk JC, Kipnis J. The role of microglia in brain maintenance: implications for Rett syndrome. Trends Immunol 2013;34(3):144–50.
[69] Andoh-Noda T, Akamatsu W, Miyake K, Matsumoto T, Yamaguchi R, Sanosaka T, et al. Differentiation of multipotent neural stem cells derived from Rett syndrome patients is biased toward the astrocytic lineage. Mol Brain 2015;8:31.
[70] Ananiev G, Williams EC, Li H, Chang Q. Isogenic pairs of wild type and mutant induced pluripotent stem cell (iPSC) lines from Rett syndrome patients as in vitro disease model. PLoS One 2011;6(9):e25255.
[71] Weaver IC, Cervoni N, Champagne FA, D'Alessio AC, Sharma S, Seckl JR, et al. Epigenetic programming by maternal behavior. Nat Neurosci 2004;7(8):847–54.
[72] McGowan PO, Sasaki A, D'Alessio AC, Dymov S, Labonté B, Szyf M, et al. Epigenetic regulation of the glucocorticoid receptor in human brain associates with childhood abuse. Nat Neurosci 2009;12(3):342–8.
[73] Murgatroyd C, Patchev AV, Wu Y, Micale V, Bockmühl Y, Fischer D, et al. Dynamic DNA methylation programs persistent adverse effects of early-life stress. Nat Neurosci 2009;12(12):1559–66.
[74] Lumey LH. Decreased birthweights in infants after maternal in utero exposure to the Dutch famine of 1944-1945. Paediatr Perinat Epidemiol 1992;6(2):240–53.
[75] Painter RC, de Rooij SR, Bossuyt PM, Simmers TA, Osmond C, Barker DJ, et al. Early onset of coronary artery disease after prenatal exposure to the Dutch famine. Am J Clin Nutr 2006;84(2):322–7.
[76] St Clair D, Xu M, Wang P, Fang Y, Zhang F, Zheng X, et al. Rates of adult schizophrenia following prenatal exposure to the Chinese famine of 1959–1961. JAMA 2005;294(5):557–62.
[77] Gluckman PD, Seng CY, Fukuoka H, Beedle AS, Hanson MA. Low birthweight and subsequent obesity in Japan. Lancet 2007;369(9567):1081–2.
[78] Lillycrop KA, Phillips ES, Jackson AA, Hanson MA, Burdge GC. Dietary protein restriction of pregnant rats induces and folic acid supplementation prevents epigenetic modification of hepatic gene expression in the offspring. J Nutr 2005;135(6):1382–6.
[79] Lillycrop KA, Phillips ES, Torrens C, Hanson MA, Jackson AA, Burdge GC, et al. Feeding pregnant rats a protein-restricted diet persistently alters the methylation of specific cytosines in the hepatic PPAR alpha promoter of the offspring. Br J Nutr 2008;100(2):278–82.
[80] Heijmans BT, Tobi EW, Stein AD, Putter H, Blauw GJ, Susser ES, et al. Persistent epigenetic differences associated with prenatal exposure to famine in humans. Proc Natl Acad Sci USA 2008;105(44):17046–9.
[81] Tobi EW, Lumey LH, Talens RP, Kremer D, Putter H, Stein AD, et al. DNA methylation differences after exposure to prenatal famine are common and timing- and sex-specific. Hum Mol Genet 2009;18(21):4046–53.
[82] Breitling LP, Yang R, Korn B, Burwinkel B, Brenner H. Tobacco-smoking-related differential DNA methylation: 27K discovery and replication. Am J Hum Genet 2011;88(4):450–7.
[83] Shenker NS, Polidoro S, van Veldhoven K, Sacerdote C, Ricceri F, Birrell MA, et al. Epigenome-wide association study in the European Prospective Investigation into Cancer and Nutrition (EPIC-Turin) identifies novel genetic loci associated with smoking. Hum Mol Genet 2013;22(5):843–51.

[84] Ivorra C, Fraga MF, Bayón GF, Fernández AF, Garcia-Vicent C, Chaves FJ, et al. Redon JLurbe EDNA methylation patterns in newborns exposed to tobacco in utero. J Transl Med 2015;13:25.
[85] Richmond RC, Simpkin AJ, Woodward G, Gaunt TR, Lyttleton O, McArdle WL, et al. Prenatal exposure to maternal smoking and offspring DNA methylation across the lifecourse: findings from the avon longitudinal study of parents and children (ALSPAC). Hum Mol Genet 2015;24(8):2201–17.
[86] Simons E, To T, Moineddin R, Stieb D, Dell SD. Maternal second-hand smoke exposure in pregnancy is associated with childhood asthma development. J Allergy Clin Immunol Pract 2014;2(2):201–7.
[87] Anway MD, Cupp AS, Uzumcu M, Skinner MK. Epigenetic transgenerational actions of endocrine disruptors and male fertility. Science 2005;308(5727):1466–9.
[88] Manikkam M, Tracey R, Guerrero-Bosagna C, Skinner MK. Plastics derived endocrine disruptors (BPA, DEHP and DBP) induce epigenetic transgenerational inheritance of obesity, reproductive disease and sperm epimutations. PLoS One 2013;8(1):e55387.
[89] Franklin TB, Russig H, Weiss IC, Gräff J, Linder N, Michalon A, et al. Epigenetic transmission of the impact of early stress across generations. Biol Psychiatry 2010;68(5):408–15.
[90] Imura H. Life course health care and preemptive approach to noncommunicable diseases. Proc Jpn Acad Ser B Phys Biol Sci 2013;89(10):462–73.
[91] Miller JL, Lynn CH, Shuster J, Driscoll DJ. A reduced-energy intake, well-balanced diet improves weight control in children with Prader–Willi syndrome. J Hum Nutr Diet 2013;26(1):2–9.
[92] Schlumpf M, Eiholzer U, Gygax M, Schmid S, van der Sluis I, l'Allemand D. A dailycomprehensive muscle training programme increases lean mass and spontaneous activity in children with Prader–Willi syndrome after 6 months. J Pediatr Endocrinol Metab 2006;19(1):65–74.
[93] Angulo MA, Castro-Magana M, Lamerson M, Arguello R, Accacha S, Khan A. Final adult height in children with Prader–Willi syndrome with and without human growth hormone treatment. Am J Med Genet A 2007;143A(13):1456–61.
[94] Sode-Carlsen R, Farholt S, Rabben KF, Bollerslev J, Schreiner T, Jurik AG, et al. One year of growth hormone treatment in adults with Prader–Willi syndrome improves body composition: results from a randomized, placebo-controlled study. J Clin Endocrinol Metab 2010;95(11):4943–50.
[95] Myers SE, Whitman BY, Carrel AL, Moerchen V, Bekx MT, Allen DB. Two years of growth hormone therapy in young children with Prader–Willi syndrome: physical and neurodevelopmental benefits. Am J Med Genet A 2007;143A(5):443–8.
[96] Siemensma EP, Tummers-de Lind van Wijngaarden RF, Festen DA, Troeman ZC, van Alfen-van der Velden AA, Otten BJ, et al. Beneficial effects of growth hormone treatment on cognition in children with Prader–Willi syndrome: a randomized controlled trial and longitudinal study. J Clin Endocrinol Metab 2012;97(7):2307–14.
[97] Osório J. Growth and development: growth hormone therapy improves cognition in children with Prader–Willi syndrome. Nat Rev Endocrinol 2012;8(7):382.
[98] Höybye C, Thorén M, Böhm B. Cognitive, emotional, physical and social effects of growth hormone treatment in adults with Prader–Willi syndrome. J Intellect Disabil Res 2005;49(Pt 4):245–52.
[99] Li W, Liu H, Yu M, Zhang X, Zhang Y, Liu H, Wilson JX, Huang G. Folic acid alters methylation profile of JAK-STAT and long-term depression signaling pathways in Alzheimer's disease models. Mol Neurobiol 2015;.
[100] Hasan A, Mitchell A, Schneider A, Halene T, Akbarian S. Epigenetic dysregulation in schizophrenia: molecular and clinical aspects of histone deacetylase inhibitors. Eur Arch Psychiatry Clin Neurosci 2013;263(4):273–84.
[101] Schmauss C. An HDAC-dependent epigenetic mechanism that enhances the efficacy of the antidepressant drug fluoxetine. Sci Rep 2015;5:8171.

CHAPTER

EPIGENETIC CAUSES OF INTELLECTUAL DISABILITY—THE FRAGILE X SYNDROME PARADIGM

6

E. Tabolacci, G. Neri
Institute of Genomic Medicine, Catholic University, Rome, Italy

CHAPTER OUTLINE

6.1 INTRODUCTION

The exact definition of translational research is still a matter of debate. If we were to ask 10 researchers to define the concept, we would probably obtain 10 different definitions. A paper entitled "The Meaning of Translational Research and Why It Matters," by S.H. Woolf, defines translational research within the realm of medical sciences, as the "bench to bedside" enterprise of harnessing knowledge from basic sciences to produce new drugs, devices, and treatment options for patients [1]. As a relatively new research discipline, translational research includes aspects of both basic science and clinical research, requiring skills and resources that are not readily available in a basic laboratory or clinical setting. It is for these reasons that translational research is more effective in specialized academic departments or in dedicated research centers. Translational research includes two areas of translation. One is the process of applying discoveries generated during research in the laboratory, and in preclinical studies, to the development of trials and studies in humans. The second area concerns research aimed at enhancing the adoption of best practices for public health. Translational research is characterized by stages (T1 through T4): T1, translation to humans; T2, translation to patients; T3, translation to practice; T4, translation to population health.

Basic research, different from translational research, is the systematic study directed toward greater knowledge of the fundamental aspects of phenomena and is performed above and beyond practical implications. Its goal is to improve our understanding of nature and its laws. Critics of translational

Neuropsychiatric Disorders and Epigenetics. http://dx.doi.org/10.1016/B978-0-12-800226-1.00006-X

research point to examples of important medical remedies that arose as fortuitous discoveries within the mainstream of basic research, such as penicillin and benzodiazepines. Therefore, basic research comes first in improving our understanding of basic biological facts (e.g., the function and structure of DNA), setting the ground for the development of applied medical research, which may or may not lead to the discovery of new cures. Examples of failed translational research in the pharmaceutical industry include the failure of anti-aβ therapeutics in Alzheimer disease. Other problems have stemmed from the widespread irreproducibility thought to exist in translational research literature [2].

Even with these reservations in mind, the importance of translational medicine in the therapy of cancer is real and its success in developing targeted therapy for certain types of tumor is undeniable. Among mendelian disorders the first application of this approach was in phenylketonuria (PKU), an autosomal recessive inborn error of metabolism due to phenylalanine hydroxylase (PAH) deficiency, causing hyperphenylalaninemia and its clinical consequences. The current primary treatment of PKU is the limitation of dietary protein intake, which in the long-term may be associated with poor compliance and other health problems due to malnutrition. The only alternative therapy currently approved and effective in around 30% of PKU patients is the supplementation of tetrahydrobiopterin (BH4), the cofactor of PAH. There is still need to assess the actual tolerance to phenylalanine in PKU patients to ameliorate quality of life, improve nutritional status, avoiding unnecessarily restricted diets, and to interpret the effects of new therapies (for a review see Ref. [3]). But no one can deny that the story of PKU treatment is a successful story.

To complicate this scenario, one must also consider the complexity of gene expression mechanisms, largely realized through epigenetic modifications. Although the unique sequence of the four nucleotides of the genetic code is the blueprint that distinguishes one person from another, the epigenetic information can be seen as erasable annotations penciled between the lines of the DNA sequence, and allowing to distinguish one cell type from another during different stages of embryogenesis and differentiation. The first article indicating the role of epigenetic disregulation (i.e., aberrant DNA methylation) in the course of tumor development and progression was published more than 30 years ago [4]. While the role of epigenetics in cancer is now well-established, leading to novel therapeutic strategies targeting epigenetic alterations, its involvement in intellectual disability (ID) is less well defined, with some notable exceptions. An example is Rett syndrome (RTT), in which a disruption of MECP2 protein causes a severe neurological disorder with features of autism. It was recently shown that MECP2 represses gene expression by binding to methylated CA sites within long genes, and that in neurons lacking MECP2, decreasing the expression of long genes attenuates RTT-associated cellular deficits. These findings suggest that mutations in *MECP2* may cause neurological dysfunction by specifically disrupting long gene expression in the brain [5].

As of today, there are 14,719 publications cited in Pubmed, 4,867 of which are reviews, for the search "Epigenetics," whereas only 119, 58 of which are reviews, are cited for the search "Epigenetics and Intellectual disability." None of these latter articles is more than 10-year-old, demonstrating that the interest for the role of epigenetics in the pathogenesis of ID is still in its infancy, but increasing, thanks also to the introduction of new high-throughput technologies (NGS, etc.), some of which are specifically intended for the study of epigenetic changes (methylome, ChIP-on-chip, etc.). In a significant percentage of patients with congenital diseases resolved causative DNA mutation has not been found, suggesting that yet other mechanisms could play important roles in their etiology. Alterations of the "native" epigenetic imprint are likely to represent one such mechanism. Epigenetics, that is, heritable changes superimposed on the nucleotide sequence, has already been shown to play a key role in embryonic development, X-inactivation, and cell differentiation in mammals. There is, for instance, growing evidence for a contribution of epigenetics to memory formation and cognition [6], suggesting a role in the etiology of mental impairment. Disturbance of the epigenetic profile due to direct alterations at specific genomic regions, or failure of the epigenetic machinery due to malfunction of one of its components, has been demonstrated in cognitive derangements in a number of neurological disorders [7]. It is therefore tempting to speculate that the cognitive deficit in a significant percentage of patients with unexplained ID results from epigenetic modifications. Moreover, a number of disorders of the epigenetic machinery are mendelian disorders where there is disruption of the various components of the epigenetic machinery (writers, erasers, readers, and remodelers) and are thus expected to have widespread downstream epigenetic consequences [8]. In these cases, neurological dysfunction and, in particular, ID appears to be a common phenotype, in association with other features typical of each disorder. The specificity of some of these features raises the question whether specific cell types are particularly sensitive to the loss of epigenetic regulation. Most of these disorders demonstrate dosage sensitivity, as loss of a single allele appears to be sufficient to cause the observed phenotypes. Although the pathogenic sequence is unknown in most cases, there are several examples where disrupted expression of downstream target genes accounts for a substantial portion of the phenotype. Interestingly, in two of these disorders, Rubinstein–Taybi and, Kabuki syndrome the postnatal rescue of markers of the neurological dysfunction by histone deacetylase inhibitors suggests that in some cases the intellectual impairment may be treatable [9,10].

We will focus our attention on fragile X syndrome as a paradigmatic condition where epigenetic mechanisms induce gene silencing and where detailed knowledge of these mechanisms may lead to the discovery of molecular targets for new specific drugs.

6.2 FRAGILE X SYNDROME

Fragile X syndrome (FXS; OMIM #300624) is the most common cause of inherited and monogenic ID. In a recent metaanalysis the frequency of affected males was reduced from 1:4,000 to 1.4:10,000 and that of affected females from 1:7,000 to 0.9:10,000 [11].

The term "fragile X" derived from the presence of the folate-sensitive fragile site FRAXA on the long arm of chromosome X (Xq27.3-q28) in affected males [12]. The syndrome was also known as Martin–Bell syndrome from the authors who described the first family with 11 males affected with ID and a "big face and jaw," showing an apparent X-linked mode of inheritance [13]. Subsequently, it was demonstrated that the Martin–Bell syndrome and FXS are one and the same entity [14]. Molecular analysis of the FRAXA region led to the cloning of the *FMR1* gene, which contains in its 5' UTR a polymorphic CGG triplet repeat [15]. The syndrome is almost exclusively caused by an expansion of this triplet repeat beyond 200 units, followed by methylation of the cytosines, including those of the CpG island in the promoter region upstream (methylated full mutation). This massive methylation blocks the transcription of the gene, preventing the production of the FMRP protein (loss-of-function mutation), even though the coding sequence of the gene remains intact. In the carrier individuals (both males and females) the CGG triplets number between 50 and 200 and are not methylated (premutation).

FXS belongs to a family of disorders called FRAXopathies, which includes other clinical conditions characterized by CGG triplet instability, namely Fragile X Tremor Ataxia Syndrome (FXTAS; OMIM #300623) and Fragile X Premature Ovarian Insufficiency (FXPOI; OMIM #300624) [16].

While FXS results from a loss of *FMR1* expression, FXTAS results form a toxic gain-of-function of the *FMR1* transcript [17].

The phenotype of affected individuals is mainly characterized by cognitive impairment. In general, the cognitive, behavioral and physical phenotype varies according to sex, with hemizygous males being more severely affected than heterozygous females [18]. In addition to moderate ID, manifestations of FXS include several physical signs and behavioral findings, such as distinct facial features, with long face, large ears, and prominent jaw; macroorchidism; hypotonia; hand flapping; perseveration of speech; anxiety; poor eye contact; and social shyness. A general overview of the physical findings at different ages is reported in Fig. 6.1. Two common findings, joint hypermobility and mitral valve prolapse, are probably related to a connective tissue dysfunction [19,20]. Up to 67% of fragile X boys meet either the autism disorder (AD) or the autism spectrum disorder (ASD) criteria on at least one of the diagnostic tests [21]. Attention deficit/hyperactivity disorder (ADHD) has also been often reported in FXS boys [22]. Seizures are present in around 40% of cases and tend to be partial and to regress with time and are amenable to treatment [23]. Hypersensitivity to sensory stimuli, especially audiogenic ones, and hyperarousal are also frequently found [24]. About 5–10% of children with FXS present with a Prader–Willi like phenotype (hyperphagia, lack of satiation after meals, and hypogonadism or delayed puberty) [25]. It has been suggested that this specific subgroup has lowered expression of a gene located on chromosome 15 in the 15q11–q13 region, encoding the cytoplasmic *FMR1* interacting

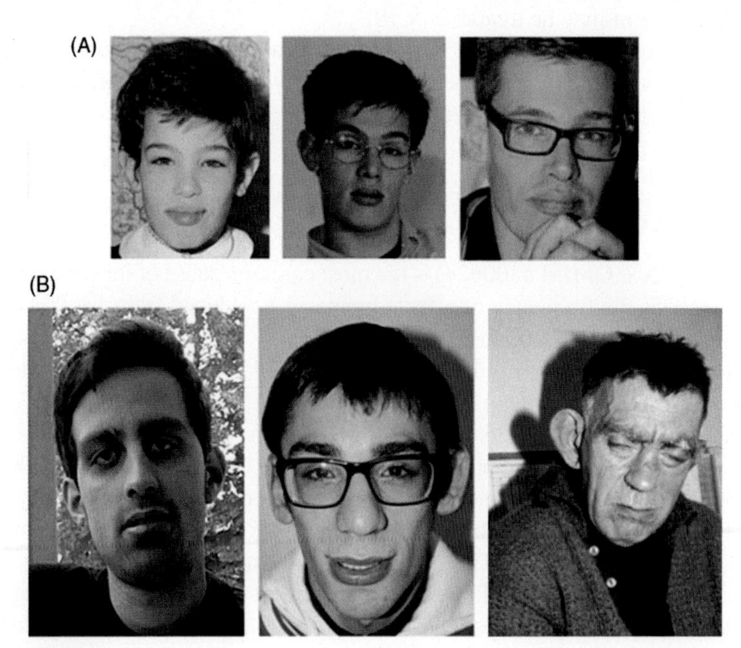

FIGURE 6.1 FXS Phenotype

In (A), the same affected patient is shown at different ages. Note the changing facial phenotype is between the ages of 6 and 26 years in the same individual. In (B), two young men in their early twenties, with more pronounced facial traits, and an older man with full facial manifestations: tall forehead, elongated face, midface hypoplasia, long philtrum and chin, large ears. The accentuation of the phenotype with age largely depends on the persistent muscular hypotonia.

protein 1 (CYFIP1) [26]. In postmortem brains of FXS individuals and in a knockout mouse model cortical spine morphology appears immature, with long, thin spines much more common than the stubby and mushroom-shaped spines that are typical of normal development. In affected individuals there is also a higher density of spines along dendrites, suggesting a possible failure of synapse elimination. While variously misshapen spines are characteristic of a number of ID syndromes, the overabundance of spines seen in FXS is unusual [27]. Overall, affected females usually present with a milder phenotype due to partial dosage compensation by the unaffected X chromosome [28].

Hagerman et al. [29] described the first cases of FXTAS, reporting five men with a *FMR1* premutation, containing from 78 to 98 CGG repeats, who presented in their sixth decade with progressive intentional tremor, parkinsonism, cognitive decline, generalized brain atrophy on MRI and impotence. Males, rarely females, over 50 years of age may develop this neurodegenerative disorder, characterized histologically by intranuclear inclusions both in neurons and in astrocytes [30]. The penetrance and severity of this condition are directly correlated to the size of the CGG expansion [31].

FMR1 premutation in female carriers appears to be a risk factor for premature ovarian insufficiency (POI), defined as menopause at an age less than 40 years [32]. Both penetrance and the age of onset of FXPOI are correlated with the size of premutation in the range of 80–120 repeats, while greater premutations have lower risk [33]. The pathogenesis of FXPOI is still unknown.

6.3 THE *FMR1* GENE AND THE FMRP PROTEIN

The *FMR1* gene, responsible for FXS, was identified in 1991 through positional cloning [15]. It is located on the long arm of the X chromosome (in Xq27.3) and it contains 17 exons spanning 38 kb [34]. The *FMR1* gene produces a 4.4 kb mRNA with an open reading frame (ORF) of 1.9 kb, plus minor isoforms resulting from alternative splicing. The human *FMR1* gene is expressed in many fetal cells and ubiquitously in adult tissues, with higher abundance in brain and testis, according to the phenotype [35,36]. The main product of *FMR1* expression, FMRP, is a 632 aa protein, with a molecular weight spanning from 70 to 80 kDa [37]. FMRP can be primarily classified as an RNA-binding protein that regulates (mostly inhibiting) translation, transport, and stability of target mRNAs, particularly those associated with neuronal development. FMRP may bind both coding and noncoding (nc) RNAs, including the brain cytoplasmic RNA BC1/BC200 [38], as well as microRNAs [39]. FMRP can shuttle from nucleus to cytoplasm and back and can form homodimers and interacts with several cytoplasmic and nuclear proteins involved in mRNA metabolism and cytoskeleton-remodeling proteins via its N-terminal and central regions. FMRP interactions and functions may be modulated by its posttranslational modifications, such as phosphorylation [40]. The functional domains of FMRP are depicted in Fig. 6.2.

The *FMR1* promoter contains approximately 56 CpG sites (CpG island), major consensus sites for USF1/USF2 (E-Box) and α-Pal/Nrf-1, which are both positive regulatory proteins of *FMR1* transcription [42,43] and two GC-boxes with affinity for Sp1/Sp3 proteins, which were shown to have a direct positive role in *FMR1* expression [44]. Like other GC-rich human promoters, it lacks the canonical TATA box, but it includes three initiator-like (InR) sequences, localized about 130 nucleotides (nt) upstream the CGG stretch. The transcription of the *FMR1* gene is initiated from one of these three transcription start sites within an approximately 50 nt region; the size of the CGG repeat may act as a downstream enhancer/modulator of transcription: as the size of the CGG repeat expands, initiation shifts to the upstream sites [45].

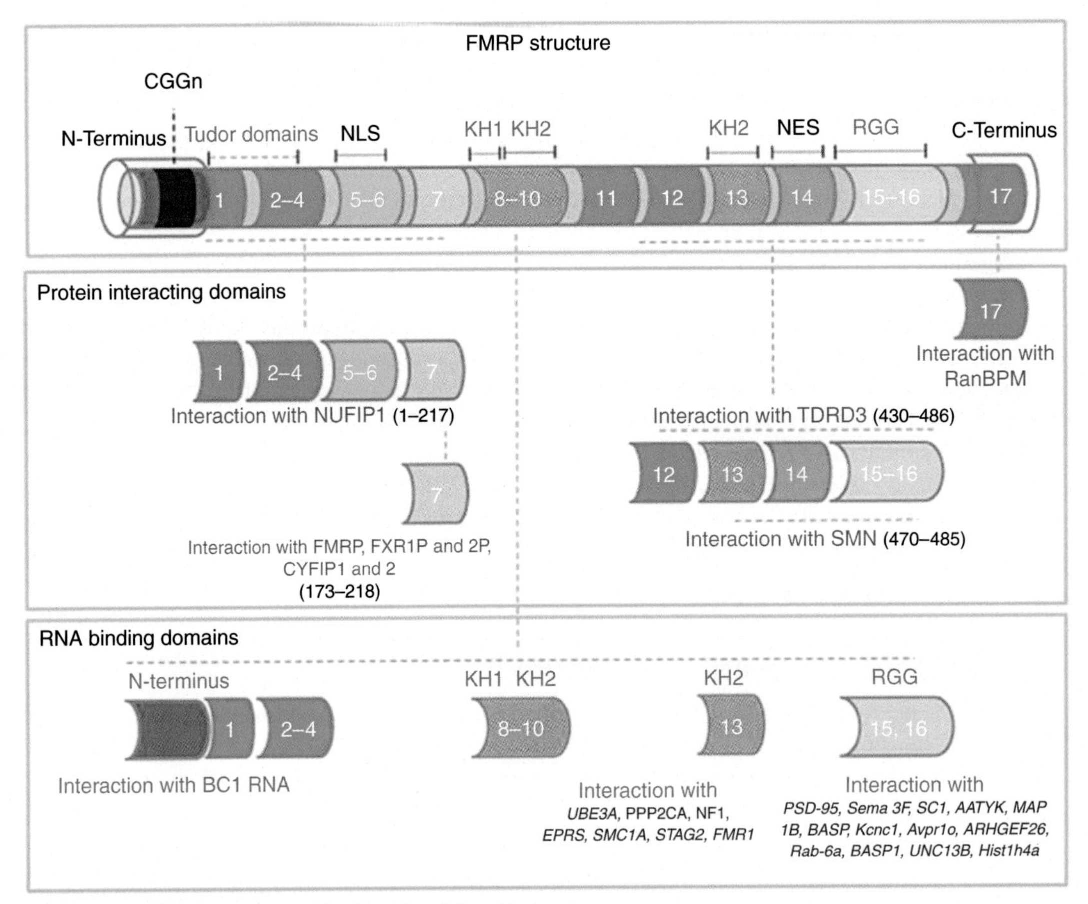

FIGURE 6.2 FMRP Structure and its Functional Domains

The red box at the N-terminus of exon 1 indicates the CGG triplet repeat within the 5′ UTR of the mRNA. The four RNA binding domains are: the N-terminus, the two K homology domains (KH1 and KH2) and the RGG box (upper panel). The interaction between FMRP and other proteins, such as NUFIP1, CYFIP1, CYFIP2, FXR1P, FXR2P, TDRD3, and SMN is indicated (middle panel). The nuclear localization signal *(NLS)* and the nuclear export signal *(NES)*, as well as the FMRP interaction through RNA binding domains and the RNA/mRNA targets directly bound are also indicated (lower panel).

From Fernández E, Rajan N, Bagni C. The FMRP regulon: from targets to disease convergence. Front Neurosci 2013;7:191 [41].

Furthermore, a methylation boundary region located around 650–800 nucleotides upstream of the CGG repeat was recently described, with normal alleles presenting a zone of transition between upstream methylated CpGs and downstream unmethylated ones all the way down to the *FMR1* promoter, allowing normal transcription. This methylation boundary appears to be lost in FXS alleles, which are fully methylated throughout this region. The boundary is also conserved in the mouse genome, even if human and mouse are only 46.7% identical in the 5’ region upstream the *FMR1* gene. The methylation boundary region contains binding sites for various nuclear proteins, including CTCF (CCCTC-binding factor), which is the only known insulator present in the region. This binding is likely to prevent methylation spreading toward the *FMR1* promoter [46].

The polymorphic CGG triplet repeat is located between the promoter region and exon 1, in the 5' UTR of the gene. Based on the CGG expansion range three main classes of alleles are described: normal, with 5–55 CGGs (29 or 30 repeats being the most common allele); premutation (PM), with 56–200 repeats; full mutation (FM), with over 200 repeats, the latter leading to DNA hypermethylation and epigenetic silencing of the gene and consequent loss of its protein product, FMRP (Fig. 6.3). This loss-of-function mutation is the likely cause of the FXS phenotype.

Those three classes of alleles correlate with a different transmission pattern through generations: normal alleles are stable during the intergeneration transmission even if Sullivan et al. [47] demonstrated that transmission of the repeat through males appears to be less stable than that through females, at the common- and intermediate-size level (common $\leq$ 39; intermediate = 40–59 repeats). PM alleles become progressively more unstable and can easily expand to a full mutation during maternal meiosis. In this latter case the paternally transmitted alleles appear to be much more stable than the maternally

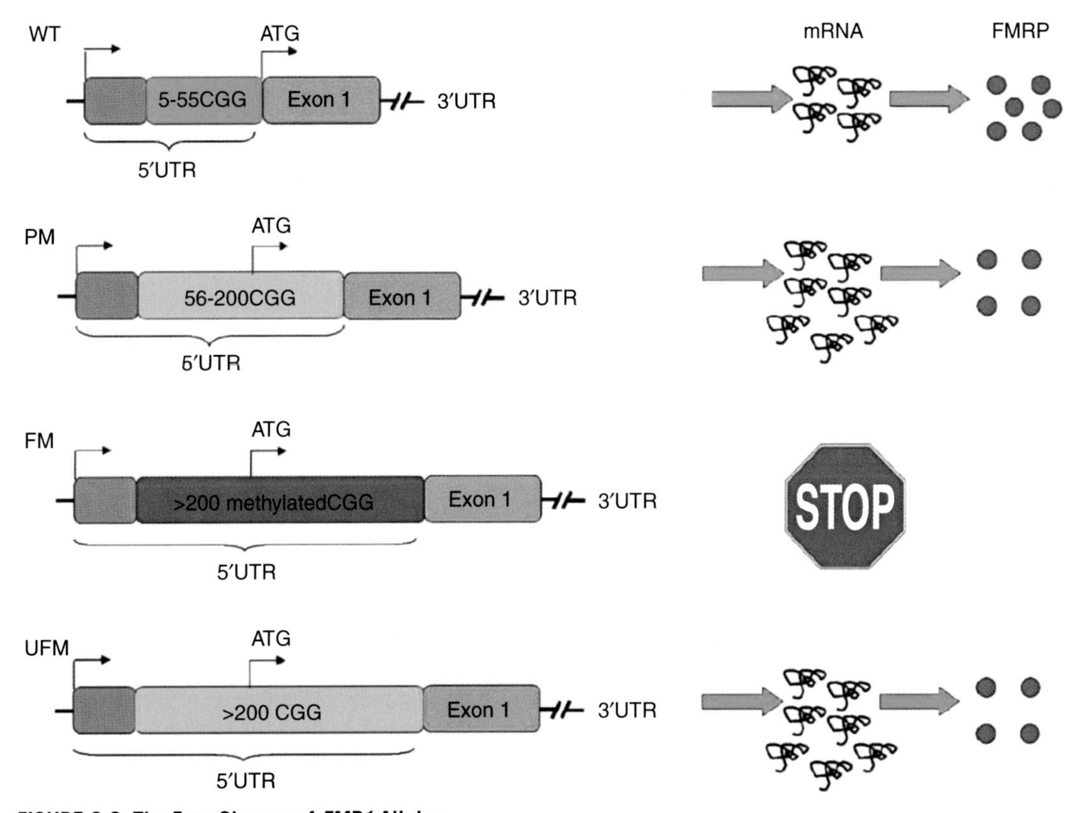

FIGURE 6.3 The Four Classes of *FMR1* Alleles

From top to bottom: there are indicated normal *(WT)*, premutation *(PM)*, full mutation *(FM)*, and unmethylated full mutation alleles *(UFM)*. The arrows indicate the transcriptional and the translational start sites, respectively.

From Pirozzi F, Tabolacci E, Neri G. The FRAXopathies: definition, overview, and update. Am J Med Genet A 2011;155A:1803–16 [16].

transmitted ones [47]. So far, the smallest CGG expansion capable of a complete switch from PM to FM in a single maternal meiosis was found to contain 56 repeats [48]. Until now, no case has been reported of direct expansions from normal to full mutation in FXS families [49]. FM alleles are also unstable, tending to expand both germinally and somatically.

Allelic instability depends not only on the number of CGG repeats but also on their sequence. In normal alleles the CGG sequence is typically interrupted by one AGG triplet every 9–10 CGGs [34]. The role of the AGG triplets is to keep the CGG tract stable during DNA replication, and their absence is among the events that favor the expansion [50]. As a consequence, the presence of the AGG triplets lowers the risk for maternal premutations to be passed down as full mutations. It has been reported that the presence of two interspersed AGG trinucleotides leads to a 60% decrease of such risk, in women carrying 70–80 CGG repeats [51].

Because the number of triplets expands over the generations, the number of affected individuals increases accordingly. This phenomenon was called the Sherman paradox [52]. However, a real anticipation does not exist in FXS, like in other unstable triplet disorders, there being instead an "all or nothing" effect, a full mutation, no matter how large, being required for the expression of the classical FXS phenotype.

In a small number of FXS individuals, variants of the *FMR1* gene were identified (missense mutations, small deletions, introns, and 3'-UTR variants) [53]. The most severe missense mutation, Ile304Asn (I304N), was reported in an individual with a severe manifestation of FXS [54]. The mouse model in which the endogenous *Fmr1* gene harbors the I304N mutation mimic the symptoms of Fragile X syndrome [55]. Mutated FMRP lost its RNA binding capacity and its levels were reduced in the brain particularly at the time when synapses are forming postnatally.

6.4 DIFFERENT TRANSCRIPTS AT THE *FMR1* LOCUS

Vast genomic regions are transcribed but not translated and many of the resulting transcripts, known as noncoding RNAs (ncRNAs), are enriched in the brain [56]. Long ncRNAs (lncRNAs), whose length exceeds 200 nucleotides, perform a wide range of functions, including modulation of transcription or of the epigenetic landscape of their loci of origin. LncRNAs can be transcribed from both strands of the gene, sense and antisense. In 2007 Ladd et al. identified a novel transcript, *FMR1-AS1*, transcribed in the antisense orientation with respect to *FMR1*. Similar to *FMR1*, the *FMR1-AS1* transcript is silenced in FM alleles and upregulated in PM carriers. *FMR1-AS1* transcript is alternatively spliced, polyadenylated, and exported into the cytoplasm. Antisense transcription appears to be driven by two alternative promoters: one is the canonical *FMR1* bidirectional promoter and the second is located in the second intron of the gene. Notably, the latter was identified as a major promoter for *FMR1-AS1* in PM cells. The corresponding antisense transcript probably spans the CGG repeat of the gene. Moreover, normal and PM alleles spliced a 9.7 kb intron corresponding to *FMR1* intron 1, using the complementary splice donor and acceptor of *FMR1*, which represents a nonconsensus CT to AC splice site. PM alleles show a specific alternative splicing in intron 2 that also used a nonconsensus CT-AC splice site. Interestingly, Ladd et al. [57] also identified an antisense transcript in the mouse that overlaps the murine *Fmr1* gene, suggesting a conserved cellular function for *FMR1-AS1*.

In 2008 Khalil et al. identified another ncRNA, called *FMR4*, transcribed in the antisense direction with respect to the *FMR1* locus, expressed in several adult tissue cells, particularly in frontal cortex and hippocampus. *FMR4* is affected by the CGG expansion, with an expression profile similar to *FMR1* mRNA: it is absent in FM alleles and slightly overexpressed in PM alleles, compared to normal control alleles. Transfection experiments performed in HEK293 cells demonstrated that the two RNAs (*FMR4* and *FMR1* sense) are nonfunctionally linked; in fact neither siRNA-induced gene silencing of *FMR4* and *FMR1*-sense, nor overexpression of *FMR4* was able to modify the expression of the opposite transcript [58].

Very recently two additional noncoding transcripts were identified at the *FMR1* locus: *FMR5* and *FMR6* [59]. *FMR5* is a sense lncRNA transcribed 1 kb upstream of the *FMR1* transcription start site (in the methylated region upstream of the methylation boundary), whereas *FMR6* is an antisense transcript that overlaps exons 15, 16, and 17 as well as the 3' UTR of the gene. *FMR5* is expressed in several human brain regions from unaffected, as well as FM and PM individuals. *FMR6* was silenced in both FM and PM carriers. Moreover, the splicing sites in *FMR6* correspond exactly to those of *FMR1* for the introns present between the exons 15–16 and 16–17, recognizing noncanonical consensus sites for the splicing, as demonstrated for the splicing of *FMR1-AS1* [57].

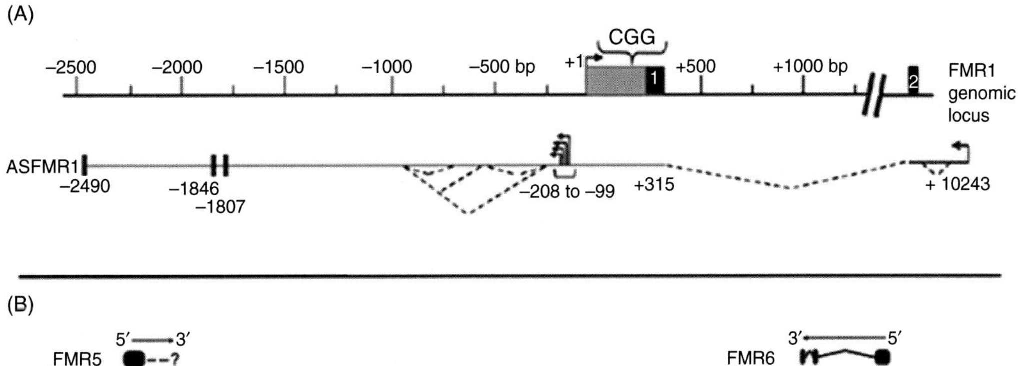

FIGURE 6.4 Scheme of ncRNAs at the *FMR1* Locus

The antisense transcript, *FMR1-AS1*, has two transcriptional start sites: one shared with *FMR1*-mRNA (−208/−99) and the other located 10 kb downstream the transcriptional start site of *FMR1* (+10243), in intron 2, including the CGG triplets (A). The *FMR5* sense transcript is located in the promoter region, while the antisense *FMR6* transcript maps at the 3' UTR, including exons 15, 16 and 17 (B).

Modified from Ladd PD, Smith LE, Rabaia NA, Moore JM, Georges SA, Hansen RS, et al. An antisense transcript spanning the CGG repeat region of FMR1 *is upregulated in premutation carriers but silenced in full mutation individuals. Hum Mol Genet 2007;16D:3174–87; Lanni S, Goracci M, Borrelli L, Mancano G, Chiurazzi P, Moscato U, et al. Role of CTCF protein in regulating* FMR1 *locus transcription. PLoS Genet 2013;9D;e1003601 [57,63].*

All ncRNAs at the *FMR1* locus are schematically depicted in Fig. 6.4. Although the exact role of these antisense transcripts is not yet elucidated, it is possible that they act as regulators of *FMR1* transcription. In other genes, the lncRNAs seem to act as a scaffold for the assembly of proteins necessary for heterochromatin formation or as a guide for the recruitment of silencing complexes [60]. It is also possible that the repeats in the *FMR1* locus are targeted by Polycomb group proteins (PcGs), homed to the locus by lncRNA acting either in *cis*, as in the case of the *Kcnq1-Kcnq1ot1* gene cluster [61], or in *trans*, as in the case of the *Hox-HOTAIR* gene clusters [62]. Since most PcGs targets are G-C-rich, the CGG•CCG-repeats may be particularly prone to be silenced by these complexes.

A new scenario was opened very recently, when it was demonstrated for the first time that specific lncRNAs bind to DNA methyltransferase 1 (*DNMT1*); these RNAs represent a new class of transcripts, that is, *DNMT1*-interacting RNAs. DNMT1 is a DNA methyltransferase involved in the maintenance of cytosine methylation at every cell cycle. Using the *CEBPA* gene (implicated in hematological malignancies) as a model, it was shown that the DNA methylation levels of the locus are inversely correlated with the levels of a lncRNA of *CEBPA* (*ecCEBPA*). The interaction between *ecCEBPA* and DNMT1 prevents *CEBPA* methylation and results in robust *CEPBA*-mRNA production. In RNA immuno-precipitation sequencing it was demonstrated that such functional DNMT1–RNA association occurs at numerous gene loci, including *FMR1*. This could be a starting point to explore the mechanism of such interactions in FXS cells in view of a possible targeted-therapeutic approach to FXS [64].

6.5 EPIGENETIC MODIFICATIONS OF THE *FMR1* LOCUS AND THE ENIGMA OF UNMETHYLATED FULL MUTATION ALLELES

Although silencing of the *FMR1* gene seems to require the described epigenetic regulation in all of its complexity, the critical silencing mechanism is not fully understood [65]. It is possible to categorize the epigenetic status of *FMR1* primarily from its transcriptional status, which is derived from its epigenetic modifications and chromatin conformation. Silenced alleles present a heterochromatic, nonpermissive configuration, while transcribed alleles are characterized by a more "open," permissive, euchromatic status. Generally, switching from active transcription to transcriptional silencing is a direct consequence of CGG repeat expansion over 200 units and its consequent epigenetic modifications [66,67]. Most common changes are: cytosine methylation of the expanded CGG sequence and of the upstream CpG island, representing the most common epigenetic mechanism used to switch off gene transcription; deacetylation or hypoacetylation of histones 3 and 4 (H3 and H4), mediated by histone deacetylases (HDACs); demethethylation of lysine 4 on histone 3 (H3K4), reversing its methylated status, established in normal alleles by HMT Set9, in association with coactivator complexes [68]; methylation of lysine 9 on histone 3 (H3K9), which determines the subsequent recruitment of chromodomain proteins, such as heterochromatin protein 1 (HP1) [69] and trimethylation of lysine 27 on histone 3 (H3K27), leading in turn to the recruitment of other Polycomb group proteins and ultimately to the silencing of target genes [70]. Kumari and Usdin [71] also reported increased methylation of lysine 20 on histone 4 (H4K20) near the CGG expansion.

All these epigenetic changes result in heterochromatin-mediated gene silencing, which has long been recognized as the cause of FXS. In contrast to the hypoacetylation of FM alleles, PM alleles have from 1.5- to 2-fold the normal levels of acetylated H3 and H4 [72], responsible for their increased transcription [73]. These epigenetic modifications confer an initially more open chromatin structure to the *FMR1* promoter and RNAs transcribed from premutated CGG expansions tend to form hairpins. These hairpins may account for the stalling of the 40S ribosomal subunits, thought to be responsible for the translation deficit in PM alleles [74], favoring in turn the use of additional promoters or affecting the expression of chromatin modifying proteins [45]. The paradox of overexpression of PM alleles and the lack of expression of FM alleles remains unresolved. The overall picture is further complicated by the existence of rare unmethylated full mutation (UFM) alleles, identified in individuals with apparently normal intelligence belonging to FXS families [75–77] (Fig. 6.5). These individuals are carriers of CGG expansions of more than 200 repeats, completely devoid of cytosine methylation. The CpG island of the promoter is likewise unmethylated, while the histone marks are similar to those of a normal control allele (histones 3 and 4 are acetylated, H3K4 is methylated and H3K27 is dimethylated) with the exception of H3K9 which remains partially methylated. These rare alleles represent the status of FXS cells before full mutations are silenced, at around 11 weeks of gestation [78]. In vitro, the *FMR1* gene was found to be expressed in embryonic stem cells (ESC) from a human FXS embryo and to undergo transcriptional silencing after ESC differentiation. Notably, during differentiation of FX-ESCs, H3K9

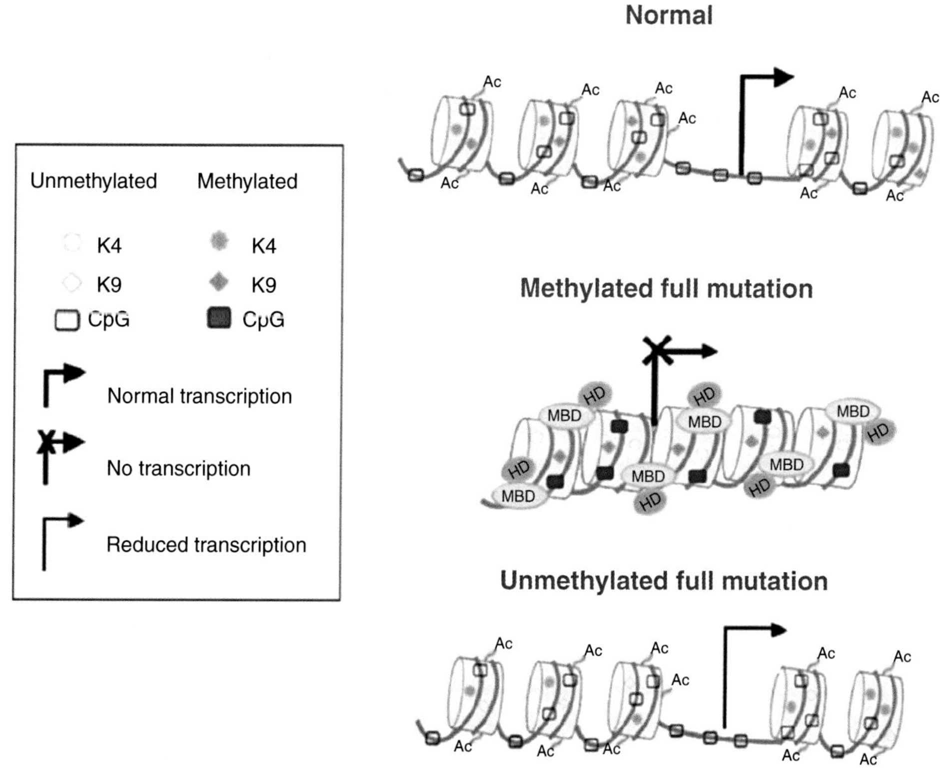

FIGURE 6.5 Major Epigenetic Modifications at the *FMR1* Locus

In normal (*WT*) and unmethylated full mutation (*UFM*) alleleles a permissive euchromatinic configuration is present, while in methylated full mutation the heterochromatinic configuration does not allow transcription. *MBD*, Methyl-binding domain protein; *HD*, histone deacethylases; *Ac*, acetyl group bound to histones.

dimethylation on the *FMR1* promoter was detected before the occurrence of DNA methylation [79]. On the other hand, *FMR1* cannot be reverted to an active status in induced pluripotent stem (iPS) cells derived from FXS fibroblasts, retaining DNA methylation and histone modifications typical of inactive heterochromatin, especially the H3K9 dimethylation [80].

The DNA binding protein CTCF was recently considered as a possible regulator of *FMR1* transcription [63]. Binding sites for CTCF within the *FMR1* locus had already been identified [57]. Lanni et al. [63] confirmed and refined these findings, localizing four CTCF binding sites in the methylation boundary, promoter, exon 1 and intron 2, respectively. The latter site coincides with one of the transcriptional start site of *FMR1-AS1*. For the first time it was shown that these four sites bind CTCF in UFM cells, both lymphoblasts and fibroblasts, with a binding level similar to that of normal control cells. CTCF binding is absent in methylated FM alleles. Notably, pharmacological demethylation with 5-aza-2-deoxycytidine (5-azadC) of FXS cells does not restore CTCF binding to the *FMR1* gene. CTCF depletion with siRNA causes a reduction of both *FMR1* and *FMR1-AS1* transcription, which however does not appear to be caused by remethylation of the *FMR1* promoter, both in WT and UFM cell lines. The antisense transcript *FMR1-AS1* in UFM cell lines were found to be higher compared to normal controls, similar to what happens with the sense transcript in the PM alleles [57,63].

Recently, it was shown that *FMR1* silencing is mediated by the mRNA in human ESCs [81]. According to this study, the *FMR1* mRNA hybridizes to the complementary CGG repeat of the gene to form RNA:DNA duplex. Disrupting the interaction of the mRNA with the CGG repeat seems to prevent silencing of the promoter, supporting a mechanism of RNA-directed gene silencing. However, it should be noted that the human ESCs employed in this work are partially methylated and that, in any event, this mechanism does not explain the existence of UFM individuals, who preserve the *FMR1* transcription in presence of CGG expansion.

6.6 TRANSLATIONAL STRATEGIES

Two different approaches can be attempted to cure FXS: either to normalize the defective synaptic functions, or to restore *FMR1* expression. FXS-iPS cells were obtained and stabilized by different groups worldwide and neurons derived from FXS-iPS cells represent a potentially useful cellular model to recapitulate the spine dysmorphogenesis in FXS [79,80,82]. Fewer and shorter neurites were observed in iPS-derived neurons from FXS, as observed in *Fmr1* knockout mice and in human postmortem brain tissues [82]. Encouraging results led recently to a host of clinical trials, all aimed at correcting the synaptic defect in FXS individuals. Most of these stemmed from the discovery of excessive metabotropic glutamate receptor (GluR) signaling at synapses lacking Fmrp [83]. Endogenous (via glutamate) or exogenous (via dihydroxyphenyl-glycine, DHPG) activation of mGluRs lead to a state called long-term depression (LTD), which induces local protein synthesis and degradation in the postsynaptic space [84]. The mGluR-dependent stimulation might in turn explain some of the neurological manifestation of FXS, such as epilepsy, behavioral abnormalities and dendritic spine dysmorphology. Fig. 6.6 depicts a schematic representation of the so-called "mGluR-theory." Treatment of *fmr1* KO mice with the mGluR5 antagonists 2-methyl-6-(phenylethynyl)-pyridine (MPEP) and fenobam effectively reverse audiogenic seizures and open field hyperactivity. In a single pilot study in which FXS patients were treated with fenobam, Berry-Kravis et al. [85] show reduced hyperactivity and anxiety. In a study performed on *Fmr1* KO mice, compound AFQ056 (Novartis), a subtype-selective inhibitor of mGluR5, rescues the

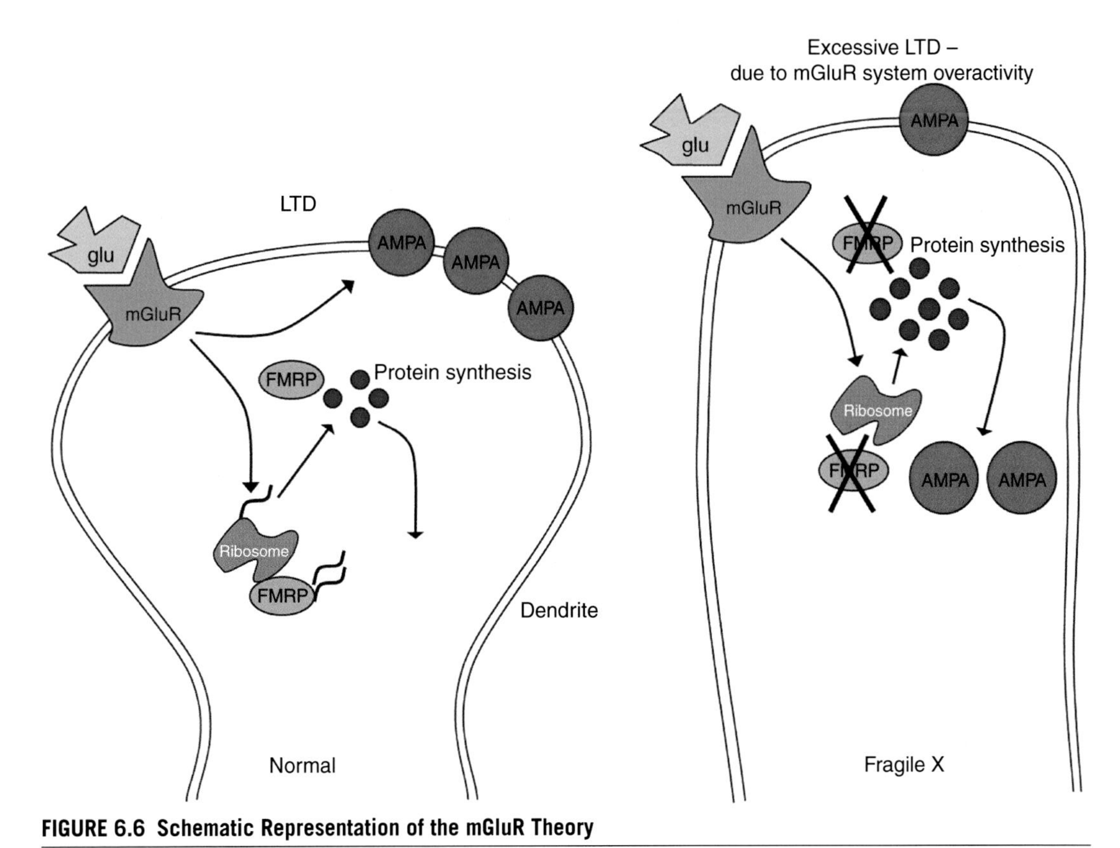

FIGURE 6.6 Schematic Representation of the mGluR Theory

In normal dendrites FMRP inhibits translation of preexisting mRNAs after mGluR stimulation, with normal ionotropic glutamatergic receptor (AMPA) internalization. FXS dendritic spines appear thinner compared to normal because mGluR stimulation by glutamate causes an excess of local protein synthesis with exaggerated LTD due to an increased AMPA receptor internalization activity.

prepulse inhibition to startle response, while cultured hippocampal neurons showed shortened dendritic spines [86]. A double-blind, placebo-controlled two-period crossover study of AFQ056 in 30 FXS males aged 18–35 years results in significant amelioration of hyperactivity in a subgroup of FXS patients who received the drug [87]. Interestingly, the responding patients were found to have complete methylation of the *FMR1* gene, while in nonresponders methylation was incomplete. The authors argued that AFQ056 may be an effective drug for the treatment of a subgroup of FXS patients with the fully methylated gene and therefore with complete lack of FMRP. The hypothesis that AFQ056 may affect DNA methylation was not supported by an in vitro study, specifically addressing this question [88]. Unfortunately, these encouraging preliminary results were not confirmed by subsequent trials with AFQ056 (http://www.fraxa.org/novartis-discontinues-development-mavoglurant-afq056-fragile-x-syndrome).

With the increasing knowledge accumulating on the many functions of the FMRP protein, especially at the synaptic level, many other pharmacological trials were performed to compensate for the altered function of specific neuronal receptors (reviewed in Ref. [89]). Considering the large number of mRNAs targeted by FMRP and the various dysregulated pathways, the GABAergic pathway was considered a possible target [90]. Acamprosate, ganaxolone, gaboxadol and riluzole are all GABA-A agonists used in open label clinical trials. While few clinical benefits where observed on seizures and behavioral problems, adverse side effects were also noted. Agonists of GABA-B receptors have been studied both in *fmr1* KO mice and in FXS patients. In a double-blind, placebo-controlled trial with 63 FXS patients (55 males) aged 6–39 years arbaclofen, a GABA-B receptor agonist, did not show any effect on the primary outcome measure (the ABC-irritability subscale) [91]. Seaside Therapeutics decide to discontinue a large-scale trial with Arbaclofen in adolescents and adults with FXS due to lack of positive results (http://www.clinicaltrials.gov/ct2/results?term=fragile+x&pg=1).

Aside from the mGluR and GABA receptor pathways, other potential targets which may compensate for the absence of FMRP were taken into account. Minocycline treatment, which is thought to act via matrix metalloproteinases 9, was found to correct hippocampal dendritic spine deficits in *Fmr1* KO mice [92]. In an open-label trial with minocycline in 20 FXS patients' aged 13–32 years, improvements were observed in the ABC-irritability subscale, global clinical improvements and visual analog scale for behavior [93].

Lovastatin (a HMG-CoA reductase inhibitor) was able to inhibit the expression of audiogenic seizures in *Fmr1* KO mice [94]. This compound inhibits the Ras-ERK1/2 signaling pathway that controls the protein synthesis in the mouse hippocampus, thus correcting the excessive protein synthesis in mice lacking Fmrp [95]. Its use in FXS patients has not yet been tested, but it will be of extreme interest to discover the effects on the FXS phenotype, given that lovastatin is an approved drug in people with hypercholesterolemia.

The second therapeutic approach to cure FXS is based on the possibility of reversing the epigenetic marks that maintain the mutated *FMR1* gene in a silent state. The concept of reversibility is supported by the existence of those already mentioned UFM individuals. Who for some yet unknown reasons are unable to methylate a fully expanded CGG tract. DNA demethylation can be induced with 5-azacytidine (5-azaC) or, more efficiently, with 5-aza-2-deoxycytidine (5-azadC) that is incorporated into DNA as an analog of deoxycitidine during cell replication and irreversibly blocks DNA methyltransferases [96]. In 1998 we first achieved in vitro reactivation of the *FMR1* full mutation by treating fragile X lymphoblastoid cells with 5-azadC [97], detecting the presence of *FMR1* mRNA as well as FMRP in a fraction of treated cells. The low efficiency of mRNA translation, due to the CGG expansion [98], probably accounts for the observed discrepancy between mRNA and protein levels. Combined treatment with 5-azadC and various histone deacetylase (HDAC) inhibitors (butyrate, phenylbutyrate and trichostatin A) resulted in a synergistic effect on *FMR1* reactivation, even though HDAC inhibitors alone were unable to induce reactivation [99], suggesting that DNA methylation is dominant over histone hypoacetylation in the transcriptional regulation *FMR1* locus [100], as also reported for other heavily methylated genes [101]. By investigating histone modifications in the promoter, exon 1 and exon 16 of *FMR1* by chromatin immunoprecipitation (ChIP) before and after pharmacological treatment of FXS lymphoblasts with 5-azadC, it was shown that this compound induces both histone acetylation and increased methylation of H3K4, while only partly reducing H3K9 methylation [67]. These epigenetic changes led to restoration of a euchromatinic configuration of the *FMR1* promoter effectively transforming a methylated into an unmethylated full mutation. Importantly, our unpublished observations suggest that 5-azadC induces DNA demethylation only in selected genomic regions, leaving other loci unaffected, for example, those containing imprinted genes, such as the Prader–Willi region in chromosome 15q11 and the Beckwith–Wiedemann region in chromosome 11p13.

An obvious concern that arises when considering the clinical use of 5-azadC is its toxicity. In fact, while 5-azaC and 5-azadC are generally well tolerated by patients affected with hematological malignancies [102], the effects of a long-term treatment are unknown. A second obstacle is the apparent requirement for cell division for 5-azadC to be effective. Interestingly, at least two reports suggest that 5-azadC may require minimal or no incorporation into DNA to effectively reduce the activity of the maintenance DNA methyltransferase DNMT1 [103,104]. Furthermore, Bar-Nur et al. [105] treated FXS-iPS cells and their derived neurons with 5-azaC and observed a robust *FMR1* reactivation. Although it was a pilot study, these authors demonstrated that an epigenetic intervention in target cells is possible. As to the objection that drugs like 5-azadC or HDAC inhibitors may have unspecific and unwanted genome-wide effects, it is interesting to note that a microarray screening of 10,814 genes in human colorectal cancer by Suzuki et al. [106] showed that a very limited set of genes are actually transcriptionally upregulated by treatment with 5-azadC (51 genes) and/or trichostatin A (23 genes).

The feasibility of reactivating the mutant *FMR1* gene was confirmed by further experiments, showing a (modest) reactivating effect of valproic acid (VPA), which acts as histone acetylator without DNA demethylation [107]. (Valproic acid has been shown to induce DNA demethylation in the brain of mice and in human progenitor cells in vitro [108].) An open-label clinical trial with VPA provided encouraging preliminary results with a decrease in hyperactivity disorder [109]. Similar findings were previously obtained in a clinical trial with L-acetylcarnitine (ALC) [110], a natural compound that can efficiently increase histone acetylation, but is not sufficient to cause *FMR1* reactivation when used alone in vitro [67].

Among rare genetic disorders, FXS appears to be more amenable than others to an effective pharmacological treatment. It is strictly monogenic, virtually all patients having the same mutation; the mutation does not affect the open reading frame of the gene but rather its (reversible) epigenetic status; the pathogenic mechanism is relatively well elucidated; the phenotype (ID and behavioral problems) ranges from mild to moderate and it does not normally include structural defects of tissues or organs. However, as testified by the clinical trials described earlier, even under ideal conditions, the effective and durable correction of a genetic defect continues to be a tremendous challenge, still requiring a wider basic knowledge of the pathophysiology underlying each disease.

ABBREVIATIONS

ADHD Attention deficit hyperactivity disorder
ALC Acetyl-L-carnitine
ASD Autism spectrum disorders
5-azaC 5-azacitidine (or vidaza)
5-azadC 5-aza-2′-deoxycytidine (or decitabine)
DNMT1 DNA Methyltranserase 1
ID Intellectual disability
iPS Induced pluripotent stem cell
FM Full mutation of the FMR1 gene (CGG sequence over 200 triplets, which is methylated)
FMR1 Fragile X Mental Retardation 1
FMRP Fragile X mental retardation protein
FRAXA Fragile site A on chromosome X, folic acid type (in which was cloned the FMR1 gene responsible for FXS)
FXPOI Fragile X premature ovarian insufficiency (phenotype associated with PM)

FXS	Fragile X syndrome
FXTAS	Fragile X tremor ataxia syndrome (phenotype associated with PM)
GABA	Gamma-aminobutyric acid
hESC	Human embryonic stem cell
HDAC	Histone deacetylase
lncRNA	long non-coding RNA (over 200 nucleotides)
mGluR	Metabotropic glutamate receptor
miRNA	MicroRNA
ncRNA	Non-coding RNA
ORF	Open reading frame (DNA coding for protein)
PM	Premutation of the FMR1 gene (CGG sequence between 56-200 triplets)
UFM	Unmethylated full mutation (CGG sequence over 200 triplets, not methylated)
VPA	Valproic acid

REFERENCES

[1] Woolf SH. The meaning of translational research and why it matters. JAMA 2008;299:3140–8.

[2] Sung NS, Crowley WF Jr, Genel M, Salber P, Sandy L, Sherwood LM, et al. Central challenges facing the national clinical research enterprise. JAMA 2003;289:1278–87.

[3] Ho G, Christodoulou J. Phenylketonuria: translating research into novel therapies. Transl Pediatr 2014;3:49–62.

[4] Feinberg AP, Vogelstein B. Hypomethylation distinguishes genes of some human cancers from their normal counterparts. Nature 1983;301:89–92.

[5] Gabel HW, Kinde B, Stroud H, Gilbert CS, Harmin DA, Kastan NR, et al. Disruption of DNA-methylation-dependent long gene repression in Rett syndrome. Nature 2015;522:89–93.

[6] Steffen PA, Ringrose L. What are memories made of? How polycomb and trithorax proteins mediate epigenetic memory. Nat Rev Mol Cell Biol 2014;15:340–56.

[7] Fahrner JA, Bjornsson HT. Mendelian disorders of the epigenetic machinery: tipping the balance of chromatin states. Annu Rev Genom Hum Genet 2014;15:269–93.

[8] Bjornsson HT. The Mendelian disorders of the epigenetic machinery. Genome Res 2016;25:1473–81.

[9] Bjornsson HT, Benjamin JS, Zhang L, Weissman J, Gerber EE, Chen YC, et al. Histone deacetylase inhibition rescues structural and functional brain deficits in a mouse model of Kabuki syndrome. Sci Transl Med 2014;6. 256ra135.

[10] Murata T, Kurokawa R, Krones A, Tatsumi K, Ishii M, Taki T, et al. Defect of histone acetyltransferase activity of the nuclear transcriptional coactivator CBP in Rubinstein–Taybi syndrome. Hum Mol Genet 2001;10:1071–6.

[11] Hunter J, Rivero-Arias O, Angelov A, Kim E, Fotheringham I, Leal J. Epidemiology of fragile X syndrome: a systematic review and meta-analysis. Am J Med Genet A 2014;164A:1648–58.

[12] Lubs HA. A marker X chromosome. Am J Hum Genet 1969;21:231–44.

[13] Martin JP, Bell J. A pedigree of mental defect showing sex-linkage. J Neurol Phychiatry 1943;6:154–7.

[14] Richards BW, Sylvester PE, Brooker C. Fragile X-linked mental retardation: the Martin–Bell syndrome. J Ment Defic Res 1981;25:253–6.

[15] Verkerk AJ, Pieretti M, Sutcliffe JS, Fu YH, Kuhl DP, Pizzuti A, et al. Identification of a gene (*FMR-1*) containing a CGG repeat coincident with a breakpoint cluster region exhibiting length variation in fragile X syndrome. Cell 1991;65:905–14.

[16] Pirozzi F, Tabolacci E, Neri G. The FRAXopathies: definition, overview, and update. Am J Med Genet A 2011;155A:1803–16.

[17] Hagerman PJ, Hagerman RJ. Fragile X-associated tremor/ataxia syndrome (FXTAS). Ment Retard Dev Disabil Res Rev 2004;10:25–30.

[18] Crawford DC, Acuña JM, Sherman SL. *FMR1* and the fragile X syndrome: human genome epidemiology review. Genet Med 2001;3:359–71.

[19] Opitz JM, Westphal JM, Daniel A. Discovery of a connective tissue dysplasia in the Martin–Bell syndrome. Am J Med Genet 1984;17:101–9.

[20] Pyeritz RE, Stamberg J, Thomas GH, Bell BB, Zahka KG, Bernhardt BA. The marker Xq28 syndrome ("Fragile-X SYndrome") in a retarded man with mitral valve prolapse. Johns Hopkins Med J 1982;151:231–7.

[21] Clifford S, Dissanayake C, Bui QM, Huggins R, Taylor AK, Loesch DZ. Autism spectrum phenotype in males and females with fragile X full mutation and premutation. J Autism Dev Disord 2007;37:738–47.

[22] Sullivan K, Hatton D, Hammer J, Sideris J, Hooper S, Ornstein P, et al. ADHD symptoms in children with FXS. Am J Med Genet A 2006;140:2275–88.

[23] Musumeci SA, Colognola RM, Ferri R, Gigli GL, Petrella MA, Sanfilippo S, et al. Fragile-X syndrome: a particular epileptogenic EEG pattern. Epilepsia 1988;29:41–7.

[24] Loesch DZ, Huggins RM, Hagerman RJ. Phenotypic variation and FMRP levels in fragile X. Ment Retard Dev Disabil Res Rev 2004;10:31–41.

[25] Nowicki ST, Tassone F, Ono MY, Ferranti J, Croquette MF, Goodlin-Jones B, et al. The Prader–Willi phenotype of fragile X syndrome. J Dev Behav Pediatr 2007;28:133–8.

[26] Irwin SA, Galvez R, Greenough WT. Dendritic spine structural anomalies in fragile-X mental retardation syndrome. Cereb Cortex 2000;10:1038–44.

[27] Bennetto L, Pennington BF, Porter D, Taylor AK, Hagerman RJ. Profile of cognitive functioning in women with the fragile X mutation. Neuropsychology 2001;15:290–9.

[28] Hagerman RJ, Leehey M, Heinrichs W, Tassone F, Wilson R, Hills J, et al. Intention tremor, parkinsonism, and generalized brain atrophy in male carriers of fragile X. Neurology 2001;57:127–30.

[29] Greco CM, Hagerman RJ, Tassone F, Chudley AE, Del Bigio MR, Jacquemont S, et al. Neuronal intranuclear inclusions in a new cerebellar tremor/ataxia syndrome among fragile X carriers. Brain 2002;125: 1760–71.

[30] Leehey MA, Berry-Kravis E, Goetz CG, Zhang L, Hall DA, Li L, et al. *FMR1* CGG repeat length predicts motor dysfunction in premutation carriers. Neurology 2008;70:1397–402.

[31] Allingham-Hawkins DJ, Babul-Hirji R, Chitayat D, Holden JJ, Yang KT, Lee C, et al. Fragile X premutation is a significant risk factor for premature ovarian failure: the International Collaborative POF in Fragile X study—preliminary data. Am J Med Genet 1999;83:322–5.

[32] Gleicher N, Weghofer A, Oktay K, Barad D. Relevance of triple CGG repeats in the *FMR1* gene to ovarian reserve. Reprod Biomed Online 2009;19:385390.

[33] Eichler EE, Holden JJ, Popovich BW, Reiss AL, Snow K, Thibodeau SN, et al. Length of uninterrupted CGG repeats determines instability in the *FMR1* gene. Nat Genet 1994;8:88–94.

[34] Kaufmann WE. Neurobiology of Fragile X syndrome: from molecular genetics to neurobehavioral phenotype. Microsc Res Tech 2002;57:131–4.

[35] Lozano R, Rosero CA, Hagerman RJ. Fragile X spectrum disorders. Intractable Rare Dis Res 2014;3: 134–46.

[36] Xie W, Dolzhanskaya N, LaFauci G, Dobkin C, Denman RB. Tissue and developmental regulation of fragile X mental retardation 1 exon 12 and 15 isoforms. Neurobiol Dis 2009;35:52–62.

[37] Verheij C, de Graaff E, Bakker CE, Willemsen R, Willems PJ, Meijer N, et al. Characterization of *FMR1* proteins isolated from different tissues. Hum Mol Genet 1995;4:895–901.

[38] Zalfa F, Adinolfi S, Napoli I, Kühn-Hölsken E, Urlaub H, Achsel T, et al. Fragile X mental retardation protein (FMRP) binds specifically to the brain cytoplasmic RNAs BC1/BC200 via a novel RNA-binding motif. J Biol Chem 2005;280:33403–10.

[39] Muddashetty R, Bassell GJ. A boost in microRNAs shapes up the neuron. EMBO J 2009;28:617–8.

[40] Cheever A, Ceman S. Translation regulation of mRNAs by the fragile X family of proteins through the microRNA pathway. RNA Biol 2009;6:175–8.

[41] Fernández E, Rajan N, Bagni C. The FMRP regulon: from targets to disease convergence. Front Neurosci 2013;7:191.

[42] Eichler EE, Richards S, Gibbs RA, Nelson DL. Fine structure of the human *FMR1* gene. Hum Mol Genet 1993;2:1147–53.

[43] Verkerk AJ, de Graaff E, De Boulle K, Eichler EE, Konecki DS, Reyniers E, et al. Alternative splicing in the fragile X gene *FMR1*. Hum Mol Genet 1993;2:1348.

[44] Kumari D, Gabrielian A, Wheeler D, Usdin K. The roles of Sp1, Sp3, USF1/USF2 and NRF-1 in the regulation and three-dimensional structure of the Fragile X mental retardation gene promoter. Biochem J 2005;386:297–303.

[45] Beilina A, Tassone F, Schwartz PH, Sahota P, Hagerman PJ. Redistribution of transcription start sites within the *FMR1* promoter region with expansion of the downstream CGG-repeat element. Hum Mol Genet 2004;13:543–9.

[46] Naumann A, Hochstein N, Weber S, Fanning E, Doerfler W. A distinct DNA-methylation boundary in the 5'-upstream sequence of the *FMR1* promoter binds nuclear proteins and is lost in fragile X syndrome. Am J Hum Genet 2009;85:606–16.

[47] Sullivan AK, Crawford DC, Scott EH, Leslie ML, Sherman SL. Paternally transmitted *FMR1* alleles are less stable than maternally transmitted alleles in the common and intermediate size range. Am J Hum Genet 2002;70:1532–44.

[48] Fernandez-Carvajal I, Lopez Posadas B, Pan R, Raske C, Hagerman PJ, Tassone F. Expansion of an *FMR1* grey-zone allele to a full mutation in two generations. J Mol Diagn 2009;11:306–10.

[49] Oostra BA, Chiurazzi P. The fragile X gene and its function. Clin Genet 2001;60:399–408.

[50] Ludwig AL, Raske C, Tassone F, Garcia-Arocena D, Hershey JW, Hagerman PJ. Translation of the *FMR1* mRNA is not influenced by AGG interruptions. Nucleic Acids Res 2009;37:6896–904.

[51] Yrigollen CM, Durbin-Johnson B, Gane L, Nelson DL, Hagerman R, Hagerman PJ, et al. AGG interruptions within the maternal *FMR1* gene reduce the risk of offspring with fragile X syndrome. Genet Med 2012;14:729–36.

[52] Sherman SL, Jacobs PA, Morton NE, Froster-Iskenius U, Howard-Peebles PN, Nielsen KB, et al. Further segregation analysis of the fragile X syndrome with special reference to transmitting males. Hum Genet 1985;69:289–99.

[53] Collins SC, Bray SM, Suhl JA, Cutler DJ, Coffee B, Zwick ME, et al. Identification of novel *FMR1* variants by massively parallel sequencing in developmentally delayed males. Am J Med Genet A 2010;152A:2512–20.

[54] De Boulle K, Verkerk AJ, Reyniers E, Vits L, Hendrickx J, Van Roy B, et al. A point mutation in the *FMR-1* gene associated with fragile X mental retardation. Nat Genet 1993;3:31–5.

[55] Zang JB, Nosyreva ED, Spencer CM, Volk LJ, Musunuru K, Zhong R, et al. A mouse model of the human Fragile X syndrome I304N mutation. PLoS Genet 2009;5:e1000758.

[56] Djebali S, Davis CA, Merkel A, Dobin A, Lassmann T, Mortazavi A, et al. Landscape of transcription in human cells. Nature 2012;489:101–8.

[57] Ladd PD, Smith LE, Rabaia NA, Moore JM, Georges SA, Hansen RS, et al. An antisense transcript spanning the CGG repeat region of *FMR1* is upregulated in premutation carriers but silenced in full mutation individuals. Hum Mol Genet 2007;16:3174–87.

[58] Khalil AM, Faghihi MA, Modarresi F, Brothers SP, Wahlestedt C. A novel RNA transcript with antiapoptotic function is silenced in fragile X syndrome. PLoS One 2008;3:e1486.

[59] Pastori C, Peschansky VJ, Barbouth D, Mehta A, Silva JP, Wahlestedt C. Comprehensive analysis of the transcriptional landscape of the human *FMR1* gene reveals two new long noncoding RNAs differentially expressed in Fragile X syndrome and Fragile X-associated tremor/ataxia syndrome. Hum Genet 2014;133:59–67.

[60] Wang X, Song X, Glass CK, Rosenfeld MG. The long arm of long noncoding RNAs: roles as sensors regulating gene transcriptional programs. Cold Spring Harb Perspect Biol 2011;3:a003756.
[61] Pandey RR, Mondal T, Mohammad F, Enroth S, Redrup L, Komorowski J, et al. *Kcnq1ot1* antisense noncoding RNA mediates lineage-specific transcriptional silencing through chromatin-level regulation. Mol Cell 2008;32:232–46.
[62] Rinn JL, Kertesz M, Wang JK, Squazzo SL, Xu X, Brugmann SA, et al. Functional demarcation of active and silent chromatin domains in human *HOX* loci by noncoding RNAs. Cell 2007;129:1311–23.
[63] Lanni S, Goracci M, Borrelli L, Mancano G, Chiurazzi P, Moscato U, et al. Role of CTCF protein in regulating *FMR1* locus transcription. PLoS Genet 2013;9:e1003601.
[64] Di Ruscio A, Ebralidze AK, Benoukraf T, Amabile G, Goff LA, Terragni J, Figueroa ME, De Figueiredo Pontes LL, Alberich-Jorda M, Zhang P, Wu M, D'Alò F, Melnick A, Leone G, Ebralidze KK, Pradhan S, Rinn JL, Tenen DG. DNMT1-interacting RNAs block gene-specific DNA methylation. Nature 2013;503:371–6.
[65] Tabolacci E, Chiurazzi P. Epigenetics, fragile X syndrome and transcriptional therapy. Am J Med Genet A 2013;161A:2797–808.
[66] Coffee B, Zhang F, Warren ST, Reines D. Acetylated histones are associated with *FMR1* in normal but not fragile X-syndrome cells. Nat Genet 1999;22:98–101.
[67] Tabolacci E, Pietrobono R, Moscato U, Oostra BA, Chiurazzi P, Neri G. Differential epigenetic modifications in the *FMR1* gene of the fragile X syndrome after reactivating pharmacological treatments. Eur J Hum Genet 2005;13:641–8.
[68] Nishioka K, Chuikov S, Sarma K, Erdjument-Bromage H, Allis CD, Tempst P, et al. Set9, a novel histone H3 methyltransferase that facilitates transcription by precluding histone tail modifications required for heterochromatin formation. Genes Dev 2002;16:479–89.
[69] Li Y, Kirschmann DA, Wallrath LL. Does heterochromatin protein 1 always follow code? Proc Natl Acad Sci USA 2002;99(Suppl 4):16462–9.
[70] Sparmann A, van Lohuizen M. Polycomb silencers control cell fate, development and cancer. Nat Rev Cancer 2006;6:846–56.
[71] Kumari D, Usdin K. The distribution of repressive histone modifications on silenced *FMR1* alleles provides clues to the mechanism of gene silencing in fragile X syndrome. Hum Mol Genet 2010;19:4634–42.
[72] Todd PK, Oh SY, Krans A, Pandey UB, Di Prospero NA, Min KT, et al. Histone deacetylases suppress CGG repeat-induced neurodegeneration via transcriptional silencing in models of fragile X tremor ataxia syndrome. PLoS Genet 2010;6:e1001240.
[73] Wang Z, Taylor AK, Bridge JA. *FMR1* fully expanded mutation with minimal methylation in a high functioning fragile X male. J Med Genet 1996;33:376–8.
[74] Handa V, Saha T, Usdin K. The fragile X syndrome repeats form RNA hairpins that do not activate the interferon-inducible protein kinase, PKR, but are cut by Dicer. Nucleic Acids Res 2003;31:6243–8.
[75] Smeets HJ, Smits AP, Verheij CE, Theelen JP, Willemsen R, van de Burgt I, et al. Normal phenotype in two brothers with a full *FMR1* mutation. Hum Mol Genet 1995;4:2103–8.
[76] Pietrobono R, Tabolacci E, Zalfa F, Zito I, Terracciano A, Moscato U, et al. Molecular dissection of the events leading to inactivation of the *FMR1* gene. Hum Mol Genet 2005;14:267–77.
[77] Tabolacci E, Moscato U, Zalfa F, Bagni C, Chiurazzi P, Neri G. Epigenetic analysis reveals a euchromatic configuration in the *FMR1* unmethylated full mutations. Eur J Hum Genet 2008;16:1487–98.
[78] Willemsen R, Bontekoe CJ, Severijnen LA, Oostra BA. Timing of the absence of *FMR1* expression in full mutation chorionic villi. Hum Genet 2002;110:601–5.
[79] Eiges R, Urbach A, Malcov M, Frumkin T, Schwartz T, Amit A, et al. Developmental study of fragile X syndrome using human embryonic stem cells derived from preimplantation genetically diagnosed embryos. Cell Stem Cell 2007;1:568–77.
[80] Urbach A, Bar-Nur O, Daley GQ, Benvenisty N. Differential modeling of fragile X syndrome by human embryonic stem cells and induced pluripotent stem cells. Cell Stem Cell 2010;6:407–11.

[81] Colak D, Zaninovic N, Cohen MS, Rosenwaks Z, Yang WY, Gerhardt J, et al. Promoter-bound trinucleotide repeat mRNA drives epigenetic silencing in fragile X syndrome. Science 2014;343:1002–5.
[82] Sheridan SD, Theriault KM, Reis SA, Zhou F, Madison JM, Daheron L, et al. Epigenetic characterization of the *FMR1* gene and aberrant neurodevelopment in human induced pluripotent stem cell models of fragile X syndrome. PLoS One 2011;6:e26203.
[83] Bear MF, Huber KM, Warren ST. The mGluR theory of fragile X mental retardation. Trends Neurosci 2004;27:370–7.
[84] Gladding CM, Collett VJ, Jia Z, Bashir ZI, Collingridge GL, Molnár E. Tyrosine dephosphorylation regulates AMPAR internalisation in mGluR-LTD. Mol Cell Neurosci 2009;40:267–79.
[85] Berry-Kravis E, Hessl D, Coffey S, Hervey C, Schneider A, Yuhas J, et al. A pilot open label, single dose trial of fenobam in adults with fragile X syndrome. J Med Genet 2009;46:266–71.
[86] Levenga J, Hayashi S, de Vrij FM, Koekkoek SK, van der Linde HC, Nieuwenhuizen I, et al. AFQ056, a new mGluR5 antagonist for treatment of fragile X syndrome. Neurobiol Dis 2011;42:311–7.
[87] Jacquemont S, Curie A, des Portes V, Torrioli MG, Berry-Kravis E, Hagerman RJ, et al. Epigenetic modification of the *FMR1* gene in fragile X syndrome is associated with differential response to the mGluR5 antagonist AFQ056. Sci Transl Med 2011;3:64ra1.
[88] Tabolacci E, Pirozzi F, Gomez-Mancilla B, Gasparini F, Neri G. The mGluR5 antagonist AFQ056 does not affect methylation and transcription of the mutant *FMR1* gene *in vitro*. BMC Med Genet 2012;13:13.
[89] Bagni C, Oostra BA. Fragile X syndrome: from protein function to therapy. Am J Med Genet A 2013;161A:2809–21.
[90] D'Hulst C, Kooy RF. The GABAA receptor: a novel target for treatment of fragile X? Trends Neurosci 2007;30:425–31.
[91] Berry-Kravis EM, Hessl D, Rathmell B, Zarevics P, Cherubini M, Walton-Bowen K, et al. Effects of STX209 (arbaclofen) on neurobehavioral function in children and adults with fragile X syndrome: a randomized, controlled, phase 2 trial. Sci Transl Med 2012;4:152ra127.
[92] Bilousova TV, Dansie L, Ngo M, Aye J, Charles JR, Ethell DW, et al. Minocycline promotes dendritic spine maturation and improves behavioural performance in the fragile X mouse model. J Med Genet 2009;46:94–102.
[93] Paribello C, Tao L, Folino A, Berry-Kravis E, Tranfaglia M, Ethell IM, et al. Open-label add-on treatment trial of minocycline in fragile X syndrome. BMC Neurol 2010;10:91.
[94] Auerbach BD, Osterweil EK, Bear MF. Mutations causing syndromic autism define an axis of synaptic pathophysiology. Nature 2011;480:63–8.
[95] Osterweil EK, Chuang SC, Chubykin AA, Sidorov M, Bianchi R, Wong RK, et al. Lovastatin corrects excess protein synthesis and prevents epileptogenesis in a mouse model of fragile X syndrome. Neuron 2013;77:243–50.
[96] Jackson-Grusby L, Laird PW, Magge SN, Moeller BJ, Jaenisch R. Mutagenicity of 5-aza-2'-deoxycytidine is mediated by the mammalian DNA methyltransferase. Proc Natl Acad Sci USA 1997;94:4681–5.
[97] Chiurazzi P, Pomponi MG, Willemsen R, Oostra BA, Neri G. *In vitro* reactivation of the *FMR1* gene involved in fragile X syndrome. Hum Mol Genet 1998;7:109–13.
[98] Feng Y, Zhang F, Lokey LK, Chastain JL, Lakkis L, Eberhart D, et al. Translational suppression by trinucleotide repeat expansion at *FMR1*. Science 1995;268:731–4.
[99] Chiurazzi P, Pomponi MG, Pietrobono R, Bakker CE, Neri G, Oostra BA. Synergistic effect of histone hyperacetylation and DNA demethylation in the reactivation of the *FMR1* gene. Hum Mol Genet 1999;8:2317–23.
[100] Pietrobono R, Pomponi MG, Tabolacci E, Oostra B, Chiurazzi P, Neri G. Quantitative analysis of DNA demethylation and transcriptional reactivation of the *FMR1* gene in fragile X cells treated with 5-azadeoxycytidine. Nucleic Acids Res 2002;30:3278–85.
[101] Cameron EE, Bachman KE, Myöhänen S, Herman JG, Baylin SB. Synergy of demethylation and histone deacetylase inhibition in the re-expression of genes silenced in cancer. Nat Genet 1999;21:103–7.

[102] Gnyszka A, Jastrzebski Z, Flis S. DNA methyltransferase inhibitors and their emerging role in epigenetic therapy of cancer. Anticancer Res 2013;33:2989–96.

[103] Ghoshal K, Datta J, Majumder S, Bai S, Kutay H, Motiwala T, et al. 5-Aza-deoxycytidine induces selective degradation of DNA methyltransferase 1 by a proteasomal pathway that requires the KEN box, bromo-adjacent homology domain, and nuclear localization signal. Mol Cell Biol 2005;25:4727–41.

[104] Patel K, Dickson J, Din S, Macleod K, Jodrell D, Ramsahoye B. Targeting of 5-aza-2'-deoxycytidine residues by chromatin-associated DNMT1 induces proteasomal degradation of the free enzyme. Nucleic Acids Res 2010;38:4313–24.

[105] Bar-Nur O, Caspi I, Benvenisty N. Molecular analysis of FMR1 reactivation in fragile-X induced pluripotent stem cells and their neuronal derivatives. J Mol Cell Biol 2012;4:180–3.

[106] Suzuki H, Gabrielson E, Chen W, Anbazhagan R, van Engeland M, Weijenberg MP, et al. A genomic screen for genes upregulated by demethylation and histone deacetylase inhibition in human colorectal cancer. Nat Genet 2002;31:141–9.

[107] Tabolacci E, De Pascalis I, Accadia M, Terracciano A, Moscato U, Chiurazzi P, et al. Modest reactivation of the mutant *FMR1* gene by valproic acid is accompanied by histone modifications but not DNA demethylation. Pharmacogenet Genom 2008;18:738–41.

[108] Dong E, Chen Y, Gavin DP, Grayson DR, Guidotti A. Valproate induces DNA demethylation in nuclear extracts from adult mouse brain. Epigenetics 2010;5(8):730–5.

[109] Torrioli M, Vernacotola S, Setini C, Bevilacqua F, Martinelli D, Snape M, et al. Treatment with valproic acid ameliorates ADHD symptoms in fragile X syndrome boys. Am J Med Genet A 2010;152A:1420–7.

[110] Torrioli MG, Vernacotola S, Peruzzi L, Tabolacci E, Mila M, Militerni R, et al. A double-blind, parallel, multicenter comparison of L-acetylcarnitine with placebo on the attention deficit hyperactivity disorder in fragile X syndrome boys. Am J Med Genet A 2008;146A:803–12.

CHAPTER

7 EPIGENETICS OF ATTENTION-DEFICIT HYPERACTIVITY DISORDER

N. Perroud*,, S. Weibel*, J.-M. Aubry*,**, A. Dayer*,****

**Department of Mental Health and Psychiatry, Service of Psychiatric Specialties, University Hospitals of Geneva, Geneva, Switzerland; **Department of Psychiatry, University of Geneva, Geneva, Switzerland*

CHAPTER OUTLINE

7.1 INTRODUCTION

Attention-deficit hyperactivity disorder (ADHD) is the most common psychiatric disorder in childhood with a prevalence ranging from 5% to 8% [1]. It persists in adulthood in 70% of the cases with a prevalence of 4–6% in adults [2]. ADHD is usually considered as a neurodevelopmental disorder with an early onset, although recent studies have suggested that de novo cases may arise later in adulthood [3]. The disorder is characterized by attention deficit, impulsivity, and hyperactivity possibly linked to a delay in brain maturation [4]. These symptoms result in impairment of learning performance in school and in multiple domains of personal, professional, and social life with considerable socioeconomical costs in childhood and adulthood [5]. Due to its chronic course, a majority of the cases in adults suffer from psychiatric comorbidities, such as depression and substance use disorders [6].

Several etiological factors have been identified in ADHD including genetic and environmental ones. However, despite the amount of studies performed in the field, biomarkers have not been identified for understanding the underlying mechanism and development of the disorder. Therefore, it is difficult to find efficient strategies to prevent the emergence of the disorder. More recently epigenetics has given

Neuropsychiatric Disorders and Epigenetics. http://dx.doi.org/10.1016/B978-0-12-800226-1.00007-1

hope in the search for prevention and new therapies for psychiatric disorders, such as ADHD. The rational for the investigation of the epigenome in ADHD comes from the fact that even if ADHD is one of the most heritable psychiatric disorders, environment clearly plays a role in the emergence and worsening of the disorder possibly by modifying gene expression.

7.1.1 THE ROLE OF THE ENVIRONMENT

Although the heritability of ADHD is estimated to be around 70% [7], the genetic and neurobiological mechanisms underlying the disorder are still poorly understood [8]. For instance, no clear genetic risk factors have emerged from numerous candidate gene and genome-wide association studies performed to date [9]. The ultimate objective of identifying genes to help diagnose the disorder and develop novel therapies has yet to be reached. ADHD, like other neuropsychiatric disorders, is considered to be a complex disease with high heterogeneity between cases which may explain this inconsistency in the results of genetic studies. The complex ADHD phenotype may possibly be better explained by taking into account environmental factors as done in gene-environment interaction studies.

The link between exposure to early environmental factors and ADHD is strongly supported by animal studies. Indeed, in utero or postnatal exposure to a variety of toxins and poor nutritional environments has been shown to induce a wide range of behavioral symptoms in animal models mimicking those observed in ADHD with cumulative effects on the developing brain [10–14]. For instance, prenatal exposure to alcohol has been shown to cause attention deficit in rats [10], postnatal exposure to subtoxic doses of pesticides is associated with behavioral hyperactivity in rats [11], and prenatal exposure to nicotine with increased locomotion and impulsivity in mice offspring [13].

Following these lines of evidence in animals, a variety of environmental factors have been associated with an increased risk for ADHD in humans, including:

1. Distal risk factors occurring before the emergence of the disorder mainly during the prenatal period, such as exposure to smoking [15], alcohol [16], recreational drugs [17], toxic compounds, such as pesticides like polychlorinated biphenyls (PCBs) [18] and hexachlorobenzene [19], maternal stress [20], glucocorticoids [21], viral infection [22], poor maternal diet [23], or just after birth, such as extreme birth weights (too low or too large) [24]. All of these factors are thought to permanently alter the development of the dopaminergic system which is a key neurobiological mechanism underlying ADHD. As distal risk factors they are believed to confer a risk for the condition at some time in the future.
2. Proximal risk factors occurring either when the disorder is imminent and may thus serve as a trigger for its emergence, or while the disorder is already established and thought to contribute to the persistence and worsening of the disease, such as childhood maltreatment and difficulties interacting with parents [25–29].

There might thus be a window of exposure to environmental factors which seems to be important, with those occurring during the immediate neonatal period having the strongest effect on the development of the disorder while those occurring late in life being more linked to the persistence of it. In addition, it has been hypothesized that certain types of environmental factors may be associated with specific ADHD subtypes [30,31]. Based on these findings, researchers have attempted to link these environmental factors to ADHD through epigenetics.

7.1.2 WHY EPIGENETICS IN ADHD?

Epigenetics may be particularly appropriate for linking environmental risk factors with biological mechanisms and disease phenotype. Epigenetics refers to reversible changes to genomic function that are independent of DNA sequence. Among epigenetic modifications, DNA methylation at CpG sites is thought to be a key epigenetic modification [32]. Methylation in the promoter regulatory region of genes typically represses gene expression. Other common epigenetic processes affect chromatin structure and include, among others, histone acetylation, methylation, and phosphorylation. Studies have shown that epigenetic processes can be influenced by exposure to several environmental factors, such as perinatal stress, social environment, toxins, drugs, and childhood trauma [33–36]. Exposure to prenatal environmental risk factors, such as those found in ADHD as well as psychosocial environment during key developmental periods in early life have been shown to epigenetically alter gene expression in animals [19,37–39]. For instance, lead exposure, commonly considered as a risk factor for the development of ADHD [40,41], has been associated in rats with increased histone acetylation in the hippocampus and increased locomotor activity (evoking the hyperactivity found in ADHD) [42]. Another example is that rodents being born large for gestational age, another common risk factor for ADHD, displayed global DNA hypomethylation across all brain regions as adults [43]. It is thus likely that, in humans, environmental factors may exert their effects on behavior through long-lasting epigenetic changes to the genome, directly altering gene expression and ultimately the phenotype.

7.2 RESULTS OF EPIGENETIC STUDIES OF ADHD

7.2.1 EPIGENETICS AND DISTAL RISK FACTORS

To our knowledge, only one study looked at distal risk factors, although indirectly, in terms of epigenetics and ADHD at the level of candidate genes. Van Mil et al. [44] found in cord blood samples assessed at birth in 426 children from a large population-based cohort of Dutch national origin, lower DNA methylation levels of the dopamine receptor D4 gene (*DRD4*) and the serotonin transporter gene (*SLC6A4*) in children developing ADHD symptoms when they were 6 years old. More specifically, DNA methylation levels were negatively associated with ADHD symptom scores in both genes among seven genes selected for their possible involvement in the etiology of ADHD.

The result involving *DRD4* is of particular interest as previous studies found increased *DRD4* expression in animal models of ADHD, thus in agreement with the observed hypomethylation of this gene in the Van Mil et al. study [45]. Based on the effective response to psychostimulants, which mainly increase dopamine levels by inhibiting reuptake of this neurotransmitter, dysregulation of the dopaminergic system within mesocortical and frontostriatal pathways is thought to be one of the main dysfunctional systems found in ADHD [46]. Genes coding for key proteins involved in the regulation of the dopaminergic system, such as the dopamine transporter gene (*DAT1*) or *DRD4* have thus been studied in ADHD [47–49] not only as having a direct effect on the phenotype, but also as genes interacting with environmental factors, such as prenatal smoking with some convincing results. Indeed, the risk of developing ADHD has been shown to be stronger in carriers of particular variants within these two key genes [50,51]. For instance, *DAT1* has been shown to interact with prenatal nicotine and alcohol exposure to increase the risk of ADHD [50,52]. Thus, the findings of study by the Van Mil et al. add to the growing body of evidence linking the dopaminergic system to ADHD. In addition they might indirectly explain how in animals, and maybe also in humans, exposure to maternal toxins, such as nicotine affects in terms of behavioral and metabolic symptoms not only the first generation of offspring but also the next generation [53]. That is, epigenetic modifications on *DRD4* and other genes during the prenatal period may set the stage for intergenerational transmission of a future environment, characterized by impulsive-ADHD parents often smoking and taking other risks for the next generation of offspring, the latter being thus at additional risk for ADHD [54]. This might be a way to understand the recent observations that epigenetic effects are transgenerational [55].

7.2.2 EPIGENETICS AND PROXIMAL RISK FACTORS

There are more studies that directly or indirectly have looked at proximal, rather than distal, risk factors for ADHD, namely looking at epigenetic differences between children with ADHD and control children. Xu et al. [56] looked at the promoter methylation level of genes involved in the dopaminergic system (*DAT1, DRD4, DRD5*) as well as their expression in 50 children with ADHD and 50 non-ADHD control children (from peripheral blood). In parallel, they investigated the expression profiles of the following genes encoding proteins involved in the epigenetic machinery: two genes encoding histone acetyltransferases (*p300* and *MYST4*), one encoding histone deacetylase type1 (*HDAC1*), and one encoding the methyl-CpG binding protein MeCP2.

Expression of all three dopaminergic genes was found to be lower in ADHD patients compared to controls. This was associated, for the *DRD4* gene only, with an increased methylation level at one CpG in the promoter region of the gene. This methylation change of a promoter CpG site did negatively affect the transcription levels suggesting that this CpG directly modulated the dopaminergic system and the possible development of ADHD. Concerning the histone-modifying genes the authors found that *HDAC1*'s expression was increased for ADHD children, compatible with a decreased histone acetylation level and associated gene silencing, consistent with the findings in schizophrenia [57]. The authors also found that *MeCP2* mRNA levels were significantly decreased for ADHD boys only, suggesting a gender effect at the histone levels which might possibly explain the gender difference (more boys than girls affected by the disorder) observed in children. The latter result was previously shown in a study performed by Nagarajan et al. [58].The authors, investigating brain samples of subjects suffering from several neuropsychiatric disorders including ADHD and autism, found a significantly lower *MeCP2* expression in ADHD samples compared to controls which was linked to increased *MeCP2* promoter methylation at least in the autism samples in the frontal cortex compared to controls. This is in agreement with animal studies showing that in rodents prenatal exposure to alcohol was associated with increased locomotor activity, attention deficit and impulsivity, with decreased expression of *MeCP2* in both prefrontal cortex and striatum [59]. The reduced *MeCP2* expression is thought to induce several neurological consequences possibly by regulating expression of key neurodevelopmental genes, such as the brain derived neurotrophic factor gene. There have been very few studies investigating histone acetylation in ADHD in humans although preliminary results in animals have linked epigenetic modifications of the histones associated with adverse environmental exposure and hyperactivity [42]. Clearly, additional studies are needed in this field.

Park et al. [60] investigated the methylation level of the serotonin transporter gene (*SLC6A4*), in the peripheral blood of 102 ADHD children aged 6–15 years, and looked at association with clinical characteristics and regional thickness of the cerebral cortex using MRI. Their results showed that higher methylation of the *SLC6A4* promoter region was associated with more hyperactive-impulsive symptoms and more commission errors as measured by a neuropsychological test (continuous performance test). In addition, Park et al. observed a negative correlation between *SLC6A4* methylation levels and cortical thickness in the right occipito-temporal region. We might thus hypothesize based on these findings that lower expression of the gene coding for the serotonin transporter due to hypermethylation of its promoter region might be a vulnerability risk factor for ADHD. More specifically, hypermethylation of the promoter of the serotonin transporter gene may indicate increased severity of the disorder characterized by higher behavioral disinhibition, more hyperactive-impulsive symptoms and more commission errors, possibly linked to abnormal development of specific brain regions. This is in agreement with previous studies showing that in children more attention problems and hyperactivity are associated with a thinner cerebral cortex [61]. Of note, the results of Park et al.'s study were in a different direction from the results found by Van Mil et al. One hypothesis suggested by the authors is that DNA methylation status determined with peripheral blood may be influenced by a variety of proximal environmental factors, such as childhood maltreatment not found in cord blood samples and that prenatal *SLC6A4* hypomethylation linked to prenatal stress increases the risk of ADHD whereas childhood *SLC6A4* hypermethylation, related to postnatal adversities, is associated with severity of ADHD and possibly its persistence in adulthood.

The serotoninergic system is of interest in the association between early life adversities and vulnerability to psychiatric disorders including ADHD [14,62]. Indeed, several studies have convincingly shown that the *SLC6A4* interacts with childhood maltreatment in order to increase adulthood psychopathologies [63,64]. Among the serotonin transporter polymorphisms the most studied one is the *5-HTTLPR*, a repeat polymorphism in the promoter region of the *SLC6A4*, which was shown to moderate the effect of stress on severity and neurobiological correlates of ADHD [65]. Moreover, a metaanalysis showed that *5-HTTLPR* may confer risk to ADHD with modest effect [66]. In addition, epigenetic changes at the serotonin 3A receptor gene (*5-HT3AR*) were found to be associated with adult ADHD supporting the involvement of the serotoninergic system in this disorder [67].

Few studies have investigated the link between childhood maltreatment and adult ADHD and its severity, most of the studies focusing on early-life stress and childhood ADHD. Perroud et al. [67] found in a study comparing adult subjects suffering from ADHD, borderline personality disorder, or bipolar disorder that DNA methylation of the gene coding for *5-HT3AR* mediated the effect of childhood maltreatment on the clinical severity of the disorder. In addition the methylation status of the *5-HT3AR* gene was significantly different between the three disorders suggesting that these serotonin receptors play a role in the etiology of ADHD. The 5-HT3ARs are of interest because they regulate several cellular processes involved in the formation of cortical circuits, including interneuron migration and pyramidal neuron dendritic morphology [68]. Further studies are needed to confirm the contribution of the *5-HT3AR* in ADHD and the impact of childhood maltreatment on its methylation status.

7.2.3 METHYLOME-WIDE ASSOCIATION STUDIES

Besides candidate gene studies, more recent investigations have focused on the epigenetic modifications at the whole genome level. Wilmot et al. [69] reported the first methylome-wide study in ADHD in children aged between 7 and 12 years. The study was performed on salivary DNA using the Illumina 450K Human Methylation array in a discovery sample of 92 boys (half control and half ADHD) and a confirmation sample of 10 ADHD boys and 10 controls. Although two genes were found to be associated with ADHD, namely, the vasoactive intestinal peptide receptor2 (*VIPR2*) and the myelin transcription factor 1-like (*MYT1L*) which is involved in myelin and nervous system formation, only *VIPR2* passed the correction for multiple testing in the confirmation sample and was subsequently confirmed through bisulfite sequencing. The result showed a lower methylation level of this gene in ADHD children compared to controls. *VIPR2* is a gene expressed in the caudate nucleus, a key ADHD-relevant brain area, and its underexpression is associated with hyperactivity in animals [70]. In addition, in their study, Wilmot et al. in an enrichment analysis, found an involvement of pathways related to inflammatory processes via oxidation and antioxidation. This is of interest as several studies have recently involved the protective role of antioxidant supplementation in ADHD [71]. Finally the authors showed the possible role of cholinergic neurotransmission in ADHD supporting the recent interest in nicotinic receptors and treatments targeting these receptors as possible novel therapies for ADHD in animal and human trials [72].

7.3 DISCUSSION

The investigation of epigenetic modifications in ADHD is a promising tool for the understanding of the environmental effects on the emergence and persistence of the disorder. Although several genes have been identified at the epigenetic level as potential candidates for ADHD including genes associated with neurotransmission (*DRD4*, *SLC6A4, 5-HT3AR*), with histone regulation (*HDAC1, MeCP2*), or with brain development (*VIPR2*), many questions remain unanswered. For instance, it is still unclear what are the underlying biological mechanisms involved in the persistence of the disorder in adulthood. Most studies focused on children and on perinatal environments although it has been shown that DNA methylation may be influenced by several postnatal factors during childhood, such as toxins, drugs, medication, social stress and childhood maltreatment [33,34,39], and only studies in adults can take into account all these environmental factors. An effort to better define specific effects of pre- and postnatal environments on ADHD should definitely be done as this might help determine which biological correlates are involved in the etiology of the disorder and which are more specific to the persistence of ADHD in adulthood. Disentangling which genes are associated with proximal and distal risk factors might also help better define and help in the understanding of this heterogeneous disorder. This issue has recently been raised with a publication by Moffitt et al. [3] suggesting that de novo cases might emerge only in adulthood. To our knowledge, only one study has investigated epigenetics in adult ADHD, and replication is needed to validate this first finding involving the *5-HT3AR* gene [67].

Although childhood maltreatment has been extensively investigated in terms of epigenetic modifications in several other psychiatric disorders, this has not been the case for ADHD and further research is clearly needed in this field. For instance, convincing evidence has linked epigenetic modifications of the gene coding for the glucocorticoid receptor (*NR3C1*), childhood maltreatment and adulthood psychiatric disorders, such as borderline personality disorder [73]. As borderline personality disorder shares several features with ADHD, these studies suggest that environmental factors may durably influence the HPA axis leading to the development of symptoms, such as impulsivity and emotion dysregulation, key ADHD dimensions. Moreover, disturbed HPA axis may not only be linked to proximal environmental factors but also to distal ones as shown in animal studies. Indeed, studies in rodents have found that prenatal maternal stress is associated with increased activity of the HPA axis with direct effect on offspring behavior [74,75]. Thus, investigating more thoroughly the role of epigenetics in the regulation of the HPA axis may shed light on the effects of a whole range of environmental factors from distal to proximal ones.

The paucity of studies investigating epigenetics in ADHD is surprising given the major role played by the environment. For instance, although being the most prevalent psychiatric disorder in childhood and one for which we have the most convincing evidence that distal factors play an important role, only one study has truly investigated the link between prenatal factors, epigenetics, and emergence of ADHD. Also poorly investigated is the complex link between genetic polymorphisms and DNA methylation. For instance, in the Van Mil et al. study [44], the authors found that the *DRD4*-48 bp VNTR and *5-HTTLPR* were associated with lower DNA methylation levels of the adjacent regions. We also found a similar effect of the polymorphism rs1062613 on the methylation level at one CpG located 1bp upstream of the *5-HT3AR* gene [67]. Such studies are more difficult to perform as they require more power and thus a larger sample size. However, they may shed light on the interplay between genes and the environment.

Another major issue in investigating epigenetic processes in ADHD and more specially when looking at proximal environmental risk factors is the medication issue. Most patients suffering from ADHD are taking psychostimulants, such as methylphenidate. Exposure to drugs has been shown to induce changes in gene expression through epigenetic mechanisms in animals [76]. Among the genes shown to be altered in their expression and DNA methylation profile following administration of psychostimulants is the *MeCP2* gene [77,78]. Interestingly, this gene has been associated with ADHD as described above. So better investigating this gene in terms of responders versus nonresponders to treatment might help in deciphering the underlying mechanisms involved in drug response. A major limitation of most of these studies, excluding those directly investigating epigenetic processes in the brain, is that methylation levels were assessed in DNA derived from leukocytes. The precise manner in which DNA methylation affects the neural system, leading to ADHD symptoms, remains thus to be established.

7.4 CONCLUSIONS

In summary, the investigation of the epigenome in ADHD may help elucidate how the environment is at play in the disorder. Understanding the epigenetic processes involved in ADHD and in the impact of proximal and distal environmental factors on ADHD may give the possibility for new therapies and interventions. Moreover, the fact that some epigenetic traces have been shown to be dynamic and may thus be reversible offers hope that adverse environmental effects may be reversed [79]. Finally, the investigation of epigenetic processes in ADHD may also help better distinguish this disorder form other closely related disorders, such as dyslexia or autism spectrum disorder in which epigenetic processes are also involved [80].

ABBREVIATIONS

BDNF Brain-derived neurotrophic factor
DAT1 Dopamine transporter
DRD4 Dopamine receptor 4
DRD5 Dopamine receptor 5

HDAC1	Type 1 histone deacetylase
5-HT3AR	Serotonin receptor 3A
MeCP2	Methyl-CpG binding protein 2
MYT1L	Myelin transcription factor 1-like
NR3C1	Glucocorticoid receptor
SLC6A4	Serotonin transporter
VIPR2	Vasoactive intestinal peptide receptor 2

REFERENCES

[1] Polanczyk G, de Lima MS, Horta BL, Biederman J, Rohde LA. The worldwide prevalence of ADHD: a systematic review and metaregression analysis. Am J Psychiatry 2007;164(6):942–8.

[2] Kessler RC, Adler L, Barkley R, Biederman J, Conners CK, Demler O, et al. The prevalence and correlates of adult ADHD in the United States: results from the National Comorbidity Survey Replication. Am J Psychiatry 2006;163(4):716–23.

[3] Moffitt TE, Houts R, Asherson P, Belsky DW, Corcoran DL, Hammerle M, et al. Is adult ADHD a childhood-onset neurodevelopmental disorder? Evidence from a four-decade longitudinal cohort study. Am J Psychiatry 2015;172(10):967–77.

[4] Shaw P, Eckstrand K, Sharp W, Blumenthal J, Lerch JP, Greenstein D, et al. Attention-deficit/hyperactivity disorder is characterized by a delay in cortical maturation. Proc Natl Acad Sci USA 2007;104(49):19649–54.

[5] Leibson CL, Katusic SK, Barbaresi WJ, Ransom J, O'Brien PC. Use and costs of medical care for children and adolescents with and without attention-deficit/hyperactivity disorder. JAMA 2001;285(1):60–6.

[6] Harpin VA. The effect of ADHD on the life of an individual, their family, and community from preschool to adult life. Arch Dis Child 2005;90(Suppl. 1):i2–7.

[7] Faraone SV, Perlis RH, Doyle AE, Smoller JW, Goralnick JJ, Holmgren MA, et al. Molecular genetics of attention-deficit/hyperactivity disorder. Biol Psychiatry 2005;57(11):1313–23.

[8] Faraone SV, Mick E. Molecular genetics of attention deficit hyperactivity disorder. Psychiatr Clin North Am 2010;33(1):159–80.

[9] Romanos M, Freitag C, Jacob C, Craig DW, Dempfle A, Nguyen TT, et al. Genome-wide linkage analysis of ADHD using high-density SNP arrays: novel loci at 5q13.1 and 14q12. Mol Psychiatry 2008;13(5): 522–30.

[10] Hausknecht KA, Acheson A, Farrar AM, Kieres AK, Shen RY, Richards JB, et al. Prenatal alcohol exposure causes attention deficits in male rats. Behav Neurosci 2005;119(1):302–10.

[11] Holene E, Nafstad I, Skaare JU, Sagvolden T. Behavioural hyperactivity in rats following postnatal exposure to sub-toxic doses of polychlorinated biphenyl congeners 153 and 126. Behav Brain Res 1998;94(1):213–24.

[12] LeSage MG, Gustaf E, Dufek MB, Pentel PR. Effects of maternal intravenous nicotine administration on locomotor behavior in preweanling rats. Pharmacol Biochem Behav 2006;85(3):575–83.

[13] Paz R, Barsness B, Martenson T, Tanner D, Allan AM. Behavioral teratogenicity induced by nonforced maternal nicotine consumption. Neuropsychopharmacology 2007;32(3):693–9.

[14] Mill J, Petronis A. Pre- and perinatal environmental risks for attention-deficit hyperactivity disorder (ADHD): the potential role of epigenetic processes in mediating susceptibility. J Child Psychol Psychiatry 2008;49(10):1020–30.

[15] Langley K, Rice F, van den Bree MB, Thapar A. Maternal smoking during pregnancy as an environmental risk factor for attention deficit hyperactivity disorder behaviour. A review. Minerva Pediatr 2005;57(6):359–71.

[16] Mick E, Biederman J, Faraone SV, Sayer J, Kleinman S. Case-control study of attention-deficit hyperactivity disorder and maternal smoking, alcohol use, and drug use during pregnancy. J Am Acad Child Adolesc Psychiatry 2002;41(4):378–85.

[17] Accornero VH, Amado AJ, Morrow CE, Xue L, Anthony JC, Bandstra ES. Impact of prenatal cocaine exposure on attention and response inhibition as assessed by continuous performance tests. J Dev Behav Pediatr 2007;28(3):195–205.

[18] Jacobson JL, Jacobson SW. Prenatal exposure to polychlorinated biphenyls and attention at school age. J Pediatr 2003;143(6):780–8.

[19] Ribas-Fito N, Torrent M, Carrizo D, Julvez J, Grimalt JO, Sunyer J. Exposure to hexachlorobenzene during pregnancy and children's social behavior at 4 years of age. Environ Health Perspect 2007;115(3):447–50.

[20] O'Connor TG, Heron J, Golding J, Beveridge M, Glover V. Maternal antenatal anxiety and children's behavioural/emotional problems at 4 years. Report from the avon longitudinal study of parents and children. Br J Psychiatry 2002;180:502–8.

[21] French NP, Hagan R, Evans SF, Mullan A, Newnham JP. Repeated antenatal corticosteroids: effects on cerebral palsy and childhood behavior. Am J Obstet Gynecol 2004;190(3):588–95.

[22] Murphy TK, Kurlan R, Leckman J. The immunobiology of Tourette's disorder, pediatric autoimmune neuropsychiatric disorders associated with Streptococcus, and related disorders: a way forward. J Child Adolesc Psychopharmacol 2010;20(4):317–31.

[23] Neugebauer R, Hoek HW, Susser E. Prenatal exposure to wartime famine and development of antisocial personality disorder in early adulthood. JAMA 1999;282(5):455–62.

[24] Grissom NM, Reyes TM. Gestational overgrowth and undergrowth affect neurodevelopment: similarities and differences from behavior to epigenetics. Int J Dev Neurosci 2013;31(6):406–14.

[25] Kvist AP, Nielsen HS, Simonsen M. The importance of children's ADHD for parents' relationship stability and labor supply. Soc Sci Med 2013;88:30–8.

[26] Prada P, Hasler R, Baud P, Bednarz G, Ardu S, Krejci I, et al. Distinguishing borderline personality disorder from adult attention deficit/hyperactivity disorder: a clinical and dimensional perspective. Psychiatry Res 2014;217(1–2):107–14.

[27] Ouyang L, Fang X, Mercy J, Perou R, Grosse SD. Attention-deficit/hyperactivity disorder symptoms and child maltreatment: a population-based study. J Pediatr 2008;153(6):851–6.

[28] Wymbs BT, Pelham WE Jr, Gnagy EM, Molina BS. Mother and adolescent reports of interparental discord among parents of adolescents with and without attention-deficit hyperactivity disorder. J Emot Behav Disord 2008;16(1):29–41.

[29] Biederman J, Petty CR, Clarke A, Lomedico A, Faraone SV. Predictors of persistent ADHD: an 11-year follow-up study. J Psychiatr Res 2011;45(2):150–5.

[30] Grizenko N, Shayan YR, Polotskaia A, Ter-Stepanian M, Joober R. Relation of maternal stress during pregnancy to symptom severity and response to treatment in children with ADHD. J Psychiatry Neurosci 2008;33(1):10–6.

[31] Park S, Cho SC, Kim JW, Shin MS, Yoo HJ, Oh SM, et al. Differential perinatal risk factors in children with attention-deficit/hyperactivity disorder by subtype. Psychiatry Res 2014;219(3):609–16.

[32] Hochberg Z, Feil R, Constancia M, Fraga M, Junien C, Carel JC, et al. Child health, developmental plasticity, and epigenetic programming. Endocr Rev 2011;32(2):159–224.

[33] Rampon C, Jiang CH, Dong H, Tang YP, Lockhart DJ, Schultz PG, et al. Effects of environmental enrichment on gene expression in the brain. Proc Natl Acad Sci USA 2000;97(23):12880–4.

[34] Bollati V, Baccarelli A, Hou L, Bonzini M, Fustinoni S, Cavallo D, et al. Changes in DNA methylation patterns in subjects exposed to low-dose benzene. Cancer Res 2007;67(3):876–80.

[35] Szyf M. The genome- and system-wide response of DNA methylation to early life adversity and its implication on mental health. Can J Psychiatry 2013;58(12):697–704.

[36] Roth TL, Lubin FD, Funk AJ, Sweatt JD. Lasting epigenetic influence of early-life adversity on the BDNF gene. Biol Psychiatry 2009;65(9):760–9.

[37] Chang HS, Anway MD, Rekow SS, Skinner MK. Transgenerational epigenetic imprinting of the male germline by endocrine disruptor exposure during gonadal sex determination. Endocrinology 2006;147(12):5524–41.

[38] Szyf M, Weaver I, Meaney M. Maternal care, the epigenome and phenotypic differences in behavior. Reprod Toxicol 2007;24(1):9–19.
[39] Weaver IC, Cervoni N, Champagne FA, D'Alessio AC, Sharma S, Seckl JR, et al. Epigenetic programming by maternal behavior. Nat Neurosci 2004;7(8):847–54.
[40] Braun JM, Kahn RS, Froehlich T, Auinger P, Lanphear BP. Exposures to environmental toxicants and attention deficit hyperactivity disorder in US children. Environ Health Perspect 2006;114(12):1904–9.
[41] Nigg JT, Knottnerus GM, Martel MM, Nikolas M, Cavanagh K, Karmaus W, et al. Low blood lead levels associated with clinically diagnosed attention-deficit/hyperactivity disorder and mediated by weak cognitive control. Biol Psychiatry 2008;63(3):325–31.
[42] Luo M, Xu Y, Cai R, Tang Y, Ge MM, Liu ZH, et al. Epigenetic histone modification regulates developmental lead exposure induced hyperactivity in rats. Toxicol Lett 2014;225(1):78–85.
[43] Vucetic Z, Kimmel J, Totoki K, Hollenbeck E, Reyes TM. Maternal high-fat diet alters methylation and gene expression of dopamine and opioid-related genes. Endocrinology 2010;151(10):4756–64.
[44] van Mil NH, Steegers-Theunissen RP, Bouwland-Both MI, Verbiest MM, Rijlaarsdam J, Hofman A, et al. DNA methylation profiles at birth and child ADHD symptoms. J Psychiatr Res 2014;49:51–9.
[45] Zhang K, Tarazi FI, Baldessarini RJ. Role of dopamine D(4) receptors in motor hyperactivity induced by neonatal 6-hydroxydopamine lesions in rats. Neuropsychopharmacology 2001;25(5):624–32.
[46] Biederman J, Faraone SV. Attention-deficit hyperactivity disorder. Lancet 2005;366(9481):237–48.
[47] Hawi Z, Segurado R, Conroy J, Sheehan K, Lowe N, Kirley A, et al. Preferential transmission of paternal alleles at risk genes in attention-deficit/hyperactivity disorder. Am J Hum Genet 2005;77(6):958–65.
[48] Martel MM, Nikolas M, Jernigan K, Friderici K, Waldman I, Nigg JT. The dopamine receptor D4 gene (DRD4) moderates family environmental effects on ADHD. J Abnorm Child Psychol 2011;39(1):1–10.
[49] Brown K. Neuroscience. New attention to ADHD genes. Science 2003;301(5630):160–1.
[50] Becker K, El-Faddagh M, Schmidt MH, Esser G, Laucht M. Interaction of dopamine transporter genotype with prenatal smoke exposure on ADHD symptoms. J Pediatr 2008;152(2):263–9.
[51] Neuman RJ, Lobos E, Reich W, Henderson CA, Sun LW, Todd RD. Prenatal smoking exposure and dopaminergic genotypes interact to cause a severe ADHD subtype. Biol Psychiatry 2007;61(12):1320–8.
[52] Brookes KJ, Mill J, Guindalini C, Curran S, Xu X, Knight J, et al. A common haplotype of the dopamine transporter gene associated with attention-deficit/hyperactivity disorder and interacting with maternal use of alcohol during pregnancy. Arch Gen Psychiatry 2006;63(1):74–81.
[53] Holloway AC, Cuu DQ, Morrison KM, Gerstein HC, Tarnopolsky MA. Transgenerational effects of fetal and neonatal exposure to nicotine. Endocrine 2007;31(3):254–9.
[54] Maughan B, Taylor A, Caspi A, Moffitt TE. Prenatal smoking and early childhood conduct problems: testing genetic and environmental explanations of the association. Arch Gen Psychiatry 2004;61(8):836–43.
[55] Bohacek J, Mansuy IM. Molecular insights into transgenerational nongenetic inheritance of acquired behaviours. Nat Rev Genet 2015;16(11):641–52.
[56] Xu Y, Chen XT, Luo M, Tang Y, Zhang G, Wu D, et al. Multiple epigenetic factors predict the attention deficit/hyperactivity disorder among the Chinese Han children. J Psychiatr Res 2015;64:40–50.
[57] Chase KA, Gavin DP, Guidotti A, Sharma RP. Histone methylation at H3K9: evidence for a restrictive epigenome in schizophrenia. Schizophr Res 2013;149(1–3):15–20.
[58] Nagarajan RP, Hogart AR, Gwye Y, Martin MR, LaSalle JM. Reduced MeCP2 expression is frequent in autism frontal cortex and correlates with aberrant MECP2 promoter methylation. Epigenetics 2006;1(4):e1–e11.
[59] Kim P, Park JH, Choi CS, Choi I, Joo SH, Kim MK, et al. Effects of ethanol exposure during early pregnancy in hyperactive, inattentive and impulsive behaviors and MeCP2 expression in rodent offspring. Neurochem Res 2013;38(3):620–31.
[60] Park S, Lee JM, Kim JW, Cho DY, Yun HJ, Han DH, et al. Associations between serotonin transporter gene (SLC6A4) methylation and clinical characteristics and cortical thickness in children with ADHD. Psychol Med 2015;45(14):3009–17.

[61] Mous SE, Muetzel RL, El Marroun H, Polderman TJ, van der Lugt A, Jaddoe VW, et al. Cortical thickness and inattention/hyperactivity symptoms in young children: a population-based study. Psychol Med 2014;44(15):3203–13.
[62] Laucht M, Treutlein J, Schmid B, Blomeyer D, Becker K, Buchmann AF, et al. Impact of psychosocial adversity on alcohol intake in young adults: moderation by the LL genotype of the serotonin transporter polymorphism. Biol Psychiatry 2009;66(2):102–9.
[63] Karg K, Burmeister M, Shedden K, Sen S. The serotonin transporter promoter variant (5-HTTLPR), stress, and depression meta-analysis revisited: evidence of genetic moderation. Arch Gen Psychiatry 2011;68(5):444–54.
[64] Caspi A, Hariri AR, Holmes A, Uher R, Moffitt TE. Genetic sensitivity to the environment: the case of the serotonin transporter gene and its implications for studying complex diseases and traits. Am J Psychiatry 2010;167(5):509–27.
[65] van der Meer D, Hoekstra PJ, Zwiers M, Mennes M, Schweren LJ, Franke B, et al. Brain correlates of the interaction between 5-HTTLPR and psychosocial stress mediating attention deficit hyperactivity disorder severity. Am J Psychiatry 2015;172(8):768–75.
[66] Gizer IR, Ficks C, Waldman ID. Candidate gene studies of ADHD: a meta-analytic review. Hum Genet 2009;126(1):51–90.
[67] Perroud N, Zewdie S, Stenz L, Adouan W, Bavamian S, Prada P, et al. Methylation of serotonin receptor 3a in ADHD, borderline personality, and bipolar disorders: link with severity of the disorders and childhood maltreatment. Depress Anxiety 2015;33(1):45–55.
[68] Murthy S, Niquille M, Hurni N, Limoni G, Frazer S, Chameau P, et al. Serotonin receptor 3A controls interneuron migration into the neocortex. Nat Commun 2014;5:5524.
[69] Wilmot B, Fry R, Smeester L, Musser ED, Mill J, Nigg JT. Methylomic analysis of salivary DNA in childhood ADHD identifies altered DNA methylation in VIPR2. J Child Psychol and Psychiatry 2016;57(2):152–60.
[70] Sheward WJ, Naylor E, Knowles-Barley S, Armstrong JD, Brooker GA, Seckl JR, et al. Circadian control of mouse heart rate and blood pressure by the suprachiasmatic nuclei: behavioral effects are more significant than direct outputs. PloS One 2010;5(3):e9783.
[71] Widenhorn-Muller K, Schwanda S, Scholz E, Spitzer M, Bode H. Effect of supplementation with long-chain omega-3 polyunsaturated fatty acids on behavior and cognition in children with attention deficit/hyperactivity disorder (ADHD): a randomized placebo-controlled intervention trial. Prostaglandins Leukot Essent Fatty Acids 2014;91(1–2):49–60.
[72] Bain EE, Apostol G, Sangal RB, Robieson WZ, McNeill DL, Abi-Saab WM, et al. A randomized pilot study of the efficacy and safety of ABT-089, a novel alpha4beta2 neuronal nicotinic receptor agonist, in adults with attention-deficit/hyperactivity disorder. J Clin Psychiatry 2012;73(6):783–9.
[73] Perroud N, Paoloni-Giacobino A, Prada P, Olie E, Salzmann A, Nicastro R, et al. Increased methylation of glucocorticoid receptor gene (NR3C1) in adults with a history of childhood maltreatment: a link with the severity and type of trauma. Transl Psychiatry 2011;1:e59.
[74] Talge NM, Neal C, Glover V. Early stress, translational research, prevention science network: fetal and neonatal experience on child and adolescent mental health. Antenatal maternal stress and long-term effects on child neurodevelopment: how and why? J Child Psychol Psychiatry 2007;48(3–4):245–61.
[75] Teicher MH, Andersen SL, Polcari A, Anderson CM, Navalta CP, Kim DM. The neurobiological consequences of early stress and childhood maltreatment. Neurosci Biobehav Rev 2003;27(1–2):33–44.
[76] Novikova SI, He F, Bai J, Cutrufello NJ, Lidow MS, Undieh AS. Maternal cocaine administration in mice alters DNA methylation and gene expression in hippocampal neurons of neonatal and prepubertal offspring. PloS One 2008;3(4):e1919.
[77] Host L, Dietrich JB, Carouge D, Aunis D, Zwiller J. Cocaine self-administration alters the expression of chromatin-remodelling proteins; modulation by histone deacetylase inhibition. J Psychopharmacol 2011;25(2):222–9.

[78] Anier K, Malinovskaja K, Aonurm-Helm A, Zharkovsky A, Kalda A. DNA methylation regulates cocaine-induced behavioral sensitization in mice. Neuropsychopharmacology 2010;35(12):2450–61.
[79] Perroud N, Salzmann A, Prada P, Nicastro R, Hoeppli ME, Furrer S, et al. Response to psychotherapy in borderline personality disorder and methylation status of the BDNF gene. Transl Psychiatry 2013;3:e207.
[80] Kern JK, Geier DA, Sykes LK, Geier MR, Deth RC. Are ASD and ADHD a continuum? A comparison of pathophysiological similarities between the disorders. J Atten Disord 2015;19(9):805–27.

CHAPTER

8 THE EPIGENETICS OF BRAIN AGING AND PSYCHIATRIC DISORDERS

H. Gong*, X. Xu,†**

**The Key Laboratory of Geriatrics, Beijing Institute of Geriatrics, Beijing Hospital, National Center of Gerontology, Beijing, China; **Max Planck Institute for Biology of Ageing, Cologne, Germany; †Department of Anesthesiology, Yale University School of Medicine, New Haven, CT, United States*

CHAPTER OUTLINE

8.1 EPIGENETICS

The term "epigenetics" was first coined by Conrad Waddington in 1940s to describe interactions of genes with their environment during development [1]. It has developed from a phenomenon to an immensely studied branch of science. Broadly, epigenetics encompasses conformational changes in DNA and/or chromatin without altering the basic genetic code that regulates the intricate molecular

Neuropsychiatric Disorders and Epigenetics. http://dx.doi.org/10.1016/B978-0-12-800226-1.00008-3

machinery through which the spatio-temporal dynamics of gene expression are implemented. These mechanisms primarily involve DNA methylation (DNAm), histone posttranslational modifications (HPTMs), and posttranscriptional regulation by noncoding RNAs (ncRNAs), such as microRNAs (miRNAs) and long noncoding RNAs (lncRNAs).

8.1.1 DNA METHYLATION

8.1.1.1 5′-Methylcytosine

DNA methylation (DNAm) is the best known epigenetic mark. It is widespread in a spectrum of species, including plants, rodents and humans, and plays a key role in maintaining genome stability and regulating gene expression [2–4]. DNAm involves the addition of a methyl group to the fifth carbon of a cytosine residue to form 5′-methylcytosine (5mC) by DNA methyltransferases (Dnmts), most frequently in the context of CpG dinucleotides [2–4] (Fig. 8.1). This biochemical process creates a relatively stable covalent modification that is traditionally understood to repress gene transcription by promoting closed chromatin states through the recruitment of transcriptional repressors and limiting DNA accessibility to transcriptional machinery [3,5]. CpGs are not uniformly distributed in the genome and tend to be enriched in CpG islands, which are stretches of DNA roughly 1000 base pairs long that have a higher CpG density than the rest of the genome [2].

The dynamics of genome-wide DNAm are regulated by Dnmts, including Dnmt1, Dnmt3a, and Dnmt3b (Fig. 8.1B,C). Dnmt1 is highly expressed in mammalian tissues including the brain. It is responsible for the maintenance of DNA methyl states during DNA replication, whereas Dnmt3a and Dnmt3b perform de novo DNAm during development and other pathophysiological conditions [3,5,6]. All three Dnmts are extensively involved in the development of an embryo and their expression is much reduced by the time cells reach terminal differentiation. This suggests that the DNAm pattern is relatively stable in postmitotic cells, such as neurons and cardiocytes. However, in the mature mammalian brain, postmitotic neurons still express substantial levels of Dnmts, raising the possibility that Dnmts and DNAm may play a pivotal role in the brain [5]. Indeed, the loss of Dnmt1 and Dnmt3a in the adult brain leads to cognitive deficits in mice; in humans, mutations in Dnmt1 are associated with a form of neurodegenerative disease [3]. These studies suggest that impairment of DNAm may be a fundamental mechanism in regulating mouse learning, memory and cognition.

DNAm might play an important biologic role was first suggested by Griffith and Mahler in 1969 speculating that DNAm could provide a basis for long-term memory in the brain [7]. It has been shown that the presence of 5mC at CpG sites affects the binding of transcription factors, and consequently gene expression [3,5]. DNAm regulates gene expression by recruiting corepressor complexes [e.g., histone deacetylases (HDACs) and histone methyltransferases] that can sterically hinder the transcriptional machinery or modifty nucleosome structure [5]. Such complexes involve several DNA methyl-binding domain proteins (MBDs), which are required for normal cell growth and development. MBDs are expressed at higher levels in brain than in any other tissues, and many MBDs are important for normal neuronal development and function [5,6], such as methyl CpG binding protein2 (MeCP2). MeCP2 is recognized as a transcriptional repressor, and dysfunction of MeCP2 leads to the neurodevelopmental disorder Rett syndrome [8]. The effects of DNAm on gene expression are complex and may vary according to genomic location. Numerous studies suggest that methylation occurring within CpG-rich regions near the transcription start-site of a gene tends to have a repressive effect on gene expression across tissues. Gene body DNAm is associated with a higher level of gene expression in dividing cells.

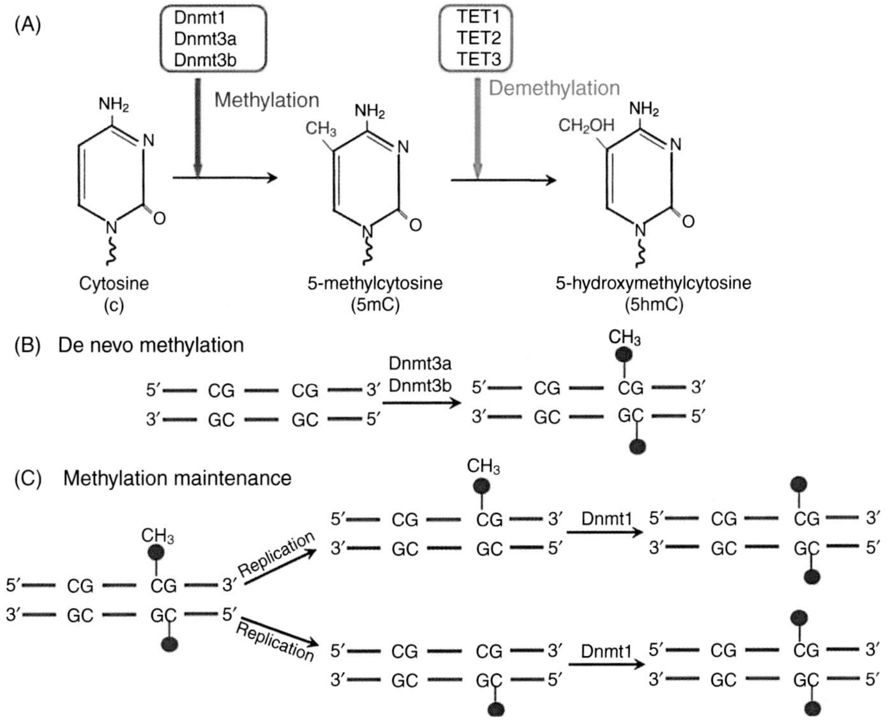

FIGURE 8.1 DNA Modification Pathways

(A) A family of DNA methyltransferases (Dnmts) catalyzes the transfer of a methyl group to the fifth carbon of cytosine (C) residue to form 5-methylcytosine (5mC). To the methyl group of 5mC can be added a hydroxyl group mediated by Tet enzymes to generate 5-hydroxymethylcytosine (5hmC). (B) Dnmt3a and Dnmt3b are de novo Dnmts and transfer methyl groups (red) onto naked DNA. (C) Dnmt1 is the maintenance Dnmt and maintains DNA methylation pattern during replication. When DNA undergoes semiconservative replication, the parental DNA stand (green) retains the original DNA methylation pattern. Dnmt1 precisely replicates the original DNA methylation pattern by adding methyl groups (red) onto the newly formed daughter strand (purple). The majority of methylation occurs at cytosines that are followed by a guanine (i.e., CpG motifs).

However, in the murine frontal cortex, gene body methylation of non-CpG sites is negatively correlated with gene expression [9]. The existence of variability in DNAm independent of the underlying nucleotide sequence suggests that epigenetic modifications can modulate the impact of genetic variation (i.e., genotypes) on biological processes including brain function.

In order to maintain or alter genome-wide DNA methylation patterns in cells, including neurons, in response to environmental changes, both DNA methylation and demethylation have to be active. The process of DNA demethylation is much less understood. Several studies imply the key DNA demethylation enzymes are ten-eleven translocation (Tet) family of enzymes Tet1, Tet2, and Tet3. Tet enzymes are involved in both global and locus-specific DNA demethylation [10–13]. Demethylation often occurs during the process of hippocampal learning, memory, and adult neurogenesis, which impacts the olfactory system [14]. Mutant Tet1 animals exhibit abnormal hippocampal long-term depression and impaired memory extinction. Learning and memory require neuronal activity, which can strengthen synaptic connections and weaken other synaptic connection strengths through a process of synaptic plasticity [14].

8.1.1.2 5′-Hydroxymethylcytosine

5′-Hydroxymethylcytosine (5hmC) is the oxidized form of the canonical 5mC and was identified in mammalian brain tissue and stem cells [12,15]. Tet enzymes add a hydroxyl group onto the methyl group of 5mC to form 5hmC [12,13] (Fig. 8.1A). Such as 5mC, 5hmC may also regulate gene expression in neurons since the conversion of 5mC to 5hmC impairs the binding of MeCP2 [16].

5hmC accounts for ~40% of modified cytosine in the brain, which is typically 5–10 times higher than in any other tissues, and has been implicated in DNAm–related synaptic plasticity [15]. In neuronal cells, 5hmC markedly increases from the early postnatal stage to adulthood, suggesting a strong correlation between 5hmC and neurodevelopment. Interestingly, 5hmC is markedly depleted on the X chromosome during postnatal neurodevelopment and aging. Functionally, 5hmC is associated with actively transcribed genes in adult cerebellum and is acquired in developmentally activated genes. 5hmC–regulated regions are dynamically changed during neurodevelopment and aging. 5hmC is enriched throughout gene bodies in the brain, whereas 5hmC is also present in ES cells in the bodies of active genes, although to a lesser degree than that found in the brain. The overall abundance of 5hmC is negatively correlated with MeCP2 dosage [17]. These findings suggest that 5hmC-mediated epigenetic regulation is critical in neurodevelopment and aging, as well as in human neurological disorders.

8.1.2 HISTONE POSTTRANSLATIONAL MODIFICATIONS

Nuclear DNA in eukaryotic cells is densely packed into chromatin. The nucleosome is the fundamental unit of chromatin and it is composed of an octamer of the four core histones (H3, H4, H2A, and H2B) around which 147 base pairs of DNA are wrapped (Fig. 8.2). Numerous types of posttranslational modifications of the amino terminal-tail of histones alter chromatin compaction to create more "open" states (euchromatin, which is transcriptionally permissive) versus "closed" states (heterochromatin, which is transcriptionally repressive) [18,19]. There are over 60 different residues on histones where modifications have been detected. Such modifications, including lysine (K), arginine (R) or histidine methylation, K acetylation, serine (S), threonine (T), or tyrosine (Y) phosphorylation, ubiquitination, adenosine diphosphate (ADP)-ribosylation, crotonylation, hydroxylation, proline isomerization, and K SUMOylation etc. together constitute the histone code [20] (Fig. 8.2). The small covalent modifications, such as acetylation, methylation, and phosphorylation of histones, have been widely explored [18].

Histone acetylation is associated with transcriptional activation by negating the positive charge of lysine residues in the histone tail. This process is controlled by histone acetyltransferases and HDACs, each of which comprises multiple enzyme classes whose expression and activity are exquisitely regulated (reviewed in Ref. [18]). Currently, there are 18 known HDACs in humans, generally subdivided into four classes: class I (HDACs 1, 2, 3, and 8), class IIa (HDACs 4, 5, 7, and 9), class IIb (HDACs 6 and 10), class III [sirtuins (SIRTs) 1, 2, 3, 4, 5, 6, and 7], and class IV (HDAC11) [20]. The different HDACs undertake many different tasks, either by effecting changes in gene expression or by directly regulating protein function. The expression of different HDACs is also highly region- and cell-type-specific. For instance, while HDAC2 is expressed in most brain regions, it is predominantly active in mature neurons and weakly or not present in progenitor and glial cells [20]. Histone methylation has

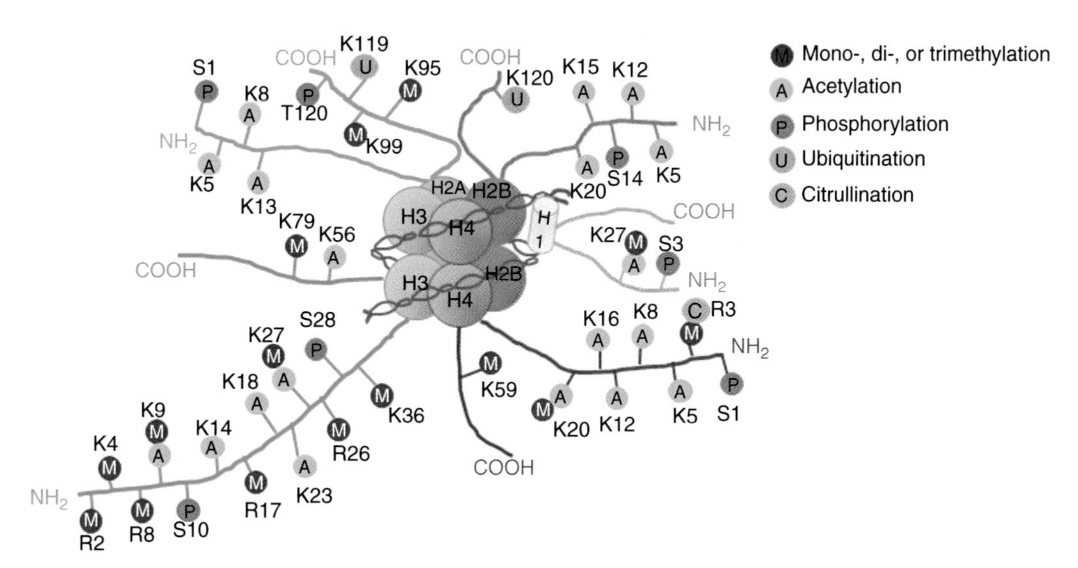

FIGURE 8.2 Schematic of the Common Histone Modifications

Histones pack and order DNA into nucleosomes. Each nucleosome contains two subunits and each subunit is composed of a copy of histones H2A, H2B, H3, and H4, known as the core histones. The linker histone H1 acts as a stabilizer of the internucleosome DNA. Histone modifications including methylation, acetylation, phosphorylation, and ubiquitination, etc., occur at specific amino acids along histone tails. *K*, lysine; *R*, arginine; *S*, serine; *T*, threonine.

been associated with both transcriptional activation and repression, depending on the particular residue and the extent of methylation. Methylation at lysines or arginine has three different forms: mono-, di-, or trimethyl for lysines, and mono- or di- (asymmetric or symmetric) methyl for arginines (reviewed in Ref. [18]). Several families of histone methyltransferases can methylate lysine and arginine residues, and equally diverse histone demethylases can reverse this reaction. There are infinite possible HPTMs patterns to form an epigenetic "histone code" which can alter interactions among histones, between histones and DNA, or facilitate the recruitment of additional chromatin-modifying proteins [14]. The actual effect and interplay between these HPTMs are complex, and depend on the type of histone protein and the specific amino acid that is modified, and a combination of certain HPTMs can even have a function that is different from that of individulized HPTMs [20]. Posttranslational modifications of nuclear histone tails represent one of basic molecular epigenetic mechanisms that alter chromatin structure and accessibility of DNA, and influence gene expression and potentially cellular/organismal phenotypes [3]. Moreover, modifications on histones are dynamic and rapidly changing. Acetylation, methylation, phosphorylation, and deimination can appear and disappear on chromatin within minutes of a stimulus arriving at the cell surface [18]. So the modifications can be adjusted through reversible modifications of their N-terminal tails.

8.1.3 NONCODING RNAs

Increasing attention has focused on a variety of ncRNAs that are important in epigenetic regulation. ncRNAs are functional RNA molecules which are transcribed from genomic DNA, but are not translated into protein. Accumulating ncRNAs were identified and assigned into many different types. The ncRNAs are briefly categorized into two size groups. A group of ncRNAs with a size < 200 nucleotides are the small ncRNAs (sncRNAs), including microRNAs (miRNAs), small interfering RNAs (siRNAs), small nuclear RNAs (snRNAs), small nucleolar RNAs (snoRNAs), small Cajal body-specific RNAs (scaRNAs), piwi-interacting RNAs (piRNAs), splice junction-associated RNAs (spliRNAs), small modulatory RNAs (smRNAs), repeat-associated small interfering RNAs (rasiRNAs), transcription initiation RNAs (tiRNAs), promoter-associated short RNAs (PASRs), transcription start site-associated RNAs (TSSa-RNAs), promoter upstream transcripts (PROMPTS), ribosomal RNAs (rRNAs), transfer RNAs (tRNAs), and small double-stranded RNAs (dsRNAs) [20,21]. The other group of ncRNAs with a size greater than 200 nucleotides is called long ncRNAs (lncRNAs), and which include a recently discovered, poorly conserved, and abundant class of heterogeneous regulatory ncRNAs. Intergenic ncRNAs (lincRNAs), natural antisense transcripts (NATs), ncRNA expansion repeats, promoter-associated RNAs (PARs), and enhancer RNAs (eRNAs), are all lncRNAs [20,21]. While pleiotropic ncRNAs, for example, miRNAs, can target large numbers of genes and signaling pathways simultaneously, there are also ncRNAs, such as NATs, which can hybridize to a limited and precise subset of candidates [22]. Additionally, some lncRNAs have been proposed to direct epigenetic enzymes to their target sites, while others are thought to bind and sequester other epigenetic players, such as Dnmts, thereby hampering their activity [20]. ncRNAs can participate in epigenetic regulation by recruiting DNA- or chromatin-modifying enzymes, or by directly modifying other RNA molecules or RNA–protein complexes. Numerous ncRNA classes are now implicated in central nervous system function and play an important role in memory formation and other forms of normal or pathological neuroplasticity [22]. However, owing to the highly dynamic and cell specific nature of these epigenetic processes, relatively little is known about their specific roles in vivo.

The best characterized ncRNAs are the miRNAs, which regulate a variety of cellular processes through the posttranscriptional repression of gene expression. Like HPTMs and DNAm, expression of miRNAs can alter the transcriptional potential of a gene in the absence of any change to the DNA sequence, and thus can be considered an epigenetic phenomenon. Much is already known about miRNA biogenesis and function [23]. Mature single-stranded miRNAs, of about 22 nucleotides, are generated through a series of well-orchestrated cleavage steps, including the nuclear processing of primary miRNAs (pri-miRNAs) into precursor miRNAs (pre-miRNAs) by the DGCR8/Drosha complex, the cytoplasmic processing of pre-miRNAs into imperfectly paired miRNA duplexes by Dicer, and the preferential incorporation of one strand (the "guide" miRNA strand) onto the RNA-induced silencing complex (RISC) [23,24]. Mature miRNAs bind to complementary sequences on target mRNAs, and this induces mRNA degradation, destabilization, translational silencing, or combinations thereof thus silencing gene expression [24]. Thousands of miRNAs have been discovered in many organisms, and there are currently 1881 precursors and 2588 mature human miRNAs registered in the fast growing miRBase [25]. More than half of the protein-coding genes are predicted, to be regulated by miRNAs and each miRNA can potentially manipulate the expression of up to hundreds of target genes. Furthermore, one miRNA can regulate the expression of several genes within a specific biological or cellular pathway [23,26]. miRNAs are positioned as possible "master regulators" of many cellular processes. miRNAs are ubiquitously expressed throughout the brain and govern all major neuronal pathways. Abnormalities in miRNA expression and miRNA-mediated gene regulation have been observed in a variety of human diseases, such as psychiatric disorders, cancer, heart disease, and viral infection [27–29]. Over half of the miRNAs identified to-date are highly or exclusively expressed in the brain, and many of them have been implicated in many important aspects of neuronal function [26,28,29]. Adult brain miRNAs function as endogenous "hubs" for the fine-tuning of target gene expression and thereby affect the structure and function of neuronal networks [22,26].

lncRNAs have been increasingly appreciated as an integral component of gene regulatory networks. lncRNAs actually make up the majority of the mammalian transcriptome. There are 10,000–50,000 lncRNA genes that have been annotated in the human genome to date [30]. The brain is one of the richest sources of lncRNAs since a striking 40% (equivalent to 4,000–20,000 lncRNA genes) are expressed specifically in the brain, where they show precisely regulated temporal and spatial expression patterns [31]. Their expression is dynamically regulated during development and in response to neuronal activity and is also often highly restricted to specific brain regions in adult mice, such as the hippocampus or particular cortical domains [30]. A large portion of lncRNAs are produced by Pol II and undergo 5′capping and polyadenylation, thus they are molecularly indistinguishable from mRNAs. In most cases where the functions of lncRNAs have been relatively well characterized, their prime roles lie at the regulation of gene expression and epigenetic processes in the nucleus. In accordance with this, lncRNAs are highly enriched in the nucleus, many of which are tightly associated with the chromatin fraction. Genome-wide features of their origin and expression patterns ascribe a prominent role for lncRNAs to the regulation of protein-coding genes, and also suggest a potential link to many human diseases [32]. Many lncRNAs are regulators of gene expression, genome stability and modulators of neural function and dysfunction [21]. Recent studies have begun to unravel the intricate regulatory mechanism of lncRNAs occurring at multiple levels, including transcriptional repression by recruiting the repressive chromatin modifying complexes to create a repressive chromatin state, gene activation by establishing transcriptionally competent chromatin structure at their target gene loci (reviewed in Refs. [30,32,33]).

8.1.4 INTEGRATION OF MULTIPLE EPIGENETIC MODIFICATIONS

The epigenetic processes associated with DNA methylation/demethylation, HPTMs, and ncRNAs do not act independently, but closely interact to form a complex, multilayered regulatory system that can dynamically fine-tune gene expression.

DNAm works with HPTMs and miRNAs to regulate transcription. Dnmts cooperate with histone-modifying enzymes involved in adding and/or stripping histone markers in order to impose a repressive state on a gene region [5]. There is an interesting interplay between Dnmt3a-dependent DNAm and Polycomb-group (PcG)-dependent H3K27me3 marks. Dnmt3a activity at nonpromoter regions correlates with increased expression of neurogenic genes, by interfering with PcG binding and H3K27me3-mediated gene repression. In contrast, Dnmt3a activity at promoter regions inhibits gene expression. Additionally, Dnmt inhibitors block changes in H3 acetylation which is associated with memory formation. Furthermore, deficits in memory and hippocampal synaptic plasticity induced by Dnmt inhibitors can be reversed by pretreatment with an HDAC inhibitor [34]. MeCP2, binding preferentially to fully methylated DNA, is associated with both HDAC machinery and histone methyltransferases to alter specific HPTMs [35]. Thus, DNAm acts in concert with HPTMs to regulate gene expression, through interference with transcription factor binding and chromatin compaction.

It is also conceivable that HPTMs influence the DNAm pattern, indicating a bidirectional relationship between histone and DNA modifications. DNAm patterns are established and maintained by specific combinations of chromatin modifications. Consistent with this hypothesis, elevated histone acetylation can trigger DNA demethylation and thereby gene expression in vitro [5]. Conversely, HDACs are known to interact with Dnmts and inhibit gene expression through the induction of DNAm [20], whereas transcription factors that recruit histone acetyltransferases can trigger demethylation of DNA. Likewise, HDAC inhibitors are capable of inducing DNA demethylation [35]. Taken together, these results reveal a complex relationship between HPTMs and DNAm.

There are additional complex interactions between miRNAs and other components of the epigenetic machinery. Some miRNAs regulate the expression of proteins involved in epigenetic regulation, and the expression of various miRNAs themselves is also subject to factors, such as DNAm and HPTMs. For example, miR-184, involved in the regulation of proliferation and differentiation of neural stem cells, is surrounded by CpG islands attracting MBD1, which can suppress its expression as described earlier [20].

8.2 EPIGENETICS OF BRAIN AGING

Brain aging is characterized by the gradual decline in different aspects of cognitive performance, brain structure, and brain function. Memory impairment in normal aging, meaning memory decline not associated with disease, is a conserved feature of growing older across invertebrates, rodents, monkeys, and humans [3]. In the United States it is expected that by 2050, ~12% of people over the age of 65 will suffer from moderate-to-severe memory impairment [36]. Cognition decline is often correlated with age-dependent deterioration of synaptic function in brain regions crucial for memory formation and consolidation, such as the hippocampus and prefrontal cortex (PFC). Human PFC plays a critical role in complex cognitive behaviors, personality, decision making, and orchestration of thoughts and actions. The hippocampus is an extremely important component in the brain and is closely associated with the cerebral cortex for learning, memory, and cognitive functions. Two major functions of the hippocampus are the storage and interpretation of spatial information and mediation of consolidation of short-term memory into long-term memory. Unlike neurodegenerative disease that profoundly impacts memory (i.e., Alzheimer's disease), hippocampus-dependent memory deficits that emerge during normal aging can occur in the absence of massive neuronal death. Instead, subtle changes in the connectional and functional integrity of key hippocampal circuits appear to underlie memory impairment in older individuals. In the context of aging, altered transcriptional regulation of genes that promote or are necessary for synaptic plasticity is associated with memory impairment in aged rodents. Alongside a focus on negative outcomes, there is increasing recognition that memory decline is not an inevitable consequence of aging as some older individuals maintain normal memory abilities that match those of younger individuals across the lifespan [37]. Identifying the neurobiological mechanisms that impact differential cognitive outcomes with age is critical.

The neurobiological processes underlying age-related learning and memory deficits include aberrant changes in gene transcription that eventually affect the plasticity of the aged brain. Changes in gene expression in neurons were thought to take place during brain aging, and analysis of regions of the hippocampus and frontal cortex by microarray has confirmed this [38,39]. The molecular mechanisms underlying these changes in gene expression and the regulation are largely unknown. Over the past few years, evidence has accumulated that epigenetic mechanisms may be critically involved in mediating age related changes of the brain. Epigenetic modifications of chromatin structure contribute to the experience-dependent synaptic plasticity in the hippocampus that underlies memory formation.

Recent research also demonstrates that dysfunctional epigenetic regulation of experience-dependent gene transcription in the hippocampus is associated with memory impairment in aged rodents [36].

8.2.1 DNA METHYLATION IN BRAIN AGING

DNAm levels are particularly promising biomarkers of aging and this implies a profound effect on DNAm levels in most human tissues and cell types. In the central nervous system, DNAm is critical for proper postnatal neurodevelopment and undergoes age-dependent changes in the adult brain [17].

An overall decline in DNAm has been associated with cell and tissue aging including brain [3]. For example, DNAm in the PFC shows unique temporal patterns across life. The fastest changes occur during the prenatal period, slowing down markedly after birth and continuing to slow further with aging [40]. However, the effects of aging are complex, with some evidence pointing to age-related decreases in global DNAm, together with increased methylation at CpG islands across multiple brain regions in humans [40]. The enrichment of methylation at CpG sites tends to occur more frequently among functionally related gene transcripts, including gene classes that regulate DNA binding and transcription factors. This age-related aggregation of methylation might contribute to transcriptional abnormalities reported in the aged brain. Consistent with this possibility, altered methylation of *Arc* DNA in the CA1 and dentate gyrus of the hippocampus in aged rats is associated with decreased *Arc* transcription and spatial memory impairment [41]. Evidence suggesting that DNAm influences differential cognitive outcomes in aging derives from a targeted study examining methylation in the promoter regions of *Gabra5*, *Hspa5*, and *Syn1* previously implicated in age-related cognitive decline in the Long-Evans rat [42]. The overall results reveal an increase in the number of methylated sites across all three genes, but only in relation to chronological age and not cognitive status. The loss of Dnmt1 and Dnmt3a in the adult brain leads to cognitive deficits in mice [43]. Transient over-expression of Dnmt3a2, a Dnmt3a isoform, in mouse hippocampus restores age-associated cognitive deficits. Moreover, inhibition of hippocampal Dnmt3a2 expression by RNAi leads to cognitive behavioral deficits in the young mouse [43]. Mutant Tet1 animals exhibit abnormal hippocampal long-term depression and impaired memory extinction. In humans, mutations in Dnmt1 are associated with a form of neurodegenerative disease. These results suggest that impairment of DNAm plays a crucial role and is a fundamental mechanism operative in regulating mouse learning, memory and cognition [3].

Both developmental programming and age-dependent alterations of 5hmC occur in the mammalian brain. A recent study found that 5hmC loci may not only be maintained, but may be further acquired with age at specified loci in hippocampus and cerebellum [17]. In the mouse hippocampus, global 5hmC content increases during aging in the absence of 5mC decrease, suggesting that 5hmC acts as an epigenetic marker and not simply as an intermediary in DNA demethylation [44]. Besides, 5hmC levels inversely correlate with MeCP2 dosage, a protein encoded by a gene in which mutations cause Rett syndrome [17]. These findings suggest that 5hmC-mediated epigenetic regulation is critical in neurodevelopment, aging and human diseases.

8.2.2 HISTONE POSTTRANSLATIONAL MODIFICATIONS IN BRAIN AGING

Apart from widespread changes in the neuronal DNAm profile throughout the lifespan, the neuronal histone code also undergoes age-related alterations (reviewed in Ref. [20]). Our study shows that in mouse brain, alterations of HPTMs are extensive and include both acetylation and methylation of histones H3 and H4 at both K and R [45]. In the senescence-accelerated prone mouse 8 (SAMP8) brain, it was also shown that many histone methylation marks are altered with age [46]. Recent reports indicate that chromatin remodeling via histone acetylation plays a crucial role in regulating synaptic and cognitive function in aging and neurodegenerative brains [47]. Increasing histone acetylation by HDAC inhibition enhances gene transcription and improves hippocampal long-term potentiation or memory functions in several experimental models of neurological diseases [34,47,48]. This indicates that dysregulation of chromatin acetylation may be involved in certain forms of cognitive impairment [49,50]. Age-related impairment in aged mice with memory deficits is selectively associated with decreased learning-induced histone acetylation at lysine 12 on histone H4 (H4K12) in the hippocampus. Reduced acetylation of H4K12 is also associated with blunted learning-induced hippocampal gene transcription [36]. The finding that HDAC2 expression increases with age in the mouse hippocampus is consistent with findings of decreased acetylation levels [51]. Conversely, levels of the class III HDAC Sirt1 decline with age in mouse and rat, a change not limited to the brain but also observed in other tissues and senescent cells [52,53]. Reduced levels of SIRT1 are associated with increased levels of H4K16ac in vitro [20]. These data suggest that alterations in histone acetylation are site- and gene-specific during brain aging.

H3K4me3, an active mark for transcription, is upregulated in hippocampus 1 h following contextual fear conditioning [54]. The levels of H3K4me3 increase with rat hippocampal aging and correlate with failed object learning. However, young adult rats, whose hippocampi gain increased H3K4me3 levels by treatment with a histone demethylase inhibitor, exhibit similar memory deficits as detected in aged rats. Population genetic studies show that mutations in EZH2, a histone H3 methyltransferase that targets H3K27me3, cause Weaver syndrome with intellectual disability. Thus, emerging findings strongly suggest a role for histone marks as important in modulating brain learning and cognitive function.

8.2.3 ncRNAs IN BRAIN AGING

Increasing evidence shows that ncRNA-driven processes, such as R-loop formation and toxic RNA accumulation represent neurotoxic mechanisms in brain aging and neurodegeneration [21]. miRNAs have been linked to processes associated with brain aging, brain function decline, and neurodegenerative diseases. In a recent research study, the majority of miRNAs expressed in the study decline in relative abundance in the aged brain, in agreement with trends observed in other miRNA studies in aging tissues and organisms. Many aging-associated miRNAs target genes in the insulin signaling pathway, a central node of aging associated genetic networks [55]. In another study, a link between aberrant miRNA expression and age-related decline in mitochondrial respiration rates was forged. Seventy miRNAs were upregulated in the aging mouse brain, 27 of which are implicated in the downregulation of the mitochondrial complexes III, IV, and F0F1-ATPase, that are all pivotal to the oxidative phosphorylation process [56]. Not only have widespread changes in miRNA expression been reported, but the importance of certain miRNAs specific to the aging brain and their roles in the development of neurodegenerative disease have been shown [55,57,58]. These miRNAs may therefore regulate aging-related functions in the brain. Some miRNAs may promote brain aging and/or initiate neurotoxic processes that contribute to specific neurodegenerative diseases (reviewed in Ref. [21]). miR-144 has a strong positive correlation with aged brains in the cortex and cerebellum of humans, chimpanzees, and rhesus macaques [58]. The age-associated miR-144 may contribute to declining brain function via the downregulation of longevity/protective factors, such as ataxin-1 (encoded by the gene mutated in spinocerebellar ataxia 1) [58]. The miR-34 family of miRNAs is another important determinant of brain aging. SIRT1, a target of miR-34a, correlates inversely with age-associated increases in brain and circulating levels of this miRNA [59].

lncRNAs constitute diverse classes of sense- and antisense transcripts that are abundantly expressed in the mammalian central nervous system in cell type- and developmental stage-specific manners. They are implicated during brain development, differentiation, neuronal plasticity, and other cognitive functions [60]. However, only a few had been found to be directly linked to the aging process in brain. One such lncRNA is termed LINC-RBE (long intergenic noncoding-rat brain expressed transcript), which is expressed distinctly and differentially in cortex, hippocampus and cerebellum of the rat, three interconnected brain compartments involved in learning, memory formation, and other cognitive functions, in an age-dependent manner [61]. LINC-RBE is expressed in an age-dependent manner with significantly higher levels of expression in the brains of adult (16 weeks) compared with both immature (4 weeks) and old (70 weeks) rats. Moreover, the expression pattern of LINC-RBE shows distinctive association with specific neuroanatomical regions, cell types and subcellular compartments of the rat brain in an age-related manner [61]. Thus, its expression increases from immature stages through adulthood and declines further in old age. Similar with LINC-RBE, both sense and antisense transcripts of LINC-RSAS (repeat-rich sense-antisense transcript) are expressed in cortical, hippocampal and cerebellar regions of the rat brain in both cell type-specific and age-related manners [62]. Both LINC-RBE and LINC-RSAS can be upregulated by all-trans retinoic acid (atRA), a vitamin A-derivative involved in development, differentiation, neurogenesis, synaptic transmission and other cognitive functions of the brain, such as learning and memory [60,62]. Another example is TUG1 (Taurine Up-Regulated 1), a growth regulator induced by p53. The expression level of TUG1 is highly elevated with increasing age in human subependymal zone (SEZ), the largest reservoir of newly formed neurons and glia cells in the adult brain, and its level is found to be reduced in patients with age-related neurodegenerative conditions, for example, Huntington's disease. TUG1 regulates cell growth and proliferation by modulating HOXB7 expression and could also play a role in senescence and aging by regulating p53-mediated inhibition of cell growth and proliferation [63]. These findings shed light on research of the link between lncRNAs and aging process in brain.

Some other types of ncRNAs are also involved in brain aging. For instance, several small nucleolar RNAs (snoRNAs) are differentially expressed in aged rat brains, and so related microprocessor action may also be important for the maintenance of proper snoRNA regulation during aging [21]. Similarly, telomere small RNAs (tel-sRNAs) might be involved in age-associated cognitive disorders because they modulate telomere maintenance and cellular senescence. As illustrated earlier, ncRNAs and aging are interconnected. ncRNAs and related factors impact the nervous system in various positive and negative manners depending on the precise underlying molecular processes [21].

8.3 EPIGENETICS OF PSYCHIATRIC DISORDERS

Psychiatric disorders, such as schizophrenia, depression and drug addiction impose an ever-increasing burden on society. Psychiatric disorders are multifaceted illnesses with complex etiologies involving chronic alterations in the structure and function of neural circuits. While genetic factors are important in the etiology of these disorders, relatively high rates of discordance among identical twins clearly indicate the importance of additional contributing mechanisms. Furthermore, epidemiological studies report associations between environmental factors, mainly exposure to psychological or physiological stressors, and psychiatric morbidity [22]. Such exposure to environmental insults induces stable changes in gene expression, neural circuit function, and ultimately behavior. Moreover, these maladaptations appear distinct depending on whether exposure to the environmental insults occurs developmentally or in the adult [64]. Thus, a complex interaction between genetic predisposition and environmental factors is suggested to be the cause of mental disorders. Environmental events, such as maternal care, also have marked influences on brain function through epigenetic mechanisms [65], indicating that many brain disorders (including those mediated by the environment) likely involve epigenetic modifications. If epigenetic dysregulation truly plays a part in the pathogenesis of psychiatric disorders, then it would be fair to expect changes in the activity and/or expression of epigenetic modifiers and alterations in their corresponding substrates. Substantial evidence for this reasoning was provided by the discovery of neurological disorders caused by mutations in genes encoding epigenetic players [66]. Besides, growing evidence suggests that epigenetic modifications in certain brain regions and neural circuits represent a key mechanism through which environmental factors interact with an individual's genetic constitution to affect risk of psychiatric conditions throughout life.

8.3.1 DNA METHYLATION

Mounting evidence indicates that altered patterns of DNAm are associated with many psychiatric disorders. An important feature of DNAm in germ cells is that it can be altered by environmental factors, such as toxins, stress, and/or aging at specific genes and these alterations remain across generations [19]. Cocaine use modulates Dnmt3a expression within the nucleus accumbens (NAc) and enhances spine formation. Repeated cocaine use also increases MeCP2 levels that, in turn, increase Bdnf expression. Another example, earlylife stress, such as maternal neglect in rodent models or childhood abuse in humans, results in increased methylation within the promoter of the glucocorticoid receptor and a decrease in its expression [5]. Furthermore, aberrant DNAm patterns are observed in a wide variety of psychiatric disorders, such as schizophrenia and bipolar disorder [5,67]. Extensive changes in DNA methylation have been observed in schizophrenia and bipolar disorder, and this may contribute to the pathogenesis of these disorders. A recent DNA methylome study identified numerous changes in DNA methylation at differentially methylated regions (DMRs) in schizophrenia and bipolar disorder. It also demonstrated that DNA methylation alterations in schizophrenia and bipolar disorder relative to normal subjects are highly brain-region-specific: in the BA9 region, both schizophrenia and bipolar disorder subjects show more hypomethylated DMRs, while in contrast, in the BA24 region, more hypermethylated DMRs are found [68]. A study of monozygotic twins discordant for psychosis found that DMRs involved in known pathways for psychiatric disorders and brain development were overrepresented [69]. It was noted that among more than 100 different markers examined, reelin is one of the most abnormal markers in the context of schizophrenia and bipolar illness [70]. Reelin mRNA and protein levels are reduced by ~50% in various cortical structures of postmortem brain from schizoprenia or bipolar disorder patients. In cortices of schizophrenia patients, there is an increased methylation at the promoter of the gene encoding reelin [71]. Dnmt1 is overexpressed in the cerebral cortex of patients with schizophrenia and bipolar disorder [72]. In one of the largest studies of postmortem human brain tissue, it was found that developmentally associated changes in DNAm are significantly enriched for genomic regions that confer genetic risk for schizophrenia. In addition, several thousand individual

CpGs demonstrated small, but statistically significant, differences in DNAm levels between adult patients with schizophrenia and controls. These changes did not appear confounded by cellular composition or smoking [73]. It is possible that aberrant DNAm in schizophrenia might be the consequence of altered normal developmental trajectories triggered either by dysregulation of methyltransferase activity and/or the involvement of environmental genetic factors affecting DNAm status [40]. It has been shown that changes in DNA methyltransferase activity may be environmentally induced. It was reported that Dnmt3a and 3b are induced 24 h following acute cocaine administration in the nucleus accumbens of mice [74]. Furthermore, in the cortex of mice, the expression of Dnmt1 and DNA methylation is reduced by nicotine and α4β2 nicotinic acetylcholine receptors (nAChR) agonists [67]. As a high percentage of schizophrenia patients smoke cigarettes, this raises another potential source of variability among patients.

8.3.2 HISTONE POSTTRANSLATIONAL MODIFICATIONS

Social defeat stress in rodents causes changes in both histone methylation and acetylation [75]. Acute and chronic stress promotes HPTMs leading to repression or activation of genes related to memory and other processes. Significant and dynamic changes in repressive histone methylation were observed in upstream gene regulatory regions in both chronic social defeat stress and protracted social isolation, with ~20% overlap [76]. Chronic social defeat stress downregulates the histone methyltransferase, G9a and G9a-like protein, which catalyze H3K9 methylation (forming H3K9me2), a major repressive mark in NAc. H3K27me3, another repressive histone mark, is increased upstream of the Rac1 promoter. Rac1 influences characteristic dendritic spine changes in defeated mice [77]. H3K27me3 is also implicated in the ability of chronic stress to suppress Bdnf expression in the hippocampus [75]. In hippocampus, the repressive histone mark H3K9me3 is dramatically induced by restraint stress at repetitive DNA elements, thus influencing genomic instability. Interestingly, a mark of gene activation, H3K4me3, is elevated at the synapsin gene family in PFC of depressed humans [78]. A recent study demonstrated that chronic social defeat stress induces a repressive chromatin remodeling complex in NAc. Induction of this repressive complex at suppressed genes correlates with lower levels of activating histone marks (e.g., H3M4me3 and H4K16ac) and increased levels of certain repressive histone marks (e.g., H3K9me2), thus emphasizing the coordinated nature of epigenetic regulation [79]. However, thus far, genome-wide analysis of HPTMs in the human brain in psychiatric disorders is sparse. This is a high priority for future research. The methylation of H3K4 and H3K27 has been identified as a stable epigenetic mark that appear to be well-preserved in postmortem tissue for a certain period after death, as evidenced by a lack of correlation with tissue pH and autolysis times that were within a range (6–30 h) representative for most of the specimens stored in brain banks [80]. Such stability makes these modifications suitable for studies of psychosis in postmortem brain. However, not all histone modifications appear to be as stable during the postmortem interval, such as H3K9, H3K14, and H4K12 [81].

8.3.3 NONCODING RNAs

It is now abundantly clear that neurobehavioral phenotypes are epigenetically controlled by ncRNAs. Growing evidence indicates that distinct neuronal ncRNA mechanisms, particularly miRNAs, likely influence the development of psychiatric disease. miRNAs are ubiquitously expressed throughout the brain and govern all major neuronal pathways. In view of the vital role of miRNAs in the brain, it is not surprising that there is substantial research supporting the dysregulation of miRNAs in psychiatric disorders [82]. Recent studies have revealed that patients with psychiatric disorders have altered microRNA (miRNA) expression profiles in the circulation and in brain (reviewed in Refs. [22,26,82]). One postmortem study found that schizophrenia is associated with an increase in cortical miRNA expression [83]. More specifically, although the mature and pre-miRNA species are increased (particularly of miR-181b and miR-26b), there were no significant differences in transcription of the source pri-miRNA. In accordance with this phenomenon, the expression of Drosha and DGCR8, essential contributors to pri-miRNA processing, are upregulated, which is probably the ultimate cause of aberrant miRNA levels [83]. In major depressive disorder subjects, a global downregulation of miRNA levels was found in the PFC. Furthermore, the polymorphism DGCR8 rs3757 is associated with increased risk of suicidal tendency and an improved response to antidepressant treatment, whereas AGO1 rs636832 shows decreased risk of suicidal tendency, suicidal behavior, and recurrence of suicide [84]. Thus, polymorphisms in miRNA processing genes may influence both depression risk and treatment. Furthermore, animal studies have shown that manipulating the levels of particular miRNAs in the brain alters behavior [26].

Although miRNAs are arguably the most extensively characterized class of ncRNA in neurons, long ncRNAs (lncRNA) are increasingly implicated in central nervous system functions. Susceptibility genes associated with psychiatric symptoms are also subject to lncRNA modulation. For instance, lncRNA Gomafu is downregulated in cortical tissue of schizophrenia patients and interferes with Disrupted in Schizophrenia 1 (DISC1) splicing, resulting in splice variants linked to schizophrenia, which is in accordance with previous reports that genetic variations in the *DISC1* gene are consistently linked with schizophrenia-associated behaviors [85]. Additionally, *DISC1* is regulated by its lncRNA antisense transcript DISC2, which is also associated with schizophrenia as well as other psychiatric disorders [22]. Also, the lncRNA antisense transcript coded by LOC285758 has been implicated in patients who are violent suicide completers. The increase in this lncRNA is likely triggered by depression or other psychiatric disorders [86].

8.4 INTERVENTION AND PHARMACOLOGY OF EPIGENETICS IN BRAIN AGING AND PSYCHIATRIC DISORDERS

Epigenetics is not only heritable but also reversible. Therefore, strategies aimed at reversing age-associated epigenetic alterations may lead to the development of a novel therapeutic intervention that can delay aging or alleviate symptoms of devastating, age-associated diseases.

8.4.1 INTERVENTION

Caloric restriction (CR), without undernutrition, appears to be a promising strategy to extend the life span and counteract detrimental age-related alterations in a fashion that is evolutionarily conserved from yeast to primates and humans [87], although studies in humans are very limited and with mixed results [88]. Dietary restriction (DR) of caloric intake and enhanced levels of endogenous and exogenous antioxidants are approaches that are potentially able to mitigate age-related deterioration of the brain. The beneficial effects include, in mammals, the attenuation of age-associated cognitive impairment and neurodegeneration [89]. More specifically, synaptic plasticity was shown to be enhanced by DR, as evidenced by increased long-term potentiation [90]. Besides CR, rapamycin is the first drug intervention to reliably increase mammalian lifespan by 10% or more [91]. The link between aging and disease by rapamycin treatment has been carefully discussed [92]. Interestingly, rapamycin treatment suppresses brain aging in senescence-accelerated OXYS rats [93], and also produces an improvement in cognitive functions that normally decline with age in mice [94]. The results from our study in mouse brain unexpectedly demonstrated that both DR and rapamycin could restore, at least partially, the age-related alterations in histone methylation levels [45]. This allows us to suggest a novel and beneficial epigenetic mechanism for age-interventions. Additionally, age and DR or rapamycin exhibit similar effects on the overall level of several HPTMs, such as H3K18ac, H3K4me2, and H3K4me [45]. This implies that those HPTMs may play dual regulatory roles in mediating both age and age-interventions. HPTMs, such as H4K16ac and H3K56ac are stable with age but regulated by DR or rapamycin [45]. The overall alterations of HPTM in brain suggest changes of HPTM related-enzymes by these interventions. In accordance with this, a recent study showed that, independent from genotype, CR prevents the age-related increase of HDAC2 in the hippocampus, particularly in the CA3 and CA1-2 subregions. Furthermore, HDAC2 correlates positively with 5mC while these markers were shown to colocalize in the nucleus of hippocampal cells [51]. Interestingly, in mouse cerebellar Purkinje cells, aging is associated with an increase of 5mC and 5hmC, and these age-related increases are mitigated by CR, and the ratio between 5mC and 5hmC decreases with age and CR treatment, suggesting that CR has a stronger effect on DNAm than DNA hydroxymethylation [95]. Aforementioned, DNAm is implicated in age-related changes in gene expression as well as in cognition and Dnmt3a is essential for memory formation and underlying changes in neuronal and synaptic plasticity. CR indeed attenuates age-related changes in Dnmt3a in mouse hippocampus [96]. These findings enforce the notion that aging is closely connected to marked epigenetic changes, affecting multiple brain regions, and that CR is an effective means to prevent or counteract deleterious age-related epigenetic alterations.

Physical exercise improves the efficiency of the capillary system and increases oxygen supply to the brain, thus enhancing metabolic activity and oxygen intake in neurons, and increases neurotrophin levels and resistance to stress. Regular exercise and an active lifestyle during adulthood have been associated with reduced risk and protective effects for mild cognitive impairment. Recent studies have examined the epigenetic impact of exercise in brain. For example, in a rodent study, epigenetic changes in the hippocampus and cerebral cortex have been correlated with an environmental enrichment that includes voluntary exercise, which increases synaptic integrity and neuroplasticity in the brain, while improving memory, learning and stress response [97]. This study clearly indicates that a lifestyle intervention can improve cognitive functions through epigenetic mechanisms. Another study has revealed that regular physical exercise induces epigenetic modifications at the dentate gyrus, which may regulate gene expression responses involved in neuroplastic and cognitive responses to stressful events. These behavioral responses to exercise were found to correlate with changes in the levels of H3K14ac and H3S10p [98]. Aside from histones, DNAm is significantly increased in the hypothalamus of rats by physical exercise. Physical exercise can also increase global DNAm in the hippocampus, cortex, and hypothalamus and decrease expression of the *Dnmt1* gene in the hippocampus and hypothalamus of rats that undergo repeated restraint stress. These findings indicate that physical exercise affects DNAm of the hypothalamus and might modulate epigenetic responses evoked by repeated restraint stress in the hippocampus, cortex, and hypothalamus [99]. Although the experimental data that link physical exercise or CR to epigenetics is still limited, insight into the epigenetic mechanisms involved in the aging process and their modulation through lifestyle interventions such as

CR and physical exercise might open new avenues for the development of preventive and therapeutic strategies to treat aging-related diseases.

8.4.2 PHARMACOLOGY

A final critical step is to explore the use of epigenetic drugs as potential therapies for brain diseases. Unlike genetic mutations or SNPs that cannot be reversed without gene therapy, epigenetic marks that accumulate in brain during aging and psychiatric disorders are reversible and can be modulated and possibly corrected through classical pharmacology.

Studies using pharmacological tools provide an additional, independent window on the role of histone acetylation in learning and memory. Commonly used HDAC inhibitors (HDACis), including valproic acid (VPA), trichostatin A, sodium and phenyl butyrate, and suberoylanilide hydroxamic acid (SAHA), target multiple HDACs [66]. Despite their broad specificity, HDACis reportedly enhance electrophysiological signatures of hippocampal plasticity (i.e., long-term potentiation), and improve long-term memory when administered prior to learning [19,36]. For brain diseases, HDACis can mimic the effects of antidepressants and alleviate cognitive and neurological defects in animals [100]. Administration of the HDACi sodium butyrate also reportedly rescues age-related memory impairment assessed by a novel object recognition procedure in rats, further supporting the efficacy of HDACi for cognitive decline in normal aging [101]. Remarkably, restoration of H4K12 acetylation levels with administration of the HDACi SAHA broadly reinstates experience-dependent gene transcription and ameliorates the impaired fear memory observed in aged mice [47]. Long-term treatment with VPA, an anticonvulsant mood-stabilizing drug, has been shown to have neuroprotective effects on neurons. VPA may also directly or indirectly induce DNA demethylation through HDAC inhibition [102]. Although still at a preclinical stage, derivatives of these drugs could, in the future, relieve or cure symptoms of complex neuropsychiatric disorders in patients. Before such treatments can be seriously envisaged, however, major progress in basic research is necessary to identify epigenetic targets more precisely to allow for the development of more selective drugs. For instance, an novel antidepressant candidate acetyl-l-carnitine (LAC) corrects mGlu2 deficits by increasing acetylation of histone H3 lysine 27 (H3K27) bound to *Grm2* promoter gene as well as acetylation of the NF-κB p65 subunit and exert fast antidepressant responses [103].

Like the HDACis, a subset of histone methyltransferase inhibitors are in clinical trials for cancer treatment and are likely to be explored in the context of neurological disease in the near future. The small molecule BIX-01294, an interesting candidate, is an inhibitor for the histone H3K9-specific methyltransferases G9a and Glp [104]. When administered directly into the ventral striatum–a key structure in the brain's addiction circuitry, this drug strongly enhances the development of reward behaviors in mice exposed to the stimulant cocaine [105]. The mechanism at least partially involves the inhibition of G9a- and Glp-mediated repressive chromatin remodeling at the promoters of *Bdnf, Cdk5, Arc*, and other genes that function as key regulators for spine density and synaptic connectivity in the mouse brain [105]. Whole forebrain overexpression of Setdb1, a histone methyltransferase that catalyzes H3K9me3, reduces depression-like behavior [106].

Several DNAm inhibitors, including the cytidine analogs 5-azacytidine and zebularine and nucleoside analogs that sequester Dnmt after being incorporated into DNA, are approved or are in preclinical and clinical trials for the treatment of cancer. Interestingly, Dnmt inhibitors become powerful modulators of reward and addiction behaviors and disrupt synaptic plasticity and hippocampal learning and memory when administered directly into the brains of mice and rats [107].

8.5 CONCLUSIONS

As an emerging field, epigenetic investigations into brain aging, as well as, on psychiatric disorders have just started to attract attention from both academics and pharmaceutical industries. Epigenetics has shown promising clues in dissecting the secrets to brain aging and for elucidating the etiology and pathology of psychiatric disorders. More intriguingly, interventions and drugs targeting epigenetic factors, DNAm, HPTM, and ncRNA, are not only providing possible cures for brain symptoms, but also assisting a mechanistic understanding of the etiology and pathophysiology of brain aging and psychiatric diseases. Hopefully, more investigations on the epigenetic hypothesis of brain aging and psychiatric disorders will be carried out. These studies could provide new insights to prevent and/or treat age-related and/or psychiatric disease-associated deficits in cognition and behavior in the near future.

ABBREVIATIONS

CR Caloric restriction
DISC Disrupted in schizophrenia
DNAhm DNA hydroxymethylation
DNAm DNA methylation
Dnmt DNA methyltransferase
DR Dietary restriction
5hmC 5′-Hydroxymethylcytosine
HDAC Histone deacetylase
HDACi Histone deacetylase inhibitor
HPTM Histone posttranslational modification
K Lysine
5mC 5′-Methylcytosine
MeCP Methyl CpG binding protein
NAc Nucleus accumbens
NATs Natural antisense transcripts
ncRNA Noncoding RNA
PFC Prefrontal cortex
R Arginine
RISC RNA-induced silencing complex
SAHA Suberoylanilide hydroxamic acid
TET Ten-eleven translocation
VPA Valproic acid

GLOSSARY

Rapamycin Also called sirolimus, it is an immunosuppressant produced by the bacterium Streptomyces hygroscopicus

Rhesus macaque An old world monkey also called the Rhesus monkey. Its scientific name is *Macaca mulatta*

REFERENCES

[1] Van Speybroeck L. From epigenesis to epigenetics: the case of C. H. Waddington. Ann NY Acad Sci 2002;981:61–81.

[2] Weber M, Hellmann I, Stadler MB, Ramos L, Paabo S, Rebhan M, et al. Distribution, silencing potential and evolutionary impact of promoter DNA methylation in the human genome. Nat Genet 2007;39:457–66.

[3] Xu X. DNA methylation and cognitive aging. Oncotarget 2015;6:13922–32.

[4] Qian H, Xu X. Reduction in DNA methyltransferases and alteration of DNA methylation pattern associate with mouse skin ageing. Exp Dermatol 2014;23:357–9.

[5] Moore LD, Le T, Fan G. DNA methylation and its basic function. Neuropsychopharmacology 2013;38:23–38.

[6] Robison AJ, Nestler EJ. Transcriptional and epigenetic mechanisms of addiction. Nat Rev Neurosci 2011;12:623–37.

[7] Griffith JS, Mahler HR. DNA ticketing theory of memory. Nature 1969;223:580–2.

[8] Chahrour M, Zoghbi HY. The story of Rett syndrome: from clinic to neurobiology. Neuron 2007;56:422–37.

[9] Xie W, Barr CL, Kim A, Yue F, Lee AY, Eubanks J, et al. Base-resolution analyses of sequence and parent-of-origin dependent DNA methylation in the mouse genome. Cell 2012;148:816–31.

[10] Ito S, Shen L, Dai Q, Wu SC, Collins LB, Swenberg JA, et al. Tet proteins can convert 5-methylcytosine to 5-formylcytosine and 5-carboxylcytosine. Science 2011;333:1300–3.

[11] He YF, Li BZ, Li Z, Liu P, Wang Y, Tang Q, et al. Tet-mediated formation of 5-carboxylcytosine and its excision by TDG in mammalian DNA. Science 2011;333:1303–7.

[12] Tahiliani M, Koh KP, Shen Y, Pastor WA, Bandukwala H, Brudno Y, et al. Conversion of 5-methylcytosine to 5-hydroxymethylcytosine in mammalian DNA by MLL partner TET1. Science 2009;324:930–5.

[13] Ito S, D'Alessio AC, Taranova OV, Hong K, Sowers LC, Zhang Y. Role of Tet proteins in 5mC to 5hmC conversion, ES-cell self-renewal and inner cell mass specification. Nature 2010;466:1129–33.

[14] Keverne EB, Pfaff DW, Tabansky I. Epigenetic changes in the developing brain: effects on behavior. Proc Natl Acad Sci USA 2015;112:6789–95.

[15] Kriaucionis S, Heintz N. The nuclear DNA base 5-hydroxymethylcytosine is present in Purkinje neurons and the brain. Science 2009;324:929–30.

[16] Valinluck V, Tsai HH, Rogstad DK, Burdzy A, Bird A, Sowers LC. Oxidative damage to methyl-CpG sequences inhibits the binding of the methyl-CpG binding domain (MBD) of methyl-CpG binding protein 2 (MeCP2). Nucleic Acids Res 2004;32:4100–8.

[17] Szulwach KE, Li X, Li Y, Song CX, Wu H, Dai Q, et al. 5-hmC-mediated epigenetic dynamics during postnatal neurodevelopment and aging. Nat Neurosci 2011;14:1607–16.

[18] Kouzarides T. Chromatin modifications and their function. Cell 2007;128:693–705.

[19] Bohacek J, Gapp K, Saab BJ, Mansuy IM. Transgenerational epigenetic effects on brain functions. Biol Psychiatry 2013;73:313–20.

[20] Lardenoije R, Iatrou A, Kenis G, Kompotis K, Steinbusch HW, Mastroeni D, et al. The epigenetics of aging and neurodegeneration. Prog Neurobiol 2015;131:21–64.

[21] Szafranski K, Abraham KJ, Mekhail K. Non-coding RNA in neural function, disease, and aging. Front Genet 2015;6:87.

[22] Kocerha J, Dwivedi Y, Brennand KJ. Noncoding RNAs and neurobehavioral mechanisms in psychiatric disease. Mol Psychiatry 2015;20:677–84.

[23] Hammond SM. An overview of microRNAs. Adv Drug Deliv Rev 2015;87:3–14.

[24] Bartel DP. MicroRNAs: genomics, biogenesis, mechanism, and function. Cell 2004;116:281–97.

[25] Griffiths-Jones S, Grocock RJ, van Dongen S, Bateman A, Enright AJ. miRBase: microRNA sequences, targets and gene nomenclature. Nucleic Acids Res 2006;34:D140–144.

[26] Issler O, Chen A. Determining the role of microRNAs in psychiatric disorders. Nat Rev Neurosci 2015;16:201–12.

[27] Gong H, Liu CM, Liu DP, Liang CC. The role of small RNAs in human diseases: potential troublemaker and therapeutic tools. Med Res Rev 2005;25:361–81.

[28] Beveridge NJ, Cairns MJ. MicroRNA dysregulation in schizophrenia. Neurobiol Dis 2012;46:263–71.

[29] Xu B, Karayiorgou M, Gogos JA. MicroRNAs in psychiatric and neurodevelopmental disorders. Brain Res 2010;1338:78–88.

[30] Briggs JA, Wolvetang EJ, Mattick JS, Rinn JL, Barry G. Mechanisms of long non-coding RNAs in mammalian nervous system development, plasticity, disease, and evolution. Neuron 2015;88:861–77.

[31] Derrien T, Johnson R, Bussotti G, Tanzer A, Djebali S, Tilgner H, et al. The GENCODE v7 catalog of human long noncoding RNAs: analysis of their gene structure, evolution, and expression. Genome Res 2012;22:1775–89.

[32] Schaukowitch K, Kim TK. Emerging epigenetic mechanisms of long non-coding RNAs. Neuroscience 2014;264:25–38.

[33] Holoch D, Moazed D. RNA-mediated epigenetic regulation of gene expression. Nat Rev Genet 2015;16:71–84.

[34] Vecsey CG, Hawk JD, Lattal KM, Stein JM, Fabian SA, Attner MA, et al. Histone deacetylase inhibitors enhance memory and synaptic plasticity via CREB: CBP-dependent transcriptional activation. J Neurosci 2007;27:6128–40.

[35] Day JJ, Sweatt JD. Epigenetic mechanisms in cognition. Neuron 2011;70:813–29.

[36] Spiegel AM, Sewal AS, Rapp PR. Epigenetic contributions to cognitive aging: disentangling mindspan and lifespan. Learn Mem 2014;21:569–74.

[37] Fandakova Y, Lindenberger U, Shing YL. Maintenance of youth-like processing protects against false memory in later adulthood. Neurobiol Aging 2015;36:933–41.

[38] Xu X, Zhan M, Duan W, Prabhu V, Brenneman R, Wood W, et al. Gene expression atlas of the mouse central nervous system: impact and interactions of age, energy intake and gender. Genome Biol 2007;8:R234.

[39] Xu X. Single cell transcriptome study in brain aging. Single Cell Biol 2012;1:e111.

[40] Numata S, Ye T, Hyde TM, Guitart-Navarro X, Tao R, Wininger M, et al. DNA methylation signatures in development and aging of the human prefrontal cortex. Am J Hum Genet 2012;90:260–72.

[41] Penner MR, Roth TL, Chawla MK, Hoang LT, Roth ED, Lubin FD, et al. Age-related changes in Arc transcription and DNA methylation within the hippocampus. Neurobiol Aging 2011;32:2198–210.

[42] Haberman RP, Quigley CK, Gallagher M. Characterization of CpG island DNA methylation of impairment-related genes in a rat model of cognitive aging. Epigenetics 2012;7:1008–19.

[43] Oliveira AM, Hemstedt TJ, Bading H. Rescue of aging-associated decline in Dnmt3a2 expression restores cognitive abilities. Nat Neurosci 2012;15:1111–3.

[44] Chen H, Dzitoyeva S, Manev H. Effect of aging on 5-hydroxymethylcytosine in the mouse hippocampus. Restor Neurol Neurosci 2012;30:237–45.

[45] Gong H, Qian H, Ertl R, Astle CM, Wang GG, Harrison DE, et al. Histone modifications change with age, dietary restriction and rapamycin treatment in mouse brain. Oncotarget 2015;6:15882–90.

[46] Wang CM, Tsai SN, Yew TW, Kwan YW, Ngai SM. Identification of histone methylation multiplicities patterns in the brain of senescence-accelerated prone mouse 8. Biogerontology 2010;11:87–102.

[47] Peleg S, Sananbenesi F, Zovoilis A, Burkhardt S, Bahari-Javan S, Agis-Balboa RC, et al. Altered histone acetylation is associated with age-dependent memory impairment in mice. Science 2010;328:753–6.

[48] Francis YI, Fa M, Ashraf H, Zhang H, Staniszewski A, Latchman DS, et al. Dysregulation of histone acetylation in the APP/PS1 mouse model of Alzheimer's disease. J Alzheimers Dis 2009;18:131–9.

[49] Strahl BD, Allis CD. The language of covalent histone modifications. Nature 2000;403:41–5.

[50] Greer EL, Shi Y. Histone methylation: a dynamic mark in health, disease and inheritance. Nat Rev Genet 2012;13:343–57.

[51] Chouliaras L, van den Hove DL, Kenis G, Draanen M, Hof PR, van Os J, et al. Histone deacetylase 2 in the mouse hippocampus: attenuation of age-related increase by caloric restriction. Curr Alzheimer Res 2013;10:868–76.

[52] Gong H, Pang J, Han Y, Dai Y, Dai D, Cai J, et al. Age-dependent tissue expression patterns of Sirt1 in senescence-accelerated mice. Mol Med Rep 2014;10:3296–302.
[53] Quintas A, de Solis AJ, Diez-Guerra FJ, Carrascosa JM, Bogonez E. Age-associated decrease of SIRT1 expression in rat hippocampus: prevention by late onset caloric restriction. Exp Gerontol 2012;47:198–201.
[54] Gupta S, Kim SY, Artis S, Molfese DL, Schumacher A, Sweatt JD, et al. Histone methylation regulates memory formation. J Neurosci 2010;30:3589–99.
[55] Inukai S, de Lencastre A, Turner M, Slack F. Novel microRNAs differentially expressed during aging in the mouse brain. PLoS One 2012;7:e40028.
[56] Li N, Bates DJ, An J, Terry DA, Wang E. Up-regulation of key microRNAs, and inverse down-regulation of their predicted oxidative phosphorylation target genes, during aging in mouse brain. Neurobiol Aging 2011;32:944–55.
[57] Wei YN, Hu HY, Xie GC, Fu N, Ning ZB, Zeng R, et al. Transcript and protein expression decoupling reveals RNA binding proteins and miRNAs as potential modulators of human aging. Genome Biol 2015;16:41.
[58] Persengiev S, Kondova I, Otting N, Koeppen AH, Bontrop RE. Genome-wide analysis of miRNA expression reveals a potential role for miR-144 in brain aging and spinocerebellar ataxia pathogenesis. Neurobiol Aging 2011;32(2316):e2317–2327.
[59] Li X, Khanna A, Li N, Wang E. Circulatory miR34a as an RNAbased, noninvasive biomarker for brain aging. Aging 2011;3:985–1002.
[60] Kour S, Rath PC. All-trans retinoic acid induces expression of a novel intergenic long noncoding RNA in adult rat primary hippocampal neurons. J Mol Neurosci 2015;.
[61] Kour S, Rath PC. Age-dependent differential expression profile of a novel intergenic long noncoding RNA in rat brain. Int J Dev Neurosci 2015;47:286–97.
[62] Kour S, Rath PC. Age-related expression of a repeat-rich intergenic long noncoding RNA in the rat brain. Mol Neurobiol 2016;.
[63] Barry G, Guennewig B, Fung S, Kaczorowski D, Weickert CS, Long Non-Coding RNA. Expression during aging in the human subependymal zone. Front Neurol 2015;6:45.
[64] Pena CJ, Bagot RC, Labonte B, Nestler EJ. Epigenetic signaling in psychiatric disorders. J Mol Biol 2014;426:3389–412.
[65] McGowan PO, Sasaki A, D'Alessio AC, Dymov S, Labonte B, Szyf M, et al. Epigenetic regulation of the glucocorticoid receptor in human brain associates with childhood abuse. Nat Neurosci 2009;12:342–8.
[66] Narayan P, Dragunow M. Pharmacology of epigenetics in brain disorders. Br J Pharmacol 2010;159:285–303.
[67] Grayson DR, Guidotti A. The dynamics of DNA methylation in schizophrenia and related psychiatric disorders. Neuropsychopharmacology 2013;38:138–66.
[68] Xiao Y, Camarillo C, Ping Y, Arana TB, Zhao H, Thompson PM, et al. The DNA methylome and transcriptome of different brain regions in schizophrenia and bipolar disorder. PLoS One 2014;9:e95875.
[69] Dempster EL, Pidsley R, Schalkwyk LC, Owens S, Georgiades A, Kane F, et al. Disease-associated epigenetic changes in monozygotic twins discordant for schizophrenia and bipolar disorder. Hum Mol Genet 2011;20:4786–96.
[70] Torrey EF, Barci BM, Webster MJ, Bartko JJ, Meador-Woodruff JH, Knable MB. Neurochemical markers for schizophrenia, bipolar disorder, and major depression in postmortem brains. Biol Psychiatry 2005;57:252–60.
[71] Grayson DR, Jia X, Chen Y, Sharma RP, Mitchell CP, Guidotti A, et al. Reelin promoter hypermethylation in schizophrenia. Proc Natl Acad Sci USA 2005;102:9341–6.
[72] Veldic M, Guidotti A, Maloku E, Davis JM, Costa E. In psychosis, cortical interneurons overexpress DNA-methyltransferase 1. Proc Natl Acad Sci USA 2005;102:2152–7.
[73] Jaffe AE, Gao Y, Deep-Soboslay A, Tao R, Hyde TM, Weinberger DR, et al. Mapping DNA methylation across development, genotype and schizophrenia in the human frontal cortex. Nat Neurosci 2016;19:40–7.
[74] Anier K, Malinovskaja K, Aonurm-Helm A, Zharkovsky A, Kalda A. DNA methylation regulates cocaine-induced behavioral sensitization in mice. Neuropsychopharmacology 2010;35:2450–61.

[75] Tsankova NM, Berton O, Renthal W, Kumar A, Neve RL, Nestler EJ. Sustained hippocampal chromatin regulation in a mouse model of depression and antidepressant action. Nat Neurosci 2006;9:519–25.

[76] Wilkinson MB, Xiao G, Kumar A, LaPlant Q, Renthal W, Sikder D, et al. Imipramine treatment and resiliency exhibit similar chromatin regulation in the mouse nucleus accumbens in depression models. J Neurosci 2009;29:7820–32.

[77] Golden SA, Christoffel DJ, Heshmati M, Hodes GE, Magida J, Davis K, et al. Epigenetic regulation of RAC1 induces synaptic remodeling in stress disorders and depression. Nat Med 2013;19:337–44.

[78] Cruceanu C, Alda M, Nagy C, Freemantle E, Rouleau GA, Turecki G. H3K4 tri-methylation in synapsin genes leads to different expression patterns in bipolar disorder and major depression. Int J Neuropsychopharmacol 2013;16:289–99.

[79] Sun H, Damez-Werno DM, Scobie KN, Shao NY, Dias C, Rabkin J, et al. ACF chromatin-remodeling complex mediates stress-induced depressive-like behavior. Nat Med 2015;21:1146–53.

[80] Huang HS, Matevossian A, Jiang Y, Akbarian S. Chromatin immunoprecipitation in postmortem brain. J Neurosci Methods 2006;156:284–92.

[81] Akbarian S, Huang HS. Epigenetic regulation in human brain-focus on histone lysine methylation. Biol Psychiatry 2009;65:198–203.

[82] Geaghan M, Cairns MJ. MicroRNA and posttranscriptional dysregulation in psychiatry. Biol Psychiatry 2015;78:231–9.

[83] Beveridge NJ, Gardiner E, Carroll AP, Tooney PA, Cairns MJ. Schizophrenia is associated with an increase in cortical microRNA biogenesis. Mol Psychiatry 2010;15:1176–89.

[84] He Y, Zhou Y, Xi Q, Cui H, Luo T, Song H, et al. Genetic variations in microRNA processing genes are associated with susceptibility in depression. DNA Cell Biol 2012;31:1499–506.

[85] Barry G, Briggs JA, Vanichkina DP, Poth EM, Beveridge NJ, Ratnu VS, et al. The long non-coding RNA Gomafu is acutely regulated in response to neuronal activation and involved in schizophrenia-associated alternative splicing. Mol Psychiatry 2014;19:486–94.

[86] Punzi G, Ursini G, Shin JH, Kleinman JE, Hyde TM, Weinberger DR. Increased expression of MARCKS in post-mortem brain of violent suicide completers is related to transcription of a long, noncoding, antisense RNA. Mol Psychiatry 2014;19:1057–9.

[87] Fontana L, Partridge L, Longo VD. Extending healthy life span—from yeast to humans. Science 2010;328:321–6.

[88] Mattison JA, Roth GS, Beasley TM, Tilmont EM, Handy AM, Herbert RL, et al. Impact of caloric restriction on health and survival in rhesus monkeys from the NIA study. Nature 2012;489:318–21.

[89] Maalouf M, Rho JM, Mattson MP. The neuroprotective properties of calorie restriction, the ketogenic diet, and ketone bodies. Brain Res Rev 2009;59:293–315.

[90] Hori N, Hirotsu I, Davis P, Carpenter D. Long-term potentiation is lost in aged rats but preserved by calorie restriction. Neuroreport 1992;3:1085–8.

[91] Harrison DE, Strong R, Sharp ZD, Nelson JF, Astle CM, Flurkey K, et al. Rapamycin fed late in life extends lifespan in genetically heterogeneous mice. Nature 2009;460:392–5.

[92] Blagosklonny MV. Rapamycin extends life- and health span because it slows aging. Aging 2013;5:592–8.

[93] Kolosova NG, Vitovtov AO, Muraleva NA, Akulov AE, Stefanova NA, Blagosklonny MV. Rapamycin suppresses brain aging in senescence-accelerated OXYS rats. Aging 2013;5:474–84.

[94] Majumder S, Caccamo A, Medina DX, Benavides AD, Javors MA, Kraig E, et al. Lifelong rapamycin administration ameliorates age-dependent cognitive deficits by reducing IL-1beta and enhancing NMDA signaling. Aging Cell 2012;11:326–35.

[95] Lardenoije R, van den Hove DL, Vaessen TS, Iatrou A, Meuwissen KP, van Hagen BT, et al. Epigenetic modifications in mouse cerebellar Purkinje cells: effects of aging, caloric restriction, and overexpression of superoxide dismutase 1 on 5-methylcytosine and 5-hydroxymethylcytosine. Neurobiol Aging 2015;36:3079–89.

[96] Chouliaras L, van den Hove DL, Kenis G, Dela Cruz J, Lemmens MA, van Os J, et al. Caloric restriction attenuates age-related changes of DNA methyltransferase 3a in mouse hippocampus. Brain Behav Immun 2011;25:616–23.
[97] Fischer A, Sananbenesi F, Wang X, Dobbin M, Tsai LH. Recovery of learning and memory is associated with chromatin remodelling. Nature 2007;447:178–82.
[98] Chandramohan Y, Droste SK, Arthur JS, Reul JM. The forced swimming-induced behavioural immobility response involves histone H3 phospho-acetylation and c-Fos induction in dentate gyrus granule neurons via activation of the *N*-methyl-D-aspartate/extracellular signal-regulated kinase/mitogen- and stress-activated kinase signalling pathway. Eur J Neurosci 2008;27:2701–13.
[99] Kashimoto RK, Toffoli LV, Manfredo MH, Volpini VL, Martins-Pinge MC, Pelosi GG, et al. Physical exercise affects the epigenetic programming of rat brain and modulates the adaptive response evoked by repeated restraint stress. Behav Brain Res 2016;296:286–9.
[100] Covington HE III, Maze I, LaPlant QC, Vialou VF, Ohnishi YN, Berton O, et al. Antidepressant actions of histone deacetylase inhibitors. J Neurosci 2009;29:11451–60.
[101] Reolon GK, Maurmann N, Werenicz A, Garcia VA, Schroder N, Wood MA, et al. Posttraining systemic administration of the histone deacetylase inhibitor sodium butyrate ameliorates aging-related memory decline in rats. Behav Brain Res 2011;221:329–32.
[102] Gottlicher M, Minucci S, Zhu P, Kramer OH, Schimpf A, Giavara S, et al. Valproic acid defines a novel class of HDAC inhibitors inducing differentiation of transformed cells. EMBO J 2001;20:6969–78.
[103] Nasca C, Xenos D, Barone Y, Caruso A, Scaccianoce S, Matrisciano F, et al. L-acetylcarnitine causes rapid antidepressant effects through the epigenetic induction of mGlu2 receptors. Proc Natl Acad Sci USA 2013;110:4804–9.
[104] Kubicek S, O'Sullivan RJ, August EM, Hickey ER, Zhang Q, Teodoro ML, et al. Reversal of H3K9me2 by a small-molecule inhibitor for the G9a histone methyltransferase. Mol Cell 2007;25:473–81.
[105] Maze I, Covington HE III, Dietz DM, LaPlant Q, Renthal W, Russo SJ, et al. Essential role of the histone methyltransferase G9a in cocaine-induced plasticity. Science 2010;327:213–6.
[106] Jiang Y, Matevossian A, Huang HS, Straubhaar J, Akbarian S. Isolation of neuronal chromatin from brain tissue. BMC Neurosci 2008;9:42.
[107] Jakovcevski M, Akbarian S. Epigenetic mechanisms in neurological disease. Nat Med 2012;18:1194–204.

CHAPTER

EPIGENETICS AND DOWN SYNDROME

A.D. Dekker*,, P.P. De Deyn*,**, M.G. Rots†**

**Department of Neurology and Alzheimer Research Center, University Medical Center Groningen, University of Groningen, Groningen, The Netherlands; **Laboratory of Neurochemistry and Behavior, Institute Born-Bunge, University of Antwerp, Wilrijk, Belgium; †Department of Pathology and Medical Biology, University Medical Center Groningen, University of Groningen, Groningen, The Netherlands*

CHAPTER OUTLINE

9.1 INTRODUCTION: EPIGENETICS HAS BEEN LARGELY NEGLECTED IN DOWN SYNDROME

With an incidence of approximately 1 in 650–1000 live births, Down syndrome (DS) is the most common genetic cause of intellectual disability [1]. In 1866, the British physician John Langdon Down described various recurring symptoms of the "Mongolian type of idiocy" that he observed among more than 10% of the children that he treated for cognitive impairment [2]. The cause of what became known as DS remained unclear for almost a century until the late 1950s when Lejeune et al. discovered its origin: trisomy 21 [3]. Over 95% of DS cases is a whole-chromosome trisomy due to meiotic nondisjunction, that is, a failed separation of one of the paired chromosomes [4,5]. The three copies of the human chromosome 21 (HSA21) lead to various complications, such as the characteristic facial appearance and the intellectual disability associated with impaired linguistic skills and diminished learning and memory capacities [6].

Neuropsychiatric Disorders and Epigenetics. http://dx.doi.org/10.1016/B978-0-12-800226-1.00009-5

In addition to the congenital intellectual disability, individuals with DS face accelerated aging, including early-onset dementia due to Alzheimer's disease (AD). By the time they reach 60–70 years of age, 50–70% of the DS population has developed AD compared to 11% of those aged 65+ in the general population [7,8]. This strongly increased risk for AD in DS has been predominantly attributed to the triplication of the HSA21-encoded amyloid precursor protein (APP) gene. Consequently, this yields higher levels of APP protein and its secretase-splicing product amyloid-β (Aβ), the main constituent of the characteristic extraneuronal amyloid plaques in AD [9]. Despite the fact that 95% of DS cases is due to a full trisomy 21, the DS population is characterized by an enormous variability in the type and the severity of clinical features [10]. This phenotypical variability is strikingly illustrated by the observation that the onset of clinical dementia symptoms in DS differs tremendously. Remarkably, 30–50% of the DS individuals do not develop dementia symptomatology, despite the full-blown AD-like neuropathology that is present in practically all DS individuals aged 40 years and older [7,11,12].

The complete DNA sequence of HSA21 was elucidated in 2000 [13]. Since then, many researchers have investigated the overexpressed protein-encoding genes and their effects on learning and memory. Despite increased understanding of the possible underlying genetic mechanisms, explaining the aforementioned variability among the DS population remains a scientific challenge [14,15]. Although the triplication of HSA21 would theoretically lead to a 1.5-fold increase in gene transcription, gene expression studies suggested otherwise. For instance, analysis of HSA21 gene expression in DS lymphoblastoid cells showed that only 22% of the analyzed genes had expression levels closely matching this level, compared to control individuals. In particular, 7% had an amplified expression (significantly higher than 1.5), 56% had an expression level that was significantly lower than 1.5, and 15% of the genes had highly variable expression profiles between subjects [16].

Similar results were obtained using the most widely used Ts65Dn mouse model of DS. Ts65Dn mice carry an additional chromosome, consisting of a duplicated part of the mouse chromosome 16 that is translocated to a small segment of the mouse chromosome 17 [17]. As a consequence, Ts65Dn mice are trisomic for about 50% of the genes on HSA21 [18]. However, it was demonstrated that many of these genes have transcript levels that significantly deviate from the theoretical 1.5-fold increase [4,19,20]. For instance, Lyle et al. reported that not more than 37% of the genes in Ts65Dn matched the theoretical expression level of 1.5 [20]. Accordingly, certain genes are more dosage sensitive than others, thereby contributing in varying extents to the DS phenotypes [4]. Although various studies have tried to identify the crucial phenotype-determining genes, the underlying cause of the gene expression variation has been largely neglected.

Conceivably, epigenetic mechanisms play a role in gene expression regulation and as such might play a crucial role in the development of cognitive deficits in DS. Epigenetic mechanisms, including DNA methylation, posttranslational histone modifications and histone core variants, regulate gene expression without affecting the DNA itself. Importantly, epigenetic marks are reversible and thus offer a huge therapeutic potential to alleviate or cure certain genetic deficits.

As described in previous chapters, an increasing body of evidence illustrates the role of epigenetic mechanisms in synaptic plasticity, learning and memory, and intellectual disabilities. Surprisingly, epigenetic mechanisms have been hardly investigated in DS. Most DS studies have focused on genomic aspects, neglecting the mounting evidence that demonstrates the contribution of epigenetics to impaired learning and memory. Importantly, epigenetic therapy is already in use for cancer, which may provide novel possibilities for cognition-enhancing treatment in DS as well. To our knowledge, no studies so far have investigated epigenetic therapy in mouse models of DS. Classical pharmacological treatment has not been successful yet in diminishing cognitive deficits in DS [21]. Epigenetic therapy offers potentially important new avenues. This chapter summarizes and evaluates the limited knowledge on epigenetics in the neurobiology of DS, and discusses the huge potential of epigenetic therapy to reverse dysregulated gene expression.

9.2 EPIGENETIC MECHANISMS AFFECT LEARNING AND MEMORY

In the human genome, cytosines preceding a guanine (CpG) are frequently methylated into 5-methylcytosine using the methyl group of S-adenosylmethionine (SAM) as donor [22]. DNA methylation, generally associated with the formation of heterochromatin and repressed gene expression, is involved in the process of memory formation: increased DNA methylation of memory suppressor genes and diminished DNA methylation of memory promoting genes [22,23]. The presence of high levels of DNA methyltransferases (DNMTs), as well as methyl-CpG-binding proteins in neurons suggests a role for DNA methylation in neuronal functioning [24]. Indeed, the expression patterns of DNMTs change depending on the stage of neurodevelopment. Furthermore, Rett syndrome, a progressive neurodevelopmental disorder in females that results in intellectual disability, is caused by a single mutation in the methyl-CpG-binding protein 2 (MECP2) that recognizes methylated DNA [25]. Moreover, neuronal plasticity was found to depend on neuronal activity-induced hydroxymethylation of certain critical genes [26]. Hydroxymethylation, an early intermediate of DNA demethylation, is mediated by the ten-eleven translocation (TET) protein family [27]. In fact, a recent study showed that *Tet1* knock-out mice presented deficits in hippocampal neurogenesis and impaired learning and memory [28].

Next to DNA methylation, posttranslational histone modifications also relate to learning and memory. Whether a certain (combination of) histone mark(s) is associated with stimulated or repressed gene transcription depends on the type and the position of a modification and the presence of particular effector proteins [29]. Histone-modifying enzymes establish (writers) or remove (erasers) particular histone marks, for example, acetylation (generally associated with gene expression) is increased by histone acetyltransferases (HATs) and reduced by histone deacetylases (HDACs). Acetylation has been strongly associated with promoting synaptic plasticity and memory formation, while histone deacetylation was associated with memory deficits [30]. Deregulated acetylation of H4K12 has been related to memory impairment in aged mice (16 months), which was overcome by administration of HDAC inhibitors [31]. In agreement, HATs are crucial for memory formation, as is demonstrated by the intellectual disability in Rubinstein–Taybi syndrome that is caused by a loss of function mutation in the CBP/P300 HAT [32]. Furthermore, histone lysine methylation is associated with learning-dependent synaptic plasticity and hippocampus-dependent long-term memory formation [33].

Due to the genetic base pair mutations in epigenetic factors, Rett syndrome and Rubinstein–Taybi syndrome are classified as chromatin diseases [34]. Such mutations have not been documented in DS, but DS may be regarded, in part, as a chromatin disease as well. A growing body of evidence has, indeed, demonstrated that the triplication of HSA21, via the subsequent overexpression of various genes, directly dysregulates cellular epigenetic mechanisms in DS (Table 9.1). In turn, these disrupted epigenetic processes are associated with altered gene expression profiles and thus provide obvious candidates that might contribute to cognitive deficits in DS.

Despite the demonstrated involvement of epigenetics in learning and memory processes, only a few epigenetic studies have been conducted in DS. To the extent that it is known, the subsequent sections

Table 9.1 Aberrant Epigenetic Mechanisms due to Overexpressed HSA21-Linked Proteins in DS

HSA21 Product	Class of Gene Expression[a]	Primary Function	Downstream Epigenetic Effector	Epigenetic Consequence	References
DNMT3L	Unknown	DNA methyltransferase	DNMT3A, DNMT3B	DNA methylation and histone deacetylation	[36,37]
CBS	Class I	Homocysteine conversion	SAM depletion	DNA methylation and histone methylation	[38,39]
DYRK1A	Class I	Kinase	SIRT1 (HDAC)	Histone deacetylation	[40]
			CREB and CBP/P300 (HAT)	Histone acetylation	[32,41]
			SWI/SNF complex	Histone modifications	[42]
BRWD1	Class IV	Transcription regulator	SWI/SNF complex	Histone modifications	[43]
RUNX1	Class III	Transcription factor	SWI/SNF complex	Histone modifications	[44]
ETS2	Class III	Transcription factor	CBP/P300 (HAT)	Histone acetylation	[45]
H2AFZP	Unknown	Histone variant	Unknown	Unknown	[24,46]
H2BFS	Class III	Histone variant	Unknown	Unknown	[47,48]
CHAF1B	Class III	Constitutive chromatin protein	Multiprotein complex with MBD1 and HP1	Methylation-mediated transcriptional repression	[49]
HMGN1	Class I	Constitutive chromatin protein	CBP/P300 (HAT)	Histone acetylation	[50]
			MECP2	Activated or repressed gene transcription	[51]

[a]*Gene expression classification based on Ref. [16]. Due to the triplication of HSA21 in DS, an increased expression level of 1.5-fold is expected in DS compared to non-DS controls. Four classes of genes were reported with expression levels that were around 1.5 (class I), significantly higher than 1.5 (class II), significantly lower than 1.5 (class III), or highly variable (class IV).*
Adapted from Dekker AD, De Deyn PP, Rots MG. Epigenetics: the neglected key to minimize learning and memory deficits in Down syndrome. Neurosci Biobehav Rev 2014;45C:72–84 [35], with permission of Elsevier.

discuss the contribution of each of the three epigenetic mechanisms to cognitive deficits in DS, particularly focusing on epigenetic alterations due to any overexpressed HSA21 gene product (Table 9.1).

9.3 ALTERED DNA METHYLATION IS ASSOCIATED WITH DS

DNA methylation is conducted by four enzymes with different functions: DNMT1 maintains the methylation marks after DNA replication, DNMT3A and 3B are mainly responsible for de novo DNA methylation [52,53], and DNMT3L has no methyltransferase activity, but mediates transcriptional gene repression [36] and stimulates the methylation activity of DNMT3A and 3B by direct binding [37,54]. DNMT3L is especially interesting in the context of DS, since it is encoded on HSA21 [47].

Previously, it was shown that female *Dnmt3l* knock-out mice presented specific hypomethylation of maternally imprinted genes, suggesting that DNMT3L together with DNMT3A/3B mediates de novo DNA methylation of these maternally imprinted genes [55–57]. Opposed to a such a knock-out, overexpression of DNMT3L in DS likely affects DNA methylation patterns as well, potentially contributing to the cognitive deficits in DS.

Although DNA methylation patterns in DS have not been extensively investigated, studies have indicated that DNA methylation is different in DS individuals compared to the general population. To our knowledge, the first report on the presence of differential DNA methylation in DS was published in 2001, demonstrating increased genome-wide hypermethylation of lymphocyte DNA in DS children with a full trisomy 21, compared to their euploid siblings [58]. In agreement, Chango et al. identified six DNA fragments that were hypermethylated in eight DS subjects compared to eight healthy controls. However, the applied technology did not allow for determination of the specific DNA sequence [59].

Then, in 2010, Kerkel et al. performed a high throughput screen for differentially methylated genes in DS using DNA that was extracted from total peripheral white blood cells and isolated T-lymphocytes [60]. Compared to non-DS controls, a range of stable, gene-specific alterations in CpG methylation patterns was observed, which was independent of the differential cell counts. Strikingly, these genes were found on autosomes other than HSA21, indicating the influence of an additional copy of HSA21 on the epigenetic marks on other chromosomes. Many of the differentially methylated genes are involved in the development and functioning of white blood cells, which are supportive of the fact that DS is characterized by immune system deficiencies, among others, resulting in the high frequency of infections [61].

Recently, Bacalini et al. confirmed most of these differentially methylated regions (DMRs) in another DS cohort. They studied DNA methylation profiles of peripheral white blood cells obtained from 29 DS subjects, as well as from their mothers and non-DS siblings to reduce the effect of genetic and environmental confounding factors. Correction for differential cell counts between the three study groups was implemented in the statistical analysis. Seven of the DMRs described by Kerkel et al. were also included, six of these probes were differentially methylated in the new DS cohort as well. Kerkel et al. reported DMRs on autosomes other than HSA21. Bacalini et al. also found a genome-wide distribution of DMRs, but the DMRs were especially enriched on HSA21. The reported DMRs primarily related to four functions: embryonic development, neuronal development, hematopoiesis (including the runt-related transcription factor 1 gene (*RUNX1*), discussed latter), and chromatin modulation (including *TET1,* and *KDM2B,* encoding a histone lysine demethylase) [62]. Once more, this illustrates the role of DNA methylation in (neuro)development, possibly contributing to the intellectual disability in DS.

A first association between differential DNA methylation and a measure of cognitive functioning in DS was established by Jones et al. in 2013. They examined DNA obtained from cheek swabs of ten adult DS individuals and ten age-matched, healthy controls. Three thousand and three hundred CpGs were reported with DNA methylation levels that differed more than 10% between both groups. In accordance with Kerkel et al. [60] but in contrast to Bacalini et al. [62] no enrichment on HSA21 was observed. Cognitive function was briefly assessed using the Dalton Brief Praxis test and subsequently correlated with the DNA methylation results. Five differentially methylated probes correlated with cognitive functioning, indicating the relation between cognitive impairment due to trisomy 21 and altered DNA methylation. Two of those probes were observed in the *TSC2* gene, which has been associated with the tau neuropathology of AD—the second major cognitive deficit in DS [63].

Interestingly, reduced levels of the methyl donor SAM have been described in DS. In contrast to the aforementioned reports on hypermethylation in DS, this suggests a reduced cellular methylation

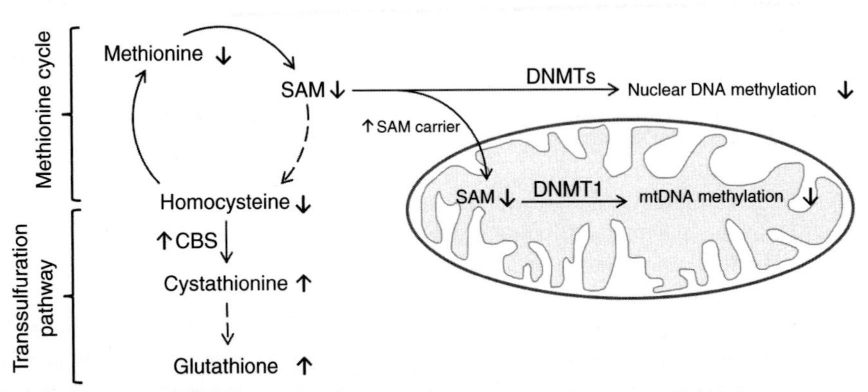

FIGURE 9.1 Schematic Illustration of the Cystathionine β-Synthase (CBS)-Induced Depletion of S-Adenosylmethionine (SAM) and Its Effects on Nuclear and Mitochondrial DNA Methylation.

Simplified: CBS is encoded on HSA21 and therefore overexpressed in DS, leading to the increased conversion of homocysteine into cystathionine. Accordingly, less homocysteine is available for conversion into the SAM-precursor methionine, yielding decreased SAM levels and a subsequently reduced DNA methylation capacity. Although hypomethylated mtDNA was indeed observed [38], various studies reported hypermethylated nuclear DNA in DS [38,58,59]. Interrupted lines indicate that intermediary components are omitted. *mtDNA*, Mitochondrial DNA.

Reprinted from Dekker AD, De Deyn PP, Rots MG. Epigenetics: the neglected key to minimize learning and memory deficits in Down syndrome. Neurosci Biobehav Rev 2014;45C:72–84 [35], with permission of Elsevier

capacity [38,58]. The decreased SAM levels are attributed to overexpression of the HSA21-encoded cystathionine β-synthase (CBS) in DS. CBS is a central enzyme in the one-carbon metabolism, catalysing the conversion of homocysteine into cystathionine. This is the first step in the trans-sulfuration pathway, which results in the synthesis of the antioxidant glutathione (Fig. 9.1). As a consequence, less homocysteine is available for the methionine cycle in which homocysteine is converted to methionine, the precursor of SAM [58]. Therefore, reduced SAM levels and the subsequently reduced methyl transfer to DNA is a likely mechanism underlying aberrant DNA methylation patterns in DS.

In addition to nuclear DNA methylation, SAM is also required for methylation of cytosines in mitochondrial DNA (mtDNA) by the mitochondrial DNMT1, the only catalytically active DNMT found in mitochondria so far [64,65]. Importantly, despite more than 50% increased expression of the SAM carrier, which transports SAM into the mitochondrion, significantly decreased mitochondrial SAM levels were found in lymphoblastoid cells of DS individuals compared to controls [38]. Whereas previous studies reported hypermethylation of nuclear DNA in DS [38,58,59,62], the opposite was demonstrated for mtDNA (Fig. 9.1). Compared to age-matched controls, mtDNA is hypomethylated in DS, suggesting an impaired mitochondrial methylation capacity in DS that, in turn, could lead to mitochondrial dysfunction [38]. In fact, mitochondrial dysfunction has been convincingly demonstrated in DS, including reduced expression of genes encoding mitochondrial enzymes [66,67] and impaired ATP synthesis [68]. The latter is likely to affect many epigenetic processes that require ATP as a substrate, including the ATP-dependent production of SAM for DNA methylation [39].

The case of altered DNA methylation in DS is reinforced by two studies that demonstrated global hypermethylation in placental villi samples derived from DS fetuses compared to normal villi, illustrating that epigenetic changes are already present in early development [69,70]. Indeed, DNA methylation provides a new diagnostic method for the detection of DS. In contrast to the commonly used invasive (and risky) sampling procedures to obtain fetal genetic material, a novel noninvasive prenatal testing (NIPT) method was developed a few years ago using cell-free fetal DNA in the maternal peripheral blood. This method analyzes differences in DNA methylation of HSA21 regions between the mother and her child and is based on the occurrence of fetal-specific DMRs on HSA21. Ratios of such DMRs for fetus over mother will be different for a trisomic child, which has an additional copy of the differentially methylated HSA21 locus, compared to a non-DS fetus. Comparing these methylation ratios for a combination of DMRs between normal and DS cases enabled correct noninvasive prenatal diagnosis of DS [71]. Recently, two of these DMRs were validated as potential fetal-specific epigenetic markers [72].

To which extent DNA methylation patterns are regionally altered in the DS brain, particularly brain areas involved in learning and memory, remains to be elucidated. In that context, the hydroxylation of 5-methylcytosine into 5-hydroxymethylcytosine (5hmC), an early intermediate of DNA demethylation, might be very relevant. TET proteins mediate this hydroxylation and were shown to be differently methylated in DS and downregulated in the DS placenta, which possibly contributes to the aforementioned hypermethylated regions [62,69,73,74]. Notably, 5hmC has the highest prevalence in mature neurons compared to other mammalian cells and global 5hmC expression increases with aging, particularly at genes involved in neurological disorders, such as AD [22]. The relationship between 5hmC and learning and memory is still in its infancy. However, its high expression in neurons and increased levels during aging and AD, seem to indicate a relevant role for 5hmC and TET.

In conclusion, increasing evidence indicates aberrant nuclear and mitochondrial DNA methylation in DS. Although DNA methylation is generally regarded as a stable epigenetic modification, recent findings have elucidated that intermediate DNA modifications are present, especially in neurons [22]. The reversibility of DNA methylation in neurons might offer a great potential for treatment of neurological disorders, such as the cognitive deficits in DS.

9.4 ALTERED HISTONE MODIFICATIONS ARE ASSOCIATED WITH DS

Posttranslational histone modifications form docking platforms for chromatin-associated effector proteins that alter the chromatin structure, thus affecting gene expression. Such histone modifications are not static, but rather the equilibrium of continuous addition and removal of these chemical groups by epigenetic writers and erasers [50]. The chromatin structure is also affected by constitutive chromatin proteins and by incorporation of different histone core variants.

9.4.1 POSTTRANSLATIONAL HISTONE TAIL MODIFICATIONS

Specific amino acids of histones are subjected to covalent posttranslational modifications, such as acetylation, methylation, and phosphorylation. Although these modifications have been reported to affect synaptic plasticity, learning, and memory, only one study so far analyzed a posttranslational histone modification in DS, namely trimethylation of H3 lysine 4 (H3K4me3) [75]. Letourneau et al. studied differential gene expression in fibroblasts from monozygotic twins discordant for trisomy 21. Such a pair of twins enables comparison of gene expression levels between trisomic and disomic cells without the noise of genomic variability. It was found that differential gene expression was distributed in defined domains along the chromosomes, so-called gene expression dysregulation domains (GEDDs). The authors suggested that the observed gene expression differences could be related to an altered chromatin state in the trisomic cells. Hence, DNA methylation and H3K4me3 were compared between the trisomic and disomic fibroblasts. Whereas differences in DNA methylation were not correlated to the GEDDs, the differences in H3K4me3 profiles between the twins markedly correlated with the GEDDs for nearly all chromosomes. Therefore, the altered H3K4me3-associated chromatin state relates to the altered gene expression in trisomic compared to disomic fibroblasts [75]. Further direct proof for altered histone marks in DS is lacking. However, an increasing body of evidence suggests that posttranslational histone modifications contribute to the neurological deficits observed in DS and other intellectual disabilities. So far, studies have found five HSA21 genes to influence particular histone modifications, namely *DYRK1A, ETS2, HMGN1, BRWD1,* and *RUNX1*, thus suggesting abnormal modifications in DS.

First, DYRK1A, a member of the dual specificity tyrosine-phosphorylated and regulated kinase (DYRK) family, has been implicated in the learning deficits in DS [76]. This highly conserved subfamily of protein kinases catalyzes autophosphorylation on tyrosine residues and phosphorylation of serine/threonine residues on exogenous substrates [77]. Studies in Drosophila and mice have revealed that the DYRK1A protein is necessary for normal brain development in a dose-sensitive way [78,79]. Increased DYRK1A expression in primary murine cortical neurons reduced dendritic growth and complexity [42]. Furthermore, transgenic mice that overexpressed DYRK1A showed significant impairment in cognitive flexibility and spatial learning. This indicates a causative role of DYRK1A overexpression in the intellectual disability associated with DS [80].

Interestingly, DYRK1A has a bipartite effect on histone modifications. First of all, DYRK1A directly phosphorylates the threonine residue 522 of the SIRT1 histone deacetylase, thereby promoting deacetylation and possibly deteriorating cognitive capacities [40]. In addition, DYRK1A phosphorylates the cyclic AMP response element-binding protein (CREB) at the serine residue 133, inducing the recruitment of the CREB binding protein (CBP/P300). CBP/P300 is a histone acetyltransferase that promotes CREB-mediated expression of genes [41,81]. A loss of function mutation in CBP/P300 results in the aforementioned Rubinstein–Taybi syndrome that includes intellectual disability [32,82]. Besides DYRK1A, two other HSA21 proteins influence the activity of CBP/P300: the erythroblastosis virus E26 oncogene homolog 2 (ETS2) and the nucleosome-binding high-mobility group N1 (HMGN1) [45,50]. Accordingly, it is conceivable that the HAT/HDAC balance is dysregulated in DS, causing aberrant histone acetylation patterns that affect learning and memory processes.

Apart from its direct phosphorylation effects, DYRK1A also alters gene expression via the neuron-restrictive silencer factor (NRSF, also known as REST). NRSF regulates the expression of a range of neuronal genes involved in the function of, amongst others, ion channels, neurotransmitter receptors and synapses [83–85]. NRSF represses transcription of these neuronal genes in nonneuronal cells by binding to the neuron-restrictive silencer element (NRSE). Besides nonneuronal cells, NRSF is also present in undifferentiated neuronal progenitors. However, it ceases to be expressed in differentiated neurons, thereby enabling gene expression. Therefore, NRSF has been termed as "a master negative regulator of neurogenesis" [84].

In DS, NRSF levels seem to be perturbed. For instance, decreased NRSF expression was observed in the aforementioned placental villi samples from DS fetuses compared to non-DS [69]. Moreover, various NRSF-regulated genes were repressed in neurospheres derived from fetal DS brain cells, while non-NRSF-regulated genes with similar functions were unaffected [86]. Furthermore, Canzonetta et al. reported a 30–60% reduced NRSF expression in the transchromosomic Tg*Dyrk1A* mouse model of DS, resulting in increased transcript levels of downstream targets [83]. This inverse correlation, however, was lost in another transgenic mouse model of DS that overexpressed DYRK1A [42]. In agreement, silencing the third copy of *DYRK1A* by RNA interference rescued NRSF levels, confirming the role of DYRK1A in NRSF-mediated gene regulation [83]. Importantly, DYRK1A regulates NRSF by binding to the SWI/SNF chromatin remodeling complex [42]. This complex uses ATP to mobilize nucleosomes and rearrange the chromatin structure and induces the expression of multiple other genes involved in histone modifications, for example, the histone methyl transferase L3MBTL2, the histone demethylase JARID1D and the HDAC interactor NCOR [24,42,87]. Therefore, the overexpression of DYRK1A in DS is likely to affect a range of epigenetic mechanisms, which strongly indicates that epigenetic marks are presumably altered in DS compared to the non-DS population.

The contribution of DYRK1A to learning and memory deficits in DS is further supported by findings from Altafaj et al. [88]. They administered short hairpin RNA against DYRK1A to Ts65Dn mice, resulting in normalized DYRK1A protein levels, improved synaptic plasticity and partial amelioration of the hippocampal-dependent search strategy in the Morris water maze. Another study showed that the DYRK1A inhibitor epigallocatechin-gallate (EGCG), a green tea flavonol, rescued visuospatial memory (Morris water maze) and object recognition memory (novel object recognition test) in both Ts65Dn and Tg*Dyrk1A* mice. A pilot study with EGCG-treatment (3 months) in young adults with DS, however, did not convincingly improve cognitive functioning: marginal positive effects were reported on visual memory recognition ($p = 0.04$), working memory ($p = 0.08$), and social functioning ($p = 0.05$), as compared to placebo-treated subjects [89].

In addition to DYRK1A, two other HSA21-encoded proteins interact with the SWI/SNF complex, thereby altering histone modifications and likely gene expression: BRWD1 and RUNX1. The bromodomain and WD repeat-containing 1 (BRWD1) modulates the chromatin by binding through its two bromodomains and by associating with the SWI/SNF complex [43]. Moreover, the RUNX1 forms multiprotein complexes at target gene promoters to which the SWI/SNF subunits BRG1 and INI1 bind. RUNX1 is associated with histone modifications that are typical of euchromatin, such as dimethylated H3K4 and acetylated H4 [44]. As described in Section 9.3, the *RUNX1* gene was found to be hypermethylated in DS compared to controls [62,70], suggesting altered *RUNX1* gene expression. Consequently, altered RUNX1 protein levels likely affect epigenetic marks as well.

Unfortunately, the role of the SWI/SNF complex has not been investigated in DS yet. A growing body of evidence indicates the involvement of this chromatin–remodelling complex in neurodevelopment and hence might be critical in the cognitive deficits in DS. For instance, the expression of the SWI/SNF subunit BRG1 is enriched in the brain and the spinal cord of mice embryos [90] and the dorsal neural tube of chick embryos [91]. In zebrafish, Eroglu et al. revealed that BRG1-deficiency leads to impaired neurogenesis and neural crest cell differentiation [92]. Furthermore, the contribution of the SWI/SNF complex to cognitive deficits is reinforced by the finding that the alpha-thalassemia X-linked intellectual disability (ATRX) syndrome is caused by mutations in the gene that encodes the SWI/SNF protein ATRX [93,94]. Therefore, aberrant functioning of the SWI/SNF complex due to overexpressed HSA21 products might lead to intellectual disability in a similar way.

Finally, posttranslational histone modifications might be altered due to mitochondrial dysfunction in DS. Mitochondria are the major cellular source of high energy intermediates such as acetyl-coenzyme A, nicotinamide adenine dinucleotide (NAD+), SAM, and ATP, which are respectively involved in acetylation, deacetylation, methylation, and phosphorylation of histones [39]. Aberrant mitochondrial production of these high energy intermediates in DS, for example due to the overexpression of CBS, is likely to cause alterations in posttranslational histone marks in DS. Interestingly, a recent study revealed that incubation of DS lymphoblasts and fibroblasts with EGCG counteracted mitochondrial dysfunction. In particular, EGCG-stimulated mitochondrial biogenesis and rescued ATP synthase catalytic activity and oxidative phosphorylation [95], probably restoring the levels of one or more high energy intermediates. In addition to its inhibitory effect on DYRK1A, EGCG might thus improve learning and memory by rescuing mitochondrial functioning in DS.

9.4.2 HISTONE CORE VARIANTS AND CONSTITUTIVE CHROMATIN PROTEINS

In addition to histone tail modifications, histone core variants can also influence gene expression. Incorporation of different histone variants into the nucleosomal core affects the chromatin structure [96]. Most variants have been discovered for H2A and H3 and were reported to have a diverse role in gene expression regulation. Incorporation of macroH2A, for instance, is associated with gene repression [96,97]. HSA21 encodes the H2A histone family member Z pseudogene 1 (H2AFZP) and the H2B histone family member S pseudogene (H2BFS) [24,46,47,98]. DNA hypermethylation of the *H2BFS* gene has been reported in three villi samples of DS fetuses compared to non-DS controls [70]. Whereas H2BFS has been described as a component of the nucleosomal core, it is currently unknown whether the H2AFZP encodes a protein [24,48]. The effects of both on gene regulation need to be established as well.

Besides these histone variants, two HSA21-encoded constitutive chromatin proteins contribute to nucleosome assembly: chromatin assembly factor 1B (CHAF1B) and HMGN1. The CHAF1B protein is involved in nucleosome assembly on to newly replicated DNA by recruiting H3 and H4 [99,100]. CHAF1B forms a multiprotein complex with the methyl-CpG binding protein 1 (MBD1) and heterochromatin protein 1 (HP1), which, again, demonstrates the involvement of epigenetics in DS [49]. HMGN1 affects posttranslational histone modifications, in particular it inhibits phosphorylation of H3S10 and H3S28 and enhances H3K14 acetylation via CBP/P300 [50,51], and has been described to regulate MECP2 expression. MECP2 is highly expressed in the brain and can activate or repress gene transcription [101]. Altered MECP2 activity may result in intellectual disability [51,102]. Abuhatzira et al. [51] reported transcript levels of *HMGN1* to be increased with 50% and transcript levels of *MECP2* to be decreased with 30% in brain tissue of DS patients compared to non-DS age-matched controls. In mice, it was found that altered HMGN1 protein levels resulted in histone modifications in the *MECP2* promoter and a modified chromatin structure. Therefore, overexpressed HMGN1 might disturb normal learning and memory processes through reduced MECP2 protein expression and subsequently altered epigenetic marks.

9.5 EPIGENETICS IN DS: A LINK TO ALZHEIMER'S DISEASE?

In addition to their intellectual disability, the cognitive capacities of people with DS are likely to deteriorate later in life due to a higher risk for early-onset AD. The triplication of the *APP* gene has been regarded as the main cause of this strongly increased risk. Postmortem analysis revealed that virtually all DS individuals have an extensive AD-like neuropathology from 40 years of age [12]. Despite the presence of extraneuronal Aβ plaques and intraneuronal neurofibrillary tangles of hyperphosphorylated tau protein, 30–50% of the DS individuals do not show signs of dementia, that is, cognitive decline, impaired activities of daily living, and behavioral and psychological alterations [103]. In those who develop dementia, the age at which the first clinical symptoms appear varies greatly [7]. Accordingly, the central question emerges: what causes the one to get clinically demented and the other to live free of AD symptoms until death, while the AD-like neuropathology is present in all of them?

Since the discovery of trisomy 21, more than 50 years of genetic research has not yielded convincing answers yet. We therefore argue that solely focusing on the consequences of the overexpressed *APP* is too limited to explain this considerable interindividual variability of AD in DS. Clearly, other factors determine why one DS individual becomes demented and the other not. Which factors are essential contributors to AD in DS is currently far from understood. However, it is conceivable that different expression levels of these factors determine the presence or absence of clinical dementia symptoms. Again, epigenetic mechanisms affect gene expression, and are thus potential therapeutic targets to interfere with AD in DS. Despite the scarce epigenetic studies on AD in DS, an increasing body of evidence demonstrates altered nuclear epigenetic modifications in AD in the general population (an up-to-date review is provided in Ref. [104]). Very recently, mtDNA methylation was implicated in AD as well [105].

Importantly, various overexpressed HSA21 genes have been attributed a role in the progression of AD, possibly via their downstream epigenetic effects. Therefore, various epigenetic marks are likely altered in demented DS individuals as well. For instance, besides phosphorylating APP at threonine residue 668 [106], DYRK1A is also able to phosphorylate multiple sites of tau proteins, contributing to AD pathology, as such. Indeed, these tau sites were found to be hyperphosphorylated in adult DS brains [107]. Whether DYRK1A-mediated histone (de)acetylation and the SWI/SNF complex contributes to this is currently unknown. Furthermore, various studies show altered expression of microRNAs in AD [108], for example, microRNA-125b is increased in the AD brain. In addition to the function of microRNAs in mRNA cleavage and degradation and translational repression [109,110], endogenously expressed microRNAs have been reported to mediate the condensation of heterochromatin [109,111,112]. Interestingly, microRNA-125b2 is located on HSA21 and thus overexpressed in DS [113]. Therefore, disturbed microRNA-125b2 levels in DS might relate to the increased risk for AD in DS.

This high risk for AD is part of the so-called accelerated aging in DS [114]. Recently, Horvath's group [115] used a new molecular marker of aging, referred to as the "epigenetic clock," to study human aging. The "epigenetic clock" was defined by measuring DNA methylation levels in different human tissues in various age ranges. Consequently, a specific set of 353 CpGs was selected to constitute the "epigenetic clock.". The authors suggested that DNA methylation levels of these selected CpGs can predict, or estimate, an individual's chronological age. This estimated age is referred to as the DNA methylation age. In non-DS individuals, the DNA methylation age is strongly correlated with the chronological age in multiple cell types and tissues, including neurons and glial cells [115].

To assess whether the "epigenetic clock" confirms the accelerated aging due to trisomy 21, DNA methylation datasets were analyzed, including blood samples and various brain areas [116]. Again, the DNA methylation age and chronological age were strongly correlated in non-DS control individuals, establishing a reference regression line. For DS samples, the DNA methylation age tended to be higher than this regression line, suggesting an accelerated aging effect. In the analysis of brain tissue samples, the authors included AD patients without DS (mean age = 60.1 (58–64) years) in addition to DS individuals (mean age = 49.6 (42–57) years, no information provided on clinical dementia symptoms) and controls (mean age = 50.5 (32–64) years). Considering the overall brain samples, DS individuals showed a significantly higher accelerated aging effect than AD patients, despite their lower age. Similar results were obtained in focused analyzes of frontal lobe and cerebellar samples [116]. As described before, extensive AD-like neuropathology is present in DS from the age of 40 [12], decades earlier than in the general population. Regarding the age range of the DS group, all DS individuals most likely presented this neuropathology. Conceivably, the significantly higher age acceleration in DS brain tissue samples reflects the early deposition of neuropathology in the DS brain.

In short, whereas AD in DS has hardly received attention in epigenetic studies, altered epigenetic signatures have been reported for accelerated aging in DS, and for AD patients in the general population. Consequently, the essential next step should be to epigenetically study AD in DS. Proper documentation on clinical dementia symptoms greatly matters. Since all DS individuals aged 40 years and older present the AD-like neuropathology, we need to distinguish between clinically demented and clinically nondemented DS individuals in order to study possible mechanisms causing the one to get demented at a relatively young age, while another grows old without any symptoms.

9.6 EPIGENETIC THERAPY MAY ALLEVIATE COGNITIVE DEFICITS IN DS

Currently, no treatment is available to prevent or alleviate intellectual disability or AD dementia in DS, though education and a stimulating environment may (slightly) improve cognitive capacities. Various studies have tried to develop pharmacological treatments to improve cognition in DS. In DS mouse models, such as the widely used Ts65Dn mice, promising cognitive-enhancing results have been described [117]. However, none of the investigated pharmacological treatments for DS patients have reached the market. Moreover, almost all clinical trials with drugs for AD in the general population have failed so far. The US Food and Drug Administration (FDA) has approved symptomatic AD treatment with cholinesterase inhibitors, such as donepezil. However, these drugs only provide short-term relief by improving cognition and to a certain, rather limited extent behavior, but do not interfere with the underlying neurodegeneration [118]. Interestingly, a 10-week administration of donepezil (2.5–10 mg/day) to DS children 10–17 years of age failed to show cognitive improvement [119]. To our knowledge, other AD drugs, such as those targeting the accumulation of Aβ, did not receive a market approval or are still in the process of clinical trials [118,120].

Hence, investigating new targets and therapies is important. Here, we have pointed out the major role that epigenetics plays in synaptic plasticity and learning and memory. Various aberrant epigenetic modifications contribute to intellectual disabilities, and hence, might be involved in the cognitive deficits in DS. In contrast to the triplication of HSA21, epigenetic marks are reversible, and thus ideal targets to alleviate certain features in DS. Therefore, drugs that inhibit epigenetic enzymes, so-called epidrugs, offer a promising new way of treatment as an alternative for, or to act synergistically with, classical pharmacology.

The field of epigenetics is booming, especially in cancer research, and provides new approaches to address a wide variety of diseases. Currently, the FDA has approved five epidrugs against cancer (Table 9.2): two DNMT inhibitors [5-azacytidine (Vidaza) and decitabine (Dacogen)] and three HDAC inhibitors [vorinostat (Zolinza), romidepsin (Istodax), and belinostat (Beleodaq)] [121,122]. In addition, it was demonstrated that valproic acid, which is already used against epilepsy and bipolar disorders for years, has anticancer properties as a HDAC inhibitor, and thus can be considered an

Table 9.2 Currently Approved Epidrugs

Epidrug (Active Ingredient)	Trade Name	Inhibition Target	Indication	FDA Approval[a]	EMA Approval[b]
5-Azacytidine	Vidaza	DNMT	Myelodysplastic syndromes	19-05-2004	17-12-2008
Decitabine	Dacogen	DNMT	Myelodysplastic syndromes	02-05-2006	20-09-2012
Vorinostat	Zolinza	HDAC	Cutaneous T-cell lymphoma	06-10-2006	Withdrawn
Romidepsin	Istodax	HDAC	Cutaneous T-cell lymphoma, peripheral T-cell lymphoma	05-11-2009	Refused on 12-02-2013
Belinostat	Beleodaq	HDAC	Peripheral T-cell lymphoma	03-07-2014	—
Valproic acid	Various	HDAC	Epilepsy, bipolar disorder	Various approval dates	Various approval dates

[a]FDA, US Food and Drug Administration [122].
[b]EMA, European Medicines Agency [124].
Adapted from Dekker AD, De Deyn PP, Rots MG. Epigenetics: the neglected key to minimize learning and memory deficits in Down syndrome. Neurosci Biobehav Rev 2014;45C:72–84 [35], with permission of Elsevier.

epidrug too [123]. Previously, studies showed that HDAC inhibitors may rescue memory impairments, thereby arousing many researchers' interest in histone acetylation and finding specific HDAC inhibitors to combat learning and memory deficits (for a comprehensive review see Ref. [30]). A wide array of clinical trials of epidrugs, however, is ongoing for neurological disorders, or to investigate new indications for approved epidrugs.

To treat the cognitive deficits in DS, the aforementioned DYRK1A inhibitor EGCG has received substantial attention in clinical trials (e.g., Ref. [125]). Interestingly, cancer studies currently investigate EGCG as an epidrug, since it can inhibit DNMT activity. Indeed, reexpression of genes that were silenced through promoter methylation have been observed [126,127]. Furthermore, Ramakrishna et al. [128] reported significantly lower expression levels of the presynaptic alpha-synuclein protein in brain tissue of DS individuals and Ts65Dn mice, and suggested that alpha-synuclein plays a key role in deficient synaptic activity. Subsequent EGCG treatment of Ts65Dn mice resulted, counterintuitively, in increased DNA methylation of the alpha-synuclein promotor, and increased expression of the alpha-synuclein protein in Ts65Dn mice [129]. Despite the inconsistent effect of EGCG on DNA methylation, it has become evident that EGCG does not only affect DYRK1A, but may also alter the methylation profile of DS individuals.

In spite of the promising results of the approved treatments (Table 9.2), epidrugs have genome-wide and nonchromatin effects, thereby altering a range of biological processes. Accordingly, specific targeting of genes or proteins in particular tissues is the major challenge. In combatting the cognitive deficits in DS, specific targeting of (a part of) the third copy of HSA21 in the brain is required without affecting peripheral epigenetic modifications. Recently, the complete third copy was silenced in vitro using the large noncoding RNA molecule X inactive specific transcript (XIST), which endogenously silences the second X-chromosome in females. Jiang et al. introduced an inducible *XIST* transgene into the *DYRK1A* locus in induced DS pluripotent stem cells. As a consequence, stable heterochromatin marks were observed (repressive histone marks and DNA methylation), leading to chromosome-wide transcriptional silencing. As confirmation it was shown that transcription of *DYRK1A* and *APP* was repressed [14]. Future studies should demonstrate if this method would be successful in in vivo models as well.

Such innovative approaches open new avenues to pinpoint pathways or genes underlying the DS phenotype. In this respect, the rapidly developing technology of epigenetic editing offers a targeted approach to modulate the expression of (combinations of) individual genes, such as *DYRK1A* and *APP*, thereby validating their role in DS. Epigenetic editing comprises the targeting of particular epigenetic enzymes (writers or erasers) to specific genes with the use of lab-engineered DNA binding domains that target the endogenous gene of interest [130]. Such engineered domains, including designer zinc finger proteins, transcription activation like effectors (TALEs) and the CRISPR/dCas9 platform [131], are subsequently fused to an epigenetic enzyme with desired properties, for example, a certain DNMT [132,133], DNA demethylase [134,135], or histone modifiers [136–138]. As a consequence, epigenetic modifications at the target gene are actively overwritten, causing long-term modulation of gene expression, either stimulating or repressing its expression [130,131]. Interestingly, epigenetic editing has demonstrated its promise to specifically modulate the expression of one or more neuronal genes in vivo [137–139], opening new avenues to ameliorate the cognitive deficits in DS.

The question then is, which targets? In this review we have considered epigenetic alterations due to overexpression of certain HSA21 genes (summarized in Table 9.1). Obviously, these genes would serve as a first group of targets. However, it is conceivable that non-HSA21 products influence the epigenetic mechanisms in DS as well. Unfortunately, not more than a mere handful of studies have conducted epigenetic profiling in DS. Therefore, future studies should more comprehensively investigate epigenetic marks in DS, preferentially including subgroups with and without dementia and groups of different ages.

9.7 CONCLUSIONS

DS is the most common genetic intellectual disability. Despite promising results in DS mouse models and ongoing clinical trials, no (preventive) treatment for the two major cognitive hallmarks (intellectual disability and early-onset AD) are currently available. Findings in other intellectual disabilities, including Rett syndrome and Rubinstein–Taybi syndrome, have elucidated the role of epigenetics in synaptic plasticity and learning and memory. However, epigenetics in DS has been largely neglected so far.

DS is characterized by extensive interindividual variability. The triplication of HSA21 would theoretically result in a 1.5-fold increased expression level, but transcript levels of various HSA21 genes deviate from this, thus differently contributing to the DS phenotypes. Conceivably, epigenetic mechanisms play a crucial role herein. Although the role of epigenetics in DS is currently far from understood, an increasing body of evidence indicates the involvement of DNA methylation, posttranslational histone modifications and histone core variants in DS. As summarized in Table 9.1, various overexpressed HSA21 gene products are epigenetic modulators, thereby dysregulating epigenetic mechanisms in DS. In turn, these disturbed mechanisms might contribute to the observed learning and memory deficits.

Importantly, epigenetic marks are reversible, and thus may offer great therapeutic potential to prevent or improve the cognitive symptoms in DS. Most promising is the technique of epigenetic editing in which specific epigenetic enzymes are recruited to specific genes by means of a laboratory-engineered DNA binding domain. Thereupon, epigenetic modifications are actively overwritten, causing a potentially persisting modulation of gene expression. Partial repression of overexpressed HSA21 genes can yield physiological expression levels that might alleviate the cognitive deficits in DS.

In short, it has become clear that one cannot ignore the involvement of epigenetics in intellectual disabilities, including DS. Although the number of studies on epigenetics in DS is relatively limited, these studies indicate disturbed epigenetic mechanisms due to overexpression of multiple HSA21 genes. Regarding the aforementioned cognition-enhancing therapies, future studies should identify the aberrant epigenetic marks in DS compared to the general population and between DS individuals with and without dementia, in order to subsequently overwrite these marks by using epigenetic editing.

ABBREVIATIONS

5hmC	5-Hydroxymethylcytosine
Aβ	Amyloid-β
AD	Alzheimer's disease
APP	Amyloid precursor protein
ATRX	Alpha-thalassemia X-linked intellectual disability
BRWD1	Bromodomain and WD repeat domain containing 1
CBS	Cystathionine β-synthase
CHAF1B	Chromatin assembly factor 1B
CpG	Cytosines preceding a guanine
CREB	Cyclic AMP response element-binding protein
CBP/P300	CREB binding protein
DMR	Differentially methylated region
DS	Down syndrome
DYRK1A	Dual specificity tyrosine-phosphorylated and regulated kinase 1A
EGCG	Epigallocatcchin-gallatc
ETS2	Erythroblastosis virus E26 oncogene homolog 2
FDA	US Food and Drug Administration
GEDD	Gene expression dysregulation domains
H2AFZP	H2A histone family member Z pseudogene 1
H2BFS	H2B histone family member S pseudogene
HMGN1	Nucleosome-binding high-mobility group N1
HP1	Heterochromatin protein 1
HSA21	Human chromosome 21
MECP2	Methyl-CpG-binding protein 2
MBD1	Methyl-CpG binding protein 1
mtDNA	Mitochondrial DNA
NAD+	Nicotinamide adenine dinucleotide
NIPT	Noninvasive prenatal testing
NRSF/REST	Neuron-restrictive silencer factor
NRSE	Neuron-restrictive silencer element
RUNX1	Runt-related transcription factor 1
SAM	S-adenosylmethionine
TALE	Transcription activation like effector
TET	Ten-eleven translocation

GLOSSARY

Epigenetic editing A technique to target epigenetic enzymes to genes of interest with the use of laboratory-engineered DNA binding domains

Epigenetic trait "A stably heritable phenotype resulting from changes in a chromosome without alterations in the DNA sequence" [140]

Euchromatin An accessible, relatively open chromatin state that is generally associated with active gene expression

Heterochromatin A relatively inaccessible, tightly compacted chromatin state that is generally associated with silenced gene expression

Histone core An octamer containing two copies of each histone type: histone 2A (H2A), H2B, H3, and H4

Nucleosome Approximately 147 base pairs of DNA wrapped around a histone core in 1.7 turns

ACKNOWLEDGMENTS

This chapter is an extended and updated version of, and thus strongly resembles, the authors' review article entitled "Epigenetics: The Neglected Key to Minimize Learning and Memory Deficits in Down Syndrome" published in Neuroscience and Biobehavioral Reviews in 2014 [35]. This work was supported by the Alzheimer Research Center of the University Medical Center Groningen (UMCG). MGR would like to acknowledge the networking support of the EU COST Action EpiChem CM1406.

REFERENCES

[1] Bittles AH, Bower C, Hussain R, Glasson EJ. The four ages of Down syndrome. Eur J Public Health 2007;17(2):221–5.
[2] Down JL. Observations on an ethnic classification of idiots. 1866. Ment Retard 1995;33(1):54–6.
[3] Lejeune J, Turpin R, Gautier M. Mongolism; a chromosomal disease (trisomy). Bull Acad Natl Med 1959;143(11–12):256–65.
[4] Antonarakis SE, Lyle R, Dermitzakis ET, Reymond A, Deutsch S. Chromosome 21 and down syndrome: from genomics to pathophysiology. Nat Rev 2004;5(10):725–38.
[5] Lubec G, Engidawork E. The brain in Down syndrome (TRISOMY 21). J Neurol 2002;249(10):1347–56.
[6] Lott IT, Dierssen M. Cognitive deficits and associated neurological complications in individuals with Down's syndrome. Lancet Neurol 2010;9(6):623–33.
[7] Zigman WB, Lott IT. Alzheimer's disease in Down syndrome: neurobiology and risk. Ment Retard Dev Disabil Res Rev 2007;13(3):237–46.
[8] Alzheimer's Assocation. 2015 Alzheimer's disease facts and figures. Alzheimer's Dement 2015;11(3):332–84.
[9] Ness S, Rafii M, Aisen P, Krams M, Silverman WP, Manji H. Down's syndrome and Alzheimer's disease: towards secondary prevention. Nat Rev Discov 2012;11(9):655–66.
[10] Roper RJ, Reeves RH. Understanding the basis for Down syndrome phenotypes. PLoS Genet 2006;2(3):e50.
[11] Wisniewski KE, Wisniewski HM, Wen GY. Occurrence of neuropathological changes and dementia of Alzheimer's disease in Down's syndrome. Ann Neurol 1985;17(3):278–82.
[12] Mann DMA. Alzheimer's disease and Down's syndrome. Histopathology 1988;13(2):125–37.
[13] Hattori M, Fujiyama A, Taylor TD, Watanabe H, Yada T, Park HS, et al. The DNA sequence of human chromosome 21. Nature 2000;405(6784):311–9.

[14] Jiang J, Jing Y, Cost GJ, Chiang JC, Kolpa HJ, Cotton AM, et al. Translating dosage compensation to trisomy 21. Nature 2013;500(7462):296–300.
[15] Prandini P, Deutsch S, Lyle R, Gagnebin M, Delucinge Vivier C, Delorenzi M, et al. Natural gene-expression variation in Down syndrome modulates the outcome of gene-dosage imbalance. Am J Hum Genet 2007;81(2):252–63.
[16] Ait Yahya-Graison E, Aubert J, Dauphinot L, Rivals I, Prieur M, Golfier G, et al. Classification of human chromosome 21 gene-expression variations in Down syndrome: impact on disease phenotypes. Am J Hum Genet 2007;81(3):475–91.
[17] Davisson MT, Schmidt C, Akeson EC. Segmental trisomy of murine chromosome 16: a new model system for studying Down syndrome. Prog Clin Biol Res 1990;360:263–80.
[18] Reeves RH, Irving NG, Moran TH, Wohn A, Kitt C, Sisodia SS, et al. A mouse model for Down syndrome exhibits learning and behaviour deficits. Nat Genet 1995;11(2):177–84.
[19] Kahlem P, Sultan M, Herwig R, Steinfath M, Balzereit D, Eppens B, et al. Transcript level alterations reflect gene dosage effects across multiple tissues in a mouse model of down syndrome. Genome Res 2004;14(7):1258–67.
[20] Lyle R, Gehrig C, Neergaard-Henrichsen C, Deutsch S, Antonarakis SE. Gene expression from the aneuploid chromosome in a trisomy mouse model of down syndrome. Genome Res 2004;14(7):1268–74.
[21] Braudeau J, Dauphinot L, Duchon A, Loistron A, Dodd RH, Herault Y, et al. Chronic treatment with a promnesiant GABA-A alpha5-selective inverse agonist increases immediate early genes expression during memory processing in mice and rectifies their expression levels in a Down syndrome mouse model. Adv Pharmacol Sci 2011;2011:153218.
[22] Weng YL, An R, Shin J, Song H, Ming GL. DNA modifications and neurological disorders. Neurotherapeutics 2013;10(4):556–67.
[23] Day JJ, Sweatt JD. DNA methylation and memory formation. Nat Neurosci 2010;13(11):1319–23.
[24] Sanchez-Mut JV, Huertas D, Esteller M. Aberrant epigenetic landscape in intellectual disability. Prog Brain Res 2012;197:53–71.
[25] Amir RE, Van den Veyver IB, Wan M, Tran CQ, Francke U, Zoghbi HY. Rett syndrome is caused by mutations in X-linked MECP2, encoding methyl-CpG-binding protein 2. Nat Genet 1999;23(2):185–8.
[26] Ma DK, Jang MH, Guo JU, Kitabatake Y, Chang ML, Pow-Anpongkul N, et al. Neuronal activity-induced Gadd45b promotes epigenetic DNA demethylation and adult neurogenesis. Science 2009;323(5917):1074–7.
[27] Tahiliani M, Koh KP, Shen Y, Pastor WA, Bandukwala H, Brudno Y, et al. Conversion of 5-methylcytosine to 5-hydroxymethylcytosine in mammalian DNA by MLL partner TET1. Science 2009;324(5929):930–5.
[28] Zhang RR, Cui QY, Murai K, Lim YC, Smith ZD, Jin S, et al. Tet1 regulates adult hippocampal neurogenesis and cognition. Cell Stem Cell 2013;13(2):237–45.
[29] Zentner GE, Henikoff S. Regulation of nucleosome dynamics by histone modifications. Nat Struct Mol Biol 2013;20(3):259–66.
[30] Gräff J, Tsai L-HH, Graff J. Histone acetylation: molecular mnemonics on the chromatin. Nat Rev Neurosci 2013;14(2):97–111.
[31] Peleg S, Sananbenesi F, Zovoilis A, Burkhardt S, Bahari-Javan S, Agis-Balboa RC, et al. Altered histone acetylation is associated with age-dependent memory impairment in mice. Science 2010;328(5979):753–6.
[32] Barrett RM, Wood Ma. Beyond transcription factors: the role of chromatin modifying enzymes in regulating transcription required for memory. Learn Mem 2008;15(7):460–7.
[33] Jarome TJ, Lubin FD. Histone lysine methylation: critical regulator of memory and behavior. Rev Neurosci 2013;24(4):375–87.
[34] Berdasco M, Esteller M. Genetic syndromes caused by mutations in epigenetic genes. Hum Genet 2013;132(4):359–83.

[35] Dekker AD, De Deyn PP, Rots MG. Epigenetics: the neglected key to minimize learning and memory deficits in Down syndrome. Neurosci Biobehav Rev 2014;45C:72–84.
[36] Deplus R, Brenner C, Burgers WA, Putmans P, Kouzarides T, de Launoit Y, et al. Dnmt3L is a transcriptional repressor that recruits histone deacetylase. Nucleic Acids Res 2002;30(17):3831–8.
[37] Ooi SK, Qiu C, Bernstein E, Li K, Jia D, Yang Z, et al. DNMT3L connects unmethylated lysine 4 of histone H3 to de novo methylation of DNA. Nature 2007;448(7154):714–7.
[38] Infantino V, Castegna A, Iacobazzi F, Spera I, Scala I, Andria G, et al. Impairment of methyl cycle affects mitochondrial methyl availability and glutathione level in Down's syndrome. Mol Genet Metab 2011;102(3):378–82.
[39] Wallace DC, Fan W. Energetics, epigenetics, mitochondrial genetics. Mitochondrion 2010;10(1):12–31.
[40] Guo X, Williams JG, Schug TT, Li X. DYRK1A and DYRK3 promote cell survival through phosphorylation and activation of SIRT1. J Biol Chem 2010;285(17):13223–32.
[41] Weeber EJ, Sweatt JD. Molecular neurobiology of human cognition. Neuron 2002;33(6):845–8.
[42] Lepagnol-Bestel AM, Zvara A, Maussion G, Quignon F, Ngimbous B, Ramoz N, et al. DYRK1A interacts with the REST/NRSF-SWI/SNF chromatin remodelling complex to deregulate gene clusters involved in the neuronal phenotypic traits of Down syndrome. Hum Mol Genet 2009;18(8):1405–14.
[43] Huang H, Rambaldi I, Daniels E, Featherstone M. Expression of the Wdr9 gene and protein products during mouse development. Dev Dyn 2003;227(4):608–14.
[44] Bakshi R, Hassan MQ, Pratap J, Lian JB, Montecino MA, van Wijnen AJ, et al. The human SWI/SNF complex associates with RUNX1 to control transcription of hematopoietic target genes. J Cell Physiol 2010;225(2):569–76.
[45] Sun HJ, Xu X, Wang XL, Wei L, Li F, Lu J, et al. Transcription factors Ets2 and Sp1 act synergistically with histone acetyltransferase p300 in activating human interleukin-12 p40 promoter. Acta Biochim Biophys Sin 2006;38(3):194–200.
[46] NCBI Gene. H2A histone family member Z pseudogene 1 [*Homo sapiens* (human)] [Internet], 2016. Available from: http://www.ncbi.nlm.nih.gov/gene/?term=h2afzp
[47] Gardiner KJ, Davisson MT. The sequence of human chromosome 21 and implications for research into Down syndrome. Genome Biol 2000;1(2):1–9.
[48] UniProtKB. P57053 (H2BFS_HUMAN) [Internet], 2016. Available from: http://www.uniprot.org/uniprot/P57053
[49] Reese BE, Bachman KE, Baylin SB, Rountree MR. The methyl-CpG binding protein MBD1 interacts with the p150 subunit of chromatin assembly factor 1. Mol Cell Biol 2003;23(9):3226–36.
[50] Ueda T, Postnikov YV, Bustin M. Distinct domains in high mobility group N variants modulate specific chromatin modifications. J Biol Chem 2006;281(15):10182–7.
[51] Abuhatzira L, Shamir A, Schones DE, Schaffer AA, Bustin M. The chromatin-binding protein HMGN1 regulates the expression of methyl CpG-binding protein 2 (MECP2) and affects the behavior of mice. J Biol Chem 2011;286(49):42051–62.
[52] Bestor TH, The DNA. The DNA methyltransferases of mammals. Hum Mol Genet 2000;9(16):2395–402.
[53] Margot JB, Ehrenhofer-Murray AE, Leonhardt H. Interactions within the mammalian DNA methyltransferase family. BMC Mol Biol 2003;4:7.
[54] Suetake I, Shinozaki F, Miyagawa J, Takeshima H, Tajima S. DNMT3L stimulates the DNA methylation activity of Dnmt3a and Dnmt3b through a direct interaction. J Biol Chem 2004;279(26):27816–23.
[55] Arima T, Hata K, Tanaka S, Kusumi M, Li E, Kato K, et al. Loss of the maternal imprint in Dnmt3Lmat−/− mice leads to a differentiation defect in the extraembryonic tissue. Dev Biol 2006;297(2):361–73.
[56] Bourc'his D, Xu GL, Lin CS, Bollman B, Bestor TH. Dnmt3L and the establishment of maternal genomic imprints. Science 2001;294(5551):2536–9.
[57] Hata K, Okano M, Lei H, Li E. Dnmt3L cooperates with the Dnmt3 family of de novo DNA methyltransferases to establish maternal imprints in mice. Development 2002;129(8):1983–93.

[58] Pogribna M, Melnyk S, Pogribny I, Chango A, Yi P, James SJ. Homocysteine metabolism in children with Down syndrome: in vitro modulation. Am J Hum Genet 2001;69(1):88–95.
[59] Chango A, Abdennebi-Najar L, Tessier F, Ferre S, Do S, Gueant JL, et al. Quantitative methylation-sensitive arbitrarily primed PCR method to determine differential genomic DNA methylation in Down syndrome. Biochem Biophys Res Commun 2006;349(2):492–6.
[60] Kerkel K, Schupf N, Hatta K, Pang D, Salas M, Kratz A, et al. Altered DNA methylation in leukocytes with trisomy 21. PLoS Genet 2010;6(11):e1001212.
[61] Ram G, Chinen J. Infections and immunodeficiency in Down syndrome. Clin Exp Immunol 2011;164(1):9–16.
[62] Bacalini MG, Gentilini D, Boattini A, Giampieri E, Pirazzini C, Giuliani C, et al. Identification of a DNA methylation signature in blood cells from persons with Down syndrome. Aging 2015;7(2):82–96.
[63] Jones MJ, Farre P, McEwen LM, Macisaac JL, Watt K, Neumann SM, et al. Distinct DNA methylation patterns of cognitive impairment and trisomy 21 in down syndrome. BMC Med Genom 2013;6:58.
[64] Shock LS, Thakkar PV, Peterson EJ, Moran RG, Taylor SM. DNA methyltransferase 1, cytosine methylation, and cytosine hydroxymethylation in mammalian mitochondria. Proc Natl Acad Sci USA 2011;108(9):3630–5.
[65] van der Wijst MGP, Rots MG. Mitochondrial epigenetics: an overlooked layer of regulation? Trends Genet 2015;31(7):353–6.
[66] Conti A, Fabbrini F, D'Agostino P, Negri R, Greco D, Genesio R, et al. Altered expression of mitochondrial and extracellular matrix genes in the heart of human fetuses with chromosome 21 trisomy. BMC Genom 2007;8:268.
[67] Lee SH, Lee S, Jun HS, Jeong HJ, Cha WT, Cho YS, et al. Expression of the mitochondrial ATPase6 gene and Tfam in Down syndrome. Mol Cells 2003;15(2):181–5.
[68] Valenti D, Tullo A, Caratozzolo MF, Merafina RS, Scartezzini P, Marra E, et al. Impairment of F1F0-ATPase, adenine nucleotide translocator and adenylate kinase causes mitochondrial energy deficit in human skin fibroblasts with chromosome 21 trisomy. Biochem J 2010;431(2):299–310.
[69] Jin S, Lee YK, Lim YC, Zheng Z, Lin XM, Ng DP, et al. Global DNA hypermethylation in down syndrome placenta. PLoS Genet 2013;9(6):e1003515.
[70] Eckmann-Scholz C, Bens S, Kolarova J, Schneppenheim S, Caliebe A, Heidemann S, et al. DNA-methylation profiling of fetal tissues reveals marked epigenetic differences between chorionic and amniotic samples. PLoS One 2012;7(6):e39014.
[71] Papageorgiou EA, Karagrigoriou A, Tsaliki E, Velissariou V, Carter NP, Patsalis PC. Fetal-specific DNA methylation ratio permits noninvasive prenatal diagnosis of trisomy 21. Nat Med 2011;17(4):510–3.
[72] Lim JH, Lee da E, Park SY, Kim do J, Ahn HK, Han YJ, et al. Disease specific characteristics of fetal epigenetic markers for noninvasive prenatal testing of trisomy 21. BMC Med Genom 2014;7(1):1.
[73] Guo JU, Su Y, Zhong C, Ming GL, Song H. Emerging roles of TET proteins and 5-hydroxymethylcytosines in active DNA demethylation and beyond. Cell Cycle 2011;10(16):2662–8.
[74] Guo JU, Su Y, Zhong C, Ming GL, Song H. Hydroxylation of 5-methylcytosine by TET1 promotes active DNA demethylation in the adult brain. Cell 2011;145(3):423–34.
[75] Letourneau A, Santoni FA, Bonilla X, Sailani MR, Gonzalez D, Kind J, et al. Domains of genome-wide gene expression dysregulation in Down's syndrome. Nature 2014;508(7496):345–50.
[76] Smith DJ, Stevens ME, Sudanagunta SP, Bronson RT, Makhinson M, Watabe AM, et al. Functional screening of 2 Mb of human chromosome 21q22.2 in transgenic mice implicates minibrain in learning defects associated with Down syndrome. Nat Genet 1997;16(1):28–36.
[77] Becker W, Joost HG. Structural and functional characteristics of Dyrk, a novel subfamily of protein kinases with dual specificity. Prog Nucleic Acid Res Mol Biol 1999;62:1–17.
[78] Fotaki V, Dierssen M, Alcantara S, Martinez S, Marti E, Casas C, et al. Dyrk1A haploinsufficiency affects viability and causes developmental delay and abnormal brain morphology in mice. Mol Cell Biol 2002;22(18):6636–47.

[79] Tejedor F, Zhu XR, Kaltenbach E, Ackermann A, Baumann A, Canal I, et al. Minibrain: a new protein kinase family involved in postembryonic neurogenesis in Drosophila. Neuron 1995;14(2):287–301.
[80] Altafaj X, Dierssen M, Baamonde C, Marti E, Visa J, Guimera J, et al. Neurodevelopmental delay, motor abnormalities and cognitive deficits in transgenic mice overexpressing Dyrk1A (minibrain), a murine model of Down's syndrome. Hum Mol Genet 2001;10(18):1915–23.
[81] Yang EJ, Ahn YS, Chung KC. Protein kinase Dyrk1 activates cAMP response element-binding protein during neuronal differentiation in hippocampal progenitor cells. J Biol Chem 2001;276(43):39819–24.
[82] Bartholdi D, Roelfsema JH, Papadia F, Breuning MH, Niedrist D, Hennekam RC, et al. Genetic heterogeneity in Rubinstein–Taybi syndrome: delineation of the phenotype of the first patients carrying mutations in EP300. J Med Genet 2007;44(5):327–33.
[83] Canzonetta C, Mulligan C, Deutsch S, Ruf S, O'Doherty A, Lyle R, et al. DYRK1A-dosage imbalance perturbs NRSF/REST levels, deregulating pluripotency and embryonic stem cell fate in Down syndrome. Am J Hum Genet 2008;83(3):388–400.
[84] Schoenherr CJ, Anderson DJ. The neuron-restrictive silencer factor (NRSF): a coordinate repressor of multiple neuron-specific genes. Science 1995;267(5202):1360–3.
[85] Sun YM, Greenway DJ, Johnson R, Street M, Belyaev ND, Deuchars J, et al. Distinct profiles of REST interactions with its target genes at different stages of neuronal development. Mol Biol Cell 2005;16(12):5630–8.
[86] Bahn S, Mimmack M, Ryan M, Caldwell MA, Jauniaux E, Starkey M, et al. Neuronal target genes of the neuron-restrictive silencer factor in neurospheres derived from fetuses with Down's syndrome: a gene expression study. Lancet 2002;359(9303):310–5.
[87] Lu P, Roberts CW. The SWI/SNF tumor suppressor complex: regulation of promoter nucleosomes and beyond. Nucleus 2013;4(5):374–8.
[88] Altafaj X, Martin ED, Ortiz-Abalia J, Valderrama A, Lao-Peregrin C, Dierssen M, et al. Normalization of Dyrk1A expression by AAV2/1-shDyrk1A attenuates hippocampal-dependent defects in the Ts65Dn mouse model of Down syndrome. Neurobiol Dis 2013;52:117–27.
[89] De la Torre R, De Sola S, Pons M, Duchon A, de Lagran MM, Farré M, et al. Epigallocatechin-3-gallate, a DYRK1A inhibitor, rescues cognitive deficits in Down syndrome mouse models and in humans. Mol Nutr Food Res 2014;58(2):278–88.
[90] Randazzo FM, Khavari P, Crabtree G, Tamkun J, Rossant J. Brg1: a putative murine homologue of the Drosophila brahma gene, a homeotic gene regulator. Dev Biol 1994;161(1):229–42.
[91] Schofield J, Isaac A, Golovleva I, Crawley A, Goodwin G, Tickle C, et al. Expression of Drosophila trithorax-group homologues in chick embryos. Mech Dev 1999;80(1):115–8.
[92] Eroglu B, Wang G, Tu N, Sun X, Mivechi NF. Critical role of Brg1 member of the SWI/SNF chromatin remodeling complex during neurogenesis and neural crest induction in zebrafish. Dev Dyn 2006;235(10):2722–35.
[93] Gibbons RJ, Bachoo S, Picketts DJ, Aftimos S, Asenbauer B, Bergoffen J, et al. Mutations in transcriptional regulator ATRX establish the functional significance of a PHD-like domain. Nat Genet 1997;17(2):146–8.
[94] Villard L, Toutain A, Lossi AM, Gecz J, Houdayer C, Moraine C, et al. Splicing mutation in the ATR-X gene can lead to a dysmorphic mental retardation phenotype without alpha-thalassemia. Am J Hum Genet 1996;58(3):499–505.
[95] Valenti D, De Rasmo D, Signorile A, Rossi L, de Bari L, Scala I, et al. Epigallocatechin-3-gallate prevents oxidative phosphorylation deficit and promotes mitochondrial biogenesis in human cells from subjects with Down's syndrome. Biochim Biophys Acta 2013;1832(4):542–52.
[96] Luger K, Dechassa ML, Tremethick DJ. New insights into nucleosome and chromatin structure: an ordered state or a disordered affair? Nat Rev Cell Biol 2012;13(7):436–47.
[97] Creppe C, Posavec M, Douet J, Buschbeck M. MacroH2A in stem cells: a story beyond gene repression. Epigenomics 2012;4(2):221–7.
[98] NCBI Gene. H2B histone family member S (pseudogene) [*Homo sapiens* (human)] [Internet], 2016. Available from: http://www.ncbi.nlm.nih.gov/gene/54145

[99] Kaufman PD, Kobayashi R, Kessler N, Stillman B. The p150 and p60 subunits of chromatin assembly factor I: a molecular link between newly synthesized histones and DNA replication. Cell 1995;81(7):1105–14.
[100] Verreault A, Kaufman PD, Kobayashi R, Stillman B. Nucleosome assembly by a complex of CAF-1 and acetylated histones H3/H4. Cell 1996;87(1):95–104.
[101] Brink MC, Piebes DG, de Groote ML, Luijsterburg MS, Casas-Delucchi CS, van Driel R, et al. A role for MeCP2 in switching gene activity via chromatin unfolding and HP1gamma displacement. PLoS One 2013;8(7):e69347.
[102] Samaco RC, Neul JL. Complexities of Rett syndrome and MeCP2. J Neurosci 2011;31(22):7951–9.
[103] Dekker AD, Strydom A, Coppus AMW, Nizetic D, Vermeiren Y, Naude PJW, et al. Behavioural and psychological symptoms of dementia in Down syndrome: early indicators of clinical Alzheimer's disease? Cortex 2015;73:36–61.
[104] Bennett DA, Yu L, Yang J, Srivastava GP, Aubin C, De Jager PL. Epigenomics of Alzheimer's disease. Transl Res 2015;165(1):200–20.
[105] Blanch M, Mosquera JL, Ansoleaga B, Ferrer I, Barrachina M. Altered mitochondrial DNA methylation pattern in Alzheimer disease-related pathology and in Parkinson disease. Am J Pathol 2016;186(2): 385–97.
[106] Ryoo SR, Cho HJ, Lee HW, Jeong HK, Radnaabazar C, Kim YS, et al. Dual-specificity tyrosine(Y)-phosphorylation regulated kinase 1A-mediated phosphorylation of amyloid precursor protein: evidence for a functional link between Down syndrome and Alzheimer's disease. J Neurochem 2008;104(5):1333–44.
[107] Liu F, Liang Z, Wegiel J, Hwang YW, Iqbal K, Grundke-Iqbal I, et al. Overexpression of Dyrk1A contributes to neurofibrillary degeneration in Down syndrome. FASEB J 2008;22(9):3224–33.
[108] Tan L, Yu JT, Hu N. Noncoding RNAs in Alzheimer's disease. Mol Neurobiol 2013;47(1):382–93.
[109] Morris KV. RNA-directed transcriptional gene silencing and activation in human cells. Oligonucleotides 2009;19(4):299–306.
[110] van den Berg A, Mols J, Han J. RISC-target interaction: cleavage and translational suppression. Biochim Biophys Acta 2008;1779(11):668–77.
[111] Farazi TA, Juranek SA, Tuschl T. The growing catalog of small RNAs and their association with distinct Argonaute/Piwi family members. Development 2008;135(7):1201–14.
[112] Mercer TR, Mattick JS. Structure and function of long noncoding RNAs in epigenetic regulation. Nat Struct Mol Biol 2013;20(3):300–7.
[113] Lukiw WJ. Micro-RNA speciation in fetal, adult and Alzheimer's disease hippocampus. Neuroreport 2007;18(3):297–300.
[114] Zigman WB. A typical aging in down syndrome. Dev Disabil Res Rev 2013;18(1):51–67.
[115] Horvath S. DNA methylation age of human tissues and cell types. Genome Biol 2013;14(10):R115.
[116] Horvath S, Garagnani P, Bacalini MG, Pirazzini C, Salvioli S, Gentilini D, et al. Accelerated epigenetic aging in Down syndrome. Aging Cell 2015;14(3):491–5.
[117] Wiseman FK, Alford KA, Tybulewicz VL, Fisher EM. Down syndrome-recent progress and future prospects. Hum Mol Genet 2009;18(R1):R75–83.
[118] Mangialasche F, Solomon A, Winblad B, Mecocci P, Kivipelto M. Alzheimer's disease: clinical trials and drug development. Lancet Neurol 2010;9(7):702–16.
[119] Kishnani PS, Heller JH, Spiridigliozzi GA, Lott IT, Escobar L, Richardson S, et al. Donepezil for treatment of cognitive dysfunction in children with Down syndrome aged 10-17. Am J Med Genet A 2010;152A(12):3028–35. (Division of Medical Genetics, Department of Pediatrics, Duke University Medical Center, Durham, North Carolina 27710, USA. kishn001@mc.duke.edu: Wiley-Liss, Inc.).
[120] US National Institutes of Health. Alzheimer's disease [Internet]. ClinicalTrials.gov. 2016. Available from: https://clinicaltrials.gov/ct2/results?term=Alzheimer%27s+disease&Search=Search
[121] Valdespino V, Valdespino PM. Potential of epigenetic therapies in the management of solid tumors. Cancer Manag Res 2015;7:241–51.

[122] US Food Drug Administration. Drugs@FDA: FDA Approved Drug Products [Internet], 2016. Available from: http://www.accessdata.fda.gov/scripts/cder/drugsatfda/index.cfm

[123] Papi A, Ferreri AM, Rocchi P, Guerra F, Orlandi M. Epigenetic modifiers as anticancer drugs: effectiveness of valproic acid in neural crest-derived tumor cells. Anticancer Res 2010;30(2):535–40.

[124] European Medicines Agency. European public assessment reports (EPAR) for human medicines [Internet], 2016. Available from: http://www.ema.europa.eu/ema/index.jsp?curl=pages/medicines/landing/epar_search.jsp&mid=WC0b01ac058001d124

[125] de la Torre R, de Sola S, Hernandez G, Farré M, Pujol J, Rodriguez J, et al. Safety and efficacy of cognitive training plus epigallocatechin-3-gallate in young adults with Down's syndrome (TESDAD): a double-blind, randomised, placebo-controlled, phase 2 trial. Lancet Neurol 2016;15:801–10.

[126] Fang MZ, Wang Y, Ai N, Hou Z, Sun Y, Lu H, et al. Tea polyphenol (−)-epigallocatechin-3-gallate inhibits DNA methyltransferase and reactivates methylation-silenced genes in cancer cell lines. Cancer Res 2003;63(22):7563–70.

[127] Nandakumar V, Vaid M, Katiyar SK. (−)-Epigallocatechin-3-gallate reactivates silenced tumor suppressor genes, Cip1/p21 and p16INK4a, by reducing DNA methylation and increasing histones acetylation in human skin cancer cells. Carcinogenesis 2011;32(4):537–44.

[128] Ramakrishna N, Meeker HC, Patel S, Brown TW, El Idrissi A. Regulation of α-synuclein expression in Down syndrome. J Neurosci Res 2012;90(8):1589–96.

[129] Ramakrishna N, Meeker HC, Brown WT. Novel epigenetic regulation of alpha-synuclein expression in Down syndrome. Mol Neurobiol 2014;53(1):155–62.

[130] de Groote ML, Verschure PJ, Rots MG. Epigenetic editing: targeted rewriting of epigenetic marks to modulate expression of selected target genes. Nucleic Acids Res 2012;40(21):10596–613.

[131] Jurkowski TP, Ravichandran M, Stepper P. Synthetic epigenetics-towards intelligent control of epigenetic states and cell identity. Clin Epigenet 2015;7(1):18.

[132] Rivenbark AG, Stolzenburg S, Beltran AS, Yuan X, Rots MG, Strahl BD, et al. Epigenetic reprogramming of cancer cells via targeted DNA methylation. Epigenetics 2012;7(4):350–60.

[133] Siddique AN, Nunna S, Rajavelu A, Zhang Y, Jurkowska RZ, Reinhardt R, et al. Targeted methylation and gene silencing of VEGF-A in human cells by using a designed Dnmt3a-Dnmt3L single-chain fusion protein with increased DNA methylation activity. J Mol Biol 2013;425(3):479–91.

[134] Chen H, Kazemier HG, de Groote ML, Ruiters MH, Xu GL, Rots MG. Induced DNA demethylation by targeting Ten-Eleven Translocation 2 to the human ICAM-1 promoter. Nucleic Acids Res 2014;42(3):1563–74.

[135] Maeder ML, Angstman JF, Richardson ME, Linder SJ, Cascio VM, Tsai SQ, et al. Targeted DNA demethylation and activation of endogenous genes using programmable TALE-TET1 fusion proteins. Nat Biotechnol 2013;31(12):1137–42.

[136] Falahi F, Huisman C, Kazemier HG, van der Vlies P, Kok K, Hospers GA, et al. Towards sustained silencing of HER2/neu in cancer by epigenetic editing. Mol Cancer Res 2013;11(9):1029–39.

[137] Heller EA, Cates HM, Peña CJ, Sun H, Shao N, Feng J, et al. Locus-specific epigenetic remodeling controls addiction- and depression-related behaviors. Nat Neurosci 2014;17(12):1720–7.

[138] Konermann S, Brigham MD, Trevino AE, Hsu PD, Heidenreich M, Cong L, et al. Optical control of mammalian endogenous transcription and epigenetic states. Nature 2013;500(7463):472–6.

[139] Bustos FJ, Varela-Nallar L, Aguilar R, Henriquez B, Falahi F, Rots MG, et al. Epigenetic editing of the PSD95 gene promoter impacts neuronal architecture. Neuroscience 2014, Society for Neuroscience; Washington DC.

[140] Berger SL, Kouzarides T, Shiekhattar R, Shilatifard A. An operational definition of epigenetics. Genes Dev 2009;23(7):781–3.

CHAPTER

EPIGENETICS AND MULTIPLE SCLEROSIS

10

L. Kular*, G. Castelo-Branco, M. Jagodic***

**Department of Clinical Neuroscience, Center for Molecular Medicine, Karolinska Institutet, Stockholm, Sweden;*
***Laboratory of Molecular Neurobiology, Department of Medical Biochemistry and Biophysics, Karolinska Institutet, Stockholm, Sweden*

CHAPTER OUTLINE

10.1 MULTIPLE SCLEROSIS: INTRODUCTION

Multiple sclerosis (MS) is a chronic inflammatory disease of the central nervous system (CNS) characterized by autoimmune destruction of myelin and subsequent loss of neurons. Today MS is the most common cause of nontraumatic disability in young adults in Europe and it affects around 2.5 million people world wide. The disease is most often diagnosed between 20 and 40 years of age, the peak years for education, career- and family-building, with women being affected nearly 3 times as often as men. MS leads to a marked reduction in quality of life in affected individuals and associates with extremely high costs for society. The disease is characterized by the breakdown of the blood brain barrier, infiltration of immune cells into the CNS and subsequent development of inflammatory and demyelinating lesions, mainly composed of lymphocytes and macrophages. Myelin is a lipid-rich layer produced by

Neuropsychiatric Disorders and Epigenetics. http://dx.doi.org/10.1016/B978-0-12-800226-1.00010-1

glial cells, that is, oligodendrocytes, which enwraps neuronal axons allowing fast transmission of electrical impulses within the neuronal circuitry and metabolic support to neurons [1]. Demyelination leads to slower impulse transmission and ultimately to neuronal degeneration and glial scar formation [2].

Symptoms greatly vary depending on the localization of the lesions in the CNS. Most MS patients (~85%) initially present with the relapsing-remitting form of MS (RRMS) that is, characterized by recurring episodes of acute neurological symptoms (relapses) followed by complete or often partial recovery (remission). This recovery associates with immune-regulation in the CNS and is in part attributed to recruitment of endogenous oligodendrocyte precursor cells (OPCs) to the sites of lesions, where they differentiate and remyelinate neuronal axons [1]. The initial relapsing-remitting phase is quite successfully treated with several disease-modifying therapies that broadly target the immune system and are associated with adverse effects. However, the majority of RRMS patients eventually convert to a more progressive form of MS, that is, secondary-progressive MS (SPMS), for which no treatments exist today. This stage is characterized by failed differentiation of OPCs and consequently failed remyelination, accumulating axonal damage and neuronal loss and persistent increase in neurological disability.

Besides neurological deficits (usually motor, sensory and visual), neuropsychiatric comorbidity is highly prevalent in MS patients. The prevalence of psychiatric disorders is high already at initial stages of the disease and increases further over time, partly depending on the type of disease course. In particular, depression, anxiety and to a lesser extent bipolar disorder, occur significantly more often in the MS population than in the general population, depression being the most common psychiatric comorbidity with a lifetime prevalence up to 50% [3]. A possible association between MS and other psychiatric comorbidities, such as psychoses and alcohol/substance abuse, is more uncertain, with conflicting data in the literature. In contrast, cognitive impairment is very common among MS patients, with prevalence rates ranging from 40% to 70% [4]. The most typical cognitive impairments are reductions in information processing speed and episodic memory, but often also includes impaired attention and executive functions, such as verbal fluency [4]. Importantly, neuropsychiatric comorbidity is associated with a lower quality of life, more fatigue, and decreased adherence to disease-modifying therapy in MS patients and is as such among the most important factors affecting quality of life, overall health and disability among patients with MS [3,4].

Although the cause of MS remains unknown, vast epidemiological data establish MS as a complex disease influenced by genetic and environmental factors. The involvement of genetic factors was suggested already at the end of the 19th century when it was recognized that MS often clusters in families. The risk of developing MS is higher in relatives of MS patients and it varies with the relatedness supporting a disease development due to the coinheritance of risk genes [5]. However, these data alone are not sufficient to distinguish the relative roles of genes and environment. On the other hand, significantly higher concordance rate of MS in monozygotic (MZ) twins (~25%) compared to dizygotic twins (~5%) favors genetic factors over the shared in utero and early childhood environment [6]. This is further supported by studies in adoptees [7], half-siblings [8], and conjugal MS pairs [9] demonstrating that genetic sharing, rather than the shared early or adulthood environment, underlies the familial aggregation. Heritability of MS was estimated to most likely follow a polygenic model with one locus of a moderate and many loci of modest effects. Indeed, the first and the strongest genetic risk factor was mapped to the human leukocyte antigen (HLA) class II complex already in early 1970s [10] and this association was later refined to a specific allele of the *HLA-DRB1* gene, *HLA-DRB1*15:01* [11]. With the advent of genome-wide association studies in large well-powered cohorts more than 100 non-HLA variants have been identified to predispose to MS together with multiple variants and alleles within the HLA locus itself [11,12]. The HLA class II locus molecules are expressed on antigen presenting cells (APCs) and present peptide antigens to CD4+ T lymphocytes, thereby having a critical role in the initiation of the antigen-specific immune response. It is generally accepted that the disease-associated variants primarily influence structure of the peptide-binding groove of the class II molecules that could lead to breakdown of tolerance and harmful T-cell responses against CNS antigens. Indeed, myelin-specific T cells are found with an increased frequency in MS patients compared to healthy individuals [13,14]. Myelin-specific CD4+ T cells extracted from naive primates can induce lesions similar to MS when expanded and transferred to naïve donors [15] and, an MS-like disease, experimental autoimmune encephalomyelitis (EAE), can be induced in rodents upon transfer of CNS-specific CD4+ T cells to naïve recipients [16]. Moreover, transgenic mice that express MS-associated HLA genes and T-cell receptors specific for myelin antigens develop spontaneous EAE [17]. Nevertheless, besides CD4+ cells, nearly every cell type of the adaptive and innate immune systems, including CD8+ T and B lymphocytes, monocytes, macrophages, dendritic cells, NK and NK T cells, have been implicated in the immunopathology of MS [18].

The epidemiological data, in particular modest concordance rate of MS in MZ twins [6], the strong latitude gradient [19] and the migration studies [20] suggest nonetheless an important role of environmental factors that act at the population level. The risk of developing MS has consistently been associated with Epstein–Barr virus (EBV) infection, smoking and low exposure to ultraviolet radiation (UVR) and low vitamin D levels [21]. More recently, associations with increased body mass index (BMI), night-shift work, and exposure to organic solvents, among others, have been reported [21]. It has been speculated that changes in environmental factors may account for the increasing incidence of MS observed during the past several decades [22]. Interestingly, many environmental factors seem to act particularly during certain susceptible windows encompassing childhood and adolescence, implicating some type of "cellular memory" of early exposures. This has been observed already in migration studies that demonstrated a reduced risk in migrants moving from a high- to a low-risk area during the first two decades of life [20]. Interestingly, migration from a low- to a high-risk area did not increase the risk of MS in the migrants but it did so in their children [20]. More recently specific environmental factors have been associated with adolescence, including infectious mononucleosis [23], night-shift work [21] and increased BMI [21]. However, the causal relationship and the molecular mechanisms underlying the effect of these factors on MS development are still largely unknown.

Epigenetic mechanisms lie at the interface between the genome and external signals and environmental changes are closely associated with changes of epigenetic landscapes in cells. Virtually all cells in the organism share an identical genome, since they are derived from one single cell, the zygote. Nevertheless, different cell types in the organism will exhibit diverse patterns of chemical modifications deposited in their genome, such as DNA methylation and histone posttranslational modifications (PTMs), leading to unique transcriptional profiles and subsequently to different cellular phenotypes [24]. These epigenetic modifications can be manifested at (1) specific gene loci level, modulating transcription; (2) at a cell/tissue level, with potential inheritance during the life span of the individual; (3) at the organism level, with potential transgenerational inheritance. Some of these epigenetic modifications have been described to be inherited through the cell cycle, and might persist in the absence of the original stimuli. As such, epigenetic modifications induced in certain cells during early stages in life might persist and only phenotypically manifest in later stages. Moreover, if epigenetic modifications are passed to the next generation through meiosis via the sperm, they can potentially lead to

BOX 10.1 EVIDENCE FOR THE INVOLVEMENT OF EPIGENETIC MECHANISMS IN THE PATHOGENESIS OF MS

- *Low concordance rate of MS in MZ twins* [6]: epigenetic changes that arise in nearly identical twin genomes [28,29] and get more pronounced in older twins [28] can account for discordant phenotypes.
- *Effect of parental origin on the risk of MS and EAE* [30–33]: penetrance of risk alleles and overall risk to develop disease is dependent on whether the alleles are inherited from a mother or a father, which implicates, beside genetic mechanisms, epigenetic mechanisms.
- *"Hidden heritability" in MS* [11,12]: the fact that all identified genes do not fully account for disease variance and heritability suggests, among other causes, epigenetic mechanisms.
- *Long-term impact of environmental exposures* [20,21,23,34,35]: epigenetic changes might explain an increased risk to develop MS that occurs years or decades after environmental exposures and in some instances when environmental triggers cease to exist (cellular memory).
- *Differences in DNA methylation and histone PTMs in MS patients* [36–42]: differences in epigenetic marks, such as DNA methylation and histone modifications between MS cases and relevant controls suggest the role of epigenetics in MS pathogenesis.

transgenerational inheritance of acquired traits. This epigenetic inheritance might be operational in mammalian organisms [25–27], although it is still unclear to what extent, and the underlying mechanisms are not known. In this chapter, we will focus on epigenetic modifications at a transcriptional and cellular/tissue level. Box 10.1 summarizes key observations that suggest involvement of epigenetic mechanisms in the pathogenesis of MS.

10.2 OVERVIEW OF EPIGENETIC STUDIES PERFORMED IN MS

In this chapter, we will review the outcome of studies that have investigated epigenetic changes in MS. We will focus on the most studied epigenetic modifications, namely DNA methylation and histone PTMs. The negatively charged genomic DNA in any given cell is densely compacted in the nucleus and associated with highly positively charged proteins called histones, in a structure called chromatin. Activation or repression of specific genes in the genome depends on the accessibility of the transcriptional machinery (i.e., RNA polymerase II and associated complexes) to their specific loci. This accessibility is dependent on epigenetic information, DNA methylation and histone PTMs, deposited on the loci of the individual genes, which will determine the level of compaction of the local chromatin and provide docking sites for proteins and noncoding RNAs (ncRNAs) associated with the transcriptional machinery. ncRNAs have in the last years been increasingly implicated in epigenetic regulation, but there is still very sparse data involving ncRNAs and MS. As such, in this chapter, we will focus on DNA methylation and histone PTMs.

10.2.1 DNA METHYLATION

The most intensely studied epigenetic mechanism is the covalent addition of a methyl group to cytosine primarily in the context of CpG dinucleotides. DNA methylation is acquired and faithfully maintained through cell division by the action of DNA methyltransferases (DNMTs) [43]. DNA methylation is established by the de novo methyltransferases, DNMT3A and DNMT3B, and recognized and copied on the newly synthetized daughter strand by DNMT1 during cellular division, thereby assuring propagation of the methylation patterns [43]. This model of rapid and stable DNA methylation propagation has been further revised by the recent discovery of the ten-eleven translocation family of proteins (TETs) that can oxidize 5-methylcytosine (5mC) to 5-hydroxymethyl cytosine (5hmC), a key intermediate in the process of active demethylation [44]. Importantly, recent studies have revealed that 5hmC is a stable and highly abundant modification in certain tissues, such as those in the CNS and displays a unique genomic distribution compared to 5mC. Emerging evidence further suggests distinct functional roles of methylation and hydroxymethylation. Indeed, while DNA methylation is a well-known repressive mark when associated with CpG-rich promoter regions mediating long-term transcriptional gene silencing, DNA hydroxymethylation in the active body of genes positively correlates with transcriptional activity. This growing evidence of the diversity of DNA modifications that are acquired in a tissue- and time-dependent manner and whose context-specific functional roles are still relatively unknown further increases the complexity of interpreting a given methylome in a clinical context.

Since DNA methylation can be analyzed by high-throughput methods in relatively small amounts of material, it has become the epigenetic modification of choice in clinical studies. Several studies have used these genome-wide DNA methylation technologies to assess differentially methylated positions (DMPs) and regions (DMRs) in immune cells and brains from MS patients compared to healthy controls. Details of the cohorts and methodologies employed in these studies are presented in Table 10.1. Nevertheless, it is noteworthy that conventional bisulfite-based methods (e.g., whole-genome bisulfite sequencing, Illumina 450K arrays) used in DNA methylation studies do not allow discrimination between 5mC and 5hmC.

10.2.1.1 DNA Methylation Changes in Immune Cells in MS

The first study aiming at identifying epigenetic marks of immune cells in MS has explored genome-wide methylation differences in CD4+ T cells from three MS-discordant MZ twin pairs [37]. Using reduced representation bisulfite sequencing (RRBS) covering > 1.7 million CpGs, the authors found very few, albeit large, variations in DNA methylation between discordant MZ twin pairs. However, these changes were not consistent between different twin pairs. Owing to the small sample size, the heterogeneity of the individuals, the limitations of methodology (in terms of low genome coverage, bias for CpG-rich regions, and the focus on only very large differences) relevant changes might have been missed. More recently, case-controls studies have investigated genome-wide DNA methylation alterations in CD4+ and CD8+ T cells sorted from peripheral blood of MS patients and healthy controls using the Illumina 450K array, a methodology that targets a small portion of CpGs (~485,000 CpGs representing 1.8% of total CpGs albeit covering 99% of the genes) [39,41,42]. However, these collective efforts to characterize the methylome signature of relevant cell types in MS have revealed difficulties in yielding consistent findings between clinical cohorts.

Two studies have investigated DNA methylation differences in CD4+ and CD8+ T cells in the same cohort of 30 RRMS patients and 28 healthy controls and identified 74 and 79 significant DMPs between MS patients and healthy controls, respectively, with minor overlap found between the two T cell subtypes [39,42]. CD4+ T cells from MS patients revealed a distinct differential methylation signal on chromosome 6, representing 25% of the DMP probes and localizing to the HLA class II locus, specifically the major MS risk gene *HLA-DRB1*. This *HLA-DRB1* cluster of CpGs exhibits strong hypomethylation, which appears to be at least partly associated with the major MS risk variant, the *HLA-DRB1*15:01* allele [39]. A large portion (54%) of the non-HLA DMPs found in CD4+ T cells maps to genes that have previously been reported in the context of MS, however whether these changes

Table 10.1 Genome-Wide DNA Methylation Studies in MS

Number		Gender (F:M)		Age[a]		Disease Status; Duration[a]	Treatment	Source[b]	Technology and Data Analysis	Major findings	References
Cases	Cont.	Cases	Cont.	Cases	Cont.						
30	28	6.5:1	1.2:1	na [20–57]	na [20–57]	RRMS; na	29/30 on various treatments	CD4 and CD8 T cells (90 ± 5%)	• Illumina 450K array[c] • $P < 0.05$ and Δmeth ≥ 0.1 • Subgroup analysis	• 74 and 79 significant DMPs in CD4 and CD8 cells, respectively • Minor overlap between DMPs found in CD4 and CD8 cells • CD4 cells: 66% DMPs map to genes previously implicated in MS (25% DMPs map to the HLA locus); GSEA: Antigen processing and presentation pathway • CD8 cells: DMPs do not map to genes previously implicated in MS; no GSEA	[39,42]
16	14	F	F	38.9 [27–65]	39.2 [27–60]	RRMS; 8.8 [1–35]	Untreated	Whole blood, CD4 and CD8 T cells (≥95%)	• Illumina 450K array[c] • Regression, $P < 0.05$ and Δmeth ≥ 0.05 and FDR • Probes containing SNPs removed	• No genome-wide significant DMPs detected in CD4, CD8, or whole blood • Minor overlap between CD4, CD8, and whole blood DMPs • CD8: majority of DMPs (nominal $P < 0.05$) hypermethylated • Top DMPs not shared with Graves et al. [39] and Maltby et al. [42]	[41]
28	19	1.5:1	0.6:1	52.3 [44–80]	67 [55–81]	PPMS and SPMS; 21.2 [7–45]	na	Brain NAWM	• Illumina 450K array[c] • 1 kb (+/− 500 bp) DMR, 1% FDR • Correction for age and gender • Probes containing common SNPs removed	• 539 genome-wide significant DMPs with subtle changes • Hypermethylated DMRs associated with increased transcript levels, involved in broad range processes and oligodendrocyte/neuronal function • Hypomethylated DMR correlated with decreased gene expression, associated with immune processes • Putative candidates: BCL2L2, HAGHL, NDRG1, CTSZ, LGMN	[40]
10	20	2.3:1	2.3:1	59 [48–75]	59 [45–75]	SPMS; 24.6 [8–48]	na				
MS-discordant MZ twin pairs		Ashkenazi Jewish F African-American F White M		56 39 19		SPMS; 26 RRMS; 1 RRMS; 6	na	CD4 T cells (na)	• RRBS[d] • Δmeth >0.6	• 2, 10, and 176 significant changes detected, respectively • 92% CpG clusters common to siblings within each twin pair • 2 DMPs shared by 2 MZ twin pairs (but in opposite direction)	[37]

Cont., Controls; DMP, differentially methylated position DMR, differentially methylated region; F, female; FDR, false discovery rate; GSEA, gene set enrichment analysis; Δmeth, difference in methylation; M, male; MS, multiple sclerosis; MZ, monozygotic; na, not available; NAWM, normal appearing white matter; PP, primary progressive; RR, relapsing-remitting; SP, secondary progressive.

[a] *Mean [range].*

[b] *Purity of sorted cells shown in bracket.*

[c] *~480,000 CpGs covering 99% annotated genes, representing 1.7% of all CpGs.*

[d] *Reduced representation bisulphite sequencing (RRBS), >1.7 million CpGs, enriched for CpG-rich regions.*

represent cause or a consequence of disease remains to be established. In contrast, in CD8+ T cells, the major DMPs identified at the *HLA-DRB1* gene in CD4+ T cells could not be found, and the cells displayed more discrete changes [42]. Indeed, none of CpGs identified in CD8+ T cells clustered together in DMRs or mapped to previously reported MS genes and gene set enrichment analysis did not result in prominent pathways. The most significant MS-associated CpGs fall into intergenic loci or gene bodies, whose regulatory properties and biological relevance are still unknown.

Conflicting results have been reported in a recent study aiming to characterize DNA methylation variations in whole blood, CD4+ and CD8+ T cells from a smaller albeit more homogenous cohort of 16 RRMS and 14 controls [41]. The authors could not reproduce the findings from Graves et al. [39] and Maltby et al. [42] as none of the top 40 differentially methylated CpGs identified in CD4+ and CD8+ T cells overlapped with the previously identified DMPs. Instead, no significant large-effect DNA methylation differences for CD4+ cells, CD8+ cells and whole blood were observed in this cohort. However, the distribution of hyper- and hypomethylated CpGs showed evidence for predominant hypermethylation in CD8+ T cells from MS patients when compared to controls. The proportion of hypermethylated probes was higher in recently diagnosed patients compared to earlier diagnosed ones and the hypermethylated CpGs were enriched in promoter regions (TSS1500 and first exon) compared to other genomic features (gene body and 3′-UTR). This finding diverges from results in the previously described cohort, which found equal distribution of hyper- and hypomethylated probes in each cell type.

Overall, the disparities between the studies might have resulted from differences in sample size, sample heterogeneity (gender ratio, disease duration, treatment status) of cohorts and stringency of data analyses (see Section entitled Challenges in epigenetic studies). Furthermore, none of the aforementioned studies has associated the identified changes with functional consequences such as transcriptomics. Therefore, larger studies of homogenous MS patients and controls are warranted to further elucidate the impact of smaller DNA methylation changes that may be important in MS pathogenesis.

10.2.1.2 DNA Methylation Changes in the CNS in MS

The first evidence that DNA methylation in the CNS could be involved in MS pathogenesis came from the study that investigated citrullination of myelin basic protein (MBP) [45]. MBP is a major constituent of myelin in the CNS and can show different posttranslational modifications. Citrullination of MBP by peptidyl arginine deaminase 2 (PADI2) induces structural changes of the molecule that can lead to unstable myelin that is more prone to breakdown but also to enhanced T-cell responses against modified MBP [46]. Analyses of normal appearing white matter (NAWM) of brain biopsies from MS patients revealed hypomethylation of the PADI2 promoter, which correlated with an increase in the levels of PADI2 expression [45]. The increased amount of PADI2 enzyme was further associated with higher citrullinated MBP in white matter from MS patients specifically (compared to other MS tissues or other neurological diseases) [45,47]. Interestingly, hypomethylation and upregulation of PADI2 was also found in peripheral blood mononuclear cells (PBMC) from MS patients compared to controls, although the impact in PBMC is yet to be determined [48]. Altogether, these studies suggest that locus-specific DNA methylation changes in the brain, which can be reflected in peripheral blood cells, could contribute to loss of myelin stability in the MS brain and therefore represent a putative mechanism underlying demyelination in MS.

More recently, Huynh et al. [40] investigated genome-wide DNA methylation changes and their functional relevance in pathology-free NAWM brain regions from MS patients compared to unaffected controls. The selection of tissue without any apparent ongoing inflammatory and demyelinating activity was supported by the aim to identify epigenetic changes indicative of environmental influences. Using the 450K arrays in a discovery cohort of 28 MS patients and 19 controls, this study revealed 539 significant DMRs in MS brains compared to controls. These DMRs displayed subtle changes and were preferentially found in genomic locations associated with enhancers. The authors further based their analysis on the directionality of methylation changes and reported that hypo- and hypermethylated regions occur in distinct genomic features and gene ontologies. Indeed, 40% hypomethylated DMRs were significantly enriched in regions surrounding transcription start sites, while the remaining 60% hypermethylated DMRs were preferentially distributed in gene bodies. Using transcriptome analysis (RNA-seq), the authors found a significant canonical anticorrelation between DNA methylation and expression. However, many DMRs did not associate with changes in expression and even displayed noncanonical correlation between DNA methylation and expression. Interestingly, some putative candidates showed biological relevance for brain homeostasis and susceptibility to pathological dysfunction. For example, hypermethylated DMRs and decreased transcript levels were detected for genes regulating oligodendrocyte and neuronal function, including BCL2L2, a member of the antiapoptotic BCL2 family, HAGHL, an hydrolase-like enzyme highly expressed in the brain and NDRG1, a gene involved in oligodendrocyte response to stress. Of special note is that the strategy of analyzing hypo- and hypermethylated loci separately might have been restrictive, since regulated pathways usually include both hypomethylated and hypermethylated genes. Moreover, the use of samples from mixed tissue raises the possibility that specific changes might have occurred in distinct cell types (e.g., mitochondrial dysfunction) or reflect differences in frequencies of different cell types. Finally, given the aforementioned inability to differentiate 5mC from 5hmC using this BS-based technology, hydroxymethylation might account for some of the signals detected in the brain samples. Nonetheless, this study sheds light on the existence of numerous discrete alterations in DNA methylation converging to distinct functions that likely favor imbalance of brain homeostasis and inability to repair after an insult.

10.2.2 HISTONE POSTTRANSLATIONAL MODIFICATIONS

The basic unit of chromatin is the nucleosome, which is composed of 1.67 turns of DNA (147 base pairs) wrapping an octamer complex of eight core histone proteins (dimers of histone H2A, H2B, H3, and H4). Variants of these core histones such as histone H3.3, H2AZ, among others, can replace core histones in the nucleosome and another histone, histone H1, is not part of the octamer, but binds at the entry point of the nucleosome. Histones are basic proteins that are strongly bound to DNA through electrostatic interactions, and their N-terminal peptide tails are highly prone to undergo posttranslational modifications, such as acetylation, methylation, citrullination, ubiquitination, phosphorylation, O-palmitoylation, and ADP ribosylation, among others. Several amino acids in the histone tails can be modified, for example, arginine 2 at histone H3 has been reported to be able to be monomethylated (H3R2me) or citrullinated (H3R2cit), while lysine 4 in histone H3 can be acetylated (H3K4ac) or mono, di, or tri-methylated (H3K4me1, H3K4me2, and H3K4me3). Advances in mass spectrometry have allowed a continuous expansion of the number and the type of posttranslational modifications identified at individual amino acids in histone tails. These PTMs can change the charge of the amino acid (e.g., acetylation and citrullination remove the positive charge of lysines and arginines, respectively), which in turn modulates the interaction with the underlying DNA and thus chromatin compaction. In addition, histone PTMs can constitute docking sites to which proteins can specifically bind. These chromatin "readers" present domains that specifically recognize a particular histone PTM, with for instance proteins with chromodomains recognizing histone methylation (e.g., heterochromatin protein 1, HP1 recognizing H3K9me3), and proteins with bromodomains recognizing histone acetylation.

The development of antibodies targeting specific histone PTMs and technologies as chromatin immunoprecipitation coupled with next generation sequencing have allowed the functional characterization of a myriad of histone PTMs. Histone PTMs have been shown to be present in specific regulatory regions in the genome and to be functionally linked to the transcriptional state of neighboring genes. For example, H3K4me3 and H3K27me3 are mainly localized at promoter regions, near transcription start sites, with H3K4me3 associating with transcriptionally active genes, while H3K27me3 with repressed genes. Treatment with chemical inhibitors of chromatin "writers"/"erasers" leads to global changes in gene transcription, but until recently it was not clear whether this was a causal effect. Recent developments with transcription activator-like effectors (TALEs), Zinc-finger nucleases and CRISPR/deactivated Cas9 technologies have shown that the targeted recruitment of histone modifiers to specific enhancers and/or promoters leads to the deposition of the respective histone PTMs and changes of the transcriptional status of the associated genes [49–51]. Furthermore, embryonic stem cells with genetically modified histone acetylation at specific sites of histone H3.3 also present induction of specific sets of genes [52], supporting a causal relationship between histone PTMs and transcriptional output. Different epigenetic PTMs at histones can constitute a histone code linked to specific transcriptional states. Consequently, the genome-wide pattern of histone PTM deposition can be associated with the transcriptome of any given cell.

10.2.3 HISTONE CHANGES IN IMMUNE CELLS AND IN CNS TISSUE IN MS

Different histone PTMs have been involved in transcriptional regulation in many of the cells involved in the pathology of MS, with histone acetylation being the most widely studied PTM. Acetylation of lysine residues at histone H3 leads to the removal of positive charge and has been mainly associated with decompaction of chromatin and thus transcriptional activation both at a global- and locus-specific level. Histone acetylation is catalyzed by histone acetyltransferases (HATs), while histone deacetylases (HDACs) erase this epigenetic mark. Increased global histone acetylation by inhibition of HDACs has been shown to modulate several cellular processes in a myriad of immune cells such as APCs and T cell subsets (i.e., Th1 cells, Foxp3+ Tregs, and IL-10-producing suppressive myeloid and Tr1 cells) [53]. Accordingly, treatment of EAE in mice and rats with HDAC inhibitors has been shown to lead to symptomatic amelioration, by targeting the immune system [53–57], but also the CNS [53]. Recently, the FDA-approved drug fingolimod, used for the treatment of MS, was shown to target not only the immune system, but to also inhibit HDACs in the brain, leading to a rescue in memory deficits [58]. HDACs and HATs have indeed been shown to exert several roles either inhibiting or promoting the differentiation of neural stem cells and OPCs into the oligodendrocyte lineage, through the transcriptional regulation of transcription factors driving or inhibiting this differentiation [59–64]. The role of HDACs in these processes appears to be context-specific, since they indeed depend on the developmental stage or local signaling factors such as thyroid hormone, which might modulate the mode of action of HDACs/HATs [59–64]. Histone acetylation is altered in the brains of MS patients, with decreased levels in active demyelinating, demyelinated, and remyelinating lesions in early stages of the disease, and slightly increased levels in the NAWM of the frontal cortex in chronic patients [65]. Moreover, single nucleotide polymorphisms (SNPs) in genes encoding HDACs have been suggested to predict changes in brain volume in MS patients [63]. Acetylated histones can also bind chromatin "readers" such as bromodomain proteins. Interestingly, bromodomain inhibitors have recently also been used in epigenetic-based therapies for mouse models of leukemias [66] and other diseases. Given the important role of histone acetylation in MS, it will be interesting to investigate whether these inhibitors can be used as therapeutic agents in MS.

Histone citrullination (or deamination) is another epigenetic PTM that has been mainly associated with decompaction of the chromatin, since it leads to the removal of the positive charge of arginine residues [67]. Histone citrullination is catalyzed by PADIs, which, as mentioned previously, can also citrullinate MBP and affect myelin stability. Histone H3 citrullination has been observed to be increased in the NAWM of MS patients [68], which might indicate a role for PADIs in demyelination and/or remyelination. Citrullination of arginine 8 (H3Cit8) in histone H3 has also been implicated in the immunological process in MS. H3Cit8 in PBMC prevents binding of the heterochromatin protein 1 (HP1) to neighboring H3K9me3, leading to derepression of cytokines such as TNF and IL8 and of human endogenous retrovirus [69]. H3cit8K9me3 is increased markedly in these cells in MS patients. The PADI inhibitor Cl-amidine can revert this derepression and interestingly, another PADI inhibitor, 2-chloroacetamidine (2CA), can lead to amelioration in mouse EAE, in a mechanism involving reduction of the pathogenic T cells [70]. As such, PADIs might be good targets for epigenetic therapy in MS, by modulating the autoimmune attack, but possibly also the remyelination process in the target tissue.

Recent studies suggest a putative role of different histone methylations in processes affected in MS. The active mark H3K4me3 has also been reported to be reduced in neurons from the gray matter of MS patients [71]. The H3K27me3 demethylase Jmjd3 is required for Th17 cell differentiation (but not Th1, Th2, and Treg) and Jmjd3 conditional knockout mice in T cells (Cd4-Cre) are refractory for developing symptoms upon EAE induction [72]. A Jmjd3 inhibitor, GSKJ4, has been used effectively in mouse models of pediatric glioma [73], and thus could be a potential small molecule for epigenetic therapy in MS. However, deletion of H3K27me3 methyltransferase EZH2 in leukocytes (Cd11c-Cre) also restricts EAE disease progression, most likely by regulating their migration to the site of inflammation [74]. These results might indicate that H3K27me3 represses differentiation of Th17 cells, but is required for their migration to the target tissue. Another repressive histone methylation mark, H3K9me3, is important for oligodendrocyte differentiation [75]. Interestingly, the antimuscarinic compound clemastine has been shown to promote oligodendrocyte differentiation by increasing H3K9me3 and rescuing social isolation behavior that has been associated with decreased myelination [76]. Clemestine, a FDA- approved compound, also leads to enhance remyelination in focal models of demyelination [81a], suggesting that it might have potential as a therapeutic agent in MS.

The studies mentioned earlier clearly indicate that histone PTMs play an important role in the pathology and regeneration in MS. Nevertheless, it is still unclear in which cells many of these PTMs are operational, and whether they regulate transcription in a global manner and/or if they regulate specific loci involved in MS. Genome-wide studies in MS patient samples are lacking. Such investigations, coupled with the development of specific inhibitors/activators of enzymes that catalyze the deposition or removal of the histone PTMs, and devising strategies to target them directly to the cell type of interest in MS, will be crucial for the development of epigenetic therapy and to prevent adverse effects.

BOX 10.2 CHALLENGES IN EPIGENETIC STUDIES

- *Cohort and sample confounders*: epigenetic marks in different tissues and cell types are highly diverse and influenced by genetic variation, environmental factors, therapy or on-going disease processes in a temporal manner.
- *Considerations: well-powered and carefully characterized, homogenous cohorts, cell type–specific investigations, validation and replication.*
- *Technical issues*: various genome-wide technologies are available and differ in genome and CpG coverage, required input, specificity (e.g., BS-based technologies do not discriminate between 5mC and 5hmC), quantification accuracy, analytical approaches and costs.
- *Considerations: use of relevant technique (e.g., oxBS treatment to establish true levels of 5mC), comprehensive approaches, standardized protocols, and analytical methods.*
- *Biological relevance*: (1) the functional relevance of epigenetic changes (in the context of genomic location, amplitude and directionality of changes, single versus multiple changes, cross-talk with other epigenetic mechanisms etc.) is largely unknown and (2) the causality (i.e., whether epigenetic changes are a cause or consequence of disease) is difficult to establish.
- *Considerations: (1) combined studies of transcriptome, DNA methylation, histone PTMs, and 3C-based chromatin studies, targeted functional investigation (e.g,. in vitro reporter systems and in vitro/in vivo epigenome editing); (2) inference of causality using analytical methods (causal inference approach (CIT) or Mendelian randomization), longitudinal cohorts, targeted epigenome editing.*

10.3 CHALLENGES IN EPIGENETIC STUDIES

The spatiotemporal dynamic nature of the epigenome renders investigation of epigenetic changes in clinical settings challenging [77]. Undoubtedly, the involvement of multiple cell types, both immune and CNS, in combination with limited availability of the key cellular players adds additional strains on epigenetic research in MS. Moreover, the diversity of available technologies and the lack of standardized analytical methods to investigate epigenomes further increase the difficulty of replicating findings. Therefore, cohort and sample homogeneity and analytical normalization appear as a necessary requirement for the identification of robust candidate genes and pathways that display altered epigenetic patterns in MS patients. Ultimately, the biological relevance of the identified changes is still to be understood and functional studies are warranted in order to determine the potential of epigenetics to aid better understanding of disease pathogenesis, diagnosis, prediction of disease progression, and treatment of MS patients. Key aspects that make epigenetic research in clinical settings challenging are outlined in Box 10.2.

10.4 POTENTIAL ROLES OF EPIGENETIC CHANGES IN MS

Epigenetic patterns are set during cell commitment and differentiation and ensure the proper functioning of a particular cell type. Therefore, any disturbance of the established epigenetic patterns might result in a breakdown of the homeostatic state and eventually in aberrant pathogenic behavior. For instance, data from murine models and in vitro studies have shown that inhibiting DNA methylation is sufficient to activate the expression of normally silenced immune genes, or the overexpression of costimulatory molecules [78]. Importantly, interference with DNA methylation can also convert antigen-specific CD4+ T cells into autoreactive cells (i.e., responsive of the self-HLA class II molecule), leading to autoimmune-like disease [78]. More generally, aberrant epigenetic mechanisms triggered by genetic, environmental, and stochastic signals can virtually affect any immune or CNS cell type implicated in MS and thereby pathogenic processes.

In this chapter we will hypothesize on potential roles of epigenetic mechanisms in MS development and progression. Disease development requires genetic susceptibility, environmental factors, and opportune circumstances operating together to trigger onset. A growing body of evidence suggests that epigenetic modulation may mediate risk in MS and can act as mediator of genetic and environmental (external as well as internal) factors.

10.4.1 POTENTIAL ROLE OF EPIGENETIC MECHANISMS IN MEDIATING GENETIC RISK

10.4.1.1 Genome-Wide Effects

Several well-established genomic loci that predispose for MS comprise genes that encode either enzymes involved in the epigenetic machinery or DNA binding proteins with chromatin modifying functions. It is tempting to speculate that genetic variations that affect epigenetic "writers," "erasers," or "readers" can predispose the carriers to altered epigenomes that, at least in part, contribute to MS development or progression. The most interesting example includes the MS risk locus on chromosome 4 encompassing the TET2 gene [12], which encodes dioxygenase that catalyzes the conversion of 5mC into 5hmC and plays a key role in the process of active DNA demethylation. As previously mentioned, 5hmC, which is particularly abundant in nervous tissue, has recently been established as a novel stable epigenetic modification with often opposite effects compared to 5mC. The disease-predisposing variant could act by altering levels of TET2 expression leading to changes in 5hmC levels of target genes and consequently their altered expression. This idea is supported by recently reported reduced levels of TET2 in PBMC of MS patients, although the link between the risk allele at the TET2 locus and TET2 levels has not been investigated [79]. Moreover, this study reports reduced global hydroxymethylation levels in MS patients compared to healthy controls, which supports a functional link between lower TET2 levels and decreased conversion of 5mC into 5hmC. Apart from impacting hydroxymethylation marks, lower TET2 might result in a reduced ability to demethylate and thus exert functional impact in MS through changes in DNA methylation too. Recently reported increased genome-wide locus-specific methylation observed in CD8+ T cells of MS patients compared to controls further supports this hypothesis [41], although a similar hypermethylation pattern has not be reported in another study, as previously discussed [42].

Several other genes that encode proteins involved in histone modifications and chromatin remodeling are encoded in MS risk loci [12]. The JARID2 gene on chromosome 6 encodes a protein that interacts with the Polycomb Repressive Complex 2 (PRC2) and regulates its enzymatic activity and its targeting to chromatin. PCR2 in turn establishes repressive chromatin marks on target genes by inducing histone H3K27 methylation. The NCOA1 gene on chromosome 2 encodes a transcriptional coregulator of nuclear receptors and exerts histone acetyltransferase activity. The RCOR1 gene on chromosome 14 encodes a protein also called CoREST that is an essential component of a complex which represses transcription by recruiting a variety of histone modifying enzymes, methyl-DNA binding proteins, and components of the SWI/SNF chromatin remodeling complex. Although the exact pathogenic variations and the genes and functions that they affect remain to be established in these loci, it is tempting to speculate that genome-wide changes in epigenetic states might mediate a fraction of genetic risk in MS.

10.4.1.2 Locus-Specific Effects

A growing body of evidence suggests that genetic and epigenetic modifications can interact biologically. This paradigm has been instrumental in deciphering the contribution of DNA methylation to the genetic risk predisposing to other complex diseases. Indeed, a major challenge is to decipher whether epigenetic changes are a consequence of disease or a possible cause, thereby mediating the risk. Importantly, it has been shown that a significant portion of variation in DNA methylation can be explained by the genotype [80], such that individual genotype at a given locus may result in different patterns of DNA methylation, due to allele-specific methylation. These sites are called methylation quantitative trait loci (meQTLs). DNA methylation status at a given locus may be controlled by local SNPs (that disrupt the CpG site) but also by proximal (short-range meQTL) and distal (long-range meQTL) SNPs, which affect DNA methylation in *cis* or *trans*. The mechanisms underlying the latter are still poorly understood but likely rely on physical and functional interactions between distant loci through chromatin rearrangements. Within this conceptual frame, a recent model proposes that genetic variants might regulate phenotypic variability in addition to mean phenotype and that this connection between genotype and phenotypic plasticity would be mediated epigenetically [81]. Such a mechanism would provide a basis for an epigenetic role in natural selection because the variants themselves would be transmitted genetically, but they would also allow increased phenotypic plasticity in response to a varying environment. According to this model, SNPs that are carriers for inheritance of DNA methylation may explain some of the "hidden" heritability as well as the clinical heterogeneity in complex diseases such as MS.

Given the robust genetic association of the HLA locus with susceptibility to, and heritability of, autoimmune diseases and its enrichment for highly heritable CpG methylation [82], emerging evidence supports a role for DNA methylation at the HLA locus in inheritance of genetic risk in immune diseases. Using a three-step filtering process followed by the causal inference test (CIT), several recent studies have identified DMPs or DMRs associated with diseases that are most likely to be acting as mediators of genetic risk rather that being a consequence of disease. Results from this approach suggest that DNA methylation at HLA class II region could mediate genetic susceptibility to immune-mediated diseases such as rheumatoid arthritis [83], type 1 diabetes [84] and food allergy [85]. More recently, a study investigating genetically controlled DNA methylation and gene expression in healthy individuals has revealed a subset of CpGs in the HLA class II locus with highly variable methylation that are genetically controlled by a single distal SNP (long-range meQTL) and multiple local SNPs (short-range meQTLs). Consistent with previous reports, expression of several genes included in this locus was also under genetic control. Interestingly, this distal meQTL was found to be in strong linkage disequilibrium with the major risk variant in MS, *HLA-DRB1*15:01* [86]. This finding suggests that specific *HLA-DRB1* variants would predispose for DNA methylation levels. This is further compatible with aforementioned results showing that the hypomethylation signal observed at several *HLA-DRB1* sites in CD4+ T cells from MS patients might be due to the *DRB1*15:01* haplotype, at least partly [39]. In line with this, a study investigating DNA methylation at *HLA-DRB1* and *-DRB5* from MS patients found significant differences between *HLA-DRB1*1501* hetero- and homozygotes [87]. Importantly, the highly polymorphic and multi-allelic nature of the HLA locus and *HLA-DRB1* gene, in particular, renders investigation of specific alleles difficult. Particular caution should be taken with regard to the methylation status of *HLA-DRB1* assessed by probe-based arrays. In this matter, allele-specific methods would be of great help to confirm the genotype-dependent DNA methylation status at HLA class II locus in MS. Furthermore, the functional consequence of such genetic control in MS patients is yet to be explored. One could hypothesize that *HLA-DRB1*15:01*-genetically controlled DMPs/DMRs might impact global expression of some of the neighboring genes or other outcomes such as alternative splicing or alternative promoter usage within the HLA locus. This would imply that *HLA-DRB1*15:01* predisposes for MS not only through differential amino acid structure but also by additional changes mediated by DNA methylation. Thus, these studies jointly suggest that DNA methylation at HLA class II locus could, at least partly mediate genetic risk in MS. Further investigations are required in order to decipher the putative role of DNA methylation as a mediator of genetic risk in MS.

10.4.2 POTENTIAL ROLES OF EPIGENETIC MODIFICATIONS IN MEDIATING ENVIRONMENTAL RISKS

10.4.2.1 Sunlight Exposure and Vitamin D

Epidemiological studies have established association between MS and sunlight exposure and/or vitamin D levels [21]; however, given their close link (vitamin D forms in the skin upon exposure to ultraviolet radiation from sunlight), it has been difficult to distinguish whether the effect is due to sunlight exposure, vitamin D, or a combination of both. Nevertheless, the protective effect of vitamin D, exerted through the regulation of balance between immunity and tolerance, via actions on APCs, T and B cells, has been extensively demonstrated [88]. In addition to its accepted antiinflammatory role, emerging studies have shown a direct neurotrophic and neuroprotective action [89]. Interestingly, neonatal hypovitaminosis D has been shown to influence risk of MS later in life [90]. Studies conducted in rats have reported that a developmental vitamin D deficiency leads to permanent alterations in the expression of genes involved in redox balance, mitochondrial dysfunction and subsequent impaired synaptic network in the adult brain [91]. Finally, a biological effect of neonatal hypovitaminosis D on neurotransmitter levels remained in adult rat brain and was transmitted to the offspring [92,93]. The effect is consistent with the concept of metabolic "imprinting" involved in intergenerational inheritance of traits. Collectively, these findings support the hypothesis of a long-term effect of vitamin D, which remains even after the period of exposure. This further raises the possibility that vitamin D could induce stable epigenetic modulation of a network of genes through action of its cognate nuclear vitamin D receptor (VDR) (Table 10.2).

Indeed, VDR binding sites in lymphoblastoid cell lines were significantly enriched near loci that associate with autoimmune diseases and with genes, which are differentially expressed between MS and controls [94]. Consistent with this, in vitamin D-sufficient individuals there was an enrichment of VDR binding near autoimmune disease risk loci and genes of importance for T cell differentiation and the amount of VDR binding correlated with vitamin D levels in naïve CD4+ T cells [95]. These data suggest potential mechanisms of interaction of environmental factors, in this case vitamin D, with risk genes. This is of particular relevance since a functional interaction has been demonstrated between vitamin D and the major MS risk variant *HLA-DRB1*15:01* [96]. *HLA-DRB1*15:01* bears a functional VDRE in the proximal region of the gene and binding to VDR of *HLA-DRB1* has been further shown to induce upregulation of the *HLA-DRB1*15:01* allele specifically compared to other haplotypes. Although the molecular mechanism behind this interaction in MS is not clear, one possibility is that vitamin D induces epigenetic changes in the *HLA-DRB1* locus and thereby impacts gene expression. In the context of cancer, the active form of vitamin D was shown to be able to induce locus-specific DNA methylation or demethylation in a cell type-specific manner [97]. Additionally, epigenetic effects of vitamin D are well linked to histone PTMs as the VDR has been shown to interact with several

Table 10.2 Potential Epigenetic Mechanisms Mediating Environmental Exposures in MS

Risk Factor	Effects in EAE/MS	Putative Epigenetic Mechanisms
Vitamin D [88,89,94–100]	• Suppresses antigen-specific T cells and proinflammatory T cells (Th1/Th17) and promotes anti-inflammatory T cells (Th2 and Treg) • Inhibits dendritic cell differentiation and HLA class II expression • Inhibits infiltration of T cell and monocytes in the CNS • Promotes apoptosis of immune cells • Neurotrophic, neuroprotective, promotes neurotransmission	• Epigenetic regulation of VDRE and ISRE-target genes involved in neuroinflammatory processes in lymphocytic and monocytic cells, microglia and astrocytes • Binding of VDR to the MS risk gene HLA-DRB1*15:01 in APCs and regulation of HLA expression via epigenetic mechanisms • Methylation of protein kinases C (e.g., PRKCZ) in T cells and effect on T-cell activation
EBV infection [101–103]	• Molecular mimicry between myelin and EBV peptides presented by HLA molecules and cross-reactivity of CD4 T cells • Sensitization to CNS antigen released after virus-induced bystander damage • Misdirected promotion of CD4+ T cell responses directed against oligodendrocyte-derived αβ-crystallin • Accumulation of EBV-infected autoreactive B cells in the brain	• Stable DNA methylation changes of host genes involved in immune processes in EBV-infected immune cells through regulation of specific HDACs and DNMTs by viral proteins • Epigenetic regulation of latent viral genes in situ in the brain • Control of microRNA network of the host cell and use of viral noncoding RNA to escape immune recognition and defense
Smoking [104–108]	• Enhances oxidative stress (kynurenine/tryptophan pathway) and proinflammatory production from T cells and monocytes • Inhibits Treg development through activation of RAS • Induces activation, viability and proinflammatory M1 polarization of microglia and demyelination • Induces proinflammatory cytokines and prooxidants (iNOS, NOX4, p22phox, Nrf2, AHR) in the brain • Increases expression of Nrf2 in astrocytes • Induces oxidative stress and impairment of the BBB integrity	• Hypomethylation of GPR15 gene encoding a chemoattractant receptor that regulates T-cell homing • Hypomethylation of F2RL3 (PAR4) gene encoding a protease-activated receptor family member that regulates production of ROS in platelets and MMP-9 in BBB cells, promotes immune cell recruitment in the brain (after pleural inflammation) and is increased in degenerating neurons and activated microglia • Hypomethylation of AHRR gene encoding AHR repressor, which can affect Th and Treg differentiation, B cell maturation and activity of macrophages, dendritic cells and neutrophils. AHR xenobiotic detoxification critical for immune functions in MS and AHR-related kynurenin/tryptophan pathways involved in MS
Obesity[a]	• Promotes Th2 to Th1 switch, proinflammatory cytokines and reduces Tregs • Enhances CNS infiltration, oxidative stress, neuroinflammation, and neurodegeneration • Decreases NPY-mediated immunosuppression	• Hypermethylation/downregulation of antiinflammatory adiponectin and hypomethylation/upregulation of proinflammatory leptin • Epigenetic dysregulation of genes involved in immune responses and neurotransmission (various tissues) • Differential methylation of epigenetic machinery genes (HDAC, DNMTs SIRT1)

(*Continued*)

Table 10.2 Potential Epigenetic Mechanisms Mediating Environmental Exposures in MS (*cont.*)

Risk Factor	Effects in EAE/MS	Putative Epigenetic Mechanisms
Gut microbiota[b]	• Affects Th17 differentiation and B-cell recruitment, antiinflammatory Treg, and Th2 responses • Immunoreactivity against gastrointestinal antigens	• Gut micriobiota is a major donor of cofactor of epigenetic enzyme (acetyl CoA, S-adenosylmethionine, folate) • Gut microbiota is a major source of epigenetic bioactive metabolites (e.g., SFCAs are potent HDAC inhibitors) and can regulate immune and nervous responses through the gut–brain axis • Gut microbiota composition associates with promoter DNA methylation status of genes involved in inflammation and nervous system function in blood (e.g., Uhrf1, TLR2/4)
Melatonin[c]	• Alters balance of effector and regulatory T cells • Reduces oxidative stress in immune cells • Protects from mitochondrial injury	• Regulates DNMTs and HDACs expression, gene promoter methylation, binding of MeCP2, histone PTMs, and expression of genes in the brain • Suppresses p300 HAT activity and subsequent NF-kB activation and related inflammation whereas promotes p300-mediated Nrf2 antioxidant response • Protects aging neurons and modulates energy homeostasis through HDAC SIRT1 pathway

Abbreviations: AHR, aryl hydrocarbon receptor; BBB, blood–brain barrier; DNMT, DNA methyl transferase; EAE, experimental autoimmune encephalomyelitis; EBV, Epstein–Barr virus; F2RL3, coagulation factor II (thrombin) receptor like 3; HAT, histone acetyl transferase; HDAC, histone deacetylase; ISRE, interferon stimulation response element; MeCP2, methyl CpG binding protein 2; MHC, major histocompatibility complex; Nrf2, nuclear factor (erythroid-derived 2)-like 2; PAR4, protease activated receptor 4; RAS, renin-angiotensin system;SFCA, short chain fatty acid; TLR, toll like receptor; VDR, vitamin D receptor; VDRE, vitamin D receptor element;

[a]Palavra F, Almeida L, Ambrósio AF,Reis F. Obesity and brain inflammation: a focus on multiple sclerosis. Obes Rev 2016;17:211–224.

Houde A, Légaré C, Biron S, Lescelleur O, Biertho L, Marceau S, Tchernof A, Vohl MC, Hivert MF, Bouchard L. Leptin and adiponectin DNA methylation levels in adipose tissues and blood cells are associated with BMI, waist girth and LDL-cholesterol levels in severely obese men and women. BMC Med Genet 2015;16:29.

Benton MC, Johnstone A, Eccles D, Harmon B, Hayes MT, Lea RA, Griffiths L, Hoffman EP, Stubbs RS, Macartney-Coxson D. An analysis of DNA methylation in human adipose tissue reveals differential modification of obesity genes before and after gastric bypass and weight loss. Genome Biol 2015;16:8

Heyward FD, Gilliam D, Coleman MA, Gavin CF, Wang J, Kaas G, Trieu R, Lewis J, Moulden J, Sweatt JD. Obesity weighs down memory through a mechanism involving the neuroepigenetic dysregulation of Sirt1. J Neurosci 2016;36:1324–1335.

Jacobsen MJ, Mentzel CM, Olesen AS, Huby T, Jørgensen CB, Barrès R,Fredholm M, Simar D. Altered methylation profile of lymphocytes is concordant with perturbation of lipids metabolism and inflammatory response in obesity. J Diab Res 2016;2016:8539057.

[b]Yadav SK, Mindur JE, Ito K, Dhib-Jalbut S. Advances in the immunopathogenesis of multiple sclerosis. Curr Opin Neurol 2015;28:206–219.

Paul B, Barnes S, Demark-Wahnefried W, Morrow C, Salvador C, Skibola C, Tollefsbol T. Influences of diet and the gut microbiome on epigenetic modulation in cancer and other diseases. Clin Epigenet 2015;7:112.

Stilling RM, Dinan TG, Cryan JF. Microbial genes, brain & behaviour-epigenetic regulation of the gut-brain axis. Genes Brain Behav 2014;13:69–86.

[c]Escribano BM, Colín-González AL, Santamaría A, Túnez I. The role of melatonin in multiple sclerosis, Huntington's disease and cerebral ischemia. CNS Neurol Disord Drug Targets 2014;13:1096–1119.

Kashani IR, Rajabi Z, Akbari M, Hassanzadeh G, Mohseni A, Eramsadati MK, Rafiee K, Beyer C, Kipp M, Zendedel A. Protective effects of melatonin against mitochondrial injury in a mouse model of multiple sclerosis. Exp Brain Res 2014;232:2835–2846.

Korkmaz A, Rosales-Corral S, Reiter RJ. Gene regulation by melatonin linked to epigenetic phenomena. Gene 2012;503(1):1–11.

Jenwitheesuk, A., Nopparat, C., Mukda, S., Wongchitrat, P., Govitrapong, P. Melatonin regulates aging and neurodegeneration through energy metabolism, epigenetics, autophagy and circadian rhythm pathways. Int J Mol Sci 2014;15:16848–16884.

coactivators that display lysine acetyltransferase activity [97] and VDR binding sites associate with regions of active open chromatin in cells of immune origin [94,109]. This is consistent with observed DNA methylation changes in the context of specific *HLA-DRB1* sequence, discussed in Section entitled Locus-specific effects earlier, and could provide an explanation for the reported parent-of-origin effect that mapped to *HLA-DRB1* in MS implicating gene–environment interactions [32]. Additionally, vitamin D could affect the epigenome through VDR-independent mechanisms as it has been shown that vitamin D regulates expression of a range of histone demethylase genes and chromatin remodeling factors [98], including JARID2, which is a candidate MS risk gene.

Finally, the first epigenome-wide study of sunlight exposure in CD4+ T cells from healthy individuals has revealed a novel association between DNA methylation status in the promoter region of a protein kinase C, PRKCZ, and sunlight exposure [99]. This is of special interest since variants of this gene have been showed to modify the association between serum vitamin D and risk of relapse in MS patients [100]. Protein kinase C genes are known to be regulated by vitamin D and to be involved in T-cell activation. Collectively these data raise the hypothesis that sunlight exposure/vitamin D level association with MS might rely on epigenetic mechanisms involving the PRKC family of genes.

10.4.2.2 Epstein–Barr Virus Infection

Infection with the Epstein–Barr virus has been unequivocally associated with MS [101]. However, whether EBV causes MS or represents epiphenomena is still being debated and the mechanisms by which EBV contributes to MS remain elusive. Several hypotheses, such as molecular mimicry or bystander effects, have been proposed to explain the role of EBV infection in the development of MS and rely on EBV-induced immune dysfunction (Table 10.2). The risk is higher when EBV is acquired during adolescence and early adulthood probably because infection in teenagers usually results in infectious mononucleosis, which increases the risk of developing disease compared to asymptomatic EBV infection [110]. Interestingly, prospective studies have estimated the mean interval between primary EBV infection and onset of MS to be 5.6 years (range 2.3–9.4 years) [111] and the risk remained increased for at least three decades [112].

Whereas several studies have implied a role of epigenetics in mediating effects of EBV-infection, the contribution of EBV-mediated epigenetic mechanisms in MS remains to be investigated. Studies of EBV-infected PBMC and B cells have shown that DNA methylation and histone PTMs participate in the temporal expression of viral genes [102]. For that purpose, EBV highjacks the host epigenetic machinery, which in turn, besides affecting viral genes, can also induce epigenetic alternations in the host genome. Indeed, gene promoter methylation profiling of DNA from EBV-infected PBMC at different times postinfection, has identified stable locus-specific changes in the DNA methylation pattern of genes involved in various processes including immune mechanisms [102]. Further investigation conducted in infected primary B cells has shown that transcriptional regulation of functionally related genes resulted from the deregulation of chromatin modifying and remodeling factors (i.e., MBDs, HDACs, and specific DNMTs) and subsequent DNMT-dependent aberrant methylation of host gene promoters (together with viral genes). Similarly, EBV exploits intrinsic cellular microRNA regulatory networks and, conversely, uses its viral ncRNAs to enable EBV-infected cells to escape recognition by host immune defenses, thereby allowing the virus to latently persist throughout the life of the individual [103]. Altogether, these findings are of special relevance given the implication of an increased viral load or a presentation of different EBV proteins in MS. The EBV-triggered epigenetic dysregulation of the host cell genome and inability of the cellular methylation system to control the viral genome might be of importance in MS pathogenesis.

10.4.2.3 Smoking

Both active smoking and exposure to passive smoking have repeatedly been associated with an increased risk of developing MS, disease progression and clinical disability [104]. This risk has recently been associated with a variant of *N*-acetyltransferase 1 (NAT1) gene, supporting the idea that a gene-environment interaction affects disease susceptibility [113]. In contrast to other MS risk factors, smoking increases MS risk regardless of the age of exposure, and although its effect is reversible, it lasts for a decade after smoking cessation [35]. However, while it is difficult to associate a heterogeneous mixture of compounds such as cigarette smoke with MS, it appears that lung irritation due to burnt tobacco products, causing oxidative stress and a proinflammatory response, rather than systemic tobacco use alters MS risk (Table 10.2).

The mechanisms responsible for smoking-induced local cellular damage and inflammation in the airway epithelium and subsequent inflammation in the CNS are still to be established; however, it is hypothesized that epigenetic modifications alter the expression of inflammatory genes [105]. DNA methylation in blood from smokers has been extensively studied and has yielded several replicable loci whose methylation levels associate with smoking intensity and time from cessation. The most frequently reported CpGs are within aryl-hydrocarbon receptor repressor (AHRR), coagulation factor II (thrombin) receptor-like 3 (F2RL3), G-protein coupled receptor 15 (GPR15) genes together with other loci within intergenic region on chromosomes 2 and 6 [106]. Importantly, both past smoking and prenatal exposure to smoking have been shown to induce long-term changes in DNA methylation [107]. However, as previously mentioned, analyses in whole blood requires particular attention since it represents a mixture of various cell types with different patterns of DNA methylation, and smoking is likely to alter the overall numbers and specific distribution of leukocytes. Given the roles of these candidates in immune and brain functions, especially the implication of the aryl-hydrocarbon receptor and related kynurenine pathways in autoimmune and neurodegenerative diseases such as MS [108], stable methylation changes of these genes might potentially participate in MS pathogenesis (Table 10.2). Nonetheless, the role of the identified genes as potential mediators or just indicators of smoking status in general and in MS, in particular, is still elusive and necessitates further investigation.

10.4.2.4 Other Environmental Factors

Even though the environmental risk factors described earlier show the most robust association with MS, recent studies have shed light on additional environmental factors such as the gut microbiome, body mass index and melatonin levels which could influence MS disease. Given the potent effects of modified gut flora, obesity, and disturbed seasonal and circadian melatonin levels in immune and nervous processes as well as their ability to alter the epigenome, the study of these factors might provide promising insight into MS pathogenesis (Table 10.2).

10.5 UTILITY OF EPIGENETICS IN DIAGNOSIS, PROGNOSIS, AND TREATMENT OF MS

10.5.1 EPIGENETIC PATTERNS AS A BIOMARKER

Although the role of epigenetic changes in the pathogenesis of different diseases remains to be firmly established, a growing list of such changes already represents signatures of diseases. As such, epigenetic modifications might provide useful biomarkers for diagnostic, prognostic, and therapeutic purposes. The biomarker potential of DNA methylation is especially appealing given the technical feasibility to accurately measure DNA methylation genome-wide. The fact that DNA methylation can be reliably quantified in a small amount of sample to a single-cell level, and that it is rather insensitive to sample handling is very suitable for clinical settings. Genome-wide analysis of histone modifications as biomarkers for disease might also be feasible given the recent developments of chromatin immunoprecipitation technologies, which are now compatible with low numbers of cells [114,115]. However, difficulties to standardize antibody-based methods of quantification and the fact that histone modifications are more dynamic compared to DNA methylation could hamper such use. The utility of DNA methylation marks has been predominantly explored in the cancer field owing to the early discovery of epigenetic dysregulation in cancer and large differences in DNA methylation that affect specific genes. DNA methylation profiling has shown promising results in cancer diagnosis, prognosis, identification of primary origin of tumors, detection of metastasis, and prediction of the response to chemotherapeutic drugs [116]. Of particular interest is the biomarker potential of DNA methylation using noninvasive methods, such as blood, stool, saliva or urine, over invasive methods such as using tissue biopsies. This is especially relevant for diseases where the target organ is inaccessible such as in MS. The utility of blood DNA methylation in indicating processes in distant target organs has been questioned, but recent studies demonstrate that blood DNA methylation may be utilized [117], although this remains to be established for every disease separately.

In addition to reflecting disease processes, DNA methylation can be used as a marker of environmental exposures [107], which is of relevance for complex diseases such as MS, where environmental factors play an important role. For example, blood DNA methylation has been shown to be a reliable long-term biomarker of prior exposure to smoking [118]. This could be useful for vast numbers of collected cohorts in which environmental exposures have not been documented.

Finally, while tissue heterogeneity poses challenges in interpreting mechanistic consequences of DNA methylation, analysis of methylomes in tissues can uncover heterogeneity that is relevant for disease processes [77]. For example, whole blood methylomes can indicate important differences in frequencies of cell types [83]. Furthermore, yet uncovered cell types or their activation states might be implicated after tissue methylome analysis and further refined using evolving methods to study methylomes in single-cells.

10.5.2 THERAPEUTIC METHODS BASED ON EPIGENETICS

The epigenome appears as an additional regulatory layer apposed onto the genome, that is, reversible but heritable, shaping and maintaining a certain cellular phenotype. These unique characteristics of stability and reversibility may position the epigenome as an interesting therapeutic target in disease treatment. Epigenetic therapy offers promising prospects to correct pathways that are dysregulated in disease by means of drugs or other epigenome-editing techniques.

10.5.2.1 Global Approaches

As aforementioned in Section 10.7, global inhibitors of HDACs and PADIs can lead to amelioration of symptoms in mouse models of MS, and other histone "writers"/"readers"/"erasers" are potential candidates for pharmacological epigenetic targeting in MS. Targeting the function of DNMTs is also an attractive therapeutic approach to correct aberrant changes in gene expression that are mediated by changes in DNA methylation. Azacitidine (AZA, 5-azacitidine) and its deoxy derivative, decitabine (DAC, 2′-deoxy-5-azacytidine), are used in the treatment of myelodysplastic disorders and leukemia. They are chemical analogues of the cytosine nucleoside that replace cytosine during DNA replication and block the action of DNA methyltransferase thereby leading to demethylation. Two recent studies demonstrated beneficial effects of these DNMT inhibitors in EAE. Prophylactic treatment with AZA prevented myelin oligodendrocyte protein $(MOG)_{35-55}$-EAE and histopathological changes in the CNS in C57BL/6 mice [119]. DAC displayed both prophylactic and lasting therapeutic effects in MOG_{35-55}-EAE in C57BL/6 and proteolipid protein $(PLP)_{139-151}$-EAE in SJL mice [120]. Both studies associate beneficial effect of DNMT inhibitors with an increased frequency of CD4+FoxP3+ Tregs and lower activation and proinflammatory profile of effector T cells. Furthermore, treatment with AZA directly upregulated FoxP3 expression [119], which is in line with the well-documented role of DNA methylation in the control of FoxP3 expression [121]. The ability of DNMT inhibitors to increase frequency of Tregs and enhance their function is of potential clinical relevance considering that MS patients have impaired Treg function [122].

10.5.2.2 Targeted Approaches

While identification of locus-specific epigenetic modifications and their functional relevance in MS is still ongoing, this effort will likely provide the molecular substratum for the development of new targeted methods for epigenome editing in MS. Indeed, as mentioned in the section entitled Histone posttranslational modifications earlier, the approach of "epigenome editing" has been shown to successfully deposit epigenetic modification onto a specific genomic region and, with this, favors a specific chromatin-state, alters gene expression and influences cellular processes in a stable manner [123]. In addition to the conventional methods that use zinc-finger proteins or transactivation activator-like effector technologies, the recent discovery of the CRISPR/Cas9 system, which facilitates the design of the DNA recognition domain, has provided a promising tool for epigenome editing. The CRISPR/Cas9 epimodifier system relies on the pairing of a guide RNA to the targeted DNA and delivery of a catalytically inactivated Cas9 protein (deactivated Cas9, dCas9) fused to the catalytic domain of the chromatin modifier. In line with this, one could imagine the use of a CRISPR/dCas9-Dnmt3a system targeting specific hypomethylated risk loci in MS. However, epigenome editing methodologies still show limitations with regard to their potential use in clinical settings. The CRISPR/dCas9 system reveals significant off-target activity in some applications (although it has been recently shown that Cas9 can be modified to minimize such activity [124]) and, importantly, might have immunogenic properties due its bacterial origin. Moreover, while the stability and functional efficiency of the newly introduced epimodification are still being currently addressed in vitro in cell lines, ex vivo in oocyte and in animal models, further work is required to affirm its potential in the clinical context.

An additional challenge to a gene-targeted strategy in MS would be specific targeting of the affected tissue and cells. In this regard, the specific targeting at the cellular level using drug-coated nanoparticles and drug-embedded microvesicles could be a more precise and effective way not only to enhance epigenetic drug availability but also to reduce adverse effects. Thus, personalized medicine based on targeted epigenome editing in relevant tissues promises great potential as a therapeutic approach for durable cellular reprogramming and control of MS.

10.6 CONCLUSIONS AND FUTURE PERSPECTIVES

The current concept of MS pathogenesis involves a model where environmental factors operate on a permissive genetic background to trigger the disease. The etiological factors are numerous and they independently typically confer very modest risks. However, genetic and environmental factors also

>100 risk loci

HLA genes display strongest influence

...

Genes

Environment

>10 environmental exposures

Sun exposure/vitamin D

Smoking

EBV infection

...

Epigenetics

DNA (hydroxy)methylation

Histone PTMs

...

Immune dysfunction and chronic inflammation

Breakdown of T-cell tolerance

Infiltration of CNS by immune cells

Direct and indirect immune-mediated attack on CNS

Activation of CNS innate immune cells

CNS damage and exhaustion of functional reserves

Demyelination, oligodendrocyte loss, and astrogliosis

ROS production, mitochondrial and energy deficiency

Loss of neuronal homeostasis and axonal injury

Neurodegeneration

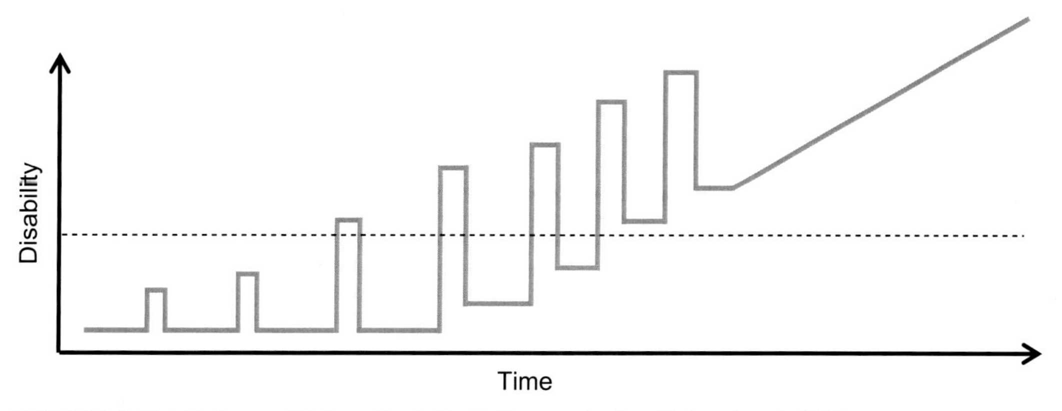

FIGURE 10.1 Risk Factors and Epigenetics in the Pathogenesis of multiple sclerosis (MS)

Genetic, environmental risk factors and epigenetic mechanisms likely interact in a tridirectional manner, that is, gene variants and environmental exposures might confer risk through epigenetic mechanisms and, reciprocally, altered epigenetic patterns may affect susceptibility to develop MS or affect disease severity in individuals burdened with risk factor(s). This complex interplay would result in aberrant cellular behavior in a tissue-specific manner at the time of exposure to the environmental factor and likely be dependent on an individual's genetic make-up. Epigenetic mechanisms might thus be involved in "cellular memory" of early exposures, both external and internal. In this way they may contribute to a cascade of immunological and neurodegenerative events that act in concert to induce brain damage in MS, and might have varying clinical relevance during the course of disease progression. Inflammatory demyelination occurs in the early stages of disease and eventually leads to clinical symptoms in a relapsing-remitting pattern. With time, depending on age and disease duration, central nervous system (CNS) damage results in exhaustion of the brain functional reserve and ultimately to sustained neurodegeneration and persistent increase in disability. Abbreviations: *CNS*, Central nervous system; *EBV*, Epstein–Barr virus; *HLA*, human leukocyte antigen; *PTMs*, posttranslational modifications; *ROS*, reactive oxygen species.

interact and these combined effects convey a dramatic increase in the risk to develop MS [21,113]. These interactions might be mediated by epigenetic mechanisms as the epigenome integrates influences from genetic and external factors at a tissue-specific level (Fig. 10.1). In line with this, we speculate that epigenetic mechanisms might mediate, at least partly, effects from genetic variants and environmental exposures, and their interactions, to convey the cellular phenotype and function. Notably, some epigenetic changes are inherited through cell division providing a plausible mechanism for "cellular memory" of altered states in chronic disease such as MS.

The first epigenome-wide studies reviewed in this chapter reveal differences in epigenetic marks between affected and healthy individuals. However, epigenetic studies remain challenging and further in depth understanding of the dynamic chromatin changes on a cellular level, and in relation to the underlying genetic sequence and environmental exposures, in well-powered and data-rich cohorts, are warranted. Such studies will lead to a better understanding of pathogenic mechanisms, but might also aid more efficient disease management in terms of robust biomarkers and novel therapeutic approaches.

ABBREVIATIONS

APCs Antigen- presenting cells
CD4+ Cluster of differentiation 4 cell (T lymphocyte subset)
CD8+ Cluster of differentiation 8 cell (T lymphocyte subset)
CRISPR Clustered Regularly Interspersed Palindromic Repeats
DMPs Differential methylated positions
DMRs Differential methylated regions
DNMTs DNA methyltransferases
EAE Experimental autoimmune encephalomyelitis
EBV Epstein–Barr virus
HATs Histone acetyltransferases
HDACs Histone deacetylases
HLA Human leukocyte antigen
MBDs Methyl-CpG-binding proteins
MBP Myelin basic protein
MOG Myelin oligodendrocyte glycoprotein
meQTL Methylated quantitative trait loci
MS Multiple sclerosis
NK cell Natural killer cell
NAWM Normal appearing white matter
OPCs Oligodendrocyte precursor cells
PADI2 Peptidyl arginine deaminase 2
PBMC Peripheral blood mononuclear cells
PTMs Posttranslational modifications
RRMS Relapsing-remitting form of multiple sclerosis
SNP Single nucleotide polymorphism
SPMS Secondary progressive multiple sclerosis
TETs Ten-eleven translocation family of proteins
Tregs Regulatory T cells, previously called suppressor T cells
VDR Vitamin D receptor
VDRE Vitamin D response element

GLOSSARY

Epigenome editing Directed alteration of chromatin marks at specific genomic loci using targeted EpiEffectors which consist of designed DNA recognition domains and catalytic domains from a chromatin modifying enzyme

Epstein–Barr virus Also called human herpes virus 4, is one of eight viruses in the herpes family, and is one of the most common viruses in humans. It is best known as the cause of infectious mononucleosis

Human leukocyte antigens Cell-surface molecules forming the major histocompatibility complex (MHC) in humans and regulate the immune response by presenting peptides and antigens to immune cells. The locus encodes a large number of polymorphic genes that belong to different classes with specific roles in defense against pathogens or graft/host tissue compatibility.

Methylation quantitative trait loci Sites in the genome where the DNA methylation status may result from the individual genotype at a locus

REFERENCES

[1] Fancy SP, Chan JR, Baranzini SE, Franklin RJ, Rowitch DH. Myelin regeneration: a recapitulation of development? Annu Rev Neurosci 2011;34:21–43.

[2] Compston A, Coles A. Multiple sclerosis. Lancet 2008;372(9648):1502–17.

[3] Marrie RA, Reingold S, Cohen J, Stuve O, Trojano M, Sorensen PS, et al. The incidence and prevalence of psychiatric disorders in multiple sclerosis: a systematic review. Mult Scler 2015;21(3):305–17.

[4] Rocca MA, Amato MP, De Stefano N, Enzinger C, Geurts JJ, Penner IK, et al. Clinical and imaging assessment of cognitive dysfunction in multiple sclerosis. Lancet Neurol 2015;14:302–17.

[5] Sadovnick AD, Baird PA, Ward RH. Multiple sclerosis: updated risks for relatives. Am J Med Genet 1988;29(3):533–41.

[6] Willer CJ, Dyment DA, Risch NJ, Sadovnick AD, Ebers GC. Twin concordance and sibling recurrence rates in multiple sclerosis. Proc Natl Acad Sci USA 2003;100(22):12877–82.

[7] Ebers GC, Sadovnick AD, Risch NJ. A genetic basis for familial aggregation in multiple sclerosis. Canadian Collaborative Study Group. Nature 1995;377(6545):150–1.

[8] Sadovnick AD, Ebers GC, Dyment DA, Risch NJ. Evidence for genetic basis of multiple sclerosis. The Canadian Collaborative Study Group. Lancet 1996;347(9017):1728–30.

[9] Ebers GC, Yee IM, Sadovnick AD, Duquette P. Conjugal multiple sclerosis: population-based prevalence and recurrence risks in offspring. Canadian Collaborative Study Group. Ann Neurol 2000;48(6):927–31.

[10] Jersild C, Svejgaard A, Fog T. HL-A antigens and multiple sclerosis. Lancet 1972;1(7762):1240–1.

[11] Sawcer S, Hellenthal G, Pirinen M, Spencer CC, Patsopoulos NA, Moutsianas L, et al. Genetic risk and a primary role for cell-mediated immune mechanisms in multiple sclerosis. Nature 2011;476(7359):214–9.

[12] Beecham AH, Patsopoulos NA, Xifara DK, Davis MF, Kemppinen A, Cotsapas C, et al. Analysis of immune-related loci identifies 48 new susceptibility variants for multiple sclerosis. Nat Genet 2013;45(11):1353–60.

[13] Olsson T, Sun J, Hillert J, Hojeberg B, Ekre HP, Andersson G, et al. Increased numbers of T cells recognizing multiple myelin basic protein epitopes in multiple sclerosis. Eur J Immunol 1992;22(4):1083–7.

[14] Cao Y, Goods BA, Raddassi K, Nepom GT, Kwok WW, Love JC, et al. Functional inflammatory profiles distinguish myelin-reactive T cells from patients with multiple sclerosis. Sci Transl Med 2015;7(287):287ra74.

[15] Mein LE, Hoch RM, Dornmair K, de Waal Malefyt R, Bontrop RE, Jonker M, et al. Encephalitogenic potential of myelin basic protein-specific T cells isolated from normal rhesus macaques. Am J Pathol 1997;150(2):445–53.

[16] Ben-Nun A, Wekerle H, Cohen IR. The rapid isolation of clonable antigen-specific T lymphocyte lines capable of mediating autoimmune encephalomyelitis. Eur J Immunol 1981;11(3):195–9.

[17] Friese MA, Jensen LT, Willcox N, Fugger L. Humanized mouse models for organ-specific autoimmune diseases. Curr Opin Immunol 2006;18(6):704–9.

[18] Dendrou CA, Fugger L, Friese MA. Immunopathology of multiple sclerosis. Nat Rev Immunol 2015;15(9):545–58.
[19] Simpson S Jr, Blizzard L, Otahal P, Van der Mei I, Taylor B. Latitude is significantly associated with the prevalence of multiple sclerosis: a meta-analysis. J Neurol Neurosurg Psychiatry 2011;82(10):1132–41.
[20] Gale CR, Martyn CN. Migrant studies in multiple sclerosis. Prog Neurobiol 1995;47(4–5):425–48.
[21] Hedstrom AK, Olsson T, Alfredsson L. The role of environment and lifestyle in determining the risk of multiple sclerosis. Curr Top Behav Neurosci 2015;26:87–104.
[22] Melcon MO, Correale J, Melcon CM. Is it time for a new global classification of multiple sclerosis? J Neurol Sci 2014;344(1–2):171–81.
[23] Handel AE, Giovannoni G, Ebers GC, Ramagopalan SV. Environmental factors and their timing in adult-onset multiple sclerosis. Nat Rev Neurol 2010;6(3):156–66.
[24] Kouzarides T. Chromatin modifications and their function. Cell 2007;128(4):693–705.
[25] Gapp K, Jawaid A, Sarkies P, Bohacek J, Pelczar P, Prados J, et al. Implication of sperm RNAs in transgenerational inheritance of the effects of early trauma in mice. Nat Neurosci 2014;17(5):667–9.
[26] Chen Q, Yan M, Cao Z, Li X, Zhang Y, Shi J, et al. Sperm tsRNAs contribute to intergenerational inheritance of an acquired metabolic disorder. Science 2016;351(6271):397–400.
[27] Sharma U, Conine CC, Shea JM, Boskovic A, Derr AG, Bing XY, et al. Biogenesis and function of tRNA fragments during sperm maturation and fertilization in mammals. Science 2016;351(6271):391–6.
[28] Fraga MF, Ballestar E, Paz MF, Ropero S, Setien F, Ballestar ML, et al. Epigenetic differences arise during the lifetime of monozygotic twins. Proc Natl Acad Sci USA 2005;102(30):10604–9.
[29] Kaminsky ZA, Tang T, Wang SC, Ptak C, Oh GH, Wong AH, et al. DNA methylation profiles in monozygotic and dizygotic twins. Nat Genet 2009;41(2):240–5.
[30] Ebers GC, Sadovnick AD, Dyment DA, Yee IM, Willer CJ, Risch N. Parent-of-origin effect in multiple sclerosis: observations in half-siblings. Lancet 2004;363(9423):1773–4.
[31] Ramagopalan SV, Herrera BM, Bell JT, Dyment DA, Deluca GC, Lincoln MR, et al. Parental transmission of HLA-DRB1*15 in multiple sclerosis. Hum Genet 2008;122(6):661–3.
[32] Chao MJ, Ramagopalan SV, Herrera BM, et al. Epigenetics in multiple sclerosis susceptibility: difference in transgenerational risk localizes to the major histocompatibility complex. Hum Mol Genet 2009;18(2):261–6.
[33] Stridh P, Ruhrmann S, Bergman P, Thessen Hedreul M, Flytzani S, et al. Parent-of-origin effects implicate epigenetic regulation of experimental autoimmune encephalomyelitis and identify imprinted Dlk1 as a novel risk gene. PLoS Genet 2014;10(3):e1004265.
[34] Hedstrom AK, Olsson T, Alfredsson L. High body mass index before age 20 is associated with increased risk for multiple sclerosis in both men and women. Mult Scler 2012;18(9):1334–6.
[35] Hedstrom AK, Hillert J, Olsson T, Alfredsson L. Smoking and multiple sclerosis susceptibility. Eur J Epidemiol 2013;28(11):867–74.
[36] Mastronardi FG, Wood DD, Mei J, Raijmakers R, Tseveleki V, Dosch HM, et al. Increased citrullination of histone H3 in multiple sclerosis brain and animal models of demyelination: a role for tumor necrosis factor-induced peptidylarginine deiminase 4 translocation. J Neurosci 2006;26(44):11387–96.
[37] Baranzini SE, Mudge J, van Velkinburgh JC, Khankhanian P, Khrebtukova I, Miller NA, et al. Genome, epigenome and RNA sequences of monozygotic twins discordant for multiple sclerosis. Nature 2010;464(7293):1351–6.
[38] Pedre X, Mastronardi F, Bruck W, Lopez-Rodas G, Kuhlmann T, Casaccia P. Changed histone acetylation patterns in normal–appearing white matter and early multiple sclerosis lesions. J Neurosci 2011;31(9):3435–45.
[39] Graves M, Benton M, Lea R, Boyle M, Tajouri L, Macartney-Coxson D, et al. Methylation differences at the HLA-DRB1 locus in CD4+ T-cells are associated with multiple sclerosis. Mult Scler 2013;20(8):1033–41.
[40] Huynh JL, Garg P, Thin TH, Yoo S, Dutta R, Trapp BD, et al. Epigenome-wide differences in pathology-free regions of multiple sclerosis-affected brains. Nat Neurosci 2014;17(1):121–30.

[41] Bos SD, Page CM, Andreassen BK, Elboudwarej E, Gustavsen MW, Briggs F, et al. Genome-wide DNA methylation profiles indicate CD8+ T cell hypermethylation in multiple sclerosis. PLoS One 2015;10(3): e0117403.
[42] Maltby VE, Graves MC, Lea RA, Benton MC, Sanders KA, Tajouri L, et al. Genome-wide DNA methylation profiling of CD8+ T cells shows a distinct epigenetic signature to CD4+ T cells in multiple sclerosis patients. Clin Epigenet 2015;7:118.
[43] Li E, Zhang Y. DNA methylation in mammals. Cold Spring Harb Perspect Biol 2014;6(5):a019133.
[44] Kriaucionis S, Tahiliani M. Expanding the epigenetic landscape: novel modifications of cytosine in genomic DNA. Cold Spring Harb Perspect Biol 2014;6(10):a018630.
[45] Mastronardi FG, Noor A, Wood DD, Paton T, Moscarello MA. Peptidyl argininedeiminase 2 CpG island in multiple sclerosis white matter is hypomethylated. J Neurosci Res 2007;85(9):2006–16.
[46] Tranquill LR, Cao L, Ling NC, Kalbacher H, Martin RM, Whitaker JN. Enhanced T cell responsiveness to citrulline-containing myelin basic protein in multiple sclerosis patients. Mult Scler 2000;6(4):220–5.
[47] Moscarello MA, Mastronardi FG, Wood DD. The role of citrullinated proteins suggests a novel mechanism in the pathogenesis of multiple sclerosis. Neurochem Res 2007;32(2):251–6.
[48] Calabrese R, Zampieri M, Mechelli R, Annibali V, Gustafierro T, Ciccarone F, et al. Methylation-dependent PAD2 upregulation in multiple sclerosis peripheral blood. Mult Scler 2012;18(3):299–304.
[49] Hilton IB, D'Ippolito AM, Vockley CM, Thakore PI, Crawford GE, Reddy TE, et al. Epigenome editing by a CRISPR-Cas9-based acetyltransferase activates genes from promoters and enhancers. Nat Biotechnol 2015;33(5):510–7.
[50] Mendenhall EM, Williamson KE, Reyon D, Zou JY, Ram O, Joung JK, et al. Locus-specific editing of histone modifications at endogenous enhancers. Nat Biotechnol 2013;31(12):1133–6.
[51] Heller EA, Cates HM, Pena CJ, Sun H, Shao N, Feng J, et al. Locus-specific epigenetic remodeling controls addiction- and depression-related behaviors. Nat Neurosci 2014;17(12):1720–7.
[52] Elsasser SJ, Ernst RJ, Walker OS, Chin JW. Genetic code expansion in stable cell lines enables encoded chromatin modification. Nat Methods 2016;13(2):158–64.
[53] Castelo-Branco G, Stridh P, Guerreiro-Cacais AO, Adzemovic MZ, Falcao AM, Marta M, et al. Acute treatment with valproic acid and L-thyroxine ameliorates clinical signs of experimental autoimmune encephalomyelitis and prevents brain pathology in DA rats. Neurobiol Dis 2014;71:220–33.
[54] Ge Z, Da Y, Xue Z, Zhang K, Zhuang H, Peng M, et al. Vorinostat, a histone deacetylase inhibitor, suppresses dendritic cell function and ameliorates experimental autoimmune encephalomyelitis. Exp Neurol 2013;241:56–66.
[55] Lv J, Du C, Wei W, Wu Z, Zhao G, Li Z, et al. The antiepileptic drug valproic acid restores T cell homeostasis and ameliorates pathogenesis of experimental autoimmune encephalomyelitis. J Biol Chem 2012;287(34):28656–65.
[56] Zhang Z, Zhang ZY, Wu Y, Schluesener HJ. Valproic acid ameliorates inflammation in experimental autoimmune encephalomyelitis rats. Neuroscience 2012;221:140–50.
[57] Camelo S, Iglesias AH, Hwang D, Due B, Ryu H, Smith K, et al. Transcriptional therapy with the histone deacetylase inhibitor trichostatin A ameliorates experimental autoimmune encephalomyelitis. J Neuroimmunol 2005;164(1–2):10–21.
[58] Hait NC, Wise LE, Allegood JC, et al. Active, phosphorylated fingolimod inhibits histone deacetylases and facilitates fear extinction memory. Nat Neurosci 2014;17(7):971–80.
[59] Ye F, Chen Y, Hoang T, Montgomery RL, Zhao XH, Bu H, et al. HDAC1 and HDAC2 regulate oligodendrocyte differentiation by disrupting the beta-catenin-TCF interaction. Nat Neurosci 2009;12(7): 829–38.
[60] Wang J, Weaver IC, Gauthier-Fisher A, Wang H, He L, Yeomans J, et al. CBP histone acetyltransferase activity regulates embryonic neural differentiation in the normal and Rubinstein-Taybi syndrome brain. Dev Cell 2010;18(1):114–25.

[61] Castelo-Branco G, Lilja T, Wallenborg K, Falcao AM, Marques SC, Gracias A, et al. Neural stem cell differentiation is dictated by distinct actions of nuclear receptor corepressors and histone deacetylases. Stem Cell Rep 2014;3(3):502–15.
[62] Shen S, Sandoval J, Swiss VA, Li J, Dupree J, Franklin RJ, et al. Age-dependent epigenetic control of differentiation inhibitors is critical for remyelination efficiency. Nat Neurosci 2008;11(9):1024–34.
[63] Inkster B, Strijbis EM, Vounou M, Kappos L, Radue EW, Matthews PM, et al. Histone deacetylase gene variants predict brain volume changes in multiple sclerosis. Neurobiol Aging 2013;34(1):238–47.
[64] Zhang L, He X, Liu L, Jiang M, Zhao C, Wang H, et al. Hdac3 interaction with p300 histone acetyltransferase regulates the oligodendrocyte and astrocyte lineage fate switch. Dev Cell 2016;36(3):316–30.
[65] Pedre X, Mastronardi F, Bruck W, Lopez-Rodas G, Kuhlmann T, Casaccia P. Changed histone acetylation patterns in normal-appearing white matter and early multiple sclerosis lesions. J Neurosci 2011;31(9):3435–45.
[66] Dawson MA, Prinjha RK, Dittmann A, Giotopoulos G, Bantscheff M, Chan WI, et al. Inhibition of BET recruitment to chromatin as an effective treatment for MLL-fusion leukaemia. Nature 2011;478(7370): 529–33.
[67] Christophorou MA, Castelo-Branco G, Halley-Stott RP, Oliveira CS, Loos R, Radzisheuskaya A, et al. Citrullination regulates pluripotency and histone H1 binding to chromatin. Nature 2014;507(7490):104–8.
[68] Mastronardi FG, Wood DD, Mei J, Raijmakers R, Tseveleki V, Dosch HM, et al. Increased citrullination of histone H3 in multiple sclerosis brain and animal models of demyelination: a role for tumor necrosis factor-induced peptidylarginine deiminase 4 translocation. J Neurosci 2006;26(44):11387–96.
[69] Sharma P, Azebi S, England P, Christensen T, Moller-larsen A, Petersen T, et al. Citrullination of histone H3 interferes with HP1-mediated transcriptional repression. PloS Genet 2012;8(9):e1002934.
[70] Moscarello MA, Lei H, Mastronardi FG, Winer S, Tsui H, Li Z, et al. Inhibition of peptidyl-arginine deiminases reverses protein-hypercitrullination and disease in mouse models of multiple sclerosis. Dis Model Mech 2013;6(2):467–78.
[71] Singhal NK, Li S, Arning E, Alkhayer K, Clements R, Sarcyk Z, et al. Changes in methionine metabolism and histone H3 trimethylation are linked to mitochondrial defects in multiple sclerosis. J Neurosci 2015;35(45):15170–86.
[72] Liu Z, Cao W, Xu L, Chen X, Zhan Y, Yang Q, et al. The histone H3 lysine-27 demethylase Jmjd3 plays a critical role in specific regulation of Th17 cell differentiation. J Mol Cell Biol 2015;7(6):505–16.
[73] Hashizume R, Andor N, Ihara Y, Lerner R, Gan H, Chen X, et al. Pharmacologic inhibition of histone demethylation as a therapy for pediatric brainstem glioma. Nat Med 2014;20(12):1394–6.
[74] Gunawan M, Venkatesan N, Loh JT, Wong JF, Berger H, Neo WH, et al. The methyltransferase Ezh2 controls cell adhesion and migration through direct methylation of the extranuclear regulatory protein talin. Nat Immunol 2015;16(5):505–16.
[75] Liu J, Magri L, Zhang F, Marsh NO, Albrecht S, Huynh JL, et al. Chromatin landscape defined by repressive histone methylation during oligodendrocyte differentiation. J Neurosci 2015;35(1):352–65.
[76] Liu J, Dupree JL, Gacias M, Frawley R, Sikder T, Naik P, et al. Clemastine enhances myelination in the prefrontal cortex and rescues behavioral changes in socially isolated mice. J Neurosci 2016;36(3):957–62.
[77] Paul DS, Beck S. Advances in epigenome-wide association studies for common diseases. Trends in Mol Med 2014;20(10):541–3.
[78] Somers EC, Richardson BC. Environmental exposures, epigenetic changes and the risk of lupus. Lupus 2014;23(6):568–76.
[79] Calabrese R, Valentini E, Ciccarone F, Gustafierro T, Bacalini MG, Ricigliano VA, et al. TET2 gene expression and 5-hydroxymethylcytosine level in multiple sclerosis peripheral blood cells. Biochim Biophys Acta 2014;1842(7):1130–6.
[80] Liu Y, Li X, Aryee MJ, Ekstrom TJ, Padyukov L, Klareskog L, et al. GeMes, clusters of DNA methylation under genetic control, can inform genetic and epigenetic analysis of disease. Am J Hum Genet 2014;94(4): 485–95.

[81] Feinberg AP, Irizarry RA. Evolution in health and medicine Sackler colloquium: stochastic epigenetic variation as a driving force of development, evolutionary adaptation, and disease. Proc Natl Acad Sci USA 2010;107(Suppl 1):1757–64.
[81a] Li Z, He Y, Fan S, Sun B. Clemastine rescues behavioral changes and enhances remyelination in the cuprizone mouse model of demyelination. Neurosci Bull 2015;31(5):617–25.
[82] McRae AF, Powell JE, Henders AK, Bowdler L, Hemani G, Shah S, et al. Contribution of genetic variation to transgenerational inheritance of DNA methylation. Genome Biol 2014;15(5):R73.
[83] Liu Y, Aryee MJ, Padyukov L, Fallin MD, Hesselberg E, Runarsson A, et al. Epigenome-wide association data implicate DNA methylation as an intermediary of genetic risk in rheumatoid arthritis. Nat Biotechnol 2013;31(2):142–7.
[84] Olsson AH, Volkov P, Bacos K, Dayeh T, Hall E, Nilsson EA, et al. Genome-wide associations between genetic and epigenetic variation influence mRNA expression and insulin secretion in human pancreatic islets. PLoS Genet 2014;10(11):e1004735.
[85] Hong X, Hao K, Ladd-Acosta C, Hansen KD, Tsai HJ, Liu X, et al. Genome-wide association study identifies peanut allergy-specific loci and evidence of epigenetic mediation in US children. Nat Commun 2015; 6:6304.
[86] Shin J, Bourdon C, Bernard M, Wilson MD, Reischl E, Waldenberger M, et al. Layered genetic control of DNA methylation and gene expression: a locus of multiple sclerosis in healthy individuals. Hum Mol Genet 2015;24(20):5733–45.
[87] Handel AE, De Luca GC, Morahan J, Handunnetthi L, Sandovnick AD, Ebers GC, et al. No evidence for an effect of DNA methylation on multiple sclerosis severity at HLA-DRB1*15 or HLA-DRB5. J Neuroimmunol 2010;223(1–2):120–3.
[88] Peelen E, Knippenberg S, Muris AH, Thewissen M, Smolders J, Tervaert JW, et al. Effects of vitamin D on the peripheral adaptive immune system: a review. Autoimmun Rev 2011;10(12):733–43.
[89] DeLuca GC, Kimball SM, Kolasinski J, Ramagopalan SV, Ebers GC. Review: the role of vitamin D in nervous system health and disease. Neuropathol Appl Neurobiol 2013;39(5):458–84.
[90] Ueda P, Rafatnia F, Baarnhielm M, Frobom R, Korzunowicz G, Lonnerbro R, et al. Neonatal vitamin D status and risk of multiple sclerosis. Ann Neurol 2014;76(3):338–46.
[91] Eyles D, Almeras L, Benech P, Patatian A, Mackay-Sim A, McGrath J, et al. Developmental vitamin D deficiency alters the expression of genes encoding mitochondrial, cytoskeletal and synaptic proteins in the adult rat brain. J Steroid Biochem Mol Biol 2007;103(3–5):538–45.
[92] Tekes K, Gyenge M, Hantos M, Csaba G. Transgenerational hormonal imprinting caused by vitamin A and vitamin D treatment of newborn rats. Alterations in the biogenic amine contents of the adult brain. Brain Dev 2009;31(9):666–70.
[93] Kesby JP, O'Loan JC, Alexander S, Deng C, Huang XF, McGrath JJ, et al. Developmental vitamin D deficiency alters MK-801-induced behaviours in adult offspring. Psychopharmacology 2012;220(3):455–63.
[94] Ramagopalan SV, Heger A, Berlanga AJ, Maugeri NJ, Lincoln MR, Burrell A, et al. A ChIP-seq defined genome-wide map of vitamin D receptor binding: associations with disease and evolution. Genome Res 2010;20(10):1352–60.
[95] Handel AE, Sandve GK, Disanto G, Berlanga-Taylor AJ, Gallone G, Hanwell H, et al. Vitamin D receptor ChIP-seq in primary CD4+ cells: relationship to serum 25-hydroxyvitamin D levels and autoimmune disease. BMC Med 2013;11:163.
[96] Ramagopalan SV, Maugeri NJ, Handunnetthi L, Lincoln MR, Orton SM, Dyment DA, et al. Expression of the multiple sclerosis-associated MHC class II Allele HLA-DRB1*1501 is regulated by vitamin D. PLoS Genet 2009;5(2):e1000369.
[97] Fetahu IS, Hobaus J, Kallay E. Vitamin D and the epigenome. Front Physiol 2014;5:164.
[98] Pereira F, Barbachano A, Singh PK, Campbell MJ, Munoz A, Larriba MJ. Vitamin D has wide regulatory effects on histone demethylase genes. Cell Cycle 2012;11(6):1081–9.

[99] Aslibekyan S, Dashti HS, Tanaka T, Sha J, Ferrucci L, Zhi D, et al. PRKCZ methylation is associated with sunlight exposure in a North American but not a Mediterranean population. Chronobiol Int 2014;31(9): 1034–40.

[100] Lin R, Taylor BV, Simpson S Jr, Charlesworth J, Ponsonby Al, Pittas F, et al. Novel modulating effects of PKC family genes on the relationship between serum vitamin D and relapse in multiple sclerosis. J Neurol Neurosurg Psychiatry 2014;85(4):399–404.

[101] Pender MP, Burrows SR. Epstein-Barr virus and multiple sclerosis: potential opportunities for immunotherapy. Clin Transl Immunol 2014;3(10):e27.

[102] Saha A, Jha HC, Upadhyay SK, Robertson ES. Epigenetic silencing of tumor suppressor genes during in vitro Epstein–Barr virus infection. Proc Natl Acad Sci USA 2015;112(37):E5199–207.

[103] Skalsky RL, Cullen BR. EBV noncoding RNAs. Curr Top Microbiol Immunol 2015;391:181–217.

[104] Weston M, Constantinescu CS. What role does tobacco smoking play in multiple sclerosis disability and mortality? A review of the evidence. Neurodegener Dis Manag 2015;5(1):19–25.

[105] Bergougnoux A, Claustres M, De Sario A. Nasal epithelial cells: a tool to study DNA methylation in airway diseases. Epigenomics 2015;7(1):119–26.

[106] Gao X, Jia M, Zhang Y, Breitling LP, Brenner H. DNA methylation changes of whole blood cells in response to active smoking exposure in adults: a systematic review of DNA methylation studies. Clin Epigenet 2015;7:113.

[107] Ladd-Acosta C. Epigenetic signatures as biomarkers of exposure. Curr Environ Health Rep 2015;2(2): 117–25.

[108] Nguyen NT, Nakahama T, Le DH, Van Son L, Chu HH, Kishimoto T. Aryl hydrocarbon receptor and kynurenine: recent advances in autoimmune disease research. Front Immunol 2014;5:551.

[109] Seuter S, Pehkonen P, Heikkinen S, Carlberg C. Dynamics of 1alpha, 25-dihydroxyvitamin D3-dependent chromatin accessibility of early vitamin D receptor target genes. Biochim Biophys Acta 2013;1829(12): 1266–75.

[110] Thacker EL, Mirzaei F, Ascherio A. Infectious mononucleosis and risk for multiple sclerosis: a meta-analysis. Ann Neurol 2006;59(3):499–503.

[111] Levin LI, Munger KL, O'Reilly EJ, Falk KI, Ascherio A. Primary infection with the Epstein–Barr virus and risk of multiple sclerosis. Ann Neurol 2010;67(6):824–30.

[112] Nielsen TR, Rostgaard K, Nielsen NM, Koch-Henriksen N, Haahr S, Sorensen PS, et al. Multiple sclerosis after infectious mononucleosis. Arch Neurol 2007;64(1):72–5.

[113] Briggs FB, Acuna B, Shen L, Ramsay P, Quach H, Bernstein A, et al. Smoking and risk of multiple sclerosis: evidence of modification by NAT1 variants. Epidemiology 2014;25(4):605–14.

[114] Lara-Astiaso D, Weiner A, Lorenzo-Vivas E, Zaretsky I, Jaitin DA, David E, et al. Immunogenetics. Chromatin state dynamics during blood formation. Science 2014;345(6199):943–9.

[115] Rotem A, Ram O, Shoresh N, Sperling RA, Goren A, Weitz DA, et al. Single-cell ChIP-seq reveals cell subpopulations defined by chromatin state. Nat Biotechnol 2015;33(11):1165–72.

[116] Heyn H, Esteller M. DNA methylation profiling in the clinic: applications and challenges. Nat Rev Genet 2012;13(10):679–92.

[117] Ma B, Wilker EH, Willis-Owen SA, Byun HM, Wong KCC, Motta V, et al. Predicting DNA methylation level across human tissues. Nucleic Acids Res 2014;42(6):3515–28.

[118] Shenker NS, Ueland PM, Polidoro S, van Veldhoven K, Ricceri F, Brown R, et al. DNA methylation as a long-term biomarker of exposure to tobacco smoke. Epidemiology 2013;24(5):712–6.

[119] Chan MW, Chang CB, Tung CH, Sun J, Suen JL, Wu SF. Low-dose 5-aza-2'-deoxycytidine pretreatment inhibits experimental autoimmune encephalomyelitis by induction of regulatory T cells. Mol Med 2014;20:248–56.

[120] Mangano K, Fagone P, Bendtzen K, Meroni PL, Quattrocchi C, Mammana S, et al. Hypomethylating agent 5-aza-2'-deoxycytidine (DAC) ameliorates multiple sclerosis in mouse models. J Cell Physiol 2014;229(12):1918–25.

[121] Polansky JK, Kretschmer K, Freyer J, Floess S, Garbe A, Baron U, et al. DNA methylation controls Foxp3 gene expression. Eur J Immunol 2008;38(6):1654–63.
[122] Viglietta V, Baecher-Allan C, Weiner HL, Hafler DA. Loss of functional suppression by CD4+CD25+ regulatory T cells in patients with multiple sclerosis. J Exp Med 2004;199(7):971–9.
[123] Kungulovski G, Jeltsch A. Epigenome editing: state of the art, concepts, and perspectives. Trends Genet 2016;32(2):101–13.
[124] Slaymaker IM, Gao L, Zetsche B, Scott DA, Yan WX, Zhang F. Rationally engineered Cas9 nucleases with improved specificity. Science 2016;351(6268):84–8.

CHAPTER

EPIGENETICS AND MIGRAINE

11

S.H. Gan, M.M. Shaik

Human Genome Centre, School of Medical Sciences, Universiti Sains Malaysia, Kubang Kerian, Kelantan, Malaysia

CHAPTER OUTLINE

11.1 INTRODUCTION

Migraine is an underrated and underreported neurovascular disorder with complex pathophysiology. It presents a high burden to both the individual and to society. Estimates suggest that approximately 10% of the world's population suffers from migraine [1]. Over the years, many theories related to migraine pathogenesis have been published [2–9]. However, the question remains as to why certain precipitators lead to migraine in some patients but not in others and the role of genetic vulnerability in selective physiological alterations is as yet unknown [10].

Various factors, such as alcohol, smoking, nutrition, stress, environmental changes, exercise, and menstrual cycles in women, have been reported to play some role in causing migraine. Cortical spreading depression (CSD) is reportedly related to migraine as it affects the brain's depolarization wave [11]. The depolarization wave propagates across the cerebral cortex when there is a decrease in cerebral blood flow [11], which occurs simultaneously with an aura (MA) [4]. In addition, an abnormal release of substance P in the trigeminal nerve, which is triggered by the depolarization of peptides, or the release of other neurotransmitters from the fifth cranial nerve explains both the hemicranial pain and the vasodilation characteristics of migraine headaches [5]. The characteristic migraine pain common to both MA and migraine without aura (MO) is postulated to occur due to the dilation of cerebral blood vessels following activation of the trigeminovascular system (TVS) [7]. The CSD can activate the TVS,

Neuropsychiatric Disorders and Epigenetics. http://dx.doi.org/10.1016/B978-0-12-800226-1.00011-3

thus providing a potential link between MA and headaches [7]. Important targets for involvement of TVS in migraine pathogenesis include biochemical factors that have the potential to disrupt both vascular endothelial function and cerebral blood flow, which can lead to CSD and/or affect the TVS [8].

Although there is a clear genetic component to migraine, it is likely that environmental factors contribute equally to the risk of developing migraine [12]. The heritable nature of migraine is influenced by a combination of environmental factors and mutations at multiple genetic loci. These factors often lead to a variable phenotypic expression, which can also occur due to chance, the influence of environmental factors or via interactions with other genes. Modulators of the frequency of migraine attacks include female sex hormones, with migraine affecting females two to three times more than males; migraine occurrence is influenced by both the menstrual cycle and pregnancy, as well as by the use of hormonal contraceptives [13]. Moreover, the menopausal condition, with reduced estrogen and progesterone production, is associated with a reduction in the frequency of migraine attacks [13]. To date, many researchers have primarily focused on the identification of genes that may cause increased susceptibility to migraine; this research may ultimately lead to an understanding of this disorder and an improvement in its therapy.

Epigenetics is one of the most rapidly developing areas in molecular biology. Epigenetics is defined as modifications of DNA other than sequence variation that carries information during cell division [14]. The best understood source of epigenetic information is DNA methylation, which is a covalent modification of cytosine maintained as CpG dinucleotides by DNA methyltransferase1. Methylation changes can arise secondary to chromatin modifications and are involved in the maintenance of boundaries between active and inactive chromatin. Additionally, DNA methylation is linked to gene silencing and chromosomal instability. In a common disease, an epigenetic framework can help provide an explanation of three important characteristics: (1) age-dependence that is not well explained by accumulated mutations, (2) the quantitative nature of a trait, and (3) the mechanism by which the environment may modulate genetic predisposition. Epigenetic alterations may influence disease phenotypes by affecting the target gene directly, regardless of sequence variation within the gene. Alternatively, the influence of epigenetic markers on disease phenotypes can occur via interaction with specific DNA sequence variants.

11.2 GENETIC VULNERABILITY TO MIGRAINE

Data from human studies have begun to identify genetic mutations, polymorphisms, and altered levels of specific proinflammatory and neuromodulatory molecules that strongly correlate with the presence of migraine as well as symptom severity. The results from a smaller number of studies have identified parameters, such as the level of the neuropeptide calcitonin gene-related peptide (CGRP), that are significantly associated with the response to specific treatments for acute migraine attacks and prophylaxis. Epigenetic mechanisms may also be involved in the development of migraine, and understanding environmentally induced genetic changes associated with this disease may eventually guide the development of therapies capable of reversing these pathophysiological changes in gene function.

DNA methylation and posttranslational histone modifications result in the heritability of traits not attributable to a change in DNA sequence [15]. Gene expression can be altered by modifying DNA methylation patterns using pharmacological techniques [16]. For example, migraine abortive and preventive drugs are effective in approximately half of patients [17–19], which may be due to the presence of functional gene variants [20]. However, the response to analgesics and triptans can also change over time, depending on the frequency of use (i.e., overuse impairs the effectiveness of abortive medication). This change in response cannot be explained by simple alterations in the genetic code, but it suggests epigenetic changes at target molecules. Such a pharmacoepigenetic model would include both DNA sequence variation and methylation, etc., as determinants of drug treatment responses [21].

It has been suggested that the methylene tetrahydrofolate reductase (MTHFR) gene, encoding an important protein of the DNA methylation cycle, is involved in migraine [22,23]. However, genome-wide association studies have pointed to other genes involved in epigenetic mechanisms [10,12,24]. Moreover, the migraine prophylactic valproic acid inhibits histone deacetylation as well as DNA methylation [25]. Finally, 17b-estradiol, which seems to be responsible for the 2–3 times higher prevalence of migraine in women compared to men, may exert at least part of its effects via epigenetic mechanisms [26,27]. Thus, DNA methylation may be involved in migraine pathophysiology. DNA methylation occurs mainly at cytosines of CpG dinucleotides found in the genome [28]. CpG island methylation changes in different types of cancer, during aging, and possibly during the menstrual cycle [29–32]. Environmental factors that might be responsible for these epigenetic mechanisms include the presence of female hormones and nutrition [33,34]. DNA methylation can differ greatly between tissues, thus methylation should be studied in the tissue of interest. For complex brain diseases, such as migraine, targeting the relevant tissue is impossible in humans. However, blood can be derived easily, and thus if the DNA methylation pattern of leukocytes is correlated with that of other tissues, it would be possible to draw conclusions from methylation studies in leukocytes.

11.3 HOMOCYSTEINE AND MIGRAINE

The MTHFR enzyme catalyzes the reduction of 5,10-methylenetetrahydrofolate to 5-methyltetrahydrofolate. The *MTHFR C677T* allele results in an amino acid change and a reduction in MTHFR activity leading to mild hyperhomocysteinemia [35]. Therefore, it is hypothesized that homocysteine-related endothelial dysfunction may be involved in the initiation and maintenance of migraine attacks. During a migraine attack, the concentration of oxygen present in the brain is reduced as a result of vasodilatation or a temporary thrombosis of cerebral blood vessels, primarily caused by the excitatory amino acid homocysteine [36]. Elevated levels of homocysteine in neurons can damage DNA, alter DNA repair functions, and/or disturb DNA methylation, all of which may lead to oxidative stress.

Folate plays an important role in the transfer of a one-carbon moiety and is an essential cofactor for the de novo biosynthesis of purines and thymidylate (Fig. 11.1) [37]. MTHFR, an intracellular coenzymatic form of folate, is pivotal for the conversion of deoxyuridylate to thymidylate and can be oxidized to 10-formyltetrahydrofolate for de novo purine synthesis [37]. Healthy subjects with the *MTHFR 677TT* genotype have been reported as having hypomethylated DNA when compared to subjects bearing the wild-type genotype, indicating an association between plasma homocysteine levels, folate levels and DNA methylation status [38]. However, subjects with the *MTHFR 1298AC* polymorphism displayed lower DNA methylation status when compared to subjects with the *677TT* genotype [39]. The difference between these polymorphisms is significant with regard to the absence of the *677CT* mutation when compared with the double wild-type genotype *677CC/1298AA* [38,40].

The *1298AA/677TT* genotype exhibited decreased genomic DNA methylation in the presence of low plasma folate levels. The *MTHFR 1298AC* polymorphism does not impair enzyme function, and

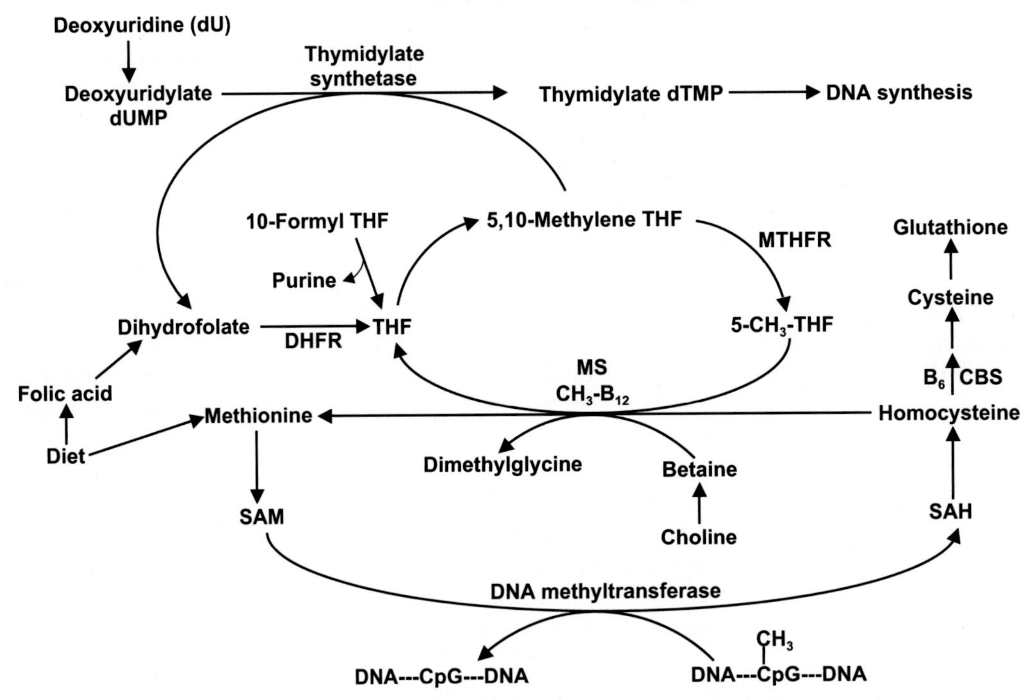

FIGURE 11.1 Homocysteine Metabolism and Role of Methionine and Vitamins in DNA Methylation

CBS, Cystathionine beta synthase; *DHFR*, dihydrofolate reductase; *dTMP*, deoxy-thymidylate mono phosphate; *dUMP*, deoxy-uridine mono phosphate; *MS*, methionine synthase; *MTHFR*, methylenetetrahydrofolate reductase; *SAM*, S-adenosyl methionine; *SAH*, S-adenosyl-L-homocysteine; *THF*, tetrahydrofolate.

thus does not significantly affect the pathway of homocysteine remethylation or biological methylation functions, indicating that the *1298AC* mutation has a minor effect on one-carbon metabolism resulting from reduced MTHFR enzyme function [41]. The *677CT* polymorphism lies at the base of the binding site of flavin adenine dinucleotide, a MTHFR cofactor, and has been shown to affect MTHFR enzyme activity more significantly than the *1298AC* variant [42,43]. Therefore, it appears that vitamin B_{12} plays an important role in MTHFR methylation [44].

11.4 ESTROGEN AND MIGRAINE

Estrogen acts by binding to the estrogen receptor (ER) in target cells, which subsequently undergoes a conformational change and binds to specific DNA sequences. This transcription complex regulates the expression of target genes within a cell. Since the ER has a unique ability to bind a wide variety of compounds with diverse structural features, many environmental toxins and plant compounds can bind to the ER, with varying affinities, and subsequently modulate estrogen activity [45,46].

Elevated intracellular concentrations of the active form of vitamin B_6 can lead to significant decreases in gene transcription response when estrogen binds to the ER. By modulating estrogen-induced gene expression in this way, vitamin B_6 can attenuate the biological effects of estrogen [47,48]. Additionally, vitamins B_6, B_{12}, and folate are important cofactors of the enzymes involved in estrogen conjugation and methylation. Therefore, decreased levels of B vitamins can disrupt estrogen detoxification and lead to increased levels of circulating estrogen [49].

Estrogen has a direct effect on neurons in the trigeminal ganglion because many such neurons express estrogen receptor alpha (ER alpha) [40,50,51]. Estrogen binds to the classic nuclear receptor ER alpha, which functions as a ligand-dependent transcription factor controlling gene expression. The nuclear form of ER alpha is a sequence-specific DNA-binding protein that recognizes *cis*-acting estrogen response elements in the promoter region of target genes. The effects of nuclear ER alpha require the synthesis of new RNA and protein. Additionally, estrogen induces rapid effects that are mediated via membrane and cytoplasmic receptors, which function by activating Mitogen activated protein (MAP) kinases, including extracellular signal-regulated kinase (ERK) [52].

The rapid effects of estrogen are important in many physiological processes, including cardiovascular protection, bone preservation, cancer cell proliferation, and neuroprotection, as well as in regulating trigeminal nociception [41,53]. Transdermal application of estrogen gel has been reported to have positive clinical effects on migraine [54]. In animals, the expression levels of serotonin transporter (SERT)-encoding mRNAs decreased after oral treatment with selective estrogen receptor agonists, such as raloxifene. In contrast, the expression of tryptophan hydroxylase (TPH), the rate-limiting enzyme in serotonin synthesis, was significantly increased in the midbrain portion of a migrainous individual [55]. Estrogen treatment has been reported to decrease serotonin uptake in a SERT and ER beta receptor-expressing rat serotonergic cell line within 15 min of administration [56]. Estrogen treatment has been reported to increase TPH expression in the neurons of the dorsal raphe nucleus of wild or ER alpha-knocked out (KO), but not in ER beta-KO, mice [57], indicating that estrogen has a role in ameliorating migraine headaches.

In a study using mice, the mRNA levels of TPH1 (the rate-limiting enzyme in serotonin synthesis) were increased more than 2-fold, and the protein levels were increased 1.4-fold during the high-estrogen stage (proestrus) of the estrous cycle when compared to the low-estrogen stage (diestrus) [50,58]. This finding is relevant for menstrual migraine because headache attacks are usually triggered during the menstrual phase when the estrogen level is decreased. Estrogen has been reported to decrease the contractile responses to serotonin primarily by inhibiting calcium influx through voltage-dependent calcium channels [59]. Estrogen-evoked antidepressant actions are also transduced by ER beta via increasing serotonin (5-HT) levels [60]. The expression of presynaptic 5-HT_{1D} autoreceptors has been reported to be significantly increased in female patients suffering from major depressive disorder but not in their male counterparts [61]. Estrogen treatment has been reported to decrease the levels of 5-HT_{1B} receptor-encoding mRNAs, where the expression of TPH2 was increased in a coordinated manner, thereby increasing serotonin synthesis and release in the forebrain regions, which are also often involved in anxiety behaviors [62]. Sex steroids exert a robust regulation on nociceptive pathways at multiple levels ranging from the expression of neurotransmitters and their cognate receptors to downstream signaling. At both the expression and functional levels, the trigeminal vascular system (TVS) is significantly modulated by sex steroids and is most likely the major determinant of the sexual dimorphism observed in migraine incidence [63].

Estrogen signaling can also target ER transcription and modulate the sensitivity of cells towards estradiol signals. Estradiol has been recognized for its ability to either enhance or decrease *ESR1* expression, depending on the cell type [64]. The effects of DNA methylation of *ESR1* [65] may be understood by examining the hypermethylation of the *ESR1* promoter CpG-island in breast cancer, which leads to the loss of *ESR1* expression in breast tumors [66]. In a nonpathological state, increased DNA methylation in the *ESR1* promoter was associated with decreased *ESR1* expression during postnatal development [67] and was also associated with estradiol exposure [68]. Changes in DNA methylation drive the physiological *ESR1* expression, which may occur in a time- and/or tissue-specific manner due to dynamic changes in circulation and local steroid hormone concentrations during development and over the lifespan of an individual. This leads us to the question of whether circulating steroid hormone concentrations, which change dynamically during development, may also play an important role in regulating *ESR1* in various other cell types and tissues. However, which factors contribute to migraine in relation to *ESR1* remains unclear. Epigenetics may provide a plausible explanation of the relationship between *ESR1* expression and estrogen levels associated with the role of vitamins in migraine patients. The epigenetic modulator 17β-estradiol is involved in the increased prevalence of migraine in women [69].

11.4.1 HOMOCYSTEINE, ESTROGEN, AND EPIGENETICS

Lower levels of homocysteine were reported among pregnant, premenopausal, and postmenopausal women who were on estrogen replacement therapy (ERT) when compared to age-matched men or postmenopausal women on ERT [70–73]. Higher estrogen status was associated with a decreased mean serum homocysteine concentration in an American population [74]. Folate deficiency due to estrogen-induced turnover among pregnant women suggested that estrogen influences folate utilization [75]. Extraembryonic coelomic fluid and amniotic fluid were found to have decreased levels of homocysteine even when high concentrations of methionine were detected, suggesting that estrogen may interact with methionine/homocysteine metabolism during early human development [76].

Estrogen directly upregulates nitric oxide synthetase (NOS) activity and increases the availability of nitric oxide (NO) [77,78] (Fig. 11.2). The formation of S-nitrosothiol inhibits the action of homocysteine [79]. Estrogen enhances the activity of glucose-6-phosphate dehydrogenase (G6PDH), which is the rate-limiting enzyme of the hexose monophosphate (HMP) shunt. This enhancement increases the activity of NADPH, which acts as a cofactor to replenish the availability of reduced glutathione (GSH) [80]. Furthermore, estrogen directly increases plasma GSH levels by reducing free radicals. This action prevents the formation of peroxynitrite ($ONOO^-$) by forming S-nitroglutathione (GSNO), which produces specific cytoprotective effects [81]. GSNO upregulates GSH synthesis and the HMP shunt, which limits the formation of $ONOO^-$ [82] (Fig. 11.2). Overall, estrogen modulates methionine biosynthesis by interfering with the transsulphuration pathway, thus diminishing homocysteine levels and preventing homocysteine accumulation.

Estrogen has both direct and indirect effects on plasma homocysteine levels and also influences the activity of homocysteine (Fig. 11.3). Estrogen also has a major effect on TVS and controls the nociceptive pathway. Homocysteine plays a major role in the biosynthesis of methionine, which is essential for the synthesis of SAM. Therefore, the metabolism of homocysteine is also important in DNA methylation. Since vitamins B_6, B_9, and B_{12} are involved in the metabolism of homocysteine, deficiencies of these vitamins may affect the methylation of the migraine-associated *MTHFR* and *ESR1* genes. Increased homocysteine levels and decreased estrogen levels may activate reactive oxygen species, which may affect endothelial dysfunction, cause direct cellular damage and activate platelets (Fig. 11.3). These findings clearly indicate the role of folate and vitamins B_6 and B_{12} in both the elevation of homocysteine and the depletion of estrogen in patients experiencing migraine.

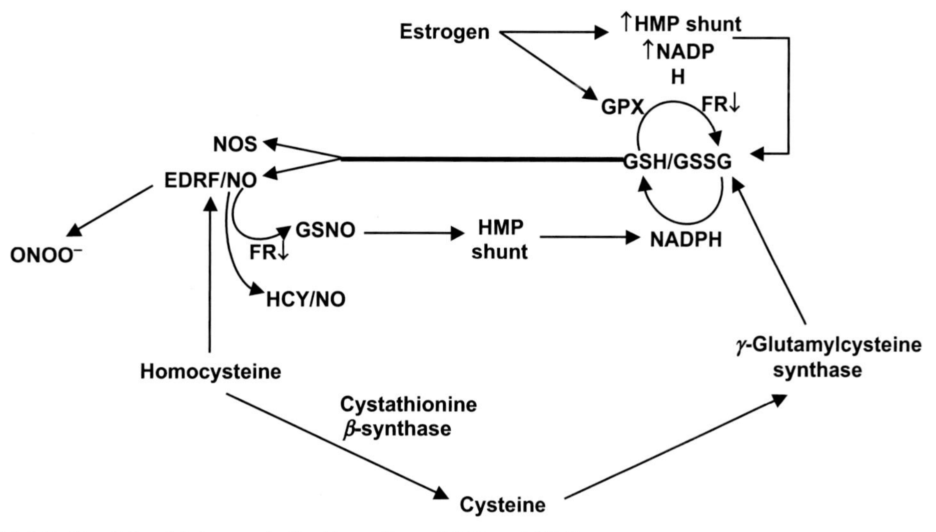

FIGURE 11.2 The Role of Estrogen in the Formation of GSH and EDRF

EDRF formation increases with increasing levels of homocysteine. EDRF levels are associated with a reduction in the formation of FR. *EDRF*, Endothelium-derived relaxing factor; *FR*, free radicals; *GSH*, glutathione sulfhydryl (reduced glutathione); *GPX*, glutathione peroxidase; *GSNO*, S-nitroglutathione; *GSSG*, glutathione disulfide (oxidized glutathione); *HCY*, homocysteine; *HMP*, hexose monophosphate; *NADPH*, nicotinamide adenine dinucleotide phosphate; *NO*, nitric oxide; *NOS*, nitric oxide synthase; *ONOO*, peroxynitrite.

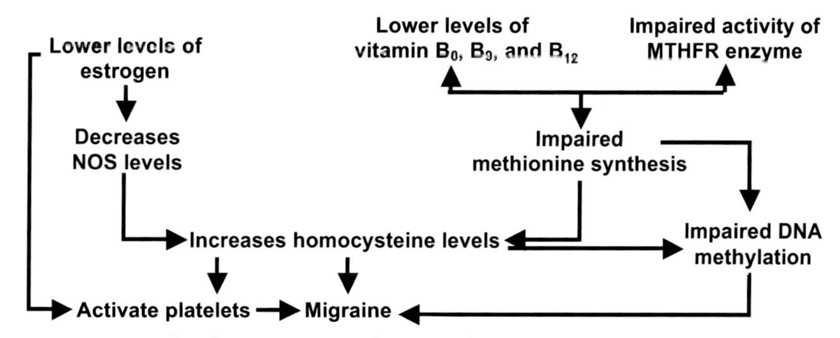

FIGURE 11.3 The Role of B Vitamins and Estrogen in Increasing the Levels of Homocysteine

The relationship between homocysteine and estrogen. *MTHFR*, Methylenetetrahydrofolate reductase; *NOS*, nitric oxide synthase.

Another important factor in migraine is the alteration of the expression of *MTHFR* and *ESR1,* which directly regulate homocysteine and estrogen levels. The methylation status of *MTHFR* and *ESR1* may be altered due to decreased formation of SAM, which may be the primary cause of altered gene expression. This relationship indicates an underlying epigenetic mechanism of *MTHFR* and *ESR1* in migraine patients; however, additional exclusive studies are needed to elucidate this mechanism.

A group of researchers recently hypothesized that platelet aggregation is one of the main reasons for migraine and stroke-induced migraine [83]. The rationale for the use of epigenetic information as a potential disease determinant stems from epigenetic heritability during cell division in a given cell lineage and the associated biological interaction with DNA sequence variation. In this context, we are particularly interested in investigating epigenetic mechanisms in migraine patients, a topic that has yet to be explored. A high-resolution microarray-based analysis of several potentially functional sites within the *ESR1* and *MTHFR* genes, which may differentially contribute to the degree of mRNA expression according to their DNA methylation patterns, is especially interesting.

Few researchers have reported the prophylactic use of vitamins B_6, B_{12}, or folic acid in migraine. These vitamins play a vital role in the metabolism of homocysteine and are also related to MTHFR gene polymorphisms. However, homocysteine and estrogen levels, together with epigenetic factors, play a vital role in *MTHFR* and *ESR1* gene polymorphism, it is important to study the role of vitamin supplementation in the methylation status of both *MTHFR* and *ESR1* in migraine patients. The following experimental protocol describes the various steps involved in investigating the possible effect of vitamin supplementation on migraine patients: During the initial phase, DNA methylation status, including gene expression levels, will be studied, along with the levels of vitamins, homocysteine and estrogen in migraine patients. The association between these factors will be investigated. Subsequently, patients will be supplemented with vitamins for 6 months. After the supplementation period, DNA methylation and gene expression, as well as the levels of vitamins, homocysteine and estrogen, will be investigated. Finally, the vitamin supplementation regime will be stopped for 3 months. All of the parameters will be reassessed after nine months of enrollment to elucidate the exact role of vitamin supplementation in maintaining levels of estrogen and homocysteine in migraine patients.

11.5 CALCITONIN AND MIGRAINE

CGRP is a neuropeptide which plays a major role in migraine pathophysiology [9,84]. Elevated plasma levels of CGRP were reported during migraine attacks [85–87], although this finding is controversial [88–90]. Importantly, individuals with migraine [91], but not healthy controls [92], develop a migraine-like headache in response to intravenous administration of CGRP, suggesting that migraine patients are especially sensitive to CGRP. CGRP receptor antagonists are effective in alleviating the pain and associated symptoms of migraine attacks [93,94]. Photophobia leading to light-aversive behavior is commonly reported by migraine patients during acute attacks and, to a lesser degree, between attacks [95]. Photophobia also accompanies other less frequent, but similarly disabling conditions, such as cluster headache and other trigeminal autonomic cephalalgias and blepharospasm [96–99]. The mechanisms underlying photophobia are currently unknown; however, it seems likely that the trigeminal system is involved [100–102].

Photophobia in blind patients suggests contributions from a nonvisual pathway [103]. In addition, trigeminal autonomic cephalalgias, a less common group of primary headache disorders, are characterized by unilateral trigeminal-mediated pain frequently associated with ipsilateral photophobia [99]. As a major neuropeptide in the trigeminal system, CGRP is likely to be involved in the development of photophobia; however, this relationship has not been directly tested to date. Epigenetic or genetic elevation of receptor activity modifying protein1 (*RAMP1*) gene levels provides a mechanism for the reported sensitivity to CGRP of migraine patients [91,92,104]. *RAMP1* is a modifier of photophobia and is known to confer species specificity to the CGRP receptor. In animal models, hypersensitivity to CGRP has been attributed to an elevated transmembrane domain of RAMP1 (hRAMP1), which also contributes to migraine susceptibility [104].

The CGRP is encoded by the calcitonin-related polypeptide alpha (*CALCA*) gene, which yields both CGRP and the hormone calcitonin (CT) as alternative splice products [105]. *CALCA* expression is normally restricted primarily to endocrine and neuronal cells and is not normally expressed by glia. However, there is increasing evidence that glia play an important role in pain conditions, possibly including migraine [106]. Neuronal expression of *CALCA* has been attributed to an 18-bp enhancer located approximately 1 kb upstream of the transcription start site [107]. Previous studies reported a CpG island extending from approximately −1.8 kb into exon 1 of human *CALCA* [108]. However, studies on the correlation of methylation status with *CALCA* expression were not clear; some "negative" tissues and cell lines (e.g., liver and small-cell lung carcinomas) actually express *CALCA*, and the restriction enzymes used only recognized a subset of CpG sites. In addition to CpG methylation, deacetylation of histones, especially H3 and H4, by histone deacetylases (HDACs) has been associated with transcriptional repression [109].

A study using rat and human cell lines as model systems reported that a CpG island near the 18-bp enhancer was hypomethylated in expressing cells, but was hypermethylated in nonexpressing cells [110]. In addition, the same study reported that histone acetylation was much higher in *CALCA* expressing cells than in nonexpressing cells. The DNA methylation inhibitor 5-aza-2′-deoxycytidine (Aza-dC) induced the CALCA gene in both nonexpressing cell lines and primary glial cultures, indicating that DNA methylation represses *CALCA* expression [110,111]. These study results demonstrate the role of DNA methylation and histone acetylation in *CALCA* expression, which can also play a vital role in migraine epigenetics.

CpG island methylation and histone H3 acetylation at the 18-bp cell-specific enhancer have been correlated with CALCA gene expression [110]. In the same study, the researchers tested the role of these epigenetic phenomena using the DNA methylation inhibitor Aza-dC and the HDAC inhibitor trichostatin A (TSA). While TSA failed to induce the CALCA gene, Aza-dC induced the gene in both human- and rat-cell lines and cultured glia. Interestingly, the combination of TSA and Aza-dC showed a synergistic effect on CALCA gene induction in glia. This finding indicates that DNA demethylation is required for the effect of histone acetylation to be manifested. Thus, it appears that CpG methylation around the 18-bp enhancer is a key determinant of cell-specific gene expression [110]. A recent study provides the first evidence that DNA methylation at the RAMP1 promoter plays a key role in migraine pathophysiology [112]. A low methylation trend was reported in migraine patients, and two CpG units were linked with a family history of migraine and female migraine. In addition, DNA methylation can be regulated by glucocorticoids [113], neonatal stress [114], vitamin D [115], and mitogen-activated protein (MAP) kinases [116]. While the regulator of CALCA demethylation has not yet been identified, there are some clues that inflammatory signals might play a role.

Neurogenic inflammation and components of the immune system have long been implicated in migraine [117]. Neurogenic inflammation with accompanying mast cell degranulation could lead to epigenetic induction. The potential mechanism by which the CALCA gene could be induced to produce pro-CT in trigeminal glia and pro-CT could be a biomarker of inflammatory activation of the trigeminal system in migraine [110]. Genetically and epigenetically, the CGRP is now firmly established as a key player in migraine pathophysiology. Clinical trials carried out during the past decade have proved that CGRP receptor antagonists are effective for treating migraine [118–120]. Despite this progress in the clinical arena, the mechanisms by which CGRP triggers migraine remain uncertain. Future studies on epistatic and epigenetic regulators of CGRP actions are expected to shed further light on CGRP actions in migraine. Targeting CGRP represents a feasible therapeutic strategy for migraine.

11.6 NEUROPSYCHIATRIC ASPECTS OF MIGRAINE

In the context of this book, psychiatric and psychological problems can be the cause and effect of migraine. Psychological factors, such as stress, or the anticipation of stress, are well known to provoke attacks of migraine. Sustained emotional tension appears to be more important than acute emotional disturbance in provoking a migraine attack, the crucial factor being the extent to which feelings are sustained, bottled up, and inadequately expressed [121]. Feelings of irritability and anxiety may last for many hours before an attack of migraine. Mental changes are common during the attack itself. Such changes include anxiety and irritability early on, with drowsiness and lethargy as the headache continues [121].

Psychiatric problems can also be the effect of attacks of migraine. Population-based studies have shown that the risk of a patient with migraine developing anxiety or depression is at least two to three times higher than that of someone without headache [121]. These psychiatric features are mainly found in patients with migraine with an aura [122]. The mechanisms underlying the development of anxiety and depression in patients with migraine are presently not well understood [122]. Psychiatric comorbidity in patients with migraine can lead to chronic substance abuse, influences the way migraine attacks progress, can affect treatment strategies of migraine attacks, and can eventually modify the outcome of the disorder.

11.7 CONCLUSIONS

The understanding of the etiology of migraine is incomplete. The role of epigenetics in migraine is a new, unexplored field that has pharmacological implications. The continued investigation and identification of genetic, epigenetic, and molecular mechanisms is likely to facilitate the goal of individualizing medicine by enabling clinicians to more accurately diagnose and treat migraine and other headache disorders.

ABBREVIATIONS

Aza-dC 5-Aza-2-deoxycytidine
CALCA Calcitonin-related polypeptide alpha
CBS Cystathionine beta synthase
CGRP Calcitonin gene related peptide
CSD Cortical spreading depression
DHFR Dihydrofolate reductase
dTMP Deoxy-thymidylate mono phosphate
dUMP Deoxy-uridine mono phosphate
EDRF Endothelium-derived relaxing factor

ESR1	Estrogen receptor 1
ER	Estrogen receptor
FR	Free radicals
GPX	Glutathione peroxidase
GSH	Glutathione sulfhydryl (reduced glutathione)
GSNO	S-nitroglutathione
GSSG	Glutathione disulfide (oxidized glutathione)
HCY	Homocysteine
HMP	Hexose monophosphate
hRAMP1	Transmembrane domain of RAMP1 protein
MA	Migraine with aura
MAP	Mitogen-activated protein
MO	Migraine without aura
MS	Methionine synthase
MTHFR	Methylenetetrahydrofolate reductase
NADPH	Nicotinamide adenine dinucleotide phosphate
NO	Nitric oxide
NOS	Nitric oxide synthase
ONOO	Peroxynitrite
RAMP1	Receptor activity modifying protein 1
SAH	S-adenosyl-L-homocysteine
SAM	S-adenosyl methionine
THF	Tetrahydrofolate
TSA	Trichostatin A
TVS	Trigeminal vascular system

GLOSSARY

Aura A subjective sensation (as of voices or lights) experienced before an attack of neurological disorders, such as epilepsy or migraine

Calcitonin A polypeptide hormone secreted by the thyroid gland that reduces the concentration of blood calcium when it rises above normal

Cephalalgia Headache, often combined with another word to indicate a specific type of headache

Triptans Drugs used to treat migraine and which act as agonists on 5-HT_1 (serotonin-1) receptors

REFERENCES

[1] World Health Organization. Atlas of Headache Disorders and Resources in the World 2011; 2011.

[2] Blau J, Harold G. Wolff: the man and his migraine. Cephalalgia 2004;24(3):215–22.

[3] Leão AAP. Spreading depression of activity in cerebral cortex. J Neurophysiol 1944;7:159–390.

[4] Olesen J, Larsen B, Lauritzen M. Focal hyperemia followed by spreading oligemia and impaired activation of rCBF in classic migraine. Ann Neurol 1981;9(4):344–52.

[5] Moskowitz MA, Reinhard JF Jr, Romero J, Melamed E, Pettibone DJ. Neurotransmitters and the fifth cranial nerve: is there a relation to the headache phase of migraine? Lancet 1979;2(8148):883–5.

[6] Moskowitz MA. The neurobiology of vascular head pain. Ann Neurol 1984;16(2):157–68.

[7] Ferrari MD. Migraine. Lancet 1998;351(9108):1043–51.
[8] Tzourio C, El Amrani M, Poirier O, Nicaud V, Bousser MG, Alperovitch A. Association between migraine and endothelin type A receptor (ETA -231 A/G) gene polymorphism. Neurology 2001;56(10):1273–7.
[9] Russo AF. Calcitonin gene-related peptide (CGRP): a new target for migraine. Annu Rev Pharmacol Toxicol 2015;55:533–52.
[10] Wessman M, Terwindt GM, Kaunisto MA, Palotie A, Ophoff RA. Migraine: a complex genetic disorder. Lancet Neurol 2007;6(6):521–32.
[11] Goadsby PJ. Current concepts of the pathophysiology of migraine. Neurol Clin 1997;15(1):27–42.
[12] Eising E, Datson NA, van den Maagdenberg AM, Ferrari MD. Epigenetic mechanisms in migraine: a promising avenue? BMC Med 2013;11(1):26.
[13] Borsook D, Erpelding N, Lebel A, Linnman C, Veggeberg R, Grant P, et al. Sex and the migraine brain. Neurobiol Dis 2014;68:200–14.
[14] Feinberg AP, Tycko B. The history of cancer epigenetics. Nat Rev Cancer 2004;4(2):143–53.
[15] Jirtle RL, Skinner MK. Environmental epigenomics and disease susceptibility. Nat Rev Genet 2007;8(4): 253–62.
[16] Peedicayil J. Epigenetic therapy-a new development in pharmacology. Indian J Med Res 2006;123(1):17.
[17] Goadsby PJ, Lipton RB, Ferrari MD. Migraine—current understanding and treatment. N Engl J Med 2002;346(4):257–70.
[18] Group SCHS. Treatment of acute cluster headache with sumatripatn. N Engl J Med 1991;325:322–6.
[19] Schürks M, Kurth T, De Jesus J, Jonjic M, Rosskopf D, Diener HC. Cluster headache: clinical presentation, lifestyle features, and medical treatment. Headache 2006;46(8):1246–54.
[20] Schürks M, Kurth T, Stude P, Rimmbach C, Jesus J, Jonjic M, et al. G protein β3 polymorphism and triptan response in cluster headache. Clin Pharmacol Therapeut 2007;82(4):396–401.
[21] Schürks M. Epigenetics in primary headaches: a new avenue for research. J Headache Pain 2008;9(3):191–2.
[22] Oterino A, Toriello M, Valle N, Castillo J, Alonso-Arranz A, Bravo Y, et al. The relationship between homocysteine and genes of folate-related enzymes in migraine patients. Headache 2010;50(1):99–168.
[23] Schürks M, Rist PM, Kurth T. MTHFR 677C> T and ACE D/I polymorphisms in migraine: a systematic review and meta-analysis. Headache 2010;50(4):588–99.
[24] Durham P, Papapetropoulos S. Biomarkers associated with migraine and their potential role in migraine management. Headache 2013;53(8):1262–77.
[25] Manev H, Uz T. DNA hypomethylating agents 5-aza-2′-deoxycytidine and valproate increase neuronal 5-lipoxygenase mRNA. Eur J Pharmacol 2002;445(1):149–50.
[26] Green CD, Han J-DJ. Epigenetic regulation by nuclear receptors. Epigenomics 2011;3(1):59–72.
[27] Imamura T. Epigenetic setting for long-term expression of estrogen receptor α and androgen receptor in cells. Horm Behav 2011;59(3):345–52.
[28] Jones PA. Functions of DNA methylation: islands, start sites, gene bodies and beyond. Nat Rev Genet 2012;13(7):484–92.
[29] Bergman Y, Cedar H. DNA methylation dynamics in health and disease. Nat Struct Mol Biol 2013;20(3): 274–81.
[30] Zhao Z, Fan L, Frick KM. Epigenetic alterations regulate estradiol-induced enhancement of memory consolidation. Proc Natl Acad Sci 2010;107(12):5605–10.
[31] Gentilini D, Mari D, Castaldi D, Remondini D, Ogliari G, Ostan R, et al. Role of epigenetics in human aging and longevity: genome-wide DNA methylation profile in centenarians and centenarians' offspring. Age (Dordr) 2013;35(5):1961–73.
[32] Guo S-W. The endometrial epigenome and its response to steroid hormones. Mol Cell Endocrinol 2012;358(2):185–96.
[33] Kvisvik EV, Stovner LJ, Helde G, Bovim G, Linde M. Headache and migraine during pregnancy and puerperium: the MIGRA-study. J Headache Pain 2011;12(4):443–51.

[34] Novensà L, Novella S, Medina P, Segarra G, Castillo N, Heras M, et al. Aging negatively affects estrogens-mediated effects on nitric oxide bioavailability by shifting ERalpha/ERbeta balance in female mice. PLoS One 2011;6(9):e25335.

[35] Lea R, Colson N, Quinlan S, Macmillan J, Griffiths L. The effects of vitamin supplementation and MTHFR (C677T) genotype on homocysteine-lowering and migraine disability. Pharmacogenet Genom 2009;19(6):422–8.

[36] Takano T, Tian GF, Peng W, Lou N, Lovatt D, Hansen AJ, et al. Cortical spreading depression causes and coincides with tissue hypoxia. Nat Neurosci 2007;10(6):754–62.

[37] Wagner C. . Biochemical role of folate in cellular metabolism. In: Bailey L, editor. Folate in health and disease. New York: Marcel Dekker Inc; 1995. p. 23–42.

[38] Friso S, Girelli D, Trabetti E, Olivieri O, Guarini P, Pignatti PF, et al. The MTHFR 1298A > C polymorphism and genomic DNA methylation in human lymphocytes. Cancer Epidemiol Biomarkers Prev 2005;14(4):938–43.

[39] Castro R, Rivera I, Ravasco P, Camilo ME, Jakobs C, Blom HJ, et al. 5,10-methylenetetrahydrofolate reductase (MTHFR) 677C-- > T and 1298A-- > C mutations are associated with DNA hypomethylation. J Med Genet 2004;41(6):454–8.

[40] Puri V, Puri S, Svojanovsky SR, Mathur S, Macgregor RR, Klein RM, et al. Effects of oestrogen on trigeminal ganglia in culture: implications for hormonal effects on migraine. Cephalalgia 2006;26(1): 33–42.

[41] Liverman CS, Brown JW, Sandhir R, Klein RM, McCarson K, Berman NE. Oestrogen increases nociception through ERK activation in the trigeminal ganglion: evidence for a peripheral mechanism of allodynia. Cephalalgia 2009;29(5):520–31.

[42] van der Put NM, Gabreels F, Stevens EM, Smeitink JA, Trijbels FJ, Eskes TK, et al. A second common mutation in the methylenetetrahydrofolate reductase gene: an additional risk factor for neural-tube defects? Am J Hum Genet 1998;62(5):1044–51.

[43] Guenther BD, Sheppard CA, Tran P, Rozen R, Matthews RG, Ludwig ML. The structure and properties of methylenetetrahydrofolate reductase from *Escherichia coli* suggest how folate ameliorates human hyperhomocysteinemia. Nat Struct Biol 1999;6(4):359–65.

[44] McKay JA, Groom A, Potter C, Coneyworth LJ, Ford D, Mathers JC, et al. Genetic and nongenetic influences during pregnancy on infant global and site specific DNA methylation: role for folate gene variants and vitamin B(12). PLoS One 2012;7(3):e33290.

[45] Cassidy A. Potential tissue selectivity of dietary phytoestrogens and estrogens. Curr Opin Lipidol 1999;10(1):47–52.

[46] Kuiper GG, Lemmen JG, Carlsson B, Corton JC, Safe SH, van der Saag PT, et al. Interaction of estrogenic chemicals and phytoestrogens with estrogen receptor beta. Endocrinology 1998;139(10):4252–63.

[47] Tully DB, Allgood VE, Cidlowski JA. Modulation of steroid receptor-mediated gene expression by vitamin B6. FASEB J 1994;8(3):343–9.

[48] Dhillon KS, Singh J, Lyall JS. A new horizon into the pathobiology, etiology and treatment of migraine. Med Hypotheses 2011;77(1):147–51.

[49] Butterworth M, Lau SS, Monks TJ. 17 beta-Estradiol metabolism by hamster hepatic microsomes. Implications for the catechol-O-methyl transferase-mediated detoxication of catechol estrogens. Drug Metab Dispos 1996;24(5):588–94.

[50] Puri V, Cui L, Liverman CS, Roby KF, Klein RM, Welch KM, et al. Ovarian steroids regulate neuropeptides in the trigeminal ganglion. Neuropeptides 2005;39(4):409–17.

[51] Bereiter DA, Cioffi JL, Bereiter DF. Oestrogen receptor-immunoreactive neurons in the trigeminal sensory system of male and cycling female rats. Arch Oral Biol 2005;50(11):971–9.

[52] Xu Y, Traystman RJ, Hurn PD, Wang MM. Neurite-localized estrogen receptor-alpha mediates rapid signaling by estrogen. J Neurosci Res 2003;74(1):1–11.

[53] Manavathi B, Kumar R. Steering estrogen signals from the plasma membrane to the nucleus: two sides of the coin. J Cell Physiol 2006;207(3):594–604.
[54] Nappi RE, Sances G, Brundu B, De Taddei S, Sommacal A, Ghiotto N, et al. Estradiol supplementation modulates neuroendocrine response to M-chlorophenylpiperazine in menstrual status migrainosus triggered by oral contraception-free interval. Hum Reprod 2005;20(12):3423–8.
[55] Bethea CL, Mirkes SJ, Su A, Michelson D. Effects of oral estrogen, raloxifene and arzoxifene on gene expression in serotonin neurons of macaques. Psychoneuroendocrinology 2002;27(4):431–45.
[56] Koldzic-Zivanovic N, Seitz PK, Watson CS, Cunningham KA, Thomas ML. Intracellular signaling involved in estrogen regulation of serotonin reuptake. Mol Cell Endocrinol 2004;226(1–2):33–42.
[57] Gundlah C, Alves SE, Clark JA, Pai LY, Schaeffer JM, Rohrer SP. Estrogen receptor-beta regulates tryptophan hydroxylase-1 expression in the murine midbrain raphe. Biol Psychiatry 2005;57(8):938–42.
[58] Berman NE, Puri V, Chandrala S, Puri S, Macgregor R, Liverman CS, et al. Serotonin in trigeminal ganglia of female rodents: relevance to menstrual migraine. Headache 2006;46(8):1230–45.
[59] Dursun N, Arifoglu C, Suer C. Relaxation effect of estradiol on different vasoconstrictor-induced responses in rat thoracal artery. J Basic Clin Physiol Pharmacol 2006;17(4):289–94.
[60] Hughes ZA, Liu F, Platt BJ, Dwyer JM, Pulicicchio CM, Zhang G, et al. WAY-200070, a selective agonist of estrogen receptor beta as a potential novel anxiolytic/antidepressant agent. Neuropharmacology 2008;54(7):1136–42.
[61] Goswami DB, May WL, Stockmeier CA, Austin MC. Transcriptional expression of serotonergic regulators in laser-captured microdissected dorsal raphe neurons of subjects with major depressive disorder: sex-specific differences. J Neurochem 2010;112(2):397–409.
[62] Hiroi R, Neumaier JF. Estrogen decreases 5-HT1B autoreceptor mRNA in selective subregion of rat dorsal raphe nucleus: inverse association between gene expression and anxiety behavior in the open field. Neuroscience 2009;158(2):456–64.
[63] Gupta S, McCarson KE, Welch KM, Berman NE. Mechanisms of pain modulation by sex hormones in migraine. Headache 2011;51(6):905–22.
[64] Ing NH, Tornesi MB. Estradiol upregulates estrogen receptor and progesterone receptor gene expression in specific ovine uterine cells. Biol Reprod 1997;56(5):1205–15.
[65] Shiota K, Kogo Y, Ohgane J, Imamura T, Urano A, Nishino K, et al. Epigenetic marks by DNA methylation specific to stem, germ and somatic cells in mice. Genes Cells 2002;7(9):961–9.
[66] Flanagan JM, Cocciardi S, Waddell N, Johnstone CN, Marsh A, Henderson S, et al. DNA methylome of familial breast cancer identifies distinct profiles defined by mutation status. Am J Hum Genet 2010;86(3):420–33.
[67] Westberry JM, Trout AL, Wilson ME. Epigenetic regulation of estrogen receptor alpha gene expression in the mouse cortex during early postnatal development. Endocrinology 2010;151(2):731–40.
[68] Kurian JR, Olesen KM, Auger AP. Sex differences in epigenetic regulation of the estrogen receptor-alpha promoter within the developing preoptic area. Endocrinology 2010;151(5):2297–305.
[69] Labruijere S, Stolk L, Verbiest M, de Vries R, Garrelds IM, Eilers P, et al. Methylation of migraine-related genes in different tissues of the rat. PloS One 2014;9(3.):e87616..
[70] Giri S, Thompson PD, Taxel P, Contois JH, Otvos J, Allen R, et al. Oral estrogen improves serum lipids, homocysteine and fibrinolysis in elderly men. Atherosclerosis 1998;137(2):359–66.
[71] Mijatovic V, Netelenbos C, van der Mooren MJ, de Valk-de Roo GW, Jakobs C, Kenemans P. Randomized, double-blind, placebo-controlled study of the effects of raloxifene and conjugated equine estrogen on plasma homocysteine levels in healthy postmenopausal women. Fertil Steril 1998;70(6):1085–9.
[72] Mijatovic V, Kenemans P, Jakobs C, van Baal WM, Peters-Muller ER, van der Mooren MJ. A randomized controlled study of the effects of 17beta-estradiol-dydrogesterone on plasma homocysteine in postmenopausal women. Obstet Gynecol 1998;91(3):432–6.
[73] van Baal WM, Smolders RG, van der Mooren MJ, Teerlink T, Kenemans P. Hormone replacement therapy and plasma homocysteine levels. Obstet Gynecol 1999;94(4):485–91.

[74] Morris MS, Jacques PF, Selhub J, Rosenberg IH. Total homocysteine and estrogen status indicators in the Third National Health and Nutrition Examination Survey. Am J Epidemiol 2000;152(2):140–8.
[75] O'Connor DL, Green T, Picciano MF. Maternal folate status and lactation. J Mamm Gland Biol Neoplasia 1997;2(3):279–89.
[76] Steegers-Theunissen RP, Wathen NC, Eskes TK, van Raaij-Selten B, Chard T. Maternal and fetal levels of methionine and homocysteine in early human pregnancy. Br J Obstet Gynaecol 1997;104(1):20–4.
[77] Konukoglu D, Serin O, Yelke HK. Effects of hormone replacement therapy on plasma nitric oxide and total thiol levels in postmenopausal women. J Toxicol Environ Health A 2000;60(2):81–7.
[78] Nakano Y, Oshima T, Matsuura H, Kajiyama G, Kambe M. Effect of 17beta-estradiol on inhibition of platelet aggregation in vitro is mediated by an increase in NO synthesis. Arterioscler Thromb Vasc Biol 1998;18(6):961–7.
[79] Morakinyo MK, Strongin RM, Simoyi RH. Modulation of homocysteine toxicity by S-nitrosothiol formation: a mechanistic approach. J Phys Chem B 2010;114(30):9894–904.
[80] Ibim SE, Randall R, Han P, Musey PI. Modulation of hepatic glucose-6-phosphate dehydrogenase activity in male and female rats by estrogen. Life Sci 1989;45(17):1559–65.
[81] Winterbourn CC, Metodiewa D. The reaction of superoxide with reduced glutathione. Arch Biochem Biophys 1994;314(2):284–90.
[82] Pechanova O, Kashiba M, Inoue M. Role of glutathione in stabilization of nitric oxide during hypertension developed by inhibition of nitric oxide synthase in the rat. Jpn J Pharmacol 1999;81(2):223–9.
[83] Borgdorff P, Tangelder GJ. Migraine: possible role of shear-induced platelet aggregation with serotonin release. Headache 2012;52(8):1298–318.
[84] Arulmani U, MaassenVanDenBrink A, Villalón CM, Saxena PR. Calcitonin gene-related peptide and its role in migraine pathophysiology. Eur J Pharmacol 2004;500(1):315–30.
[85] Goadsby P, Edvinsson L, Ekman R. Vasoactive peptide release in the extracerebral circulation of humans during migraine headache. Ann Neurol 1990;28(2):183–7.
[86] Goadsby PJ, Edvinsson L. The trigeminovascular system and migraine: studies characterizing cerebrovascular and neuropeptide changes seen in humans and cats. Ann Neurol 1993;33(1):48–56.
[87] Juhasz G, Zsombok T, Jakab B, Nemeth J, Szolcsanyi J, Bagdy G. Sumatriptan causes parallel decrease in plasma calcitonin gene-related peptide (CGRP) concentration and migraine headache during nitroglycerin induced migraine attack. Cephalalgia 2005;25(3):179–83.
[88] Tvedskov JF, Lipka K, Ashina M, Iversen HK, Schifter S, Olesen J. No increase of calcitonin gene–related peptide in jugular blood during migraine. Ann Neurol 2005;58(4):561–8.
[89] Benemei S, Nicoletti P, Capone J, Geppetti P. Pain pharmacology in migraine: focus on CGRP and CGRP receptors. Neurol Sci 2007;28(2):S89–93.
[90] Tfelt-Hansen P, Le H. Calcitonin gene-related peptide in blood: is it increased in the external jugular vein during migraine and cluster headache? A review. J Headache Pain 2009;10(3):137–43.
[91] Lassen L, Haderslev P, Jacobsen V, Iversen HK, Sperling B, Olesen J. CGRP may play a causative role in migraine. Cephalalgia 2002;22(1):54–61.
[92] Petersen KA, Lassen LH, Birk S, Lesko L, Olesen J. BIBN4096BS antagonizes human α-calcitonin gene related peptide–induced headache and extracerebral artery dilatation. Clin Pharmacol Ther 2005;77(3):202–13.
[93] Olesen J, Diener H-C, Husstedt IW, Goadsby PJ, Hall D, Meier U, et al. Calcitonin gene–related peptide receptor antagonist BIBN 4096 BS for the acute treatment of migraine. N Engl J Med 2004;350(11):1104–10.
[94] Ho TW, Ferrari MD, Dodick DW, Galet V, Kost J, Fan X, et al. Efficacy and tolerability of MK-0974 (telcagepant), a new oral antagonist of calcitonin gene-related peptide receptor, compared with zolmitriptan for acute migraine: a randomised, placebo-controlled, parallel-treatment trial. Lancet 2009;372(9656):2115–23.
[95] Mulleners W, Aurora S, Chronicle E, Stewart R, Gopal S, Koehler P. Self-reported photophobic symptoms in migraineurs and controls are reliable and predict diagnostic category accurately. Headache 2001;41(1):31–9.

[96] Goadsby PJ, Edvinsson L. Human in vivo evidence for trigeminovascular activation in cluster headache. Brain 1994;117(427):34.
[97] Bahra A, May A, Goadsby PJ. Cluster headache a prospective clinical study with diagnostic implications. Neurology 2002;58(3):354–61.
[98] Hallett M, Evinger C, Jankovic J, Stacy M. Update on blepharospasm: report from the BEBRF International Workshop. Neurology 2008;71(16):1275–82.
[99] Irimia P, Cittadini E, Paemeleire K, Cohen A, Goadsby P. Unilateral photophobia or phonophobia in migraine compared with trigeminal autonomic cephalalgias. Cephalalgia 2008;28(6):626–30.
[100] Drummond PD, Woodhouse A. Painful stimulation of the forehead increases photophobia in migraine sufferers. Cephalalgia 1993;13(5):321–4.
[101] Kowacs P, Piovesan E, Werneck L, Tatsui C, Lange M, Ribas L, et al. Influence of intense light stimulation on trigeminal and cervical pain perception thresholds. Cephalalgia 2001;21(3):184–8.
[102] Okamoto K, Thompson R, Tashiro A, Chang Z, Bereiter D. Bright light produces Fos-positive neurons in caudal trigeminal brainstem. Neuroscience 2009;160(4):858–64.
[103] Amini A, Digre K, Couldwell WT. Photophobia in a blind patient: an alternate visual pathway: case report. J Neurosurg 2006;105(5):765–8.
[104] Recober A, Kuburas A, Zhang Z, Wemmie JA, Anderson MG, Russo AF. Role of calcitonin gene-related peptide in light-aversive behavior: implications for migraine. J Neurosci 2009;29(27):8798–804.
[105] Rosenfeld MG, Amara SG, Evans RM. Alternative RNA processing: determining neuronal phenotype. Science 1984;225(4668):1315–20.
[106] Villa G, Fumagalli M, Verderio C, Abbracchio MP, Ceruti S. Expression and contribution of satellite glial cells purinoceptors to pain transmission in sensory ganglia: an update. Neuron Glia Biol 2010;6(01):31–42.
[107] Park K-Y, Russo AF. Control of the calcitonin gene-related peptide enhancer by upstream stimulatory factor in trigeminal ganglion neurons. J Biol Chem 2008;283(9):5441–51.
[108] Broad P, Symes A, Thakker R, Craig R. Structure and methylation of the human calcitonin/α-CGRP gene. Nucleic Acids Res 1989;17(17):6999–7011.
[109] Grunstein M. Histone acetylation in chromatin structure and transcription. Nature 1997;389(6649):349–52.
[110] Park K-Y, Fletcher JR, Raddant AC, Russo AF. Epigenetic regulation of the calcitonin gene–related peptide gene in trigeminal glia. Cephalalgia 2011;31(5):614–24.
[111] Robert M-F, Morin S, Beaulieu N, Gauthier F, Chute IC, Barsalou A, et al. DNMT1 is required to maintain CpG methylation and aberrant gene silencing in human cancer cells. Nat Genet 2003;33(1):61–5.
[112] Wan D, Hou L, Zhang X, Han X, Chen M, Tang W, et al. DNA methylation of RAMP1 gene in migraine: an exploratory analysis. J Headache Pain 2015;16(1):1–5.
[113] Kress C, Thomassin H, Grange T. Active cytosine demethylation triggered by a nuclear receptor involves DNA strand breaks. Proc Natl Acad Sci 2006;103(30):11112–7.
[114] Murgatroyd C, Patchev AV, Wu Y, Micale V, Bockmühl Y, Fischer D, Dynamic DNA, et al. Dynamic DNA methylation programs persistent adverse effects of early-life stress. Nat Neurosci 2009;12(12):1559–66.
[115] Kim M-S, Kondo T, Takada I, Youn M-Y, Yamamoto Y, Takahashi S, et al. DNA demethylation in hormone-induced transcriptional derepression. Nature 2009;461(7266):1007–12.
[116] Maddodi N, Bhat KM, Devi S, Zhang S-C, Setaluri V. Oncogenic BRAFV600E induces expression of neuronal differentiation marker MAP2 in melanoma cells by promoter demethylation and down-regulation of transcription repressor HES1. J Biol Chem 2010;285(1):242–54.
[117] Edvinsson L, Ho TW. CGRP receptor antagonism and migraine. Neurotherapeutics 2010;7(2):164–75.
[118] Ho TW, Ho AP, Ge YJ, Assaid C, Gottwald R, MacGregor EA, et al. Randomized controlled trial of the CGRP receptor antagonist telcagepant for prevention of headache in women with perimenstrual migraine. Cephalalgia 2016;36(2):148–61.

[119] Romero-Reyes M, Pardi V, Akerman S. A potent and selective calcitonin gene-related peptide (CGRP) receptor antagonist, MK-8825, inhibits responses to nociceptive trigeminal activation: role of CGRP in orofacial pain. Exp Neurol 2015;271:95–103.

[120] Luo G, Chen L, Conway CM, Kostich W, Macor JE, Dubowchik GM. Asymmetric synthesis of heterocyclic analogues of a CGRP receptor antagonist for treating migraine. Org Lett 2015;17(24):5982–5.

[121] Fleminger S. Cerebrovascular disorders. In: David AS, Fleminger S, Kopelman MD, Lovestone S, Mellers JDC, editors. Lishman's organic psychiatry. 4th ed. Chichester: Wiley-Blackwell; 2009. p. 473–542.

[122] Antonaci F, Nappi G, Galli F, Manzoni GC, Calabresi P, Costa A. Migraine and psychiatric comorbidity: a review of clinical findings. J Headache Pain 2011;12:115–25.

CHAPTER

THE ROLE OF EPIGENETICS IN THE PATHOPHYSIOLOGY OF EPILEPSY

12

S.M. Nam, K.-O. Cho

Department of Pharmacology, Catholic Neuroscience Institute, College of Medicine, The Catholic University of Korea, Seoul, Korea

CHAPTER OUTLINE

12.1 INTRODUCTION

Epilepsy is one of the most prevalent neurological disorders, affecting more than 50 million people worldwide regardless of age, race, or ethnic background. Each year, approximately 2.4 million people are newly diagnosed with epilepsy globally, giving it a high incidence rate [1]. Moreover, the social and economic burden of epilepsy is quite considerable. In addition to lost work productivity due to epilepsy-associated comorbidities and social discrimination, healthcare costs related to epilepsy are high. Biologically, epilepsy is characterized by a shift in the balance of excitation and inhibition toward excitation, leading to spontaneous recurrent seizures. Uncontrolled abrupt seizure episodes are basically attributed to abnormal synchronized neuronal firing, which can further cause both morphological and functional

Neuropsychiatric Disorders and Epigenetics. http://dx.doi.org/10.1016/B978-0-12-800226-1.00012-5

changes in the brain. These spontaneous repetitive seizures are believed to develop from acute brain injuries including high fever, stroke, brain tumor, and traumatic brain injury [2]. After acute injuries, the brain initiates and accumulates a variety of cellular changes, that is, neurodegeneration, aberrant neurogenesis, reactive gliosis, blood brain barrier (BBB) dysfunction, and inflammation, during the latent period when no apparent seizure events are observed. Epilepsy is a chronic condition as unprovoked recurrent seizures are generated after several months to years of a latent period [2]. By the time recurrent seizures are prominent, patients often display various neuropsychiatric manifestations including cognitive impairment and mood disturbance. Considering the fact that epilepsy is a multifaceted and chronic disease, it warrants further investigation on what are the key contributors of epileptogenesis (development of epilepsy) and/or the formation of recurrent seizures in chronic epilepsy among acute seizure-induced pathologic alterations in the brain. Furthermore, identifying common cellular mechanisms affecting all stages of epilepsy may be essential for the development of effective treatment modalities to ultimately cure epilepsy.

Epigenetics refers to heritable changes in gene expression without alterations in the DNA sequence itself: changes in phenotype with no changes in genotype. As a way for responding to environmental stimuli and maintaining homeostasis, eukaryotic cells employ various epigenetic modifications, for example, DNA methylation, histone modifications, noncoding RNAs, and chromatin remodeling. The nucleosome is a fundamental repeating unit of chromatin in the nucleus of cells, comprising 147 base pairs of DNA wrapped around an octamer of core histones (made of two molecules each of H2A, H2B, H3, and H4 histones). Nucleosomes are linked to each other by linker DNAs and histone H1 protein that condense into chromatin. Under physiological condition, nucleosomal DNA is in active equilibrium between a wrapped and unwrapped state. Epigenetic modifications, such as DNA methylation and histone modifications (acetylation, methylation, phosphorylation, or ubiquitination) control the histone core and DNA association, thereby changing the structure of chromatin that allows the accessibility of regulatory factors to DNA. Moreover, proteins hydrolyzing ATP to alter histone–DNA contacts can remodel chromatin structure so that events, such as DNA replication, DNA repair, and chromosome segregation can accurately occur. Along with DNA, histone, and chromatin modification, noncoding RNA has received recent attention as an important regulator of epigenetic mechanisms. Noncoding RNAs are functional RNA molecules that are transcribed from DNA but not translated into proteins. Categorized as miRNA, siRNA, piRNA, and lncRNA, they can modulate the expression of target genes in a transcriptional or posttranscriptional manner. With versatile utilization of different epigenetic mechanisms described earlier, cells can adapt themselves to the environment by either silencing or activating target genes. At a functional level, epigenetic modification play critical roles in modulating normal brain development, homeostasis and cognitive functions, such as learning and memory. Not surprisingly, epigenetic dysregulation is also reported to be involved in the pathophysiology of many brain disorders including epilepsy [3].

Therefore in this chapter, we will go over past and ongoing efforts to link epigenetic modifications to the pathophysiologic mechanisms of epileptogenesis and chronic epilepsy. In the first part, we will describe the general overview of epilepsy and histologic alterations induced by acute seizures. In the second half, we will discuss epigenetic alterations reported in different types of epilepsy and their impact on the pathogenesis of epilepsy.

12.2 GENERAL OVERVIEW OF EPILEPSY AND ITS BASIC MECHANISMS

12.2.1 CLASSIFICATION OF SEIZURES

Epileptogenesis in general starts from acute seizures. Unlike epilepsy, seizures are not a disease but a transient sign that can be caused by various factors. Not all seizures can trigger epileptogenic processes.

Only epileptic seizures are able to develop into epilepsy. However, available evidence to distinguish epileptic seizures from simple seizures is extremely limited, making the differentiation of the two types of seizures is almost impossible. Therefore, seizures are currently categorized based on their mode of onset, regardless of epileptogenic potential. Seizure types are basically divided into three major groups: primary generalized seizures, partial (focal) seizures, and seizures with unknown cause [4]. The criteria to classify generalized and focal seizures are in how and where the seizure activity begins. Generalized seizures rapidly engage bilaterally distributed networks of the brain, for example, generalized tonic–clonic, myoclonic, clonic, tonic, atonic, or absence seizures, whereas focal seizures are localized in discrete areas of the brain and are often confined to one cerebral hemisphere.

12.2.2 CLASSIFICATION OF EPILEPSY

Epilepsy has been defined as having at least two unprovoked seizures more than 24 h apart. However, the International League Against Epilepsy (ILAE) recently released broader diagnostic criteria for epilepsy, now encompassing additional conditions: (1) one unprovoked seizure and a probability of further seizures occurring over the next 10 years being at least 60% (similar to the general recurrence risk after two unprovoked seizures); (2) diagnosis of an epilepsy syndrome [5]. This edition reflects the heterogeneous nature of epileptic disorders sharing a common clinical feature, seizures. With the help of technological advancements in electroencephalography (EEG) and imaging modalities, ILAE proposed a hierarchical arrangement of different forms of epilepsy for the classification of epilepsy disorders [4]. Based on the etiology, seizure type, and age of onset, epilepsy can be subdivided as electroclinical syndromes, distinctive constellations, structural/metabolic epilepsies, and epilepsies of unknown cause.

Electroclinical syndromes refer to cases where clinical features, signs, and symptoms of a patient can be recognized as a distinctive clinical disorder. According to age at onset, specific EEG patterns, and seizure types, almost 30 different epilepsies are assigned in this category. For example, benign familial neonatal epilepsy, early myoclonic encephalopathy, and Ohtahara syndrome start during the neonatal period. Myoclonic epilepsy in infancy, epilepsy in infancy with migrating focal seizures, benign infantile epilepsy, benign familial infantile epilepsy, myoclonic encephalography in nonprogressive disorders, Dravet syndrome and West syndrome are frequently diagnosed in infancy. For epilepsies with childhood onset, there are febrile seizures plus, Panayiotopoulos syndrome, epilepsy with myoclonic atonic seizures, benign epilepsy with centrotemporal spikes, autosomal-dominant nocturnal frontal lobe epilepsy, late onset childhood occipital epilepsy, epilepsy with myoclonic absences, Lennox–Gastaut syndrome, epileptic encephalopathy with continuous spike-and-wave during sleep, Landau–Kleffner syndrome, and childhood absence epilepsy. From adolescence to adulthood, juvenile absence epilepsy, juvenile myoclonic epilepsy, epilepsy with generalized tonic-clonic seizures alone, progressive myoclonus epilepsies, autosomal dominant epilepsy with auditory features, and other familial temporal lobe epilepsies are often observed. Finally, there are epilepsies having less specific age relationships, such as familial focal epilepsy with variable foci and reflex epilepsies.

Distinctive constellations do not exactly meet the criteria of electroclinical syndromes but represent clinically characteristic features based on specific lesions or other causes. These include mesial temporal lobe epilepsy (TLE) with hippocampal sclerosis, epilepsies with hemiconvulsion and helmiplegia, gelastic seizures with hypothalamic hamartoma, and Rasmussen syndrome. Compared to electroclinical syndromes having strong developmental and genetic components, age at onset is not a critical factor in epilepsies that belong to distinctive constellations. Moreover, for treating this type of epilepsy, surgery can be considered in order to remove suspected epileptic foci. Particularly, in case of TLE which is the most prevalent epilepsy disorder, 50–80% of patients with drug-resistant TLE became seizure-free and the remaining showed a reduction in seizure frequency after epilepsy surgery [6].

Epilepsies secondary to specific structural or metabolic causes include genetic conditions like malformations of cortical development (i.e., hemimegalencephaly and heterotopias) and neurocutaneous syndromes (i.e., tuberous sclerosis complex and Sturge–Weber syndrome). In addition, acquired disorders, such as tumors, infection and trauma can result in structural lesions of the brain, subsequently leading to epilepsies in this class.

Finally, epilepsies that do not fit in any of the categories described earlier can be grouped as epilepsies of unknown cause. These epilepsies comprise one-third or more of all kinds of epilepsies [4]. However, very little information is currently available to subgroup these epilepsies except that they have a common clinical manifestation, seizures. Therefore, a detailed characterization of clinical features and histopathologic mechanisms of epilepsies in this category is required to understand the complex nature of epilepsy.

12.2.3 CELLULAR ALTERATIONS INDUCED BY ACUTE SEIZURES

Acute seizures can trigger pleiotropic changes in the brain. Over a period of months to years, accumulation of multiple structural alterations in the brain can damage existing neural circuits and elicit compensatory mechanisms. After all, repetitive seizure-induced histopathologic findings are believed to ultimately contribute to the development of epilepsy and recurrent seizures in the chronic stage of epilepsy. Plenty of animal and human studies on TLE have supported this idea, reporting various cellular changes after acute seizure insults [2]. Among them, we will review some of the important findings, such as neuronal death, aberrant neurogenesis, reactive astrocytosis, BBB dysfunction, and inflammation in this chapter.

12.2.3.1 Neurodegeneration in Epilepsy

Seizures can induce rapid cell death in the hilus of the dentate gyrus, followed by delayed neuronal loss in the CA1 and CA3 subregions of the hippocampus in TLE [2]. In addition to the hippocampus, the amygdala, the entorhinal cortex, the thalamus, and the cerebellum also show neuronal death. Neurodegeneration is thought to be derived from excessive neuronal excitation involving ionotropic glutamate receptors [7]. Moreover, excitotoxic damage and subsequent loss of GABAergic interneurons leads to the disinhibition of pyramidal neurons, making the system hyperexcitable and worsening the epilepsy [8]. In line with TLE, impaired excitability of inhibitory neurons by sodium channel mutation (SCN1A) is critical in Dravet syndrome. However, in contrast to the significant hippocampal neuronal loss in TLE, cerebellar Purkinje cells were the only cells found dead in Dravet syndrome [9]. In addition to the influence of neuronal loss on the recurrent seizures, granule cell death was associated with impaired memory acquisition in patients with TLE [10].

12.2.3.2 Aberrant Neurogenesis in Epilepsy

Upon acute seizure activity and subsequent neurodegeneration, endogenous neural stem cells become reactive and produce new neurons. However, unlike physiologic circumstances, seizure-induced neurogenesis is markedly dysregulated in many ways. Parent et al. [11] first reported increased proliferation of neural progenitors in the dentate gyrus after pilocarpine injection. In that study, they showed increased number of BrdU-positive cells upto 2 weeks after acute seizures, suggesting acute upregulation of hippocampal neurogenesis. Follow-up studies corroborated this seizure-induced enhanced neurogenesis utilizing retroviral labeling of newborn neurons [12], along with the acceleration of synaptic integration of seizure-born neurons. However, in the chronic phase of TLE when spontaneous recurrent seizures are prominent, hippocampal neurogenesis is decreased as the cells expressing doublecortin, a marker for newborn neuron, is significantly reduced at 5 months after kainic acid injection [13]. In addition to the temporal fluctuation of hippocampal neurogenesis, the morphology of newly generated neurons after seizures is different from normally developed cells. Seizure-generated granule neurons represented hilar basal dendrites and abnormal sprouting of the axons (mossy fibers) to the inner molecular layer [12], possibly inducing aberrant circuit formation. Additionally, seizures can disrupt the migration process of adult- and embryo-generated granule neurons since hilar ectopic granule cells and granule cell dispersion are frequently observed in models of rodent and human TLE, respectively [11,14]. The critical role of ectopic granule cells was confirmed in febrile seizure-associated epilepsy as well [15].

At a functional level, aberrant adult neurogenesis induced by acute seizures increased excitatory synaptic inputs to granule cells, leading to imbalance of excitatory and inhibitory synaptic controls. Moreover, the cause-and-effect relationships between aberrant neurogenesis and epileptogenesis were further demonstrated in a couple of papers [16,17]. Genetic disruption of hippocampal neurogenesis in just a few cells could produce spontaneous recurrent seizures, inducing epilepsy without applying chemoconvulsants [16]. In a pilocarpine mouse model, ablation of adult neurogenesis prior to acute seizures could reduce the frequency of chronic seizures, which were sustained for almost 1 year after pilocarpine treatment [17]. Together, these data indicate that seizure-born abnormal neurons are both sufficient and necessary for the development of epilepsy. With respect to epilepsy-associated comorbidities, pharmacological valproic acid (VPA) treatment could restore seizure-induced hippocampal memory impairment with the suppression of aberrant neurogenesis [18]. Furthermore, blocking adult neurogenesis using genetically engineered mice normalized epilepsy-associated memory deficits [17], providing direct evidence for the requirement of seizure-induced aberrant neurogenesis in memory impairment. In addition, chronic reduction of neurogenesis and the generation of newborn neurons with aberrant synapse formation were suggested to increase risk of depression in epilepsy. Indeed, fosB null mice showed disrupted hippocampal neurogenesis, for example, reduced proliferative activity and increased hilar ectopic cells, in accordance with spontaneous seizures and depression-like behavior [19]. In summary, these data demonstrate that seizure-induced aberrant neurogenesis is critical for chronic seizure formation, cognitive decline and depression in epilepsy.

12.2.3.3 Reactive Astrocytosis in Epilepsy

In addition to neurons, glial cells, such as astrocytes and microglia are affected by seizures. In animal models of epilepsy and the brains from patients with mesial TLE, tuberous sclerosis complex, focal cortical dysplasia, and Rasmussen syndrome, reactive astrocytes are commonly observed [20]. Their numbers are highly increased upon acute seizures with morphologic transformation into reactive hypertrophic cells, a pathological hallmark of the epileptic brain. Astrocytes are supposed to play multiple roles in the support of neurons, that is, oxygen and nutrition supply to neurons, regulation of ionic homeostasis and neurotransmission, synchronization of neuronal firing, and so on. In spite of ongoing debates, reactive astrocytosis is thought to contribute to the hyperexcitability of the brain in epilepsy. After acute seizures, glial water transport by the aquaporin-4 channel was disrupted, causing swollen astrocytes and shrinking extracellular space (ECS) volume, eventually leading to increased susceptibility to seizures [20]. Moreover, excessive extracellular K^+ concentration induced by impaired astroglial buffering could predispose to seizures [20]. These data nicely illustrate astroglial dysregulation of water and K^+ flow in epilepsy and its impact on seizure susceptibility.

Dysregulation of glutamate transport is another important mechanism of excitotoxicity in the process of chronic seizure induction. Extracellular glutamate is increased in the epileptic brain in concurrence with altered astroglial regulation of glutamate homeostasis. Indeed, astrocytic glutamate release was the initiating contributor for hypersynchronous neuronal firing activity [21]. Moreover, the expression of astroglial glutamate transporters, such as EAAT1 and EAAT2 was reduced, suggesting a decreased uptake of synaptic glutamate into astrocytes [22]. Astrocyte-mediated epileptic hyperexcitability can be further aggravated when glutamine synthetase (GS) that degrades glutamate to nontoxic glutamine is significantly downregulated in TLE [23]. When reactive astrocytosis was selectively induced using high titer injection of adeno-associated viruses, the expression of GS was markedly reduced with the impairment of inhibitory neurotransmission and the increment of network hyperexcitability [24]. Pharmacological inhibition of GS by methionine sulfoximine supported proepileptic implication of astrocytes as it induced recurrent seizures with an increment of astrocytic glutamate levels [25]. Thus, reduced astroglial GS can play a crucial role in the elevation of seizure susceptibility.

Astrocytes can release D-serine, ATP, adenosine, GABA, and tumor necrosis factor alpha (TNF-α) by increasing Ca^{2+} signals. Interestingly, selective stimulation of astroglial Ca^{2+} signaling enhanced seizure-like discharges, possibly via stimulating gliotransmission [26]. When the glial transmitter release was blocked with genetically modified dominant negative SNARE mice, the frequency of spontaneous recurrent seizures was significantly reduced with attenuated neuronal death and reactive astrocytosis, proposing proepileptogenic function of astrocyte-mediated gliotransmission [27]. Additionally, astrocytes, in concert with dendritic sprouting and new synapse formation, increased their coverage in the brain after acute seizures, thereby providing a structural basis for recurrent neuronal excitation in wider areas [28].

12.2.3.4 BBB Dysfunction in Epilepsy

The brain microvasculature changes during the course of epileptogenesis and in chronic epilepsy. In the hippocampus of patients with TLE, increased vessel density and BBB dysfunction were positively correlated with seizure frequency [29]. Vascular endothelial growth factor (VEGF) released from astrocytes was suggested as one of the main factors for triggering BBB damage and the proliferation of endothelial cells (angiogenesis) through VEGF receptor 2 on the vessels [29]. Proinflammatory cytokines, such as interleukin-1 beta (IL-1β) could also induce BBB dysfunction as treatment with anakinra, an IL-1β receptor antagonist, alleviated seizure-induced BBB breakdown [30]. Moreover, leakage of serum proteins (i.e., albumin, immunoglobulin, etc.), due to increased BBB permeability perturbed the neuronal environment, contributing to recurrent seizures, edema, and inflammatory/immune responses [31]. Therefore, many researchers are now paying attention to this vicious cycle of BBB disruption, angiogenesis, and inflammation in the hope of developing potential treatment targets for epilepsy, despite the fact that exact causal relationships among them are still poorly understood.

12.2.3.5 Inflammation in Epilepsy

From acute seizures to the establishment of chronic epilepsy, activated microglia and inflammatory cytokines have been attractive targets for treating epilepsy. Numerous papers have reported activation of microglia and astrocytes using different animal models of epilepsy and brain tissues affected by TLE, tuberous sclerosis complex, Rasmussen syndrome and focal cortical dysplasia [20]. Activated glial cells release various inflammatory mediators including IL-1β, high mobility group box-1 (HMGB1), IL-6, and TNF-α after acute seizures. Within 30 min after seizure onset, cytokine levels increased and persisted even in the chronic phase of epilepsy [20]. Among them, proepileptic roles of IL-1β are well established. Increased IL-1β and its inhibition by an IL-1 receptor antagonist facilitated and attenuated seizure induction, respectively, through the modulation of glutamatergic neurotransmission [32]. Moreover, proinflammatory cytokines increased BBB permeability, supporting dynamic interrelations among BBB breakdown, inflammation, and glial activation [32]. HMGB1 is another proepileptic mediator, which acts through a toll-like receptor (TLR) system. There is increased expression of HMGB1 and TLR4 in TLE; HMGB1 could increase both acute and chronic seizure susceptibilities in a TLR4-dependent manner [33].

IL-6 can have antiepileptic roles, as previous reports have shown that IL-6 was involved in glutamatergic receptor-mediated acute seizure aggravation [34,35]. On the other hand, TNF-α is reported to be either pro- or antiepileptic, depending on the context. For example, intracerebroventricular TNF-α infusion was sufficient to increase seizure susceptibility [36], supporting proconvulsive effects of TNF-α. Moreover, TNF-α showed a proictogenic effect by activating TNF-α receptor 1, but an opposite antiictogenic role was observed if TNF-α signaling was mediated by TNF-α receptor 2 [37]. In support of anticonvulsive effects of the TNF-α signaling pathway, modulation of TNF-α and its receptor p75 using a knockout approach showed prolonged seizures with exacerbated cell death [38]. In addition, a recent paper elegantly illustrated that microglial TNF-α production via TLR9 signaling could alleviate aberrant seizure-induced hippocampal neurogenesis, contributing to the suppression of recurrent seizures and the attenuation of hippocampal memory impairment [39].

Taken all together, glial secretions of proinflammatory and antiinflammatory mediators can alter various seizure-induced histologic and functional outcomes. As their modes of action are divergent and complex, a deeper appreciation of the underlying mechanisms will provide us valuable insights to solve unanswered questions in epilepsy.

12.2.4 MANAGEMENT OF EPILEPSY

Currently applicable treatment options for epilepsy are antiepileptic drugs (AED), electrical stimulation, diet therapy, and surgical removal of an epileptic focus. AED are the first-line treatment choice for almost all patients experiencing multiple seizures. AED are generally effective since symptomatic seizures become controllable in about two-thirds of patients with epilepsy. The basic goal of pharmacologic treatments is to recover the balance of excitation and inhibition in the brain via blocking sodium or calcium channels, and enhancing GABAergic neurotransmission. In addition, newer AED, such as levetiracetam are constantly developed, and they have different mechanisms of action [40]. Despite extensive efforts to develop better AED, current pharmacological treatments can only suppress seizures working as anticonvulsants, rather than curing the disease itself. This may be attributable to insufficient information about basic pathophysiologic mechanisms of epilepsy. Thus, more studies are required for developing true AED.

A significant number of patients (approximately 30%) are still resistant to medical therapy and suffer from persistent uncontrolled seizures. Surgical resection may be considered for some of these patients. Specifically, mesial TLE with hippocampal sclerosis and partial epilepsy with a defined lesion are the best candidates for curative resection. Indeed, surgical resection of the sclerotic hippocampus has been reported to be effective in intractable TLE, most of the patients (50–80%) undergoing epilepsy surgery becoming seizure free and 12% of the patients having reductions in seizure frequency [6]. However, it is interesting that the percentage of seizure-free patients after surgery decreases over the course of chronic follow-up, suggesting reorganization of new brain circuitry leading to relapse.

Dietary therapy can provide a new avenue in the treatment of patients with certain forms of epilepsy. The ketogenic diet and the modified Atkins diet are the most widely used ones. Both are high-fat, low-carbohydrate diets with a slight difference that the modified Atkins diet is less restrictive on protein consumption. The ketogenic diet has been shown to be effective especially for refractory childhood epilepsies, such as glucose transporter type 1 (GLUT-1) or pyruvate dehydrogenase (PDH) deficiency, infantile spasms, Rett syndrome, tuberous sclerosis complex, Dravet syndrome, and Doose syndrome [41].

Finally, stimulation techniques, such as vagal nerve stimulation (VNS) or deep brain stimulation (DBS) are the newest and rapidly advancing modalities for the treatment of epilepsy. With technological advancements, implantation of a stimulation device and the delivery of electrical currents to the peripheral and central nervous system is feasible. Although the exact mechanism remains unclear yet, VNS and DBS have been reported to reduce seizure frequency [42,43].

In conclusion, current treatment strategies for epilepsy leave much room for improvement, given that there is no true seizure prevention method and the lack of basic mechanisms providing strong rationales for each treatment option. Therefore, new approaches to view epilepsy through a different lens are required to overcome these obstacles. Epigenetics, in particular, may be a promising candidate to explain underlying molecular mechanisms of epilepsy.

12.3 EPIGENETIC MODIFICATIONS IN EPILEPSY

12.3.1 DNA METHYLATION

DNA methylation is a covalent addition of a methyl group to the cytosine nucleotide. Classically, cytosine at CpGs are the main target of DNA methylation, although we now know that DNA methylation can also occur in sites in the genome without CpGs. In general, DNA methylation inhibits transcription process through two basic mechanisms. Methylation of the cytosine within a transcription factor (TF) binding sequence motif can directly impede the binding of transcription factors, thereby silencing gene expression. The second mechanism involves special protein complexes that recognize single methylated CpG sites in a gene promoter region (regardless of TF's binding consensus sequence motif) and indirectly prevent TF binding by recruiting repressors or physically blocking TF binding. The first mechanism can be mediated by at least three DNA methyltransferases (DNMTs): DNMT1, DNMT3A, and DNMT3B. DNMT1, called a "maintenance" methyltransferase, predominantly transfers methyl groups to hemimethylated CpG sites, copying DNA methylation patterns after DNA replication. In contrast, DNMT3A and DNMT3B have an equal preference for hemimethylated and unmethylated DNAs, and are thus called "de novo" methyltransferases. A second class of proteins that can inhibit transcription by reading methyl-CpGs is methyl-CpG binding domain (MBD) superfamily proteins. Methyl-CpG-binding protein 2 (MeCP2), MBD1, MBD2, MBD3, and MBD4 are the mammalian members of this family and they are responsible for gene silencing by recruiting chromatin remodelers and other regulatory elements.

DNA demethylation is a dynamic process in that 5-methylcytosine can be replaced with a naked cytosine in either an active or a passive mode. Passive DNA demethylation is mediated by blocking DNMTs, allowing newly added cytosine to remain unmethylated. On the other hand, active DNA demethylation requires sequential enzymatic processes involving ten-eleven translocation (TET) enzymes, activation-induced cytidine deaminase/apolipoprotein B mRNA-editing enzyme complex (AID/APOBEC), and thymine DNA glycosylase (TDG). Since it is impossible to break the strong covalent bond of 5-methylcytosine directly, TET-mediated production of 5-hydroxymethylcytosine (5hmC), followed by AID/APOBEC-mediated deamination of 5hmC and TDG-mediated replacement of 5-hydroxymethyl-uracil (5hmU) with cytosine, can be an alternate way of DNA demethylation.

With regard to epilepsy, dysregulation of DNA methylation has been mainly demonstrated in TLE. In the chronic stage of a rat model of TLE, global DNA methylation was increased in the hippocampus [44]. In support of this observation, hypermethylation of the reelin promoter in TLE patients was associated with granule cell dispersion, a characteristic feature of aberrant neurogenesis [45]. Increased methylation at the reelin promoter can occur by DNMTs since DNMT1 and DNMT3A are upregulated in both experimental and human TLE [46,47]. Moreover, in a primary culture system, reelin expression was negatively regulated by DNMT1 [48]. Studies using animal models of TLE further filled the gap between epigenetic modulation of reelin expression and seizure-induced aberrant cellular alterations, showing the reduction of reelin production after acute seizures and the abnormal migration of granule neurons by pharmacologic reelin inhibition [49,50]. Collectively, it is plausible that seizures can promote hypermethylation of the reelin promoter through DNMTs, resulting in the blockade of reelin transcription and the subsequent displacement of granule neurons. In contrast to a well-established connection between DNA methylation and reelin-mediated granule cell dispersion, epigenetic regulations of other features in epilepsy have not been extensively studied. Only a few papers have suggested that DNA methylation may be associated with abnormal dendritic arborization [51] and the regulation of immediate early genes [52]. Given the commonality of DNA methylation as a cellular regulatory tool and its dysregulation in epilepsy, comprehensive evaluation of the roles of DNMTs in diverse cellular alterations shown in epilepsy will provide a better understanding of the epileptic brain controlled by epigenetics.

Another class of proteins influencing the DNA methylation process is MBD proteins that selectively bind to CpG dinucleotides in the genome and mediate transcriptional repression through interactions with histone deacetylases and corepressors. Among them, MeCP2 is a critical molecule responsible for Rett syndrome where mutation of MeCP2 is related to epileptic seizures. MeCP2 is an important regulator for the proper functioning of GABAergic interneurons in the hippocampus. Zhang et al. [53] working on hippocampal slices from MeCP2-null mice showed a diminished basal inhibitory activity in the CA3 region, which in turn rendered hyperexcitability at the network level. One more line of evidence supported this notion showing that excitatory neuron-specific deletion of MeCP2 led to a reduction of GABAergic transmission, hyperexcitation of cortical neurons, and eventually generation of absence seizures [54]. However, pathophysiologic roles of MeCP2 in epilepsy might be more complicated than expected, considering the fact that epileptic seizures are also observed in MeCP2 duplication syndrome [55]. Moreover, MeCP2 expression was increased in the temporal neocortex in patients with TLE and rats with pilocarpine injection [56]. Thus, further studies focusing on the complex regulatory mechanisms of MeCP2 in epilepsy are required.

DNA demethylation with regard to epilepsy is only a beginning to be elucidated. DNA methylation can be erased by DNA demethylases. As passive demethylation is strongly associated with the activity of DNMTs that are described earlier, we will give a couple of examples of how active demethylation plays a role in epilepsy. Growth arrest and DNA damage 45 (Gadd45) family proteins, implicated in active DNA demethylation, have been shown to be enhanced after kainic acid-evoked seizures [57]. Among three isoforms (Gadd45a, Gadd45b, and Gadd45g), upregulation of Gadd45a and Gadd45b was maintained in the chronic stage of epilepsy, with the reduced methylation of brain derived growth factor (BDNF) exon IX and the corresponding increase in BDNF mRNA expression [58]. Assessing the function of Gadd45b using electroconvulsive therapy (ECT) showed that Gadd45 deletion induced impaired proliferation of neural stem/progenitor cells and shortened dendrites of the granule neurons in the dentate gyrus [59]. In this study, it was also found that ECT-induced decreased methylation at BDNF and fibroblast growth factor 1 (FGF1) promoters shown in wild-type mice, disappeared in Gadd45b-deleted dentate gyrus as the methylation status was comparable between sham and ECT groups. Thus, the reduction of BDNF and FGF1 expression in Gadd45b knockout mice subject to ECT could lead to aberrant neurogenesis.

Other important players mediating DNA demethylation include TET enzymes. Hippocampal TET1 expression was reduced in the acute phase of epilepsy, although global 5-methylcytosine levels were inconsistent showing no difference or even reduction, surprisingly [58,60]. With technological advancements measuring precise methylation status, this discrepancy will be hopefully resolved in the near future. As emerging data suggest dynamic regulation of DNA methylation/demethylation in key pathologic features of epilepsy, there is no doubt that this field will be a fertile area for future research.

12.3.2 HISTONE MODIFICATIONS

Histone modifications target the histones. Chemical modification to the histone tails can generate synergistic or antagonistic interactions with chromatin-associated proteins. As a result, recruited proteins can alter the stability of chromatin structure and regulate transcription. Histone modifications covered in this chapter include acetylation, methylation, phosphorylation, ubiquitination, and sumoylation.

12.3.2.1 Histone Acetylation

Histone acetylation in general activates gene transcription by increasing accessibility of transcription factors to DNA. The balance between histone acetyltransferase (HAT) and histone deacetylase (HDAC) activities determines the level and turnover of acetylation. Basically, HAT families transfer an acetyl group from acetyl-CoA to the lysine residue on the histone protein. Depending on the functional similarity of catalytic domains, HATs are more defined as p300/CREB-binding protein (CBP), general control nonderepressible 5/p300-CBP-associated factor (GCN5/PCAF), steroid receptor coactivator-1 (SRC-1), and MYST (MOZ, Ybf2/Sas3, Sas2, and Tip60) families. On the other hand, HDACs can be divided into four categories: class I, II, and IV zinc-dependent HDACs versus class III nicotinamide adenine dinucleotide (NAD^+)-dependent HDACs. Class I HDACs (HDAC1, HDAC2, HDAC3, and HDAC8), class IIa HDACs (HDAC4, HDAC5, HDAC7, and HDAC9), class IIb HDACs (HDAC6 and HDAC10), and class IV HDAC (HDAC11) are all affected by trichostatin A (TSA), whereas class III HDACs (sirtuins, from SIRT1 to SIRT7) do not respond to TSA treatment. Functionally, HDACs remove acetyl groups from acetylated lysine residues on the histones, in addition to nonhistone proteins, resulting in chromatin compaction and transcription repression. Moreover, histone deacetylation and DNA methylation can be interrelated with each other for gene silencing as HDACs are recruited by MBD proteins after recognition of methylated DNA.

In the epileptic brain, histone acetylation and deacetylation are reported to be fluctuating. For example, acetylated H4 at the BDNF promoter was significantly increased at 3 h after pilocarpine-induced status epilepticus, which promoted BDNF transcription [61]. Hyperacetylated H4 induction was recapitulated in a kainic acid model where it showed concomitant induction of CBP (a HAT) and a couple of target genes, such as c-fos and c-jun [62]. Moreover, curcumin, a specific inhibitor of p300/CBP HAT, could attenuate hyperacetylation of H4 and seizure-induced upregulation of c-fos and c-jun transcription, suggesting a pivotal role of histone acetylation in the immediate early response after acute seizures.

Compared to HATs, HDACs have been vigorously studied in epilepsy with the early discovery of VPA, a well-known AED that was later identified to inhibit class I HDACs. When administered after kainic acid injection, VPA could reduce seizure-induced proliferation of progenitors and the persistence of basal dendrites by normalizing HDAC-dependent gene expression, in addition to the restoration of epilepsy-associated memory deficits [18]. Moreover, chronic VPA treatment induced the expression of the endogenous anticonvulsant neuropeptide Y [63]. In line with previous reports using VPA, class I HDACs were upregulated in TLE [64]. At a molecular level, seizure-induced H4 deacetylation at the promoter of glutamate receptor 2 (GluR2) was reversed by a HDAC inhibitor, trichostatin A (TSA) [61]. Additionally, a newer HDAC inhibitor, suberoylanilide hydroxamic acid (SAHA), showed neuroprotective effects against status epilepticus [65]. These data nicely demonstrate that HDACs can modulate critical outcomes in epilepsy.

Finally, SIRT1 in the class III HDAC family was found to be upregulated in the hippocampus after acute seizures [66]. Sirt1 activity was critical in epileptic neuronal cell death and its neuroprotective effects were mainly derived from attenuating oxidative stress. Pharmacological SIRT1 activation by resveratrol, found richly in grapes and red wine, showed neuroprotection against acute seizures and the attenuation of chronic recurrent seizures [67], supporting beneficial effects of SIRT1 activation in epilepsy.

12.3.2.2 Histone Methylation

Methylation of N-terminal tails of histone proteins attracts various chromatin-modifying enzymes. However, unlike histone acetylation, histone methylation can either promote or repress transcription depending on the site and the number of methylations. For example, H3K9me3, H3K27me3, and H4K20me3 are repressive marks of transcription, whereas H3K4me3, H3K36me, H3K79me represent a transcriptionally active state. Histone methyltransferases (HMTs) transfer methyl groups to lysine or arginine residues on the histone proteins. Some of the lysine-specific HMTs [lysine(K)-methyltransferases, KMTs] include KMT1C/EHMT2/G9a, KMT1D/EHMT/GLP, KMT2A/MLL1, KMT2B/MLL2, KMT2C/MLL3, KMT2D/MLL4, KMT2E/MLL5, KMT3A/SETD2, KMT6A/EZH2, and KMT6B/EZH1, whereas protein arginine methyltransferases (PRMTs) include PRMT1, PRMT3, PRMT4/CARM1, and PRMT5/JBP. Although many HMTs and their functional roles under physiologic condition have been identified, expression profiles and the influence of different HMTs concerning epilepsy remain largely unknown. Only one paper recently demonstrated the spatiotemporal pattern of polycomb repressive complex 2 (PRC2) after kainic acid-induced status epilepticus [66]. PRC2 containing KMT6A/EZH2 or KMT6B/EZH1 is responsible for trimethylation of H3K27, promoting chromatin compaction. In this study, the authors reported rapid downregulation of EZH1 in the hippocampus (CA1, CA3, and the dentate gyrus), as opposed to early upregulation of EZH2 in the CA3 subregion and the dentate gyrus of the hippocampus. Taken together, histone methylation appears to be a dynamic process utilizing different HMTs in a region-specific manner.

On the other hand, histone methyl marks can be erased by histone demethylases. There are six different histone lysine demethylases (KDMs) including KDM1, KDM2, KDM3, KDM4, KDM5, and KDM6. However, it is only recently that the roles of KDMs in epilepsy have begun to get attention. Interestingly, one study demonstrated that a neurospecific form of KDM1A (also known as LSD1) was reduced in an experimental model of TLE. Moreover, it suggested that neuroLSD1 might contribute to the high susceptibility against pilocarpine-induced acute seizure induction using neuroLSD1-specific null mice [68]. However, since the mechanisms underlying this phenotype are not clear at this time, further studies are needed to determine the involvement of histone demethylation by neuroLSD1 deletion. One more example of KDMs in association with epilepsy is KDM5C, also known as JARID1C. KDM5C is directly regulated by aristaless-related homeobox (ARX), which is frequently mutated in children with X-linked intellectual disability and chronic epilepsy [69]. When ARX was deleted in neural stem cells, KDM5C expression was markedly reduced with an increase in H3K4me3 [69]. As ARX mutation is found to be associated with early infantile epileptic encephalopathy, such as Ohtahara syndrome, X-linked infantile spasms, and West syndrome [70], histone demethylation may serve important molecular functions in many electroclinical syndromes.

Interestingly, KMT1C/EHMT2/G9a or KDM1 can be a part of the complex involving repressor element 1-silencing transcription factor (REST) [3]. REST can associate with KMT1C and KDM1, thereby modulating histone methylation and demethylation, respectively. In addition, REST can repress target gene expression as the histone deacetylases, HDAC1 and HDAC2, can be recruited to REST through SIN3. After status epilepticus, REST was found to be upregulated in the hippocampus [3]. Regarding the functional impact of REST, however, it is not straightforward because of conflicting results obtained by different conditional knockout approaches. In the electrical kindling model of TLE, REST showed antiepileptogenic effects, while in the pentylenetetrazol model of acute seizures, REST promoted seizure initiation. Considering multiple epigenetic partners of REST, it will be interesting to examine underlying mechanisms and the context-dependent different recruitment of corepressors, if any, in epilepsy.

12.3.2.3 Histone Phosphorylation

Histone phosphorylation via protein kinases can occur on serine (S), threonine (T), and tyrosine (Y) residues of histone proteins. It is generally thought to promote transcription activation because a repulsive force between histone and DNA is increased by adding a negatively charged phosphate group to histones.

After kainic acid and pilocarpine injection, phosphorylation of histone H3 at serine 10 was rapidly increased in the hippocampus [62], in concordance with the activation of the mitogen-activated protein kinase (MAPK) pathway [71]. Moreover, this was coupled with the induction of c-fos transcription and histone acetylation, promoting transcriptionally active euchromatin [71]. This observation was further supported by a study applying electroconvulsive seizures where the increased H3 phophoacetylation at the c-fos promoter was correlated with enhanced c-fos transcription [72]. Collectively, these data illustrate a well-established MAPK/H3 phosphorylation/c-fos pathway in epilepsy. Intriguingly, in addition to canonical histones, kainic acid-induced seizures were able to phosphorylate histone H2A.X (γ-H2AX), a histone variant that demarcates the site of DNA double strand breaks [73]. As H2A.X phosphorylation is a rapid event responding to sublethal excitotoxic injury, histone phosphorylation, such as γ-H2AX, may be a useful marker for monitoring neuronal endangerment in epilepsy.

12.3.2.4 Histone Ubiquitination

Histone ubiquitination refers to the transport of ubiquitin to the histone core proteins, such as H2A and H2B. H2A ubiquitination through PRC1 typically represses gene expression, while H2B ubiquitination can both activate and inhibit target gene expression. Interestingly, H2A and H2B ubiquitination and histone methylation are interrelated as ubiquitinized H2B is necessary for histone H3K4 methylation, whereas H2A ubiquitination blocks this methylation, resulting in chromatin compaction. With regard to epilepsy, very little information is available except for the expression profiles of the PRC1 components after acute seizure activity [66]. Several PRC1 genes including Ring1B and Bmi1 were upregulated at the hyperacute phase (1 h after kainic acid), followed by the reduction of the same genes in addition to Ring1A at later time-points (4 h, 8 h, and 24 h postkainic acid). These findings may imply PRC1-mediated immediate transcription repression by acute brain insults, accompanied by the reinitiation of transcription for compensation against seizure-mediated alterations.

12.3.2.5 Histone Sumoylation

Histone sumoylation basically involves attaching SUMO proteins to core histones, similar to the process of histone ubiquitination. Sumoylation of histones is considered to repress transcription because it can recruit histone deacetylases and other repressor complexes. When it comes to histone sumoylation in epilepsy, it is extremely unexplored. Researches focusing on the basic expression pattern or the functional role of sumoylated histones in epilepsy will provide comprehensive understanding of histone modifications after brain injury.

12.3.3 NONCODING RNAs

Noncoding RNAs (ncRNAs) are RNAs that are not translated into protein. Based on their size, ncRNAs are usually divided into small ncRNAs (sncRNAs, <200 bp) and long ncRNAs (lncRNAs, >200 bp). Since ncRNAs have been identified as critical epigenetic regulators in the nervous system, their role in epilepsy is now getting prime attention.

12.3.3.1 MicroRNAs

MicroRNAs (miRNAs) are an integral part of the RNA-induced silencing complex (RISC). miRNAs mediate posttranscriptional gene silencing by sequence-specific binding to complementary regions of the target messenger RNA (mRNA). An individual miRNA can regulate numerous different mRNAs and a single mRNA can be controlled by several miRNAs. The number of miRNAs keeps increasing with the discovery of novel miRNAs by advanced small RNA deep sequencing technology.

Recent extensive investigation has revealed that miRNA levels fluctuate in the epileptic brain and miRNA dysregulation is highly correlated with the development of, and pathological changes in, epilepsy. Indeed, blocking miRNA biogenesis by inducible dicer knockout in the forebrain showed increased seizure susceptibility, supporting this notion [74]. Moreover, discrepancies between mRNA and protein expression shown in previous studies may provide a strong rationale why posttranscriptional epigenetic regulation is critical in the pathogenesis of epilepsy. In this section, we will highlight several miRNAs influencing cell death, aberrant neurogenesis, and inflammation in epilepsy (Table 12.1).

12.3.3.1.1 miRNAs and Neuronal Death in Epilepsy

Seizure activity alters the expression of a number of miRNAs that can control neuronal death. miR-34a was increased by kainic acid–induced status epilepticus and promoted neuronal apoptosis by p53

Table 12.1 miRNAs That are Dysregulated in Epilepsy

miRNA	Source of Data	Stage of Epilepsy	Expression Change in Epilepsy	Potential Roles or Functional Effect	References
let-7	Rats (Pilo, i.p.)	Chronic (2 mo)	Down	• Known to reduce proliferation	[75,76]
miR-9	Rats (Pilo, i.p.)	Chronic (2 mo)	Up	• Known to reduce proliferation and promote neuronal differentiation	[75,77]
miR-22	Mice (KA, intraamygdala)	Acute (8 h)	Up (contralateral hippocampus)	• Repression of ionotropic P2X7 receptor-mediated IL-1β and TNF-α production • Anticonvulsive and antiepileptic effects	[78]
miR-27a	Rats (Pilo, i.p.) TLE patients	Acute (2 h) Latent (3 w) Chronic (2 mo) 5–24 yr	Up Up Up Up	• Known to increase proinflammatory cytokines (IL-1β, IL-6, TNF-α, etc.) and decrease antiinflammatory cytokine (IL-10)	[79,80]
miR-30c	Rats (Pilo, i.p.) TLE patients	Acute (2 h) Latent (3 w) Chronic (2 mo) 5–24 yr	Up Down Up Up	• Known to inhibit NFκB and the production of proinflammatory cytokines (IL-6, IL-8)	[79,81]
miR-34a	Mice (KA, intraamygdala)	Hyperacute (2 h, 6 h) Acute (24 h) Latent (3 d, 7 d)	Up — —	• Contribution to p53-dependent apoptotic cell death by caspase 3 activation at 24 h after KA	[82]
	Rats (Pilo, i.p.)	Acute (24 h) Latent (1 w, 2 w) Chronic (2 mo)	Up Up Up	• Contribution to apoptotic cell death by caspase 3 activation at 1 w after Pilo (neuroprotection by antagomir treatment)	[83]
miR-125b	Rats (Pilo, i.p.) TLE patients	Acute (2 h) Latent (3 w) Chronic (2 mo) 5–24 yr	Up — Up Up	• Known to promote reactive astrocytosis	[79,84]
miR-128	Rats (Pilo, i.p.) TLE patients	Acute (2 h) Latent (3 w) Chronic (2 mo) 5–24 yr	Down Down Down Down	• Inhibition of proliferation and promotion of neuronal differentiation-neuronal hyperexcitability and behavioral seizures by miR-128 deficiency	[76,79,85]
miR-132	Mice (KA, intraamygdala)	Acute (24 h)	Up	• Contribution to neuronal death	[86]
	rats (Pilo, i.p.)	Chronic (2 mo)	Up	• Known to promote dendritic outgrowth	[75,87]

Table 12.1 miRNAs That are Dysregulated in Epilepsy (*cont.*)

miRNA	Source of Data	Stage of Epilepsy	Expression Change in Epilepsy	Potential Roles or Functional Effect	References
miR-134	Mice (KA, i.c.v.)	Acute (24 h) Chronic (3 w)	Up Up	• Contribution to neuronal death • Proconvulsive and proepileptic effects	[88]
	rats (Pilo, i.p.)	Chronic (2 mo)	Up	• Known to promote cell proliferation and neuronal differentiation • Known to promote dendritic arborization	[75,89,90]
miR-135a	Rats (Pilo, i.p.)	Acute (2 h) Latent (3 w) Chronic (2 mo)	Up — Up	• Known to suppress cell proliferation and promote neuronal differentiation	[76,79]
	TLE patients	5–24 yr	Up		
miR-138	TLE patients	10–47.5 yr	Down	• Known to reduce dendritic spine size and to inhibit axon growth	[91–93]
miR-146a-5p	FCD/ganglioma patients	FCD: 6-17 yr ganglioma: 3–16 yr	Up	• Reduction of IRAK-1, TRAF-6, TNF-α, IL-6, HMGB1, and COX-2	[94]
miR-146a	Rats (Pilo, i.p.)	Chronic (2 mo)	Up		[75]
miR-155	Rats (Pilo, i.p.)	Acute (2 h) Latent (3 w) Chronic (2 mo)	Up — Up	• Known to upregulate TNF-α, IL-6, and microglial activation	[95,96]
miR-183	TLE patients (children)	3–7 yr	Up	• Undetermined	[79,95]
	rats (Pilo)	Acute (2 h) Latent (3 w) Chronic (2 mo)	Up — Up		
miR-184	TLE patients	5–24 yr	Up	• Promotion of neuronal survival (epileptic tolerance)	[79,97]
	mice (KA, i.p.)	seizure precondition (24 h) + SE (24 h)	Up		
miR-204	TLE patients	2–22 yr	Down	• Repression of axon guidance genes (ROBO1, GRM1, SLC1A2, GNAI2)	[98]
miR-218	TLE patients	2–22 yr	Down	• Repression of GRM1	[98]
miR-221	TLE patients	10–47.5 yr	Down	• Repression of ICAM1	[91]
miR-222	TLE patients	10–47.5 yr	Down	• Repression of ICAM1	[91]
miR-487a	TLE-HS patients	7–38 yr	Down	• Repression of ANTXR1 promoting cell adhesion and migration	[99]

D, Day; FCD, focal cortical dysplasia; h, hour; HS, hippocampal sclerosis; i.c.v., intracerebroventricular; i.p., intraperitoneal; KA, kainic acid; mo, month; Pilo, pilocarpine; TLE, temporal lobe epilepsy;w, week; yr, year.

regulation [82]. Functional analysis with a specific antagomir further supported proapoptotic miR-34a as it ameliorated caspase 3-mediated neuronal death [83]. Similarly, antagomirs targeting miR-132 and miR-134 showed a reduction of seizure-induced neuronal death in the hippocampus, with increased expression of miR-132 and miR-134 in animal models of TLE [86,88]. Thus, these miRNAs, such as miR-34a, miR-132, and miR-134 can contribute to seizure-induced cell death. On the contrary, a study from the Henshall group nicely showed that epileptic tolerance was mediated by preconditioning-induced upregulation of miR-184 in the hippocampal CA1 and CA3 regions where seizure-induced cell death is prominent [97]. The authors further confirmed the requirement of miR-184 in the acquisition of epileptic tolerance by applying miR-184 antagomir, demonstrating a neuroprotective role of this miRNA.

12.3.3.1.2 miRNAs and Aberrant Neurogenesis

Seizure-induced aberrant neurogenesis and its modulation by miRNAs have received explosive attention. As a result, a number of studies have identified multiple miRNAs working on different aspects of postseizure neurogenesis, that is, proliferation and differentiation of progenitors, dendrites and axon sprouting, and granule cell dispersion.

After acute seizures, neural progenitors differentiate into mature neurons faster than physiological neurogenesis. In epilepsy, the expression of miR-134 was increased [75]. As miR-134 promotes the proliferation of neural progenitors and their differentiation, increased miR-134 can play a facilitative role in seizure-induced neurogenesis. Moreover, let-7 miRNAs that suppress both the proliferation and the differentiation process were decreased in the chronic stage of epilepsy [75]. In addition, miR-9 and miR-135a were reported to be upregulated in epilepsy [79]. As they can also promote the differentiation of newly generated neurons at the expense of suppressed proliferation of progenitors, these data imply a proneurogenic environment induced by seizures, and possibly proepileptogenic [17]. However, the regulatory mechanism of miRNAs in postseizure neurogenesis may not be as simple as it appears. For example, under physiological conditions, induction of miR-128 has been known to promote neuronal differentiation with the suppression of the proliferative activity of progenitors, similar to miR-9 and miR-135a. However, the expression of miR-128 was reduced in an animal model and in patients with TLE [79], in contrast to miR-9 and miR135a. Since miR-128 deficiency could induce neuronal hyperexcitability and behavioral seizures by disinhibition of ion channels [85], the direct epileptogenic mechanisms involving miR-128 may be derived from ion channel dysregulation, not by promoting neurogenesis. Thus, it will be interesting to examine if miR-128 reduction in epilepsy occurs as a compensatory mechanism in an attempt to balance seizure-induced proneurogenesis.

Repetitive seizures can trigger neurite sprouting. Physiologically, neuronal dendritic complexity and spine morphogenesis are regulated by three major miRNAs: miR-132, miR-134, and miR-138. In the chronic phase of TLE when dendritic outgrowth is prominent, hippocampal expression of miR-132 and miR-134 were significantly enhanced [75], suggesting prodendritogenic roles of miR-132/134. Furthermore, silencing miR-134 could reduce dendritic spine density in hippocampal CA3 pyramidal neurons [88]. At a functional level, blocking miR-134 could increase resistance against acute and chronic seizures [88], despite no change in memory function [100]. However, in contrast to miR-132 and miR134, miR-138 was found to be decreased in human brain tissues of TLE with hippocampal sclerosis [91], making regulatory mechanisms of miRNAs on seizure-induced dendritic alterations complicated.

As for the mossy fiber sprouting, miR-204 and miR-218 can play important roles in dysregulation of axonal guidance in epilepsy. Neuronal miR-204 and miR-218 were markedly downregulated in patients with TLE, compared to controls [98]. In this study, the authors found that miR-218 was a negative regulator of roundabout guidance receptor 1 (ROBO1), metabotropic glutamate receptor 1 (GRM1), SCL1A2/EAAT2, and guanine nucleotide-binding protein Gi alpha 2 subunit (GNAI2), all of which are associated with axonal guidance and synaptic plasticity. Moreover, they also found that GRM1 was a direct target of miR-204. Therefore, downregulation of miR-204 and miR-218 in TLE may derepress these four targets, leading to axon sprouting.

Another characteristic feature of seizure-induced aberrant neurogenesis is granule cell dispersion. In order to identify critical regulators of granule cell dispersion in TLE, Zucchini et al. [99] took the microdissection approach to specifically isolate seizure-experienced granule cells with or without dispersion. When they analyzed differentially expressed miRNAs between the two groups, they found reduction of miR-487a in the scattered granule cells and concurrent increase of a potential target of miR-487a, anthrax toxin receptor 1 (ANTXR1, also known as tumor endothelial marker 8, TEM8). As ANTXR1 is an adhesion molecule promoting cellular migration, miR-487a-mediated control of ANTXR1 may be a good target for treating epilepsy.

12.3.3.1.3 miRNAs and Inflammation

Proinflammatory mediators including TNF-α and IL-1β are immediately released after acute seizures and sustained until the chronic stage of epilepsy. Among immune-related miRNAs, the expression of miR-146a and miR-155 was increased in experimental and human TLE [75,94,95]. Considering the upregulation of TNF-α and its similar expression pattern with these miRNAs, it is plausible that miR-146a and miR-155 could modulate TNF-α in epilepsy. Indeed, addition of miR-146a mimics in reactive astrocytes before challenging with proinflammatory stimuli was able to decrease the level of TNF-α and IL-6 in addition to HMGB1 and COX-2 expression, suggesting an antiinflammatory role of miR-146a [94]. Another miRNA that can influence seizure-induced inflammation is miR-22. A recent study elegantly showed underlying regulatory mechanisms controlling reduced inflammation in the contralateral hippocampus compared to the ipsilateral, after intraamygdala injection of kainic acid [78]. The authors of this study found contralateral miR-22 upregulation was responsible for the reduction of TNF-α and IL-1β expression and this was mediated by ATP-gated ionotropic P2X7 receptor (P2X7R). When they inhibited miR-22 using an antagomir approach, P2X7R-mediated TNF-α and IL-1β expression in the contralateral hippocampus where kainic acid was not directly exposed, were markedly enhanced with exacerbated acute and chronic seizures, memory impairment, and increased anxiety. Moreover, these findings were reversed by the treatment of miR-22 mimics, strongly supporting the antiinflammatory and antiepileptic roles of miR-22.

Genome-wide miRNA profiling study enables the identification of critical miRNAs targeting immune-modulatory molecules in TLE. Using an unbiased miRNA array, Kan et al. [91] reported that the expression of miR-221 and miR-222 was significantly reduced in human brain tissues of patients with TLE with hippocampal sclerosis. They also found that miR-221 and miR-222 could regulate ICAM1 expression in reactive astrocytes. Thus, the evidence suggests that the reduction of miR-221 and miR-222 in epilepsy can increase ICAM1 expression, contributing to the recruitment, accumulation, and activation of immune cells in brain.

Other inflammatory miRNAs with altered expression in epilepsy were miR-27a, miR-30c, miR-125b, and miR-183. The expression of these miRNAs was generally increased in animal models of TLE [79]. However, since the temporal expression of these miRNAs is dynamically controlled depending on the phase of TLE, further studies are required to dissect the stage-specific regulatory mechanisms of various miRNAs on inflammation in epilepsy.

12.3.3.2 Long Noncoding RNAs

Long noncoding RNAs (lncRNA) are nonprotein coding transcripts longer than 200 nucleotides. Considering approximately 25,000 protein-coding genes and 2,500 miRNAs of the human genome, it is amazing that 40% of all lncRNAs (equivalent to 4,000–20,000 lncRNA genes) are expressed only in the brain. lncRNAs can be subgrouped into intergenic, intronic, bidirectional, sense overlapping, and antisense transcripts. In the past decade, a wide range of functional implications of lncRNAs including genomic imprinting, cell differentiation and organogenesis have begun to be elucidated. Especially in the brain, lncRNAs can be a critical regulator in the process of neural differentiation and synapse formation. lncRNAs can control this process with spatiotemporal specificity via unique site-specific targeting. Epigenetically, lncRNAs can interact with chromatin-modifying proteins and subsequently affect gene transcription by modifying chromatin structure. In the nucleus, lncRNAs act as a scaffold for recruiting transcriptional regulators, such as DNMT3, PRC1/2, and H3K9 methyltransferases onto specific genomic loci. In the cytoplasm, they can control the stability, degradation, and the transcription of mRNAs by specific base pairing to a complementary target mRNA sequence.

To our knowledge, only one study has showed lncRNA expression profiles in the epileptic brain [101]. The authors screened lncRNAs dysregulated in two popular animal models of epilepsy, pilocarpine and kainic acid injection models. At 2 months after the induction of acute seizures, the authors found 118 lncRNAs commonly altered in both models. Interestingly, most of the lncRNAs were upregulated and only two were downregulated. The authors further identified two lncRNAs that were dysregulated in the same direction with the protein-coding gene located near the particular lncRNA. One of them was antisense noncoding RNA in the INK4 locus (ANRIL), known to bind its target, chromobox 7 (CBX7). Considering the facts that ANRIL binding to CBX7 promotes CBX7 activation and CBX7 consists of PRC1, histone modification with the aid of lncRNA may provide novel insights into multifactorial mechanisms in epilepsy.

Despite the fact that the association with epilepsy is less strong, several lncRNAs may play a role in epilepsy. For example, rhabdomyosarcoma 2 associated transcript (RMST) was found to be transcriptionally regulated by REST. When REST was downregulated, RMST expression went up with increased coupling of RMST and SOX2, promoting neurogenesis [102]. Moreover, HOXA transcript antisense RNA (HOTAIR) could recruit PRC2 and KDM1A-CoREST-REST complexes to chromatin, working as a scaffold for gene repression [103]. As REST is known to be upregulated after acute seizures [3], it will be interesting to assess the involvement of RMST and HOTAIR in REST-mediated epigenetic control in epilepsy. Another candidate lncRNA that might be critical in epilepsy is EVF2. LncRNA EVF2 was reported to be important for the development of GABAergic interneurons and its deficiency reduced synaptic inhibition [104]. Given that epilepsy is primarily caused by neuronal hyperexcitability, EVF2 can be an attractive focus for epilepsy research.

12.3.4 ATP-DEPENDENT CHROMATIN REMODELING COMPLEX

ATP-dependent chromatin remodeling complexes can mobilize nucleosomes for the regulation of gene expression. They can modify chromatin architecture by moving nucleosomes along the DNA (looping and sliding), evicting histones off the DNA, or facilitating the exchange of histone variants. The energy to translocate nucleosomes comes from ATP hydrolysis mediated by the conserved ATPase domain of these chromatin remodeling complexes. Based on the distinct domain structures, there are four chromatin remodeling ATPase families: switching defective/sucrose nonfermentable (SWI/SNF), imitation SWI (ISWI), chromodomain helicase DNA-binding (CHD), and inositol requiring 80 (INO80).

SWI/SNF family proteins were first discovered in yeasts. The mammalian homolog of yeast SWI/SNF is brahma-associated factor (BAF) complex that consists of 15 protein subunits. BAF complex can switch some of the subunits to serve context-dependent functions. Each BAF complex contains a subunit acting as catalytic ATPases, BRM/SMARCA2/SNF2L2 (BRM, brahma) or BRG1/SMARCA4/SNF2L4 (BRG1, brahma-related gene1). Moreover, some of the BAF complex subunits have a bromodomain that can bind to acetylated lysine residues of histones. Interestingly, SMARCA3, now known as helicase-like transcription factor (HLTF), has been shown to be critical in the antidepressant-mediated enhancement of hippocampal neurogenesis [105]. ISWI family complexes include nucleosome remodeling factor (NURF), ATP-utilizing chromatin assembly and remodeling factor (ACF), chromatin accessibility complex (CHRAC), nucleolar remodeling complex (NoRC), and the Williams syndrome transcription factor–ISWI chromatin remodeling complex (WICH). Each mammalian ISWI complex has one of the two ATPase subunits, SNF2H/SMARCA5 and SNF2L/SMARCA1. H3K4me3 is a docking site for ISWI complexes, mediating chromatin remodeling and epigenetic regulation. The third chromatin remodeler is CHD proteins. CHD proteins are classified into three subfamilies: CHD1 (CHD1, CHD2), Mi-2 (CHD3, CHD4), and CHD7 (from CHD5 to CHD9). They are characterized to have a chromodomain that serves as a module for binding to DNA, RNA or methylated H3. Similar to SWI/SNF and ISWI family complexes, CHD proteins can form a multisubunit complex in that CHD3 or CHD4, in concert with HDAC1/2 and MBD, is composed of nucleosome remodeling histone deacetylase (NURD) complex. Finally, INO80 family is a large multisubunit complex containing INO80 ATPases. It has DNA helicase activity and is known to regulate transcription, DNA replication, DNA repair, cell division, and histone exchange from H2A.Z to H2A.

Concerning epilepsy, regulatory mechanisms and the impact of chromatin remodeling ATPases on the epileptic brain are largely unknown. Currently, potential candidate genes are only identified from multiple mutation studies. For example, CHD2 mutation was found in Dravet syndrome [106], Lennox–Gastaut syndrome [107], and a neurodevelopmental disorder with an epileptic phenotype [108]. CHD4 was mutated in two classical epileptic encephalopathies, which are infantile spasms and Lennox–Gastaut syndrome [109]. Mutations of BAF47/INI1/SMARCB1 (a component of BAF complex), BRM/SMARCA2 and BRG1/SMARCA4 have been identified in Coffin–Siris syndrome [110], in which mental retardation and epileptic seizures are prominent. One possible clue to the functional impact of CHD2 modulation in epilepsy arises from a study showing that seizure-like behaviors and epileptiform discharges were recapitulated by CHD2 knockdown [106].

Interestingly, BRG1 could bind to REST through the interaction with CoREST [111]. BRG1was supposed to recognize acetylated histones by its bromodomain and in turn, promoted REST–chromatin interactions, resulting in the repression of REST-target genes. Since REST is upregulated in epilepsy and is able to recruit diverse epigenetic players, it will be interesting to figure out how REST-mediated chromatin remodeling can influence the pathology of seizures and epilepsy.

12.4 NEUROPSYCHIATRIC ASPECTS OF EPILEPSY

Patients with epilepsy are well known to experience psychiatric and behavioral problems [112]. Compared to a normal population, patients with epilepsy are likely to have a higher prevalence of psychiatric problems. Psychiatric comorbidity in patients with epilepsy can be classified first as disorders clearly attributable to the underlying brain disorder causing seizures. Here, the brain pathology itself is associated with psychiatric, cognitive, or behavioral manifestations, like in a number of syndromes affecting children which present with epilepsy, like Lennox–Gastaut syndrome. Second, there can be psychiatric disorders with an obvious temporal relationship between episodic psychiatric symptoms and seizures. These can be preictal disorders like irritability and dysphoria, ictal disorders like aura and automatisms, and postictal disorders like delirium and psychosis. Third, patients with epilepsy can develop psychiatric symptoms as a psychological reaction to the epileptic illness.

The most common psychiatric disorders in patients with epilepsy are mood disorders, followed by anxiety disorders, psychoses, and personality disorders [113]. Cognitive impairment is also common [112]. In addition, psychiatric problems can arise as adverse effects to AED, especially in children. Psychiatric comorbidity correlates with the severity and chronicity of epilepsy, and is more common in patients with some types of epilepsy like TLE and refractory epilepsy [113].

12.5 CONCLUSIONS

Understanding basic epigenetic regulation of gene expression and its dysregulation in epilepsy has great potential for the development of novel therapeutic and preventive strategies in epilepsy. Here we reviewed important pathophysiologic features in epilepsy and how each one was altered by epigenetic mechanisms, contributing to the formation of recurrent seizures, epilepsy-associated cognitive decline and mood disturbances. Since multiple epigenetic factors expressed in different cell types can influence the epileptic brain over the course of epileptogenesis and after the establishment of chronic epilepsy, the next step will be to find a common thread by assembling known facts. This will provide a general landscape of epigenetic modifications in epilepsy. Since VPA has long been used as a classical AED, but later found to play a critical role in epigenetic regulation blocking HDAC enzymes, epigenetic mechanisms in epilepsy are crucial without doubt. With recent efforts in drug development targeting epigenetics, several new drugs have been introduced, Vidaza (azacitidine) and Dacogen (decitabine) as DNMT inhibitors and Vorinostat (SAHA) and Istodax (romidepsin) as HDAC inhibitors. However, current indications of these drugs are limited to cancer patients and their application to epilepsy has not been tested yet. Another potential therapeutic agent will be antagomirs and/or miRNA mimics as a number of miRNAs were identified to mediate diverse pathologic cellular events in epilepsy.

Considering the heterogeneous and complex nature of epilepsy, a single target approach may not be enough for treating epilepsy. Despite currently available AED showing different modes of action, all of them are merely anticonvulsants without a disease-modifying capability, which further support this hypothesis. If a single target approach cannot alter the disease course, combination therapies covering multiple mechanisms simultaneously may be a new alternative modality for epilepsy. We believe epigenetic modifiers can play synergistic roles with classical (or new) AED, hopefully improving the management of patients with epilepsy. Epigenetics can influence gene expression for a long period of time, not just as a temporary inhibition of protein activities like ion channels. Taking epigenetics into consideration for treatment options may contribute to the accomplishment of our ultimate goal, that is, curing epilepsy.

ABBREVIATIONS

ACF ATP-utilizing chromatin assembly and remodeling factor
AED Antiepileptic drugs
AID Activation-induced cytidine deaminase

ANRIL Antisense noncoding RNA in the INK4 locus
ANTXR1 Anthrax toxin receptor 1
APOBEC Apolipoprotein B mRNA-editing enzyme complex
BAF Brahma-associated factor
BBB Blood brain barrier
BDNF Brain derived growth factor
BrdU Bromodeoxyuridine
BRG Brahma-related gene
BRM Brahma
CARM Coactivator-associated arginine methyltransferase
CBP CREB-binding protein
CBX7 Chromobox 7
CHD Chromodomain helicase DNA-binding
CHRAC Chromatin accessibility complex
DBS Deep brain stimulation
EAAT Excitatory amino acid transporter
EEG Electroencephalography
ECS Extracellular space
ECT Electroconvulsive therapy
EHMT Euchromatic histone-lysine *N*-methyltransferase
EVF2 Embryonic ventral forebrain-2
EZH Enhancer of zeste homolog
FGF1 Fibroblast growth factor 1
Gadd45 Growth arrest and DNA damage 45
GCN5 General control nonderepressible 5
GLP G9a-like protein
GluR2 Glutamate receptor 2
GLUT-1 Glucose transporter type 1
GNAI2 Guanine nucleotide-binding protein Gi alpha 2 subunit
GNP Gross National Product
GRM1 Metabotropic glutamate receptor 1
γ-H2AX Phosphorylated histone H2A.X
HLTF Helicase-like transcription factor
5hmU 5-Hydroxymethyl-uracil
HMGB1 High mobility group box- 1
HOTAIR HOXA transcript antisense RNA
ICAM-1 Intercellular adhesion molecule 1
IL-1β Interleukin-1 beta
ILAE International League Against Epilepsy
INO80 Inositol requiring 80
ISWI Imitation SWI
JARID Jumonji/AT-rich interaction domain (ARID)
JBP Janus kinase-binding protein
KDM Lysine demethylase
KMT Lysine(K)-methyltransferases
lncRNA Long noncoding RNAs
LSD Lysine-specific demethylase
MAPK Mitogen-activated protein kinase

MLL	Myeloid/lymphoid or mixed-lineage leukemia
MYST	MOZ, Ybf2/Sas3, Sas2, and Tip60
NAD	Nicotinamide adenine dinucleotide
NoRC	Nucleolar remodeling complex
NURD	Nucleosome remodeling histone deacetylase
NURF	Nucleosome remodeling factor
PCAF	p300-CBP-associated factor
P2X7R	P2X7 receptor
PDH	Pyruvate dehydrogenase
REST	Repressor element 1-silencing transcription factor
RISC	RNA-induced silencing complex
RMST	Rhabdomyosarcoma 2 associated transcript
ROBO1	Roundabout guidance receptor 1
SCL1A2	Solute carrier family1, member 2
SCN1A	Sodium channel, voltage gated, type I alpha subunit
SETD	SET domain-containing
SMARC	SWI/SNF-related, matrix-associated, actin-dependent regulator of chromatin
sncRNA	Small ncRNAs
SRC-1	Steroid receptor coactivator-1
SUMO	Small ubiquitin related modifier
SWI/SNF	Switching defective/sucrose nonfermentable
TDG	Thymine DNA glycosylase
TET	Ten-eleven translocation
TLE	Temporal lobe epilepsy
TLR	Toll-like receptor
TNF-α	Tumor necrosis factor alpha
VEGF	Vascular endothelial growth factor
WICH	Williams syndrome transcription factor–ISWI chromatin remodeling complex

GLOSSARY

Blood brain barrier A semipermeable membrane that separates the blood from the cerebrospinal fluid and which constitutes a barrier to the passage of cells, particles, and large molecules

Ictal Relating to or caused by a stroke or seizure

Kainic acid A central nervous system stimulant that acts by activating receptors for the excitatory neurotransmitter glutamate

Picrotoxin A central nervous system stimulant which acts by blocking GABA-activated chloride ion channels

ACKNOWLEDGMENTS

We appreciate Dr. Hoon Hur for his thoughtful comments on this manuscript. We apologize for not being able to cite all primary references due to space limitation. This work was supported by a grant of Basic Science Research Program through the National Research Foundation of Korea (NRF-2014R1A1A3049456).

REFERENCES

[1] World Health Organization. Epilepsy Fact Sheet. 2016; http://www.who.int/mediacentre/factsheets/fs999/en/

[2] Pitkanen A, Lukasiuk K. Molecular and cellular basis of epileptogenesis in symptomatic epilepsy. Epilepsy Behav 2009;14:16–25.

[3] Roopra A, Dingledine R, Hsieh J. Epigenetics and epilepsy. Epilepsia 2012;53(Suppl. 9):2–10.

[4] Berg AT, Berkovic SF, Brodie MJ, Buchhalter J, Cross JH, Boas WV, et al. Revised terminology and concepts for organization of seizures and epilepsies: report of the ILAE Commission on Classification and Terminology, 2005–2009. Epilepsia 2010;51(4):676–85.

[5] Fisher RS, Acevedo C, Arzimanoglou A, Bogacz A, Cross JH, Elger CE, et al. ILAE Official Report: a practical clinical definition of epilepsy. Epilepsia 2014;55(4):475–82.

[6] Engel J, Wiebe S, French J, Sperling M, Williamson P, Spencer D, et al. Practice parameter: temporal lobe and localized neocortical resections for epilepsy —Report of the Quality Standards Subcommittee of the American Academy of Neurology, in association with the American Epilepsy Society and the American Association of Neurological Surgeons. Epilepsia 2003;44(6):741–51.

[7] Sattler R, Tymianski M. Molecular mechanisms of glutamate receptor-mediated excitotoxic neuronal cell death. Mol Neurobiol 2001;24(1–3):107–29.

[8] Dudek FE, Hellier JL, Williams PA, Ferraro DJ, Staley KJ. The course of cellular alterations associated with the development of spontaneous seizures after status epilepticus. Prog Brain Res 2002;135:53–65.

[9] Catarino CB, Liu JY, Liagkouras I, Gibbons VS, Labrum RW, Ellis R, et al. Dravet syndrome as epileptic encephalopathy: evidence from long-term course and neuropathology. Brain 2011;134(Pt 10):2982–3010.

[10] Pauli E, Hildebrandt M, Romstock J, Stefan H, Blumcke I. Deficient memory acquisition in temporal lobe epilepsy is predicted by hippocampal granule cell loss. Neurology 2006;67(8):1383–9.

[11] Parent JM, Yu TW, Leibowitz RT, Geschwind DH, Sloviter RS, Lowenstein DH. Dentate granule cell neurogenesis is increased by seizures and contributes to aberrant network reorganization in the adult rat hippocampus. J Neurosci 1997;17(10):3727–38.

[12] Jessberger S, Zhao C, Toni N, Clemenson GD Jr, Li Y, Gage FH. Seizure-associated, aberrant neurogenesis in adult rats characterized with retrovirus-mediated cell labeling. J Neurosci 2007;27(35):9400–7.

[13] Hattiangady B, Rao MS, Shetty AK. Chronic temporal lobe epilepsy is dentate neurogenesis in the adult associated with severely declined hippocampus. Neurobiol Dis 2004;17(3):473–90.

[14] Houser CR. Granule cell dispersion in the dentate gyrus of humans with temporal lobe epilepsy. Brain Res 1990;535(2):195–204.

[15] Koyama R, Tao K, Sasaki T, Ichikawa J, Miyamoto D, Muramatsu R, et al. GABAergic excitation after febrile seizures induces ectopic granule cells and adult epilepsy. Nat Med 2012;18(8):1271–8.

[16] Pun RY, Rolle IJ, Lasarge CL, Hosford BE, Rosen JM, Uhl JD, et al. Excessive activation of mTOR in postnatally generated granule cells is sufficient to cause epilepsy. Neuron 2012;75(6):1022–34.

[17] Cho KO, Lybrand ZR, Ito N, Brulet R, Tafacory F, Zhang L, et al. Aberrant hippocampal neurogenesis contributes to epilepsy and associated cognitive decline. Nat Commun 2015;6:6606.

[18] Jessberger S, Nakashima K, Clemenson GD, Mejia E, Mathews E, Ure K, et al. Epigenetic modulation of seizure-induced neurogenesis and cognitive decline. J Neurosci 2007;27(22):5967–75.

[19] Yutsudo N, Kamada T, Kajitani K, Nomaru H, Katogi A, Ohnishi YH, et al. fosB-null mice display impaired adult hippocampal neurogenesis and spontaneous epilepsy with depressive behavior. Neuropsychopharmacology 2013;38(7):1374–5.

[20] Devinsky O, Vezzani A, Najjar S, De Lanerolle NC, Rogawski MA. Glia and epilepsy: excitability and inflammation. Trends Neurosci 2013;36(3):174–84.

[21] Tian GF, Azmi H, Takano T, Xu Q, Peng W, Lin J, et al. An astrocytic basis of epilepsy. Nat Med 2005;11(9):973–81.

[22] Sarac S, Afzal S, Broholm H, Madsen FF, Ploug T, Laursen H. Excitatory amino acid transporters EAAT-1 and EAAT-2 in temporal lobe and hippocampus in intractable temporal lobe epilepsy. APMIS 2009;117(4): 291–301.

[23] van der Hel WS, Notenboom RGE, Bos IWM, van Rijen PC, van Veelen CWM, de Graan PNE. Reduced glutamine synthetase in hippocampal areas with neuron loss in temporal lobe epilepsy. Neurology 2005;64(2):326–33.

[24] Ortinski PI, Dong J, Mungenast A, Yue C, Takano H, Watson DJ, et al. Selective induction of astrocytic gliosis generates deficits in neuronal inhibition. Nat Neurosci 2010;13(5):584–91.

[25] Wang Y, Zaveri HP, Lee TS, Eid T. The development of recurrent seizures after continuous intrahippocampal infusion of methionine sulfoximine in rats: a video-intracranial electroencephalographic study. Exp Neurol 2009;220(2):293–302.

[26] Gomez-Gonzalo M, Losi G, Chiavegato A, Zonta M, Cammarota M, Brondi M, et al. An excitatory loop with astrocytes contributes to drive neurons to seizure threshold. PLoS Biol 2010;8(4):e1000352.

[27] Clasadonte J, Dong J, Hines DJ, Haydon PG. Astrocyte control of synaptic NMDA receptors contributes to the progressive development of temporal lobe epilepsy. Proc Natl Acad Sci USA 2013;110(43):17540–5.

[28] Oberheim NA, Tian GF, Han X, Peng W, Takano T, Ransom B, et al. Loss of astrocytic domain organization in the epileptic brain. J Neurosci 2008;28(13):3264–76.

[29] Rigau V, Morin M, Rousset MC, de Bock F, Lebrun A, Coubes P, et al. Angiogenesis is associated with blood-brain barrier permeability in temporal lobe epilepsy. Brain 2007;130:1942–56.

[30] Librizzi L, Noe F, Vezzani A, de Curtis M, Ravizza T. Seizure-induced brain-borne inflammation sustains seizure recurrence and blood-brain barrier damage. Ann Neurol 2012;72(1):82–90.

[31] Morin-Brureau M, Rigau V, Lerner-Natoli M. Why and how to target angiogenesis in focal epilepsies. Epilepsia 2012;53:64–8.

[32] Marchi N, Fan Q, Ghosh C, Fazio V, Bertolini F, Betto G, et al. Antagonism of peripheral inflammation reduces the severity of status epilepticus. Neurobiol Dis 2009;33(2):171–81.

[33] Maroso M, Balosso S, Ravizza T, Liu J, Aronica E, Iyer AM, et al. Toll-like receptor 4 and high-mobility group box-1 are involved in ictogenesis and can be targeted to reduce seizures. Nat Med 2010;16(4): 413–9.

[34] De Sarro G, Russo E, Ferreri G, Giuseppe B, Flocco MA, Di Paola ED, et al. Seizure susceptibility to various convulsant stimuli of knockout interleukin-6 mice. Pharmacol Biochem Behav 2004;77(4):761–6.

[35] Peltola J, Hurme M, Miettinen A, Keranen T. Elevated levels of interleukin-6 may occur in cerebrospinal fluid from patients with recent epileptic seizures. Epilepsy Res 1998;31(2):129–33.

[36] Riazi K, Galic MA, Kuzmiski JB, Ho W, Sharkey KA, Pittman QJ. Microglial activation and TNF alpha production mediate altered CNS excitability following peripheral inflammation. Proc Natl Acad Sci USA 2008;105(44):17151–6.

[37] Weinberg MS, Blake BL, McCown TJ. Opposing actions of hippocampus TNF alpha receptors on limbic seizure susceptibility. Exp Neurol 2013;247:429–37.

[38] Balosso S, Ravizza T, Perego C, Peschon J, Campbell IL, De Simoni MG, et al. Tumor necrosis factor alpha inhibits seizures in mice via p75 receptors. Ann Neurol 2005;57(6):804–12.

[39] Matsuda T, Murao N, Katano Y, Juliandi B, Kohyama J, Akira S, et al. TLR9 signalling in microglia attenuates seizure-induced aberrant neurogenesis in the adult hippocampus. Nat Commun 2015;6:6514.

[40] Leppik IE. Three new drugs for epilepsy: levetiracetam, oxcarbazepine, and zonisamide. J Child Neurol 2002;17(Suppl. 1):S53–7.

[41] Baranano KW, Hartman AL. The ketogenic diet: uses in epilepsy and other neurologic illnesses. Curr Treat Options Neurol 2008;10(6):410–9.

[42] Amar AP, Heck CN, Levy ML, Smith T, DeGiorgio CM, Oviedo S, et al. An institutional experience with cervical vagus nerve trunk stimulation for medically refractory epilepsy: rationale, technique, and outcome. Neurosurgery 1998;43(6):1265–76. discussion 76–80.

[43] Velasco AL, Velasco F, Velasco M, Trejo D, Castro G, Carrillo-Ruiz JD. Electrical stimulation of the hippocampal epileptic foci for seizure control: a double-blind, long-term follow-up study. Epilepsia 2007;48(10):1895–903.

[44] Kobow K, Kaspi A, Harikrishnan KN, Kiese K, Ziemann M, Khurana I, et al. Deep sequencing reveals increased DNA methylation in chronic rat epilepsy. Acta Neuropathol 2013;126(5):741–56.

[45] Kobow K, Jeske I, Hildebrandt M, Hauke J, Hahnen E, Buslei R, et al. Increased reelin promoter methylation is associated with granule cell dispersion in human temporal lobe epilepsy. J Neuropathol Exp Neurol 2009;68(4):356–64.

[46] Miller-Delaney SF, Das S, Sano T, Jimenez-Mateos EM, Bryan K, Buckley PG, et al. Differential DNA methylation patterns define status epilepticus and epileptic tolerance. J Neurosci 2012;32(5):1577–88.

[47] Zhu Q, Wang L, Zhang Y, Zhao FH, Luo J, Xiao Z, et al. Increased expression of DNA methyltransferase 1 and 3a in human temporal lobe epilepsy. J Mol Neurosci 2012;46(2):420–6.

[48] Noh JS, Sharma RP, Veldic M, Salvacion AA, Jia X, Chen Y, et al. DNA methyltransferase 1 regulates reelin mRNA expression in mouse primary cortical cultures. Proc Natl Acad Sci USA 2005;102(5):1749–54.

[49] Gong C, Wang TW, Huang HS, Parent JM. Reelin regulates neuronal progenitor migration in intact and epileptic hippocampus. J Neurosci 2007;27(8):1803–11.

[50] Heinrich C, Nitta N, Flubacher A, Muller M, Fahrner A, Kirsch M, et al. Reelin deficiency and displacement of mature neurons, but not neurogenesis, underlie the formation of granule cell dispersion in the epileptic hippocampus. J Neurosci 2006;26(17):4701–13.

[51] D'Aiuto L, Di Maio R, Mohan KN, Minervini C, Saporiti F, Soreca I, et al. Mouse ES cells overexpressing DNMT1 produce abnormal neurons with upregulated NMDA/NR1 subunit. Differentiation 2011;82(1):9–17.

[52] Dyrvig M, Gotzsche CR, Woldbye DP, Lichota J. Epigenetic regulation of Dnmt3a and Arc gene expression after electroconvulsive stimulation in the rat. Mol Cell Neurosci 2015;67:137–43.

[53] Zhang L, He JW, Jugloff DGM, Eubanks JH. The MeCP2-null mouse hippocampus displays altered basal inhibitory rhythms and is prone to hyperexcitability. Hippocampus 2008;18(3):294–309.

[54] Zhang W, Peterson M, Beyer B, Frankel WN, Zhang ZW. Loss of MeCP2 from forebrain excitatory neurons leads to cortical hyperexcitation and seizures. J Neurosci 2014;34(7):2754–63.

[55] Ramocki MB, Tavyev YJ, Peters SU. The MECP2 duplication syndrome. Am J Med Genet Part A 2010;152A(5):1079–88.

[56] Tao S, Yang X, Chen Y, Wang X, Xiao Z, Wang H, et al. Up-regulated methyl CpG binding protein-2 in intractable temporal lobe epilepsy patients and a rat model. Neurochem Res 2012;37(9):1886–97.

[57] Zhu RL, Graham SH, Jin J, Stetler RA, Simon RP, Chen J. Kainate induces the expression of the DNA damage-inducible gene, GADD45, the rat brain. Neuroscience 1997;81(3):707–20.

[58] Ryley Parrish R, Albertson AJ, Buckingham SC, Hablitz JJ, Mascia KL, Davis Haselden W, et al. Status epilepticus triggers early and late alterations in brain-derived neurotrophic factor and NMDA glutamate receptor Grin2b DNA methylation levels in the hippocampus. Neuroscience 2013;248:602–19.

[59] Ma DK, Jang MH, Guo JU, Kitabatake Y, Chang ML, Pow-Anpongkul N, et al. Neuronal activity-induced Gadd45b promotes epigenetic DNA demethylation and adult neurogenesis. Science 2009;323(5917):1074–7.

[60] Kaas GA, Zhong C, Eason DE, Ross DL, Vachhani RV, Ming GL, et al. TET1 controls CNS 5-methylcytosine hydroxylation, active DNA demethylation, gene transcription, and memory formation. Neuron 2013;79(6):1086–93.

[61] Huang YF, Doherty JJ, Dingledine R. Altered histone acetylation at glutamate receptor 2 and brain-derived neurotrophic factor genes is an early event triggered by status epilepticus. J Neurosci 2002;22(19):8422–8.

[62] Sng JCG, Taniura H, Yoneda Y. Histone modifications in kainate-induced status epilepticus. Eur J Neurosci 2006;23(5):1269–82.

[63] Brill J, Lee M, Zhao S, Fernald RD, Huguenard JR. Chronic valproic acid treatment triggers increased neuropeptide Y expression and signaling in rat nucleus reticularis thalami. J Neurosci 2006;26(25):6813–22.

[64] Jagirdar R, Drexel M, Kirchmair E, Tasan RO, Sperk G. Rapid changes in expression of class I and IV histone deacetylases during epileptogenesis in mouse models of temporal lobe epilepsy. Exp Neurol 2015;273:92–104.
[65] Rossetti F, de Araujo Furtado M, Pak T, Bailey K, Shields M, Chanda S, et al. Combined diazepam and HDAC inhibitor treatment protects against seizures and neuronal damage caused by soman exposure. Neurotoxicology 2012;33(3):500–11.
[66] Reynolds JP, Miller-Delaney SFC, Jimenez-Mateos EM, Sano T, McKiernan RC, Simon RP, et al. Transcriptional response of polycomb group genes to status epilepticus in mice is modified by prior exposure to epileptic preconditioning. Front Neurol 2015;6:46.
[67] Shetty AK. Promise of resveratrol for easing status epilepticus and epilepsy. Pharmacol Therapeut 2011;131(3):269–86.
[68] Rusconi F, Paganini L, Braida D, Ponzoni L, Toffolo E, Maroli A, et al. LSD1 neurospecific alternative splicing controls neuronal excitability in mouse models of epilepsy. Cereb Cortex 2015;25(9):2729–40.
[69] Poeta L, Fusco F, Drongitis D, Shoubridge C, Manganelli G, Filosa S, et al. A regulatory path associated with X-linked intellectual disability and epilepsy links KDM5C to the polyalanine expansions in ARX. Am J Hum Genet 2013;92(1):114–25.
[70] Shoubridge C, Fullston T, Gecz J. ARX spectrum disorders: making inroads into the molecular pathology. Hum Mutat 2010;31(8):889–900.
[71] Crosio C, Heitz E, Allis CD, Borrelli E, Sassone-Corsi P. Chromatin remodeling and neuronal response: multiple signaling pathways induce specific histone H3 modifications and early gene expression in hippocampal neurons. J Cell Sci 2003;116(Pt 24):4905–14.
[72] Tsankova NM, Kumar A, Nestler EJ. Histone modifications at gene promoter regions in rat hippocampus after acute and chronic electroconvulsive seizures. J Neurosci 2004;24(24):5603–10.
[73] Crowe SL, Tsukerman S, Gale K, Jorgensen TJ, Kondratyev AD. Phosphorylation of histone H2A.X. as an early marker of neuronal endangerment following seizures in the adult rat brain. J Neurosci 2011;31(21):7648–56.
[74] Fiorenza A, Lopez-Atalaya JP, Rovira V, Scandaglia M, Geijo-Barrientos E, Barco A. Blocking miRNA biogenesis in adult forebrain neurons enhances seizure susceptibility, fear memory, and food intake by increasing neuronal responsiveness. Cereb Cortex 2016;26(4):1619–33.
[75] Song YJ, Tian XB, Zhang S, Zhang YX, Li X, Li D, et al. Temporal lobe epilepsy induces differential expression of hippocampal miRNAs including let-7e and miR-23a/b. Brain Res 2011;1387:134–40.
[76] Sempere LF, Freemantle S, Pitha-Rowe I, Moss E, Dmitrovsky E, Ambros V. Expression profiling of mammalian microRNAs uncovers a subset of brain-expressed microRNAs with possible roles in murine and human neuronal differentiation. Genome Biol 2004;5(3):R13.
[77] Shibata M, Nakao H, Kiyonari H, Abe T, Aizawa S. MicroRNA-9 regulates neurogenesis in mouse telencephalon by targeting multiple transcription factors. J Neurosci 2011;31(9):3407–22.
[78] Jimenez-Mateos EM, Arribas-Blazquez M, Sanz-Rodriguez A, Concannon C, Olivos-Ore LA, Reschke CR, et al. microRNA targeting of the P2X7 purinoceptor opposes a contralateral epileptogenic focus in the hippocampus. Sci Rep 2015;5:17486.
[79] Alsharafi W, Xiao B. Dynamic expression of microRNAs (183, 135a, 125b, 128, 30c and 27a) in the rat pilocarpine model and temporal lobe epilepsy patients. CNS Neurol Disord Drug Targets 2015;14(8): 1096–102.
[80] Xie N, Cui H, Banerjee S, Tan Z, Salomao R, Fu M, et al. miR-27a regulates inflammatory response of macrophages by targeting IL-10. J Immunol 2014;193(1):327–34.
[81] Shukla K, Sharma AK, Ward A, Will R, Hielscher T, Balwierz A, et al. MicroRNA-30c-2-3p negatively regulates NF-kappaB signaling and cell cycle progression through downregulation of TRADD and CCNE1 in breast cancer. Mol Oncol 2015;9(6):1106–19.
[82] Sano T, Reynolds JP, Jimenez-Mateos EM, Matsushima S, Taki W, Henshall DC. MicroRNA-34a upregulation during seizure-induced neuronal death. Cell Death Dis 2012;3:e287.

[83] Hu K, Xie YY, Zhang C, Ouyang DS, Long HY, Sun DN, et al. MicroRNA expression profile of the hippocampus in a rat model of temporal lobe epilepsy and miR-34a-targeted neuroprotection against hippocampal neurone cell apoptosis post-status epilepticus. BMC Neurosci 2012;13:115.
[84] Pogue AI, Cui JG, Li YY, Zhao Y, Culicchia F, Lukiw WJ. Micro RNA-125b (miRNA-125b) function in astrogliosis and glial cell proliferation. Neurosci Lett 2010;476(1):18–22.
[85] Tan CL, Plotkin JL, Veno MT, von Schimmelmann M, Feinberg P, Mann S, et al. MicroRNA-128 governs neuronal excitability and motor behavior in mice. Science 2013;342(6163):1254–8.
[86] Jimenez-Mateos EM, Bray I, Sanz-Rodriguez A, Engel T, McKiernan RC, Mouri G, et al. miRNA expression profile after status epilepticus and hippocampal neuroprotection by targeting miR-132. Am J Pathol 2011;179(5):2519–32.
[87] Magill ST, Cambronne XA, Luikart BW, Lioy DT, Leighton BH, Westbrook GL, et al. microRNA-132 regulates dendritic growth and arborization of newborn neurons in the adult hippocampus. Proc Natl Acad Sci USA 2010;107(47):20382–7.
[88] Jimenez-Mateos EM, Engel T, Merino-Serrais P, McKiernan RC, Tanaka K, Mouri G, et al. Silencing microRNA-134 produces neuroprotective and prolonged seizure-suppressive effects. Nat Med 2012;18(7):1087–94.
[89] Gaughwin P, Ciesla M, Yang H, Lim B, Brundin P. Stage-specific modulation of cortical neuronal development by Mmu-miR-134. Cereb Cortex 2011;21(8):1857–69.
[90] Schratt GM, Tuebing F, Nigh EA, Kane CG, Sabatini ME, Kiebler M, et al. A brain-specific microRNA regulates dendritic spine development. Nature 2006;439(7074):283–9.
[91] Kan AA, van Erp S, Derijck AAHA, de Wit M, Hessel EVS, O'Duibhir E, et al. Genome-wide microRNA profiling of human temporal lobe epilepsy identifies modulators of the immune response. Cell and Mol Life Sci 2012;69(18):3127–45.
[92] Siegel G, Obernosterer G, Fiore R, Oehmen M, Bicker S, Christensen M, et al. A functional screen implicates microRNA-138-dependent regulation of the depalmitoylation enzyme APT1 in dendritic spine morphogenesis. Nat Cell Biol 2009;11(6):705–16.
[93] Liu CM, Wang RY, Saijilafu, Jiao ZX, Zhang BY, Zhou FQ. MicroRNA-138 and SIRT1 form a mutual negative feedback loop to regulate mammalian axon regeneration. Genes Dev 2013;27(13):1473–83.
[94] Iyer A, Zurolo E, Prabowo A, Fluiter K, Spliet WG, van Rijen PC, et al. MicroRNA-146a: a key regulator of astrocyte-mediated inflammatory response. PLoS ONE 2012;7(9):e44789.
[95] Ashhab MU, Omran A, Kong HM, Gan N, He F, Peng J, et al. Expressions of tumor necrosis factor alpha and microRNA-155 in immature rat model of status epilepticus and children with mesial temporal lobe epilepsy. J Mol Neurosci 2013;51(3):950–8.
[96] Lippai D, Bala S, Csak T, Kurt-Jones EA, Szabo G. Chronic alcohol-induced microRNA-155 contributes to neuroinflammation in a TLR4-dependent manner in mice. PLoS ONE 2013;8(8):e70945.
[97] McKiernan RC, Jimenez-Mateos EM, Sano T, Bray I, Stallings RL, Simon RP, et al. Expression profiling the microRNA response to epileptic preconditioning identifies miR-184 as a modulator of seizure-induced neuronal death. Exp Neurol 2012;237(2):346–54.
[98] Kaalund SS, Veno MT, Bak M, Moller RS, Laursen H, Madsen F, et al. Aberrant expression of miR-218 and miR-204 in human mesial temporal lobe epilepsy and hippocampal sclerosis-convergence on axonal guidance. Epilepsia 2014;55(12):2017–27.
[99] Zucchini S, Marucci G, Paradiso B, Lanza G, Roncon P, Cifelli P, et al. Identification of miRNAs differentially expressed in human epilepsy with or without granule cell pathology. PLoS ONE 2014;9(8):e105521.
[100] Jimenez-Mateos EM, Engel T, Merino-Serrais P, Fernaud-Espinosa I, Rodriguez-Alvarez N, Reynolds J, et al. Antagomirs targeting microRNA-134 increase hippocampal pyramidal neuron spine volume in vivo and protect against pilocarpine-induced status epilepticus. Brain Struct Funct 2015;220(4):2387–99.
[101] Lee DY, Moon J, Lee ST, Jung KH, Park DK, Yoo JS, et al. Dysregulation of long noncoding RNAs in mouse models of localization-related epilepsy. Biochem Biophys Res Commun 2015;462(4):433–40.

[102] Ng SY, Bogu GK, Soh BS, Stanton LW. The long noncoding RNA RMST interacts with SOX2 to regulate neurogenesis. Mol Cell 2013;51(3):349–59.
[103] Tsai MC, Manor O, Wan Y, Mosammaparast N, Wang JK, Lan F, et al. Long noncoding RNA as modular scaffold of histone modification complexes. Science 2010;329(5992):689–93.
[104] Bond AM, VanGompel MJW, Sametsky EA, Clark MF, Savage JC, Disterhoft JF, et al. Balanced gene regulation by an embryonic brain ncRNA is critical for adult hippocampal GABA circuitry. Nat Neurosci 2009;12(8):1020–7.
[105] Oh YS, Gao P, Lee KW, Ceglia I, Seo JS, Zhang X, et al. SMARCA3, a chromatin-remodeling factor, is required for p11-dependent antidepressant action. Cell 2013;152(4):831–43.
[106] Suls A, Jaehn JA, Kecskes A, Weber Y, Weckhuysen S, Craiu DC, et al. De novo loss-of-function mutations in CHD2 cause a fever-sensitive myoclonic epileptic encephalopathy sharing features with Dravet syndrome. A J Hum Genet 2013;93(5):967–75.
[107] Lund C, Brodtkorb E, Oye AM, Rosby O, Selmer KK. CHD2 mutations in Lennox–Gastaut syndrome. Epilepsy Behav 2014;33:18–21.
[108] Chenier S, Yoon G, Argiropoulos B, Lauzon J, Laframboise R, Ahn JW, et al. CHD2 haploinsufficiency is associated with developmental delay, intellectual disability, epilepsy and neurobehavioural problems. J Neurodev Disord 2014;6(1):9.
[109] Allen AS, Berkovic SF, Cossette P, Delanty N, Dlugos D, Eichler EE, et al. De novo mutations in epileptic encephalopathies. Nature 2013;501(7466):217–21.
[110] Kosho T, Okamoto N, Int C-SS. Genotype-phenotype correlation of coffin-siris syndrome caused by mutations in SMARCB1, SMARCA4, SMARCE1, and ARID1A. Am J Med Genet Part C 2014;166(3):262–75.
[111] Ooi L, Belyaev ND, Miyake K, Wood IC, Buckley NJ. BRG1 chromatin remodeling activity is required for efficient chromatin binding by repressor element 1-silencing transcription factor (REST) and facilitates REST-mediated repression. J Biol Chem 2006;281(51):38974–80.
[112] Mellers JDC. Epilepsy. In: David AS, Fleminger S, Kopelman MD, Lovestone S, Mellers JDC, editors. Lishman's Organic Psychiatry. 4th ed. Chichester: Wiley-Blackwell; 2009. p. 309–95.
[113] Gaitatzis A, Trimble MR, Sander JW. The psychiatric comorbidity of epilepsy. Acta Neurol Scand 2004;110:207–20.

CHAPTER

EPIGENETIC DYSREGULATION IN BRAIN TUMORS AND NEURODEVELOPMENT

13

M.M. Hefti*, N. Tsankova*,,†**

*Department of Pathology, Icahn School of Medicine at Mount Sinai, New York, NY, United States; **Department of Neuroscience, Icahn School of Medicine at Mount Sinai, New York, NY, United States; †Friedman Brain Institute, Icahn School of Medicine at Mount Sinai, New York, NY, United States

CHAPTER OUTLINE

13.1 INTRODUCTION

The realization that genomic activity can be regulated through chemical modification of the DNA molecule, or "epi"-genetics, has produced great insights into the role of such mechanisms during normal cell fate development and in aberrant neoplastic growth. The wide array of neurological abnormalities seen in patients with inborn errors of epigenetic regulation, such as alpha-thalassemia X-linked mental retardation (ATRX) or Coffin–Siris syndromes, demonstrate the key role of such mechanisms in the development of the human central nervous system (CNS) [1]. The ability of

Neuropsychiatric Disorders and Epigenetics. http://dx.doi.org/10.1016/B978-0-12-800226-1.00013-7

epigenetic modifications to activate and inactivate specific genes has led to the supposition that epigenetic dysregulation in carcinogenesis may involve the deactivation of tumor suppressors and/or the activation of oncogenes [2].

While these mechanisms have been studied in many different tumor types, their relationship to CNS tumors is of particular interest given that these tumors arise in the same anatomical structures, and often at the same time, as the genetically related cognitive and neurodevelopmental disorders. Primary brain tumors are difficult to remove surgically, due to their infiltrative nature and proximity to vital neuroanatomical structures, making it critical to understand better mechanisms driving their growth in the hope of identifying potential new therapeutic targets.

Diffusely infiltrating gliomas, the most common primary brain tumors, have characteristic global DNA methylation abnormalities, and such epigenetic signatures now assist the histological classification of specific tumor subtypes [3]. Furthermore, many gliomas carry specific mutations in genes involved in chromatin remodeling, some of which appear to be early, if not primary, oncogenic drivers of tumor growth [2,3]. While mutations in epigenetic factors have not been shown to be sufficient by themselves to initiate gliomagenic transformation, CNS tumors display remarkable spatial and temporal restriction, indicating an intimate association with niche-dependent tumor precursors cells that can acquire specific mutations during predictable vulnerable periods (Fig. 13.1). This has not only revolutionized the practice of neuropathology by introducing new diagnostic and prognostic markers, but has also opened new avenues for research into these intractable and often lethal tumors.

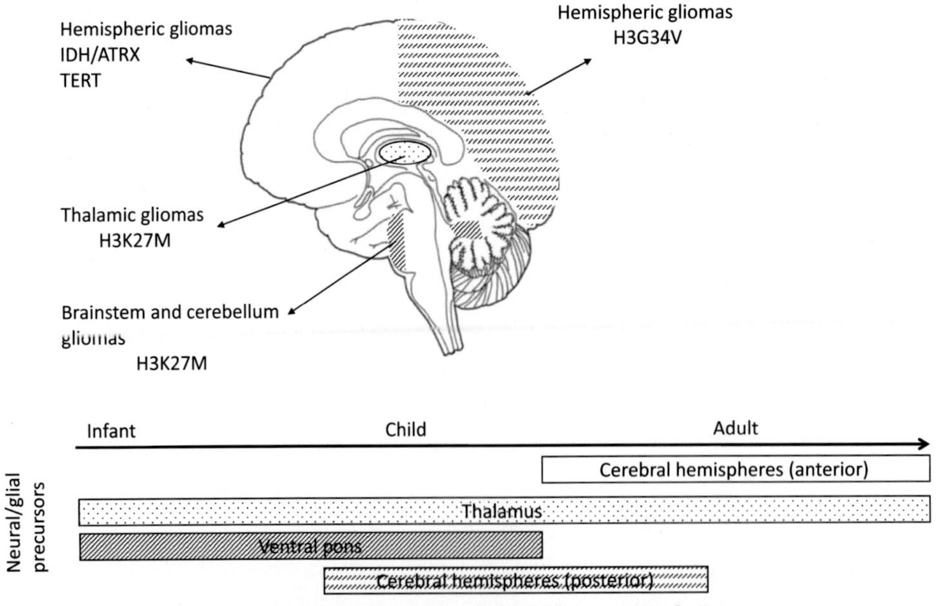

FIGURE 13.1 Spatial and Temporal Relationship Between Epimutations and Brain Tumors

13.2 EPIGENETIC DYSREGULATION IN PEDIATRIC TUMORS

13.2.1 H3K27M AND H3G34R MUTATIONS IN PEDIATRIC GLIOMAS

The most common site for pediatric solid tumors is the brain and the most common group of such tumors are those phenotypically resembling, and thought to be derived from, astrocytic cell lineages (i.e., astrocytomas). With the exception of some specific low-grade tumor subtypes, such as pilocytic astrocytomas and glioneuronal tumors, which tend to be well circumscribed and have an excellent prognosis, the majority of pediatric brain tumors are infiltrative, progressive, and ultimately fatal. A large subset of pediatric gliomas arise from the brainstem, most commonly the ventral pons, and are collectively referred to as diffuse intrinsic pontine gliomas (DIPG). These tumors are characteristic of the pediatric population, carry a dismal prognosis given their unresectable nature, and are resistant to treatment. Similarly to adult gliomas, pediatric gliomas can also arise in the cerebral hemispheres, where unique epigenetic mutations distinguish them from adult peripheral tumors (see further subsequent sections). Their highly restricted presentation, both geographically and temporally, leads to the supposition that they arise from pools of residual undifferentiated or lineage-restricted precursors cells that are seen at specific developmental stages [4] and also to great interest in identifying specific mechanisms that might trigger their development.

Recent molecular data suggests that pediatric gliomas have global epigenetic abnormalities due to recurrent hotspot mutations within genes encoding specific histone H3 variants. The majority of pediatric DIPGs show hotspot mutations in the *H3F3A* or *HIST1H3B/C* genes, encoding histone variants H3.3 or H3.1, respectively, all leading to a histone variant with a lysine (K) to a methionine (M) substitution, known as H3K27M. This mutation is uniquely present only in midline infiltrating gliomas, and occurs predominantly in children, although it is occasionally also seen in adults [5–7]. In children, these are generally located in the brainstem, while in adults they are more commonly seen in the cerebellum, spinal cord, and thalamus [6,7]. The morphology of such tumors can vary significantly, including histological features resembling classical astrocytomas, primitive neuroectodermal tumors (PNETs), and even pleomorphic xanthoastrocytomas (PXAs), among others [8]. Their exclusive midline location suggests that these tumors are derived from a midline precursor that retains or acquires an immortalized stem cell/pluripotency phenotype.

H3K27M-mutant gliomas have a profoundly altered epigenetic landscape. The mutated H3K27M histone variant binds to the repressive polycomb complex PRC2/EZH2, inhibiting its normal ability to write trimethylated H3K27me3 marks elsewhere in the genome. The resultant global hypomethylation state causes abnormal expression of several hundred genes [9], including members of the sonic hedgehog (SHH) and Myc signaling pathways, the relative proportions of which seems to carry prognosis implications [10].

The specific neuroanatomical location of these tumors is intriguing, and a subject of ongoing investigation. Evidence from animal models so far suggests that the site-specificity of pediatric tumors is likely a combination of a vulnerable precursor cell population and a specific microenvironment [11]. While such models do not fully recapitulate human tumors, they provide a tractable and readily available means of studying carcinogenesis in vivo. Several mouse studies have demonstrated that histone H3K27M mutations, induced in combination with P53 mutation and/or PDGFR overexpression in Nestin-positive brainstem precursor mouse cells, lead to the formation of high grade gliomas histologically similar to those seen in humans [12]. In these experiments, H3K27M appears to be necessary but not sufficient to produce brainstem tumors in both animal and in vitro models [12,13].

Upon transfection with H3K27M and either PDGFRA overexpression or P53 knockdown, human neural progenitor cells derived from human embryonic stem cells (hESc) cells acquire many of the characteristic traits of pediatric DIPGs, including high proliferative index, radiation resistance, and increased migration [14]. They are also able to form tumor xenografts when injected into the mouse brainstem. Interestingly however, this transformation is highly specific to a particular developmental stage. Neither more primitive hESc, nor more differentiated hESc-derived astrocytes respond to transfection with H3K27M [14].

Human neuroanatomical studies suggest the presence of a SHH-responsive Nestin and OLIG2+ neuronal precursor cell in the ventral pons during the key time period for the development of pediatric brainstem gliomas. These may represent the vulnerable in vivo population corresponding to the neuronal precursor cells used in the above-described in vitro experiments [4]. Other candidate populations include precursor cells of the 4th ventricle subventricular zone of the dorsolateral pons which, when transfected with PDGFR-containing viral vectors, can produce tumors remarkably similar to human pediatric brainstem gliomas [15].

It is also possible to grow xenografts of human pediatric brainstem gliomas in immunodeficient mice [12,16]. Xenografts of human brainstem glioma cells are best able to generate tumors when implanted into the murine brainstem rather than subcutaneously or elsewhere in the brain, suggesting the need for specific micro environmental stimuli for tumor progression [11,17]. Interestingly, some of these experiments have shown that certain human brainstem glioma-derived cell lines are capable of inducing the formation of brainstem tumors of murine rather than of human origin in mouse models [11].

While H3K27M is the most frequent and best characterized mutation in pediatric midbrain gliomas, a similar but distinct histone H3.3 mutation, H3G34R, has been discovered in pediatric gliomas arising from the peripheral hemispheres [5] (Fig. 13.1). The effect of H3G34R mutations on the global epigenetic landscape in hemispheric gliomas is less well understood compared to H3K27M mutant ones, and is a subject of ongoing studies [18]. Interestingly, H3G34R-mutant tumors show a gene expression pattern driven by abnormal methylation, similar to that of the developing forebrain [19]. Although relatively more genetically silent than adult gliomas [20], histone-mutant pediatric gliomas of either type frequently have concurrent mutations in other oncogenes and tumor suppressors, such as p53, various cyclin genes, and interestingly, ATRX, which is itself involved in chromatin remodeling [5,21,22]. In contrast to adult gliomas, ATRX loss in pediatric gliomas is not accompanied by isocitrate dehydrogenase (IDH) mutations [5].

The emerging role of histone mutations in midline gliomas is rapidly being translated into neuropathological diagnostic practice. Antibodies to both H3K27M and its affected trimethylated H3K27me3 mark are now available and correlate well with sequencing results [7]. The H3K27M antibody in particular is becoming routine for the evaluation of midline tumors at all ages and is particularly helpful in cases where only small biopsies at the edge of a tumor are available for evaluation, as is often the case in tumors located in critical areas, such as the brainstem, the thalamus, and the spinal cord (Fig. 13.2A).

13.2.2 INI-1 AND OTHER SWI/SNF COMPLEXES IN ATYPICAL TERATOID RHABDOID TUMORS

One of the first chromatin remodeling genes identified in human cancers was INI-1 (SMARCB1, SNF5, BAF47), a member of the SWI/SNF chromatin remodeling complex. First identified in malignant rhabdoid tumors [23], it was subsequently identified as a germline mutation in children with

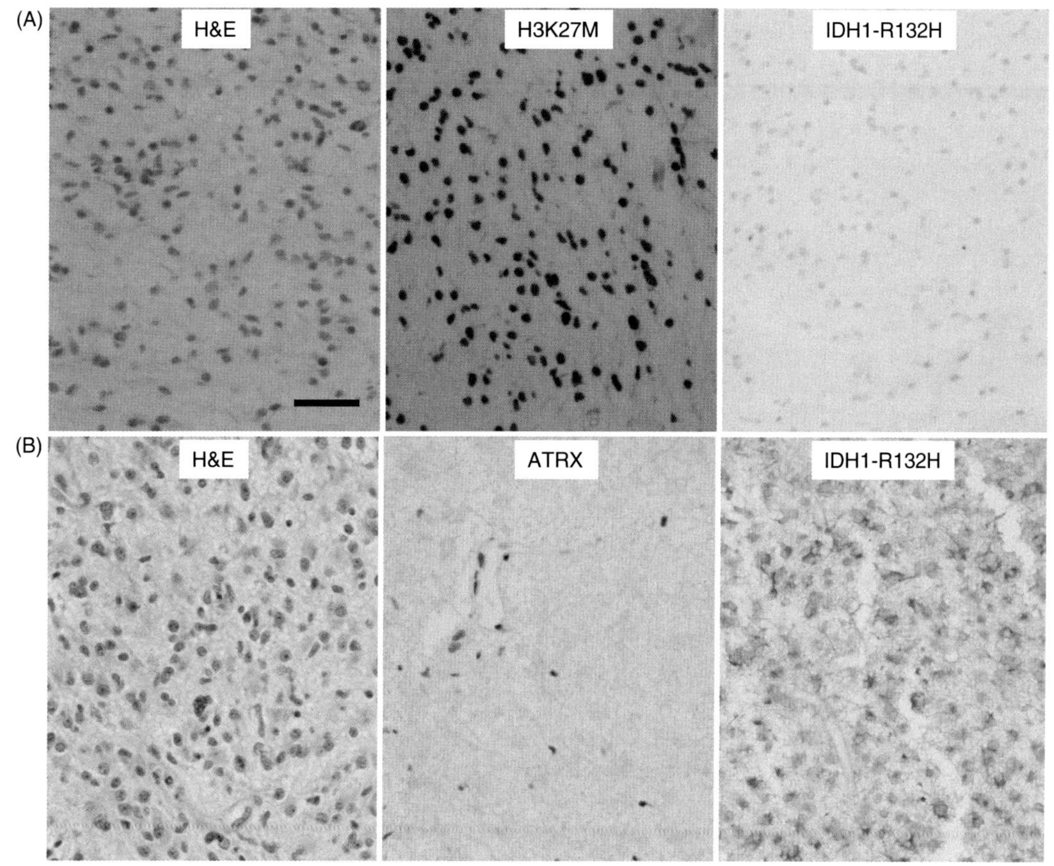

FIGURE 13.2 Analysis for H3K27M, ATRX, and IDH1 Mutations by Immunohistochemistry in Diagnostic Neuropathology

(A) Midline thalamic high-grade glioma (H&E) with positive nuclear staining for mutant H3K27M but wild-type for IDH1-R132H. (B) Anaplastic astrocytoma (H&E) with loss of ATRX nuclear staining indicating deletional mutation (notice wild-type nuclear staining in endothelial cells) and strong cytoplasmic staining for mutant IDH1-R132H protein.

multiple rhabdoid tumors [24,25] and then as a sporadic mutation in atypical teratoid rhabdoid tumors (AT/RT) [26]. With the development of reliable antibodies against INI-1, the loss of nuclear expression of INI-1 has become the defining characteristic of rhabdoid tumors in general, and AT/RTs in particular in the CNS [27,28]. Subsequently, a few AT/RT tumors have been associated with a mutation of another SWI/SNF protein, SMARCA4 (BRG1), rather than with INI-1 [29]. It is intriguing that AT/RT tumors, seen predominantly in infants, show very few, if any, other genomic alterations, and are almost exclusively driven by chromatin remodeling abnormalities due to their abnormal SWI/SNF machinery [30]. In vitro, the loss of INI-1 has been shown to prevent the expression of genes necessary for neural differentiation, while its overexpression can induce differentiation [31].

13.2.3 EMERGING EPIGENETIC MUTATIONS IN MEDULLOBLASTOMAS

Medulloblastoma is a primitive tumor arising in the cerebellum of children and young adults. With the advent of next generation sequencing, detailed genomic characterization of medulloblastomas has led to their molecular subclassification and has revealed remarkable insight into specific tumor subtype cell(s) of origin [32]. Genomic and transcriptomic data have subdivided medulloblastomas into four groups with distinct molecular drivers and clinical behavior: SHH-type, WNT-type, Group 3, and Group 4. Similar to gliomas, each subtype of medulloblastoma resembles phenotypically distinct precursor cell population(s) and at a specific age, implying distinct cellular origins [33]. SHH-type tumors, for example, arise from the lower rhombic lip at any age, but more frequently in older children and adults, while WNT-type tumors arise in the lower rhombic lip in younger children [34]. In addition to unique gene expression phenotype [35], each medulloblastoma molecular subgroup has a distinctive DNA methylation pattern. Given the difficulty in extracting RNA from paraffin embedded tissue and variable results using immunohistochemistry for substratification [36–38], DNA methylation profiling is rapidly becoming the preferred method for diagnostic subtyping of these tumors [39,40].

Aberrant methylation patterns have long been recognized in medulloblastomas [41] and more recently there is an increasing appreciation of the key role of epigenetics in their tumorigenesis [42]. In sequencing studies, approximately one-third of medulloblastomas show mutations in chromatin remodeling genes, such as MLL2/3, KDM6A and SMARCA4, although each individual mutation is relatively rare [43]. The impact of these mutations on the global epigenetic landscape and their correlation to histological subtypes and overall prognosis are beginning to emerge [44].

13.3 EPIGENETIC DYSREGULATION IN ADULT TUMORS

13.3.1 IDH MUTATIONS IN ADULT GLIOMAS

Mutations in the Krebs cycle enzyme IDH types 1 and 2 were first identified in hematologic and neurologic malignancies [45–48]. Mutations in IDH1/2 result in a gain-of-function mutant protein (IDH^{MUT}), which produces the oncometabolite 2-hydroxyglutarate, instead of the usual Krebs cycle by-product alpha-ketoglutarate produced by wild-type IDH [49]. Intriguingly, high levels of 2-hydroxyglutarate have been shown to induce global hypermethylation within the tumor epigenome [50,51], also known as the global CpG island methylator phenotype (G-CIMP), by acting as inhibitors for histone and DNA demethylase enzymes [52] (Fig. 13.3). These mutations are thought to represent early lesions in gliomagenesis [53], and are seen almost exclusively in peripheral adult diffusely infiltrating gliomas, such as grade II/III astrocytomas and oligodendrogliomas, as well as in glioblastomas (GBM) arising from an IDH-mutant astrocytoma (secondary GBM) [48,54,55] (Fig. 13.3). Interestingly, these tumors show remarkable transcriptional resemblance to oligodendrocyte progenitors residing in the subcortical white matter, based on which they have been subclassified as "proneural" gliomas [56]. While the presence of high levels of 2-hydroxyglutarate is thought to be oncogenic, the G-CIMP phenotype secondary to IDH mutations appears to be a more favorable prognostic indicator [57]. The development of a specific antibody for the most common IDH mutant protein, IDH1-R132H, has made molecular assessment for this mutation a routine part of neuropathology practice [58] (Fig. 13.2B). In fact, assessment for mutations in IDH1-R132H and ATRX are becoming a key part of the current clinical consensus guidelines for the classification of glial tumors [59].

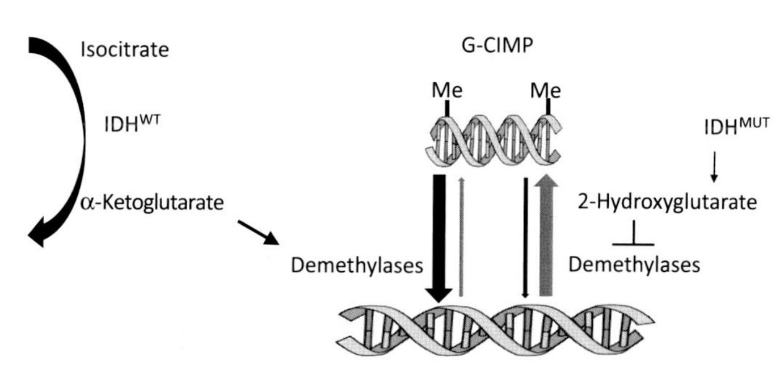

FIGURE 13.3 Proposed Mechanism for Hypermethylation in IDH-Mutant Gliomas

Wild-type IDH (IDH^{WT}) normally converts isocitrate to α-ketoglutarate as part of the Krebs cycle. The metabolite α-ketoglutarate is also a necessary cofactor for DNA and histone demethylases. In contrast, mutant IDH1 and IDH2 proteins (IDH^{MUT}) produce the alternate metabolite 2-hydroxyglutarate, which leads to inhibition of demethylases and a resultant increased global DNA and histone methylation (G-CIMP phenotype).

13.3.2 ATRX MUTATIONS IN ADULT GLIOMAS

More recently, mutations in ATRX, another key component of the SWI/SNF chromatin remodeling complex and previously implicated in the alpha-thalassemia mental retardation syndrome, have been identified in a subset of pediatric gliomas, as well as in adult lower grade (WHO Grade II and III) astrocytomas and the resultant secondary GBM (WHO Grade IV) (Table 13.1) [60]. While ATRX mutations in pediatric high-grade gliomas are invariably IDH wild-type, in adult gliomas they are almost exclusively seen in proneural IDH-mutant tumors [61,62]. ATRX combined with DAXX is involved in the incorporation of histone H3.3 variant into telomeric chromatin. The dysfunction of this complex facilitates telomere elongation by homologous recombination rather than the more common upregulation of telomerase [63]. This in turn permits cells to bypass the pro-apoptotic senescence signals triggered by telomere shortening [64].

Interestingly, like IDH mutations, but unlike other chromatin remodeling mutations, such as INI-1 or H3K27M, ATRX mutations in adults carry a relatively good prognosis [65]. They may confer susceptibility to inhibitors of the ATR protein kinase regulating homologous recombination [66]. ATRX mutations are seen in tumors with astrocytoma morphology, often along with TP53 mutations, and are mutually exclusive with codeletion of 1p/19q chromosomes characteristic of oligodendrogliomas [67]. This has been shown even in rare tumors with biphasic morphology [68]. In conjunction with IDH mutations, ATRX immunohistochemistry has permitted tumors previously classified as oligoastrocytomas to be relatively easily reclassified as one or the other histological subtypes [59,67,69–71].

13.3.3 OTHER EMERGING EPIGENETIC MUTATIONS IN ADULT GLIOMAS

Similarly to ATRX, the telomerase reverse transcriptase (TERT) promoter gene, which encodes the catalytic subunit of telomerase [72], is frequently mutated in gliomas and causes abnormalities in telomere lengthening. *TERT* promoter mutations appear to be associated with tumors arising in any tissue with a low rate of self-renewal and are mutually exclusive with ATRX mutations across all tumor types

Table 13.1 Central Nervous System Tumors and Developmental Disorders With Reported Epimutations

		Somatic Mutations in Tumors			Germline Mutations
Gene		Location	Histology	Ages	Developmental Syndromes
H3F3A/HIST1H3B/ C-H3K27M	+/− ATRX	Brainstem	Astrocytoma/ glioblastoma	Pediatric	
		Thalamus, cerebellum, spinal cord	Astrocytoma/ glioblastoma	Both	
H3F3A-H3G34R	With ATRX	Cerebral hemispheres	Astrocytoma/ glioblastoma	Both	
IDH1/2	With del(1p/19q)	Cerebral hemispheres	Oligodendroglioma	Adult	
ATRX	With IDH1/2	Cerebral hemispheres	Astrocytoma/ 2′ glioblastoma	Adult	Alpha thalassemia/X-linked mental retardation (ATRX) syndrome
TERT promoter		Cerebral hemispheres	1′ Glioblastoma, oligodendroglioma	Adult	
MLL2/3		Cerebellum	Medulloblastoma	Pediatric	
KDM6A		Cerebellum	Medulloblastoma	Pediatric	Coffin–Siris syndrome
SWI/SNF	SMARCA4	Cerebellum	Medulloblastoma	Pediatric	Nicolaides–Baraitser syndrome
	SMARCA1	Variable	AT/RT	Infant	Sotos syndrome
	INI-1	Variable	AT/RT	Infant	Rubinstein–Taiby syndrome

[72]. Unlike ATRX, TERT mutations are found almost exclusively in adult patients [73]. Interestingly, mutations within the *TERT* promoter are seen in both oligodendrogliomas, relatively less aggressive diffuse gliomas, and in primary GBM with epidermal growth factor receptor (EGFR) amplification, which are very aggressive and associated with dismal prognosis [74–76].

Bromodomain and extraterminal domain (BET) proteins, such as the BRD family (BRD2, BRD3, and BRD4) are upregulated in GBM and regulate the expression of long noncoding RNAs (lncRNAs), such as HOX transcript antisense RNA (HOTAIR) [77], which in turn activate a tumorigenic, growth-promoting gene expression profile [78]. Emerging data suggests that multiple other chromatin remodeling genes, including EZH2 [79], mixed lineage leukemia (MLL) [80], and CHD5 [81] are abnormally expressed in gliomas. How their dysregulation is related to tumorigenesis and/or tumor progression remains to be determined. It is intriguing to speculate that at least a subset of these epigenetic lesions predispose vulnerable neural/glial precursors cells toward a transformative oncogenic fate over time [3,82,83].

13.4 TUMOR CELL OF ORIGIN AND EPIGENETICS

As early as the 1920s, Bailey suggested that the majority of the brain tumors he described were the result of aberrant development of normal precursors. He hypothesized, for example, that the tumor he described as medulloblastoma was derived from a primitive cerebellar precursor cell he termed it as medulloblast [84]. While the existence of the medulloblast has been largely disproven, molecular data has shown remarkable genetic, transcriptomic, epigenetic, and functional similarities between cells derived from CNS tumors and normal CNS precursor cells, reinforcing the concept that gliomas can arise from various stem and progenitor cells in the CNS at different stages of their lineage differentiation [2,56,85,86].

Not surprisingly, given their role in the development and progression of tumors, epigenetic mechanisms play a significant role in regulating the differentiation of precursor cells, ensuring that they differentiate, and migrate appropriately to generate the complex structure of the human brain. EGFR is overexpressed in germinal matrix precursor cells and in most adult diffusely infiltrating gliomas, even in the absence of gene mutation or amplification, but is silenced in differentiated cortical neurons and glia. This expression appears to be at least in part regulated through locus-specific dynamic chromatin remodeling of H3K27 acetylation, H3K4 trimethylation, and/or H3K27 trimethylation [83]. Similarly, expression of the stem cell marker CD133 is regulated by promoter methylation and transcription factor exclusion in glioma stem cells and cell lines [87]. These and other data suggest that in parallel to genetic instability, epigenetic mechanisms may play a broader role in the development of a wider array of CNS tumors.

13.5 EMERGING EPIGENETIC THERAPIES

The emerging data on the role of epigenetics in cancer has led to considerable interest in these molecules associated with epigenetics as drug targets [3]. In vitro data suggests that tumors with ATRX mutations and the ALT phenotype are particularly susceptible to ATR inhibitors [66]. Similar data suggests that IDH mutant G-CIMP phenotype tumors may be susceptible to DNMT inhibitors [88]. HDAC inhibitors have already been tested as radiation sensitizers in glioma [89] and in vivo studies show that vorinostat (SAHA) and panobinostat may partially reverse the loss of H3K27M trimethylation in H3K27M mutant tumors [90] and are currently being tested in glioma clinical trials [91].

13.6 EPIMUTATIONS AND NEUROCOGNITIVE DYSFUNCTION VERSUS CARCINOGENESIS

One of the particularly interesting themes in tumor epigenetics is that many of the genes mutated in tumors are also seen, as germline mutations, in intellectual disability syndromes (Table 13.1). For example, somatic mutations in any one of several proteins in the SWI/SNF complex lead to tumor formation, but germline mutations in the same or similar genes can cause intellectual disability syndromes, such as Coffin–Siris and Nicolaides–Baraitser syndromes [92]. While some syndromes, such as Sotos (EZH2) or Rubinstein–Taiby (CREBBP and EP300) demonstrate both cognitive dysfunction and cancer predisposition, there are no definite reports of cancer predisposition in the Coffin–Siris syndrome [93].

How mutations in the same gene can lead to such variable phenotypes remains unclear. For example, one specific germline mutation in the SMARCB1 SNF5 (INI-1) domain (p.Arg377His) seen in Coffin–Siris syndrome [94] is also detected as a somatic mutation in meningiomas [95]. Other mutations in the same gene are reported in cases of schwannomatosis [96], rhabdoid predisposition syndrome [97], and as somatic mutations in atypical teratoid/rhabdoid tumors [98]. In some genes, such as SMARCB1, somatic mutations in cancers tend to be missense mutations, causing less dramatic changes in protein function, whereas developmental mutations tend to be truncations [92]. This is however, not the case with most other genes studied and does not explain why patients with the truncated, nonfunctional proteins do not have a profound predisposition to malignancy in addition to global neurodevelopmental dysfunction.

This complex interplay of developmental abnormalities, cancer predisposition, and somatic mutations in cancer demonstrates the key role of epigenetics in all three processes. The mechanisms by which these individual mutations have such widely varying effects remain unclear and are an active area of research in both neurodevelopment and CNS tumor biology.

13.7 NEUROPSYCHIATRIC ASPECTS OF BRAIN TUMORS

In relation to the present book, patients with brain tumors are well known to have psychiatric symptoms. Indeed, occasionally the earliest clinical manifestations may be psychiatric symptoms alone [99]. The study of psychiatric symptoms in patients with brain tumors is complicated by the fact that these symptoms can occur directly due to the tumor, and also as a psychological response to the tumor [100]. From the clinical standpoint, psychiatric symptoms are usually not of use for diagnosing the location or nature of the tumor. Neurological signs and neuroimaging are far superior in this regard [99]. Psychiatric symptoms can involve any aspect of psychological functions. Disturbance of cognitive function is the most commonly seen psychiatric change. In minor form, this may present as decreased capacity to attend and concentrate, faulty memory, and easy fatigability. More severe cognitive impairment may present in the form of dementia [99]. Anxiety, sleep disturbances, apathy, and depression are common in patients with brain tumors. There can also be psychotic symptoms in the form of hallucinations in any modality, commonly as part of an epileptic seizure, but also without epileptic seizures. Delusions may also be present, either early during the illness, or late during the illness. So far, it has not been possible to identify precise correlations between tumor location or histological type of tumor and the likelihood of psychiatric symptoms. However, patients with tumors in the frontal and temporal lobes show a somewhat higher frequency of psychiatric problems than patients with tumors involving the parietal or occipital lobes [99].

13.8 CONCLUSIONS

The recent explosion of available data on the molecular biology of brain tumors in both adults and children has increased greatly our appreciation of the role of abnormal epigenetic mechanisms in brain tumorigenesis, and is refining the classification of brain tumors for diagnostic and prognostic purposes. Although still in their early stages of development, many proposed therapies are currently being investigated to restore normal DNA methylation patterns in gliomas, making them more susceptible to treatment. The role of epigenetics in brain tumors suggests an intimate relationship between normal development and carcinogenesis in the brain, which may be also exploited to curtail tumorigenic glioma growth, such as through the induction of terminal cell fate differentiation in tumor cells. Similar or even identical mutations in the same gene(s) can lead to tumors or cognitive delay, depending on whether they are somatic or germline. While the mechanisms by which this occurs have yet to be elucidated, this preliminary data suggests a rich field for future inquiries into the roles of epigenetic regulation in both cognitive development and in tumorigenesis.

ABBREVIATIONS

AT/RT Atypical teratoid/rhabdoid tumor
ATRX Alpha-thalasaemia–related X-linked mental retardation
CNS Central nervous system
DAXX Death domain-associated protein
DIPGs Diffuse intrinsic pontine gliomas
EGFR Epidermal growth factor receptor
GBM Glioblastoma
hESc Human embryonic stem cells
IDH Isocitrate dehydrogenase
SHH Sonic hedgehog
TERT Telomerase reverse transcriptase

GLOSSARY

Astrocytoma A primary brain tumor, whose neoplastic glioma cells show morphological resemblance to astrocytes
Atypical teratoid rhabdoid tumor An uncommon, aggressive CNS tumor which in the vast majority of cases occur in children less than two years old
Glioma A malignant tumor arising from glial cells in the brain
Medulloblastoma A malignant tumor of the brain arising in the cerebellum, especially in children
Myc signaling pathway A signaling pathway that plays an important role in cell proliferation, differentiation, transformation, and apoptosis
Oligodendroglioma A primary brain tumor, whose neoplastic glioma cells show morphological resemblance to oligodendroglial cells
Oncogene A gene which has the potential of making a normal cell cancerous
Sonic hedgehog gene A gene coding for a protein called Sonic hedgehog that functions as a chemical signal essential for embryonic development
Xenograft A graft of tissue taken from a donor of one species and grafted into a recipient belonging to another species

REFERENCES

[1] Burton A, Torres-Padilla ME. Chromatin dynamics in the regulation of cell fate allocation during early embryogenesis. Nat Rev Mol Cell Biol 2014;15(11):723–34.
[2] Tsankova NM, Canoll P. Advances in genetic and epigenetic analyses of gliomas: a neuropathological perspective. J Neurooncol 2014;119(3):481–90.
[3] Yong RL, Tsankova NM. Emerging interplay of genetics and epigenetics in gliomas: a new hope for targeted therapy. Semin Pediatr Neurol 2015;22(1):14–22.

[4] Monje M, Mitra SS, Freret ME, Raveh TB, Kim J, Masek M, et al. Hedgehog-responsive candidate cell of origin for diffuse intrinsic pontine glioma. Proc Natl Acad Sci USA 2011;108(11):4453–8.
[5] Sturm D, Witt H, Hovestadt V, Khuong-Quang DA, Jones DT, Konermann C, et al. Hotspot mutations in H3F3A and IDH1 define distinct epigenetic and biological subgroups of glioblastoma. Cancer Cell 2012;22(4):425–37.
[6] Aihara K, Mukasa A, Gotoh K, Saito K, Nagae G, Tsuji S, et al. H3F3A K27M mutations in thalamic gliomas from young adult patients. Neuro Oncol 2014;16(1):140–6.
[7] Bechet D, Gielen GG, Korshunov A, Pfister SM, Rousso C, Faury D, et al. Specific detection of methionine 27 mutation in histone 3 variants (H3K27M) in fixed tissue from high-grade astrocytomas. Acta Neuropathol 2014;128(5):733–41.
[8] Solomon DA, Wood MD, Tihan T, Bollen AW, Gupta N, Phillips JJ, et al. Diffuse midline gliomas with histone H3-K27M mutation: a series of 47 cases assessing the spectrum of morphologic variation and associated genetic alterations. Brain Pathol 2015;.
[9] Bender S, Tang Y, Lindroth AM, Hovestadt V, Jones DT, Kool M, et al. Reduced H3K27me3 and DNA hypomethylation are major drivers of gene expression in K27M mutant pediatric high-grade gliomas. Cancer Cell 2013;24(5):660–72.
[10] Saratsis AM, Kambhampati M, Snyder K, Yadavilli S, Devaney JM, Harmon B, et al. Comparative multidimensional molecular analyses of pediatric diffuse intrinsic pontine glioma reveals distinct molecular subtypes. Acta Neuropathol 2014;127(6):881–95.
[11] Caretti V, Sewing AC, Lagerweij T, Schellen P, Bugiani M, Jansen MH, et al. Human pontine glioma cells can induce murine tumors. Acta Neuropathol 2014;127(6):897–909.
[12] Misuraca KL, Cordero FJ, Becher OJ. Preclinical models of diffuse intrinsic pontine glioma. Front Oncol 2015;5:172.
[13] Lewis PW, Muller MM, Koletsky MS, Cordero F, Lin S, Banaszynski LA, et al. Inhibition of PRC2 activity by a gain-of-function H3 mutation found in pediatric glioblastoma. Science 2013;340(6134):857–61.
[14] Funato K, Major T, Lewis PW, Allis CD, Tabar V. Use of human embryonic stem cells to model pediatric gliomas with H3.3K27M histone mutation. Science 2014;346(6216):1529–33.
[15] Masui K, Suzuki SO, Torisu R, Goldman JE, Canoll P, Iwaki T. Glial progenitors in the brainstem give rise to malignant gliomas by platelet-derived growth factor stimulation. Glia 2010;58(9):1050–65.
[16] Aoki Y, Hashizume R, Ozawa T, Banerjee A, Prados M, James CD, et al. An experimental xenograft mouse model of diffuse pontine glioma designed for therapeutic testing. J Neurooncol 2012;108(1):29–35.
[17] Subashi E, Cordero FJ, Halvorson KG, Qi Y, Nouls JC, Becher OJ, et al. Tumor location, but not H3.3K27M, significantly influences the blood-brain-barrier permeability in a genetic mouse model of pediatric high-grade glioma. J Neurooncol 2016;126(2):243–51.
[18] Baker SJ, Ellison DW, Gutmann DH. Pediatric gliomas as neurodevelopmental disorders. Glia 2015;879–95.
[19] Bjerke L, Mackay A, Nandhabalan M, Burford A, Jury A, Popov S, et al. Histone H3.3. mutations drive pediatric glioblastoma through upregulation of MYCN. Cancer Discov 2013;3(5):512–9.
[20] Jones C, Baker SJ. Unique genetic and epigenetic mechanisms driving paediatric diffuse high-grade glioma. Nat Rev Cancer 2014;14(10.):651–61.
[21] Wu G, Diaz AK, Paugh BS, Rankin SL, Ju B, Li Y, et al. The genomic landscape of diffuse intrinsic pontine glioma and pediatric nonbrainstem high-grade glioma. Nat Genet 2014;46(5):444–50.
[22] Schwartzentruber J, Korshunov A, Liu XY, Jones DT, Pfaff E, Jacob K, et al. Driver mutations in histone H3.3 and chromatin remodelling genes in paediatric glioblastoma. Nature 2012;482(7384):226–31.
[23] Versteege I, Sevenet N, Lange J, Rousseau-Merck MF, Ambros P, Handgretinger R, et al. Truncating mutations of hSNF5/INI1 in aggressive paediatric cancer. Nature 1998;394(6689):203–6.
[24] Biegel JA, Fogelgren B, Wainwright LM, Zhou JY, Bevan H, Rorke LB. Germline INI1 mutation in a patient with a central nervous system atypical teratoid tumor and renal rhabdoid tumor. Genes Chromosomes Cancer 2000;28(1):31–7.

[25] Taylor MD, Gokgoz N, Andrulis IL, Mainprize TG, Drake JM, Rutka JT. Familial posterior fossa brain tumors of infancy secondary to germline mutation of the hSNF5 gene. Am J Hum Genet 2000;66(4):1403–6.
[26] Biegel JA, Fogelgren B, Zhou JY, James CD, Janss AJ, Allen JC, et al. Mutations of the INI1 rhabdoid tumor suppressor gene in medulloblastomas and primitive neuroectodermal tumors of the central nervous system. Clin Cancer Res 2000;6(7):2759–63.
[27] Sigauke E, Rakheja D, Maddox DL, Hladik CL, White CL, Timmons CF, et al. Absence of expression of SMARCB1/INI1 in malignant rhabdoid tumors of the central nervous system, kidneys and soft tissue: an immunohistochemical study with implications for diagnosis. Mod Pathol 2006;19(5):717–25.
[28] Judkins AR, Mauger J, Ht A, Rorke LB, Biegel JA. Immunohistochemical analysis of hSNF5/INI1 in pediatric CNS neoplasms. Am J Surg Pathol 2004;28(5):644–50.
[29] Hasselblatt M, Gesk S, Oyen F, Rossi S, Viscardi E, Giangaspero F, et al. Nonsense mutation and inactivation of SMARCA4 (BRG1) in an atypical teratoid/rhabdoid tumor showing retained SMARCB1 (INI1) expression. Am J Surg Pathol 2011;35(6):933–5.
[30] Hasselblatt M, Isken S, Linge A, Eikmeier K, Jeibmann A, Oyen F, et al. High-resolution genomic analysis suggests the absence of recurrent genomic alterations other than SMARCB1 aberrations in atypical teratoid/rhabdoid tumors. Genes Chromosomes Cancer 2013;52(2):185–90.
[31] Albanese P, Belin MF, Delattre O. The tumour suppressor hSNF5/INI1 controls the differentiation potential of malignant rhabdoid cells. Eur J Cancer 2006;42(14):2326–34.
[32] Northcott PA, Jones DT, Kool M, Robinson GW, Gilbertson RJ, Cho YJ, et al. Medulloblastomics: the end of the beginning. Nat Rev Cancer 2012;12(12):818–34.
[33] Millard NE, De Braganca KC. Medulloblastoma. J Child Neurol 2015;.
[34] Taylor MD, Northcott PA, Korshunov A, Remke M, Cho YJ, Clifford SC, et al. Molecular subgroups of medulloblastoma: the current consensus. Acta Neuropathol 2012;123(4):465–72.
[35] Kool M, Korshunov A, Remke M, Jones DT, Schlanstein M, Northcott PA, et al. Molecular subgroups of medulloblastoma: an international meta-analysis of transcriptome, genetic aberrations, and clinical data of WNT, SHH, Group 3, and Group 4 medulloblastomas. Acta Neuropathol 2012;123(4):473–84.
[36] Ellison DW, Dalton J, Kocak M, Nicholson SL, Fraga C, Neale G, et al. Medulloblastoma: clinicopathological correlates of SHH, WNT, and non-SHH/WNT molecular subgroups. Acta Neuropathol 2011;121(3):381–96.
[37] Northcott PA, Korshunov A, Witt H, Hielscher T, Eberhart CG, Mack S, et al. Medulloblastoma comprises four distinct molecular variants. J Clin Oncol 2011;29(11):1408–14.
[38] Remke M, Hielscher T, Northcott PA, Witt H, Ryzhova M, Wittmann A, et al. Adult medulloblastoma comprises three major molecular variants. J Clin Oncol 2011;29(19):2717–23.
[39] Schwalbe EC, Williamson D, Lindsey JC, Hamilton D, Ryan SL, Megahed H, et al. DNA methylation profiling of medulloblastoma allows robust subclassification and improved outcome prediction using formalin-fixed biopsies. Acta Neuropathol 2013;125(3):359–71.
[40] Hovestadt V, Remke M, Kool M, Pietsch T, Northcott PA, Fischer R, et al. Robust molecular subgrouping and copy-number profiling of medulloblastoma from small amounts of archival tumour material using high-density DNA methylation arrays. Acta Neuropathol 2013;125(6):913–6.
[41] Fruhwald MC, O'Dorisio MS, Dai Z, Tanner SM, Balster DA, Gao X, et al. Aberrant promoter methylation of previously unidentified target genes is a common abnormality in medulloblastomas—implications for tumor biology and potential clinical utility. Oncogene 2001;20(36):5033–42.
[42] Hovestadt V, Jones DT, Picelli S, Wang W, Kool M, Northcott PA, et al. Decoding the regulatory landscape of medulloblastoma using DNA methylation sequencing. Nature 2014;510(7506):537–41.
[43] Jones DT, Northcott PA, Kool M, Pfister SM. The role of chromatin remodeling in medulloblastoma. Brain Pathol 2013;23(2):193–9.
[44] Dubuc AM, Remke M, Korshunov A, Northcott PA, Zhan SH, Mendez-Lago M, et al. Aberrant patterns of H3K4 and H3K27 histone lysine methylation occur across subgroups in medulloblastoma. Acta Neuropathol 2013;125(3):373–84.

[45] Figueroa ME, Abdel-Wahab O, Lu C, Ward PS, Patel J, Shih A, et al. Leukemic IDH1 and IDH2 mutations result in a hypermethylation phenotype, disrupt TET2 function, and impair hematopoietic differentiation. Cancer Cell 2010;18(6):553–67.
[46] Caramazza D, Lasho TL, Finke CM, Gangat N, Dingli D, Knudson RA, et al. IDH mutations and trisomy 8 in myelodysplastic syndromes and acute myeloid leukemia. Leukemia 2010;24(12):2120–2.
[47] Yan H, Bigner DD, Velculescu V, Parsons DW. Mutant metabolic enzymes are at the origin of gliomas. Cancer Res 2009;69(24):9157–9.
[48] Yan H, Parsons DW, Jin G, McLendon R, Rasheed BA, Yuan W, et al. IDH1 and IDH2 mutations in gliomas. N Engl J Med 2009;360(8):765–73.
[49] Dang L, White DW, Gross S, Bennett BD, Bittinger MA, Driggers EM, et al. Cancer-associated IDH1 mutations produce 2-hydroxyglutarate. Nature 2009;462(7274):739–44.
[50] Lu C, Ward PS, Kapoor GS, Rohle D, Turcan S, Abdel-Wahab O, et al. IDH mutation impairs histone demethylation and results in a block to cell differentiation. Nature 2012;483(7390):474–8.
[51] Turcan S, Rohle D, Goenka A, Walsh LA, Fang F, Yilmaz E, et al. IDH1 mutation is sufficient to establish the glioma hypermethylator phenotype. Nature 2012;483(7390):479–83.
[52] Noushmehr H, Weisenberger DJ, Diefes K, Phillips HS, Pujara K, Berman BP, et al. Identification of a CpG island methylator phenotype that defines a distinct subgroup of glioma. Cancer Cell 2010;17(5):510–22.
[53] Watanabe T, Nobusawa S, Kleihues P, Ohgaki H. IDH1 mutations are early events in the development of astrocytomas and oligodendrogliomas. Am J Pathol 2009;174(4):1149–53.
[54] Alentorn A, Sanson M, Idbaih A. Oligodendrogliomas: new insights from the genetics and perspectives. Curr Opin Oncol 2012;24(6):687–93.
[55] Pollack IF, Hamilton RL, Sobol RW, Nikiforova MN, Lyons-Weiler MA, LaFramboise WA, et al. IDH1 mutations are common in malignant gliomas arising in adolescents: a report from the Children's Oncology Group. Childs Nerv Syst 2011;27(1):87–94.
[56] Verhaak RG, Hoadley KA, Purdom E, Wang V, Qi Y, Wilkerson MD, et al. Integrated genomic analysis identifies clinically relevant subtypes of glioblastoma characterized by abnormalities in PDGFRA, IDH1, EGFR, and NF1. Cancer Cell 2010;17(1):98–110.
[57] Hartmann C, Hentschel B, Wick W, Capper D, Felsberg J, Simon M, et al. Patients with IDH1 wild-type anaplastic astrocytomas exhibit worse prognosis than IDH1-mutated glioblastomas, and IDH1 mutation status accounts for the unfavorable prognostic effect of higher age: implications for classification of gliomas. Acta Neuropathol 2010;120(6):707–18.
[58] Capper D, Weissert S, Balss J, Habel A, Meyer J, Jager D, et al. Characterization of R132H mutation-specific IDH1 antibody binding in brain tumors. Brain Pathol 2010;20(1):245–54.
[59] Louis DN, Perry A, Burger P, Ellison DW, Reifenberger G, von Deimling A, et al. International Society of Neuropathology—Haarlem consensus guidelines for nervous system tumor classification and grading. Brain Pathol 2014;24(5):429–35.
[60] Jiao Y, Killela PJ, Reitman ZJ, Rasheed AB, Heaphy CM, de Wilde RF, et al. Frequent ATRX, CIC, FUBP1 and IDH1 mutations refine the classification of malignant gliomas. Oncotarget 2012;3(7):709–22.
[61] Liu XY, Gerges N, Korshunov A, Sabha N, Khuong-Quang DA, Fontebasso AM, et al. Frequent ATRX mutations and loss of expression in adult diffuse astrocytic tumors carrying IDH1/IDH2 and TP53 mutations. Acta Neuropathol 2012;124(5):615–25.
[62] Abedalthagafi M, Phillips JJ, Kim GE, Mueller S, Haas-Kogen DA, Marshall RE, et al. The alternative lengthening of telomere phenotype is significantly associated with loss of ATRX expression in high-grade pediatric and adult astrocytomas: a multiinstitutional study of 214 astrocytomas. Mod Pathol 2013;26(11):1425–32.
[63] Pickett HA, Reddel RR. Molecular mechanisms of activity and derepression of alternative lengthening of telomeres. Nat Struct Mol Biol 2015;22(11):875–80.
[64] Cesare AJ, Reddel RR. Alternative lengthening of telomeres: models, mechanisms and implications. Nat Rev Genet 2010;11(5):319–30.

[65] Mangerel J, Price A, Castelo-Branco P, Brzezinski J, Buczkowicz P, Rakopoulos P, et al. Alternative lengthening of telomeres is enriched in, and impacts survival of TP53 mutant pediatric malignant brain tumors. Acta Neuropathol 2014;128(6):853–62.
[66] Flynn RL, Cox KE, Jeitany M, Wakimoto H, Bryll AR, Ganem NJ, et al. Alternative lengthening of telomeres renders cancer cells hypersensitive to ATR inhibitors. Science 2015;347(6219):273–7.
[67] Wiestler B, Capper D, Holland-Letz T, Korshunov A, von Deimling A, Pfister SM, et al. ATRX loss refines the classification of anaplastic gliomas and identifies a subgroup of IDH mutant astrocytic tumors with better prognosis. Acta Neuropathol 2013;126(3):443–51.
[68] Odia Y, Varma H, Tsankova NM. Biphasic IDH1 phenotype in a diffusely infiltrating glioma: implications for pathogenesis, treatment and prognosis. Clin Neuropathol 2015;34(5):282–7.
[69] Sahm F, Reuss D, Koelsche C, Capper D, Schittenhelm J, Heim S, et al. Farewell to oligoastrocytoma: in situ molecular genetics favor classification as either oligodendroglioma or astrocytoma. Acta Neuropathol 2014;128(4):551–9.
[70] Reuss DE, Sahm F, Schrimpf D, Wiestler B, Capper D, Koelsche C, et al. ATRX and IDH1-R132H immunohistochemistry with subsequent copy number analysis and IDH sequencing as a basis for an "integrated" diagnostic approach for adult astrocytoma, oligodendroglioma and glioblastoma. Acta Neuropathol 2015;129(1):133–46.
[71] Cai J, Yang P, Zhang C, Zhang W, Liu Y, Bao Z, et al. ATRX mRNA expression combined with IDH1/2 mutational status and Ki-67 expression refines the molecular classification of astrocytic tumors: evidence from the whole transcriptome sequencing of 169 samples samples. Oncotarget 2014;5(9):2551–61.
[72] Killela PJ, Reitman ZJ, Jiao Y, Bettegowda C, Agrawal N, Diaz LA Jr, et al. TERT promoter mutations occur frequently in gliomas and a subset of tumors derived from cells with low rates of self-renewal. Proc Natl Acad Sci USA 2013;110(15):6021–6.
[73] Koelsche C, Sahm F, Capper D, Reuss D, Sturm D, Jones DT, et al. Distribution of TERT promoter mutations in pediatric and adult tumors of the nervous system. Acta Neuropathol 2013;126(6):907–15.
[74] Eckel-Passow JE, Lachance DH, Molinaro AM, Walsh KM, Decker PA, Sicotte H, et al. Glioma groups based on 1p/19q, IDH, and TERT promoter mutations in tumors. N Engl J Med 2015;372(26):2499–508.
[75] Labussiere M, Di Stefano AL, Gleize V, Boisselier B, Giry M, Mangesius S, et al. TERT promoter mutations in gliomas, genetic associations and clinico-pathological correlations. Br J Cancer 2014;111(10).2024–32.
[76] Arita H, Narita Y, Fukushima S, Tateishi K, Matsushita Y, Yoshida A, et al. Upregulating mutations in the TERT promoter commonly occur in adult malignant gliomas and are strongly associated with total 1p19q loss. Acta Neuropathol 2013;126(2):267–76.
[77] Pastori C, Kapranov P, Penas C, Peschansky V, Volmar CH, Sarkaria JN, et al. The Bromodomain protein BRD4 controls HOTAIR, a long noncoding RNA essential for glioblastoma proliferation. Proc Natl Acad Sci USA 2015;112(27):8326–31.
[78] Tsai MC, Manor O, Wan Y, Mosammaparast N, Wang JK, Lan F, et al. Long noncoding RNA as modular scaffold of histone modification complexes. Science 2010;329(5992):689–93.
[79] Yin Y, Qiu S, Peng Y. Functional roles of enhancer of zeste homolog 2 in gliomas. Gene 2016;576(1 Pt. 2):189–94.
[80] Gallo M, Ho J, Coutinho FJ, Vanner R, Lee L, Head R, et al. A tumorigenic MLL-homeobox network in human glioblastoma stem cells. Cancer Res 2013;73(1):417–27.
[81] Wang L, He S, Tu Y, Ji P, Zong J, Zhang J, et al. Downregulation of chromatin remodeling factor CHD5 is associated with a poor prognosis in human glioma. J Clin Neurosci 2013;20(7):958–63.
[82] Flavahan WA, Drier Y, Liau BB, Gillespie SM, Venteicher AS, Stemmer-Rachamimov AO, et al. Insulator dysfunction and oncogene activation in IDH mutant gliomas. Nature 2016;529(7584):110–4.
[83] Erfani P, Tome-Garcia J, Canoll P, Doetsch F, Tsankova NM. EGFR promoter exhibits dynamic histone modifications and binding of ASH2L and P300 in human germinal matrix and gliomas. Epigenetics 2015;10(6):496–507.

[84] Ferguson S, Lesniak MS. Percival Bailey and the classification of brain tumors. Neurosurg Focus 2005;18(4):e7.
[85] Sanai N, Alvarez-Buylla A, Berger MS. Neural stem cells and the origin of gliomas. N Engl J Med 2005;353(8):811–22.
[86] Canoll P, Goldman JE. The interface between glial progenitors and gliomas. Acta Neuropathol 2008;116(5):465–77.
[87] Gopisetty G, Xu J, Sampath D, Colman H, Puduvalli VK. Epigenetic regulation of CD133/PROM1 expression in glioma stem cells by Sp1/myc and promoter methylation. Oncogene 2013;32(26):3119–29.
[88] Turcan S, Fabius AW, Borodovsky A, Pedraza A, Brennan C, Huse J, et al. Efficient induction of differentiation and growth inhibition in IDH1 mutant glioma cells by the DNMT inhibitor decitabine. Oncotarget 2013;4(10):1729–36.
[89] Krauze AV, Myrehaug SD, Chang MG, Holdford DJ, Smith S, Shih J, et al. A phase 2 study of concurrent radiation therapy, temozolomide, and the histone deacetylase inhibitor valproic acid for patients with glioblastoma. Int J Radiat Oncol Biol Phys 2015;92(5):986–92.
[90] Grasso CS, Tang Y, Truffaux N, Berlow NE, Liu L, Debily MA, et al. Functionally defined therapeutic targets in diffuse intrinsic pontine glioma. Nat Med 2015;21(6):555–9.
[91] Lee P, Murphy B, Miller R, Menon V, Banik NL, Giglio P, et al. Mechanisms and clinical significance of histone deacetylase inhibitors: epigenetic glioblastoma therapy. Anticancer Res 2015;35(2):615–25.
[92] Santen GW, Kriek M, van Attikum H. SWI/SNF complex in disorder: SWItching from malignancies to intellectual disability. Epigenetics 2012;7(11):1219–24.
[93] Kleefstra T, Schenck A, Kramer JM, van Bokhoven H. The genetics of cognitive epigenetics. Neuropharmacology 2014;80:83–94.
[94] Tsurusaki Y, Okamoto N, Ohashi H, Kosho T, Imai Y, Hibi-Ko Y, et al. Mutations affecting components of the SWI/SNF complex cause Coffin–Siris syndrome. Nat Genet 2012;44(4):376–8.
[95] Schmitz U, Mueller W, Weber M, Sevenet N, Delattre O, von Deimling A. INI1 mutations in meningiomas at a potential hotspot in exon 9. Br J Cancer 2001;84(2):199–201.
[96] Smith MJ, Wallace AJ, Bowers NL, Rustad CF, Woods CG, Leschziner GD, et al. Frequency of SMARCB1 mutations in familial and sporadic schwannomatosis. Neurogenetics 2012;13(2):141–5.
[97] Sevenet N, Sheridan E, Amram D, Schneider P, Handgretinger R, Delattre O. Constitutional mutations of the hSNF5/INI1 gene predispose to a variety of cancers. Am J Hum Genet 1999;65(5):1342–8.
[98] Bourdeaut F, Chi SN, Fruhwald MC. Rhabdoid tumors: integrating biological insights with clinical success: a report from the SMARCB1 and Rhabdoid Tumor Symposium, Paris, December 12–14, 2013. Cancer Genet 2014;207(9):346–51.
[99] Mellado-Calvo N, Fleminger S. Cerebral tumors. In: David AS, Fleminger S, Kopelman MD, Lovestone S, Mellers JDC, editors. Lishman's organic psychiatry. 4th ed. Chichester: Wiley-Blackwell; 2009.
[100] Madhusoodanan S, Ting MB, Ugur U. Psychiatric aspects of brain tumors. World J Psychiatry 2015;5:273–85.

CHAPTER

EPIGENETICS AND CEREBROVASCULAR DISEASES

14

C. Soriano-Tárraga*,,†, J. Jiménez-Conde*,**,†, J. Roquer*,**,†**

**Department of Neurology, Hospital del Mar, Barcelona, Spain; **Neurovascular Research Group, IMIM (Institut Hospital del Mar d'Investigacions Mèdiques), Barcelona, Spain; †Department of Medicine, Autonomous University of Barcelona, Barcelona, Spain*

CHAPTER OUTLINE

Neuropsychiatric Disorders and Epigenetics. http://dx.doi.org/10.1016/B978-0-12-800226-1.00014-9

14.1 CEREBROVASCULAR DISEASES—INTRODUCTION

Cardiovascular diseases are a group of diseases of the heart and blood vessels and include cerebrovascular diseases, such as stroke, coronary artery diseases, such as myocardial infarction, cardiomyopathy, cardiac arrhythmias, endocarditis, and aortic aneurysms, among others. Cerebrovascular diseases, and specifically stroke, refer to a group of conditions that affect the circulation of blood to the brain, causing limited or no blood flow to affected areas of the brain.

Stroke is one of the leading causes of disability and morbidity in many countries, at all levels of economic development. Stroke can be caused by multiple pathologies involving different vascular, systemic, genetic, and environmental factors and processes. Despite current attention to risk factors and preventive treatment, the number of stroke cases has been rising in recent decades, which is likely due to population aging [1]. Of the 17.5 million deaths due to cardiovascular disease in 2012, an estimated 7.4 million were due to heart attacks (ischemic heart disease) and 6.7 million to strokes (ischemic or hemorrhagic) [1]. Stroke symptoms depend on the area of the brain affected. However, the most common symptom is weakness or paralysis of one side of the body with partial or complete loss of voluntary movement or sensation in a leg or an arm, or weak face muscles, which can cause drooling. Speech can also be affected.

About 80% of strokes are ischemic, whereas 20% are hemorrhagic, mainly intracerebral hemorrhage (ICH). Ischemic stroke (IS) results when the blood flow of an artery to the brain is blocked by a blood clot, thrombus, or a bit of atherosclerotic plaque, such as a cholesterol and calcium deposit on the wall of the artery, or when a hemodynamic mechanism causes hypoperfusion. ICH occurs when cerebral vessels rupture, and is associated with a 3-month mortality of 40–50% and with sustained disability in more than half of survivors.

Stroke etiology has been classified in various ways. In IS, the most widespread and frequently used is based on Trial of Org 10,172 in Acute Stroke Treatment (TOAST) Classification of Ischemic Stroke Subtypes [2]. Diagnoses are based on clinical features and on data collected by tests, such as brain imaging (CT/MRI), cardiac imaging (echocardiography, etc.), duplex imaging of extracranial arteries, arteriography, and laboratory assessments of prothrombotic status. The TOAST classification system includes five categories: large-artery atherosclerosis (LAA), cardioembolism (CE), small-artery occlusion (lacunar), stroke of other determined etiology, and stroke of undetermined etiology [2].

- *Large-artery atherosclerosis*: clinical and arterial imaging findings of either significant (>50%) stenosis or occlusion of a major brain artery or branch cortical artery, presumably due to atherosclerosis.
- *Cardioembolism*: arterial occlusions, presumably due to an embolus arising in the heart. Cardiac sources are divided into high-risk and medium-risk groups based on the evidence of their relative propensities for embolism.
- *Small-artery occlusion* (in some classifications, *lacunar*): evidence of one of the traditional clinical lacunar syndromes and no evidence of cerebral cortical dysfunction. A history of diabetes mellitus or hypertension supports the clinical diagnosis. The patient should also have a normal CT or CT/MRI examination with a brainstem or subcortical hemispheric lesion of less than 1.5 cm in diameter.
- *Acute stroke of other determined etiology*: this category includes patients with rare causes of stroke, such as nonatherosclerotic vasculopathies, hypercoagulable states, hematologic disorders, genetic causes, etc.

- *Stroke of undetermined etiology/cryptogenic*: cause of stroke cannot be determined with any degree of confidence. Some patients will have no likely etiology despite an extensive evaluation; in others, no cause is found but the evaluation was superficial. This category also includes patients with two or more potential causes of stroke so that the physician is unable to make a final diagnosis.

In this chapter, we focus on the emerging evidence of the role of epigenetic factors in IS. Understanding these processes may be critical to enhance the assessment of patient risk, achieve early diagnosis, and characterize clinically relevant molecular mechanisms associated with various stroke subtypes.

14.2 VASCULAR STROKE RISK FACTORS

Numerous factors can contribute to an individual's stroke risk, and many individuals can have more than one risk factor. The Framingham study defined global stroke predictors that include age, systolic blood pressure, hypertension, diabetes mellitus, current smoking, established cardiovascular disease, atrial fibrillation (AF), and left ventricular hypertrophy [3]. Risk factors for a first stroke were classified according to their potential for being modified (nonmodifiable, modifiable, or potentially modifiable) and strength of evidence (well documented or less well documented). Nonmodifiable risk factors include age, sex, low birth weight, race/ethnicity, and genetic factors. Well documented and modifiable risk factors include hypertension, diabetes, exposure to cigarette smoke, AF, and certain other cardiac conditions, dyslipidemia, carotid artery stenosis, sickle cell disease, postmenopausal hormone therapy, poor diet, physical inactivity, obesity, and body fat distribution. Less well-documented or potentially modifiable risk factors include the metabolic syndrome, alcohol abuse, drug abuse, oral contraceptive use, sleep-disordered breathing, migraine, hyperhomocysteinemia, elevated lipoprotein(a), elevated lipoprotein-associated phospholipase, hypercoagulability, inflammation, and infection [4]. Traditional cardiovascular risk factors, such as age, sex, smoking, hypertension, cholesterol, and diabetes have remained as the keystone of risk stratification.

Although most risk factors have an independent effect, there may be important interactions between individual factors. For instance, obesity increases the likelihood of diabetes, hypertension, coronary heart disease, stroke, and certain types of cancer. Adults who are not physically active have a 20–30% increased risk of all-cause mortality compared to those who do at least 150 min of moderate-intensity physical activity per week, or equivalent, as recommended by the WHO. Regular physical activity reduces the risk of ischemic heart disease, stroke, diabetes, and breast and colon cancer. Finally, lifestyle factors, such as diet, physical activity, smoking, drinking, and socioeconomic status are important targets for primary prevention of all cardiovascular diseases, including stroke [1].

14.3 INTERMEDIATE PHENOTYPES

Intermediate states between the presence of risk factors and evident disease may serve as important indicators of cerebrovascular disease. These intermediate phenotypes, or endophenotypes, are associated with clinical cerebrovascular disease in the population, are independently and jointly heritable with stroke, and are present in individuals with and without stroke. The endophenotype concept is now broadly used in clinical genetics research. An endophenotype is not a risk factor; rather, it is a manifestation of the underlying disease. Endophenotypes, usually characterized on imaging, represent processes involved in the pathogenesis pathway toward IS. The two markers most widely used in stroke genetics are ultrasound measurements of carotid intima-media thickness (IMT) and carotid plaque for large-artery stroke and cerebral white-matter lesions (leukoaraiosis) seen on MRI as a marker for small-vessel disease. These markers must be taken into account because they might also interact with environmental factors or contribute directly to an intermediate phenotype, such as atherosclerosis. Heritability estimates range from about 30% to 60% for IMT, which means that much of the variability is due to genetic factors [5] and from 55% to 80% for cerebral white-matter lesions [6]. Due to the strong genetic contribution, researchers have started to use IMT and white-matter-lesion volumes as targets for genetic studies.

14.4 GENETIC RISK FACTORS AND STROKE

IS is a complex disorder caused by multiple genetic and environmental factors. The heritability of IS has been estimated at ~38%, varying across stroke subtypes [7,8]. Genetic predisposition differs depending on age and stroke subtype. Both twin and family history studies suggest a stronger genetic component in stroke patients younger than 70 years than in those who are older. In twins, concordance rates for the disease were reported to be about 65% greater in monozygotic than in dizygotic twins, but most twin studies have been relatively small. In case–control studies, a family history of stroke was shown to increase the risk of IS by about 75% [9,10]. Genetic factors also seem to be more important in large-vessel stroke and small-vessel stroke than in cryptogenic stroke. This finding accentuates the importance of stroke subtypes. Moreover, large-vessel stroke and myocardial infarction share similar pathological mechanisms and genetic susceptibility; further insight has come from studies on intermediate phenotypes [9]. These genetic factors contribute to conventional risk factors, such as hypertension or diabetes, and may interact with environmental factors, such as smoking habit, or contribute to triggering an intermediate phenotype, such as atherosclerosis. As a result, they may contribute to stroke latency, infarct size, functional recovery, and outcome. On the other hand, a small proportion of the overall numbers of stroke events is related to various Mendelian stroke syndromes identified in humans. Genetic predisposition to stroke can be categorized as either monogenic or polygenic. Although many studies have investigated potential risk genes for common multifactorial stroke, the most robust findings are related to monogenic disorders [9].

14.5 MONOGENIC CAUSES OF STROKE

A large number of rare monogenic disorders can cause stroke. Mendelian conditions are an important cause of stroke in young patients without known risk factors. In some disorders, stroke is the prevailing manifestation, whereas in others it is part of a wider phenotypic spectrum. Most monogenic disorders are associated with specific stroke subtypes [9]. Epidemiologic and family studies support an inherited component to stroke risk [10]. Table 14.1 provides a summary of the single gene disorders associated with IS.

Table 14.1 Summary of Single-Gene (Monogenic) Disorders Associated With Ischemic Stroke

Monogenic Disorder	Chromosome	Gene	Inheritance
CADASIL	19p13.2-p13.1	NOTCH3	AD
CARASIL	10q26.3	HTRA1	AR
COL4A1-related arteriopathy	13q34	COL4A1	AD/AR
Marfan syndrome	15q21.1	FBN1	AD
Fabry disease	X	GLA	X-linked
Sickle-cell disease	11p15.5	HBB	AR
Homocytinuria	21q22.3	CBS, others	AR
Ehlers–Danlos syndrome type IV	2q31	COL3A1	AD
Pseudoxanthoma elasticum	16p13.1	ABCC6	AR
MELAS	Mitochondrial	mtDNA	Maternal

Abbreviations: AD, autosomal dominant; AR, autosomal recessive; CADASIL, cerebral autosomal-dominant arteriopathy with subcortical infarcts and leukoencephalopathy; CARASIL, cerebral autosomal-recessive arteriopathy with subcortical infarcts and leukoencephalopathy; MELAS, mitochondrial encephalopathy, lactic acidosis, and stroke-like episodes; mtDNA, mitochondrial DNA.

14.6 STROKE AS A COMPLEX POLYGENIC DISORDER

In the vast majority of cases, stroke should be considered as a multifactorial disorder or complex trait for which no clear patterns of inheritance can be demonstrated. As with other complex traits, the genetic component of common stroke is likely to be polygenic. Many alleles may have minimal effects individually, but their wide distribution throughout the population is likely to have a large summative impact on stroke. The heterogeneity of IS poses a significant challenge for identifying genes associated with this disorder because different genes are likely to predispose to different stroke subtypes. Substantial evidence from studies in twins, families, and animal models points to a genetic risk component associated with stroke and recent genome-wide association studies (GWAS) have identified new variants associated with IS and specific genetic variants associated with IS subtypes [11–13]. In other words, genetic factors contribute to stroke prevalence. Candidate gene research and GWAS have identified genetic variants associated with IS risk. These variants often have small effects without obvious biological significance. Table 14.2 provides a list of genes associated with IS in GWAS and exome sequencing analysis. The underlying rationale for GWAS is the "common disease, common variant" hypothesis, positing that common diseases are attributable in part to allelic variants, such as single nucleotide polymorphisms (SNPs), that are present in 1–5% of the population. However, for accurate detection of these relatively small-risk genes, huge sample sizes are necessary.

Part of the genetic burden can be explained through low-frequency variants with substantial effect sizes (increasing disease risk two- to threefold), but without a clear Mendelian segregation. Exome sequencing may discover predicted protein-altering variants with a potentially large effect on IS risk. Exome sequencing has identified two novel genes and mechanisms, *PDE4DIP* and *ACOT4*, associated with increased risk for IS. In addition, *ZFHX3* and *ABCA1* were discovered to have protein-coding variants associated with IS [14].

Table 14.2 List of Genes Associated With Ischemic Stroke in Genome-Wide Association Studies and Exome Sequencing

CHR	Gene	SNP	Trait	Method	References
12p13	NINJ2	rs11833579	IS	GWAS	a,b
4q25	PITX2	rs2200733	CE	GWAS	a
16q22	ZFHX3	rs7193343	CE	GWAS	a
9p21	CDKN2B-AS1	rs2383207	LA	GWAS	a
7p21	HDAC9	rs11984041	LA	GWAS	c
6p21	Intergenic	rs556621	LA	GWAS	d
1p13.2	*TSPAN2*	rs12122341	LA	GWAS	e
9q34.2	ABO	rs505922	IS,CE, and LA	GWAS	f
1q12	PDE4DIP	rs1778155	IS	Exome sequencing	g
14q24.3	ACOT4	rs35724886	IS	Exome sequencing	g

Abbreviation: CHR, chromosome.
[a]*Traylor M, Farrall M, Holliday EG, et al. Genetic risk factors for ischaemic stroke and its subtypes (the METASTROKE collaboration): a meta-analysis of genome-wide association studies. Lancet Neurol 2012;11:951–962.*
[b]*Ikram MA, Seshadri S, Bis JC, et al. Genomewide association studies of stroke. N Engl J Med 2009;360:1718–28.*
[c]*Bellenguez C, Bevan S, Gschwendtner A, et al. Genome-wide association study identifies a variant in HDAC9 associated with large vessel ischemic stroke. Nat Genet 2012;44:328–33.*
[d]*Holliday EG, Maguire JM, Evans T-J, et al. Common variants at 6p21.1 are associated with large artery atherosclerotic stroke. Nat Genet 2012;44:1147–51.*
[e]*NINDS Stroke Genetics Network (SiGN) ISGC (ISGC). Loci associated with ischaemic stroke and its subtypes (SiGN): a genome-wide association study. Lancet Neurol 2016;15(2):174–84.*
[f]*Williams FMK, Carter AM, Hysi PG, et al. Ischemic stroke is associated with the ABO locus: the EuroCLOT study. Ann Neurol 2013;73:16–31.*
[g]*Auer PL, Nalls M, Meschia JF, et al. Rare and coding region genetic variants associated with risk of ischemic stroke: the NHLBI Exome Sequence Project. JAMA Neurol 2015;72:781–8.*

Nevertheless, stroke risk is not completely explained by these genetic factors. Most variants identified so far confer relatively small increments in risk, and explain only a small proportion of familial clustering, leading many to question how the remaining "missing heritability" can be explained. Discovering the missing heritability for IS could provide critical insights into the cause of the disease, novel pathways, and therapeutic targets [8]. Possible explanations for the missing genetic variance include gene interactions, rare genetic variants that have a large effect, and epigenetic mechanisms that bridge heredity and the environment [15].

14.7 EVIDENCE FOR EPIGENETICS IN STROKE

The genome is identical in each cell of an organism. Nevertheless, cell- and tissue-specific profiles of gene transcription and posttranscriptional RNA processing and translation are selectively regulated by epigenetic mechanisms, including DNA methylation, histone modifications, nucleosome remodeling, higher-order chromatin formation, and noncoding RNAs. Epigenomics has emerged as one of the most promising areas in addressing some of the gaps in our current knowledge of the interaction between nature and nurture in the development of cardiovascular disease [16]. The impact of epigenetic mechanisms in cardiovascular pathophysiology is now emerging as a major player in the interface between genotype and phenotype variability. This area of research has strict implications for disease development and progression, and opens up possible novel preventive strategies in cardiovascular disease. An important aspect of epigenetic mechanisms is that they are potentially reversible and may be influenced by nutritional-environmental factors and through gene–environment interactions, all of which have an important role in complex, multifactorial diseases, such as stroke. Interest in the epigenetic mechanisms involved in stroke pathophysiology has only recently gained traction. Modifier genes and environmental factors may interact to determine individual susceptibility to stroke. In addition to genetic variation, epigenetics provides a new layer of variation that might mediate the relationship between the genotype and internal and external environmental factors. This epigenetic component could help explain the marked increase in common diseases with age, as well as the frequent discordance between monozygotic twins in the incidence of diseases, such as bipolar disorder [17,18].

The involvement of epigenetic mechanisms in regulating the expression of stroke-related genes has been explored in recent human studies; for example, the role of neural cell reprogramming pathways regulated by epigenetic marks appears to have the potential for possible innovative therapeutic approaches to the damaged brain [19]. The epigenetics of stroke is at an early stage of research but offers great promise for the better understanding of stroke pathology and the potential viability of new strategies for its treatment. In this review, we describe the epigenetic studies in this field to date.

14.8 DNA METHYLATION AND STROKE

DNA methylation is the most widely studied epigenetic modification, and is a marker of genomic DNA because it adds a methyl group to the 5-carbon position of cytosine, in a 5′-CpG-3′ context. The dinucleotide, is quite rare in mammalian genomes (~1%), often clusters in regions known as CpG islands. The methylation of the CpG dinucleotide is mainly associated with gene silencing [20]. Changes in DNA methylation contribute to interindividual phenotypic variation and are associated with cancer development and other complex diseases [21,22]. Epigenetic changes measured in blood DNA of stroke patients and healthy subjects seem to predict the risk of common age-related diseases, such as coronary heart disease and stroke [23,24]. DNA methyltransferases (DNMTs) are responsible for the conversion of unmodified cytosine to 5mC, levels of which are dramatically increased after ischemic insults, contributing to ischemic brain injury. The pharmacological inhibition of DNMTs, or DNMT1 knockout, causes the brain to resist ischemic insults and promotes brain functional recovery [19]. Therefore, DNA methylation is an important factor that can affect outcomes after brain injury. Global DNA methylation has been widely used in epidemiological studies because it is cost-effective, has a high-throughput, and provides quantitative results. Aberrant genomic DNA methylation has been found in a variety of common aging-related diseases, such as atherosclerosis, cancer, hypertension, and coronary heart disease [21,25,26]. Atherosclerosis is characterized by global hypomethylation [25], and reports of low genomic methylation have been described in patients with cardiovascular disease [23,27]. These results enhance the interest in global DNA methylation as a potential biomarker of cardiovascular disease risk.

About one-third of DNA methylation occurs in repetitive elements, representing a large portion of the human genome. Long interspersed nucleotide element 1 (LINE-1) is the most abundant family of nonlong terminal repeat retrotransposons found in the genome. CpG sites in repetitive elements, like LINEs, are largely methylated in normal somatic tissue, suppressing most of their transposition activity [28]. Global DNA methylation measured in LINE-1 sequences has been considered a surrogate marker for global genome methylation [29]. LINE-1 hypomethylation in these elements increases their activity as retrotransposable sequences, which may induce genomic alterations and affect gene expression by a number of mechanisms [30].

Only two articles about stroke global methylation in LINE-1 have been published, and they report global DNA hypomethylation of stroke patients compared with healthy individuals [23,31]. Those studies showed that hypomethylation of the repetitive LINE-1 elements analyzed in blood DNA are associated with ischemic heart disease and stroke. However, a limited number of stroke patients were included in these analyses ($N = 55$ men), and stroke subtypes were not specified. Ischemic stroke is a heterogeneous disease with different etiological subtypes, resulting from differences in pathophysiology and genetic variants; consequently, it is likely its methylation profile is also specific to stroke subtype.

Another method for analyzing global DNA methylation is the luminometric methylation assay (LUMA), which measures levels of 5-mC residing in the —CCGG— motif [32]. This motif represents 8% of all CpG sites, occurs throughout the genome, and can be used as a proxy marker to estimate global DNA methylation. A recent study comparing IS patients with healthy individuals found global DNA hypomethylation of stroke patients, as well. Contrary to what might be expected, no consistent differences in global DNA methylation were observed between three different etiologies of stroke, classified according to TOAST criteria, or between large-artery atherosclerosis, small-artery disease, and cardio-aortic embolism stroke [33].

Another important factor in DNA methylation is methylenetetrahydrofolate reductase (MTHFR), which is involved in folate metabolism and in the formation of cellular reservoirs of the methyl group donor S-adenosylmethionine. Intriguingly, MTHFR deficiency causes hyperhomocysteinemia, a high level of homocysteine in the blood predisposed to endothelial cell injury, which leads to inflammation in the blood vessels, and is associated with increased risk of stroke and cardiovascular disease [34,35]. Specific MTHFR gene polymorphisms (e.g., *C677T*) are similarly associated with hyperhomocysteinemia and an increased risk of stroke, cardiovascular disease, neural developmental disorders, and a variety of other disease entities. Although the mechanisms that underlie this increased risk of stroke have not been clearly delineated, differential MTHFR activity is associated with variations in global DNA methylation levels, which suggests that the risk may, in part, be linked to the effects of DNA methylation status on the vulnerability of the brain to ischemic injury. Moreover, methylation of MTHFR significantly increases susceptibility risk for IS by mediating serum folate and vitamin B_{12} levels, but does not affect IS severity [36].

Similar to GWAS, epigenome-wide association studies (EWAS) aim to perform a genome-wide search for epigenetic variants that associate with complex phenotypes and have the potential to identify novel genes and molecular pathways in common diseases. In contrast to genetic variation, epigenetic variation can be dynamic, which has implications for EWAS methodology and design. A recent EWAS study in aortic and carotid atherosclerotic plaques showed a DNA hypermethylation profile [37]. A high-coverage epigenomics study of the human aorta and carotid arteries identified atherosclerosis-specific DNA methylation profiles that are lesion grade–independent and are therefore established relatively early during the natural history of the atheroma [37]. After early hypermethylation, the carotid plaque DNA methylome undergoes relatively minor changes with the appearance of cardiovascular symptoms, but subsequently drifts toward a widespread hypomethylation with increasing postcerebrovascular event time [38]. One pending issue is whether any DNA methylation profile is associated with plaque rupture, the most common underlying cause of the clinical complications of advanced atherosclerosis.

On the subject of pharmacoepigenetics studies, a recent epigenome-wide association study identified cg04985020 (PPM1A) associated with vascular recurrence in patients treated with aspirin [38a]. Clopidogrel is one of the most used antiplatelet drugs in patients with cardiovascular disease. However, 16–50% of patients have high on-clopidogrel platelet reactivity and an increased risk of ischemic events. A EWAS identified hypomethylation of cg03548645 (TRAF3), correlated with an increased platelet aggregation in IS patients with vascular recurrence treated with clopidogrel [38b].

14.9 5-HYDROXYMETHYLCYTOSINE AND STROKE

Recent research has implicated 5-hydroxymethylcytosine (5hmC), a DNA base derived from 5-methylcytosine (5mC) via oxidation by ten-eleven translocation (TET) enzymes, in DNA methylation-related plasticity [39]. The high levels of 5hmC in the brain suggest a significant role in neural differentiation and development [40,41]. Modifications in 5hmC contribute to many neurological disorders, among them Huntington's disease and Alzheimer's disease [42,43]. In a middle cerebral artery occlusion mouse model, the reduction of Tet2 by either a chemical (SC1) or knockout worsened ischemic brain injury. Genome-wide 5hmC profiling helped identify changes in 5hmC, which were enriched in genes associated with the cell junction and neuronal development after stroke. Alterations of 5hmC in the brain-derived neurotrophic factor (*BDNF*) gene is known to play a key role in the development and survival of neurons in the central nervous system, influencing the mRNA and protein levels of *BDNF*. These findings suggest 5hmC modifications as a potential new therapeutic target in the treatment of stroke [44].

14.10 HISTONES AND STROKE

All histones are subject to posttranscriptional modifications, several of which occur in histone tails: acetylation, methylation, phosphorylation, and ubiquitination, among others. Histone modifications have important roles in transcriptional regulation, DNA repair, DNA replication, alternative splicing, and chromosome condensation. Euchromatin is characterized by high levels of acetylation and trimethylated H3K4, H3K36, and H3K79. On the other hand, heterochromatin is characterized by low levels of acetylation and high levels of H3K9, H3K27, and H4K20 methylation. Histone modification levels are predictive for gene expression. Actively transcribed genes are characterized by high levels of H3K4me3, H3K27ac, H2BK5ac, and H4K20me1 in the promoter and H3K79me1 and H4K20me1 along the gene body. Histone acetylation is regulated by a balance between the actions of antagonistic enzymes: histone acetyltransferases (HATs) that catalyze the addition of acetyl groups to lysine residues in histone tails, and histone deacetylases (HDACs) that remove acetyl residues from histones [20,45]

Ischemia alters the expression of multiple HDACs, and these have become popular targets for preclinical neuroprotection studies in stroke [46]. A GWAS for IS in a large-vessel stroke subtype showed an association with the *HDAC9 locus*, revealing that carriers of the A-allele have about 9% increased risk of stroke. However, the underlying molecular pathways have not yet been identified [11]. During ischemic injury, a transcriptional repression associated with increased DNA methylation and histones 3 and 4 acetylation occurs in an animal model of stroke. Vast deacetylation processes occur quickly upon ischemic insult. Experimental therapeutic approaches using many HDAC inhibitors (HDACi) in IS models have revealed a decreased brain infarct volume and an increased expression of neuroprotective proteins, such as Hsp70 and Bcl-2. The administration of HDACi controls the acetylation level of H3 and H4 during the ischemic insult and induces a global transcriptional change, promoting a reduction of infarct volume and functional recovery [47,48].

SIRT1 is a class III HDAC belonging to the mammalian sirtuin family. Its activity is increased during ischemic preconditioning and it has a role in the protective mechanism of ischemic preconditioning in the mouse heart. The upregulation of SIRT1 has been described to confer protection in stroke [49]. HDACi treatment not only influences histone marks, but also modulates microRNA (miRNA) pathways poststroke [50]. These findings suggest that the whole epigenetic machinery works in concert in physiologic as well as pathologic conditions and in adaptive responses in the brain.

14.11 NONCODING RNAS (MIRNAS) AND STROKE

MicroRNAs (miRNAs) are important regulators of gene expression and play important roles in the initiation and progression of several diseases. miRNAs are short (~21–23 nucleotides long) nonprotein coding ribonucleic acids. They regulate gene expression at multiple epigenetic levels including mRNA degradation, mRNA sequestration, translational repression, and transcriptional repression [20].

Several groups have investigated the expression profiles of miRNAs following stroke in animal models and humans. These profiles exhibit variable changes in expression during the acute and recovery phases of stroke. Unfortunately, there is little overlap between the findings of these studies, which is confusing if miRNAs are truly important in the evolution of stroke and recovery [51]. The dysregulated miRNAs have been detectable even several months after stroke onset in what is usually regarded as neurologically stable patients. The peripheral blood miRNAs and their profiles could be promising biomarkers in the diagnosis and prognosis of cerebral IS [52]. miRNA-181a, for instance, increases in the ischemic brain in mice, but decreases in the penumbra area. Inhibition of miRNA-181a reduces cell death in mice models of both local and global ischemia, with Grp78 and Bcl-2 being potential target transcripts of importance. Inhibition of let7f or miRNA-1 upto 4 h after stroke appears to provide neuroprotection via IGF1 pathways, substantially reducing infarct volume. On the other hand, overexpression of miR-124 protects against stroke in vivo; however, some have reported that miR-124 actually promotes cell death by suppressing the expression of apoptosis inhibitors. Many miRNAs can target multiple different genes, which may explain some conflicting findings. Dissecting the multiple functions that these regulators can serve in the pathophysiology of stroke will require further investigation [51].

Finally, miRNAs have emerged as regulators of molecular events, such as endothelial/vascular function, erythropoiesis, hypoxia, and neural function involved in ischemia. Some miRNAs are highly expressed in the ischemic brain and can be detected in blood samples several weeks after stroke onset, suggesting their potential diagnostic use as noninvasive biomarkers [53,54]. For instance, circulating miR-30a and miR-126 levels were markedly downregulated in all patients with IS until 24 weeks, and let-7b levels were lower in patients with large-vessel atherosclerosis than in healthy volunteers but higher in other kinds of IS [54]. Recent preclinical studies have reported a neuroprotective effect of antagomirs (miRNA inhibitors) in experimental IS through modulation of several pathways including NFkB and leukocyte infiltration [47]. In summary, miRNAs represent an epigenetic mechanism up- or downregulated during ischemic insults that could be used as potential biomarkers and provide novel pharmacological targets.

14.12 MODIFIABLE STROKE RISK FACTORS AND EPIGENETICS

Unlike the small number of stroke epigenetics studies, there are a large number of studies about modifiable risk factors that contribute to individual stroke risk as a result of environmental exposure and lifestyle. As numerous factors can contribute to stroke risk and many individuals have more than one risk factor, there may be important interactions between them. We summarize later the epigenetic research in the field of modifiable risk factors that may contribute to the risk of stroke and then complete our description of the epigenetic landscape of this complex disease.

14.13 EPIGENETICS OF ATHEROSCLEROSIS

Atherosclerosis is characterized by cholesterol accumulation in the arterial wall, vascular smooth muscle cell proliferation, and deposit of extracellular matrix components, which lead to plaque rupture and/or the occlusion of blood vessels. Atherosclerosis is a multifactorial disease also associated with dietary, environmental, and lifestyle risk factors. The most widely studied epigenetic mark, DNA methylation, is a flexible regulatory modification influenced by environmental factors and may be involved in the onset and progression of atherosclerosis [24].

Differentially methylated genes participating in atherogenesis have been identified by candidate gene-based studies and epigenome-wide analysis, which showed a genome-wide increase in DNA methylation in atherosclerosis [37,38]. Global DNA hypermethylation involved unique and repetitive sequences and occurred in CpG and non-CpG contexts [37]. However, hypomethylation also has been described in human atherosclerotic lesions [25]. Moreover, a high-coverage epigenomics study of human aorta and carotid arteries identified atherosclerosis-specific DNA methylation profiles, independent of degree of lesion, that are established early in the atheroma and the DNA methylomes that distinguish symptomatic from asymptomatic plaques [37]. Differentially methylated genes identified that participate in important aorta-related processes are the following: *A2BP1, CALD1, DAAM1, EGFR, FBN2, HOXA9, HOXB3, KCNMA1, MAP1B, MIR23b, MYH10, PDGFD, PLA2G10, PLAT, RPTOR, PRKCE, PRRX1, PTK2, PXDN*, and *TSC22D1*. Further studies have identified genes epigenetically modified during the atherogenesis process, including nitric oxide synthase (*NOS*), estrogen receptor (*ER*), collagen type XV alpha 1 (*COL15A1*), vascular endothelial growth factor receptor (*VEGFR*), and ten-eleven translocation (*TET*) proteins, all of which are involved in endothelial dysfunction; genes involved in the atherosclerotic inflammatory process encode proteins, such as gamma interferon (*IFN-γ*), forkhead box 3p (*FOXP3*), and tumor necrosis factor-α (*TNF-α*); and those responsible for high cholesterol and homocysteine levels, such as p66shc, lectin-like oxLDL receptor (*LOX1*), and apolipoprotein E (*APOE*) [24].

14.14 EPIGENETICS OF HYPERTENSION

Blood pressure is a trait with an estimated heritability ranging from 30% to 70%, influenced by several biological pathways and responsive to environmental stimuli [55]. Small increments in blood pressure are associated with an increased risk of cardiovascular events [56]. GWAS have greatly contributed to the progress in understanding the role of genetic components in hypertensive disease [57,58]. However, a simple DNA-sequence approach only partially explains genome involvement in the pathogenesis of hypertension, which suggests the involvement of epigenetic mechanisms that are potentially modifiable by environmental factors.

A recent study described global DNA hypomethylation of peripheral blood mononuclear cells (PBMC) of patients affected by essential hypertension, compared to controls [59]. Candidate-gene studies have shown significant DNA methylation changes in genes related to hypertension. Hypermethylation of the promoter of *HSD11B2* in PBMC DNA of hypertensive patients is inversely related to enzyme function, disrupts cortisol degradation to cortisone, and thereby promotes hypertension [60]. Expression of Na^+-K^--2Cl cotransporter 1 (*NKCC1*), a solute carrier, is upregulated by a mechanism induced by gene promoter hypomethylation in a spontaneously hypertensive rodent model, leading to overexpression of *NKCC1*, associated with hypertension. These results encourage studies on methylation *NKCC1* status in hypertensive patients [27]. Another example is the hypermethylation of the promoter region of the norepinephrine transporter gene associated with DNA binding by MECP-2, which leads to exaggerated autonomic responsiveness, resulting in essential hypertension [61]. A recent genome-wide association and replication study in 320,251 people identified new genetic loci influencing blood pressure phenotypes (*IGFBP3, KCNK3, PDE3A, PRDM6, ARHGAP24, OSR1, SLC22A7*, and *TBX2*). The majority of the loci identified do not contain common or low-frequency coding variants to account for the association between the sentinel SNP and blood pressure. Using a 450K methylation array and mapping through bisulfite sequencing, the authors showed that these SNPs influencing blood pressure are associated in *cis* with methylation at multiple local CpG sites and that DNA methylation is associated with blood pressure [62].

14.15 EPIGENETICS OF TYPE 2 DIABETES MELLITUS

Type 2 diabetes mellitus (DM) is an established risk factor for a wide range of vascular diseases, including IS, independently of other conventional risk factors. The prevalence of DM in IS ranges from 15% to 44% [63]. DM is a complex disease, and both genetic and environmental factors contribute to its development. However, despite the depth of available research in the genetic area, all the identified DM risk variants explain only ~10% of its estimated heritability [64]. Environmental factors, such as diet, contribute to the etiology of DM and its associated microvascular and macrovascular complications. Recent studies have proposed that specific changes in the epigenome serve as markers of DM risk and are associated with DM onset and progression [65]. Global LINE-1 hypomethylation is associated with a higher risk of DM, independently of other classical risk factors [66]. Genome-wide methylation studies have shown that hypomethylation of a CpG site, cg19693031 located in the *TXNIP* gene, is related to type 2 DM. The inverse relationship between *TXNIP* methylation and HbA1c levels suggests that *TXNIP* hypomethylation is a consequence of sustained hyperglycemia levels. It has been reported that *TXNIP* expression is highly sensitive to glucose concentration, and DNA methylation may be a key mechanism to modulate its expression, and is likely to be an early biomarker for impaired glucose homeostasis [67].

14.16 EPIGENETICS OF HYPERLIPIDEMIA AND OBESITY

Total cholesterol (TC), low-density lipoprotein cholesterol (LDL-C), high-density lipoprotein cholesterol (HDL-C), and triglycerides (TG) are among the most important risk factors for cardiovascular diseases, the leading cause of death worldwide [68]. Serum lipid levels are determined by a complex interplay between environmental, lifestyle, and genetic factors. According to heritability studies, inheritance could account for up to 60% of the interindividual variability in plasma lipid concentration [69]. Genetic association studies have identified only a few well-validated genes that account for a small proportion of the heritability estimates. Approximately 35–40% of plasma lipid level heritability cannot be explained by current knowledge [70]. It is well known that dietary patterns and physical activity practices play an important role in the regulation of lipid levels [71]; on the other hand, many loci associated with lipid traits have been discovered using GWAS and linkage studies [72,73] and used to validate or identify new therapeutic targets [74]. DNA methylation at five key lipoprotein metabolism gene loci (*ABCG1*, *CETP*, *LIPC*, *LPL*, and *PLTP*) contributed to the plasma lipid level variability in familial hypercholesterolemia, independently of well-known traditional predictors [75].

On the other hand, despite the large number of epigenetic studies of obesity and metabolic diseases using animal models, there are only a few epigenetic studies of obesity in humans. Most of the human studies have explored the DNA methylation status of a few previously identified genes known to affect obesity. Several differentially methylated CpGs associated with body mass index (BMI), such as *HIF3A*, *UBASH3A*, and *TRIM3*, have been identified by EWAS [76,77]. Monozygotic (MZ) twin pairs show discordance for obesity. A genome-wide leukocyte DNA methylation study in 30 clinically healthy young adult MZ twin pairs (average within-pair BMI difference: 5.4 ± 2.0 kg/m^2) identified differentially methylated CpGs included in 23 genes known to be associated with obesity, liver fat, type 2 DM, and metabolic syndrome [78].

14.17 EPIGENETICS OF ATRIAL FIBRILLATION

Cardiac arrhythmias are caused by altered conduction properties of the heart, which may arise in response to ischemia, inflammation, fibrosis, aging, or genetic factors. AF is the most common cardiac arrhythmia, affecting 1–2% of the general population, with well-established clinical and genetic risk components. Common variants in genomic regions associated with AF (*KCNN3, PRRX1, PITX2, WNT8A, CAV1, C9orf3, SYNE2, HCN4,* and *ZFHX3* genes) have been identified by GWAS. However, the genetic variability of these risk variants does not explain the entire genetic susceptibility to AF [79].

The first evidence for epigenetic regulation of cardiac rhythm came from a study conducting microarrays on heart rhythm determinants on tissue exposed to either intermittent or chronic hypoxia, in wild-type mice. Hypoxia changed the expression profile of epigenetic modulators, such as *Hdac5*, *Mef2b*, and *Mef2c* [80]. Variable imprinting of the *KCNQ1* gene provides a possible explanation for the existence of long-QT syndrome in the absence of a coding sequence mutation in *KCNQ1*. The HDACi trichostatin A (TSA) has been shown to reduce the amount of atrial fibrosis and concomitant AF in mice [81].

The most-studied epigenetic mechanism in arrhythmias has been noncoding RNAs, mainly miRNA. miRNA expression profiles were shown to differ in right atrial disease, with 47 miRNAs being differentially expressed between disease and control states, whereas similar changes in expression could not be found in left atrial disease [82]. Multiple studies have shown that miR-208a plays an important role in action potential conduction. Overexpression of miR-208a leads to arrhythmia, cardiac fibrosis, and hypertrophy, and is a strong predictor of cardiac death [83]. Similarly, miR-328 is upregulated not only in animal models of AF but also in human tissue samples from AF patients. miRNAs have also emerged as possible therapeutic targets for the treatment of AF [81,82].

14.18 EPIGENETICS AND CIGARETTE SMOKING

Smoking is the second leading risk factor for global disease burden worldwide. Recent studies estimate that the rate of death from any cause is almost 3 times higher among current smokers, compared to never smokers [84]. The harmful health effects associated with smoking are mediated through a variety of mechanisms that are not fully understood. These include direct DNA damage, vascular dysfunction, inflammation, and platelet functionality, among others [85]. Genes for nicotine dependence and smoking behavior have been linked by GWAS to increased risk of cardiovascular, pulmonary, and malignant diseases [86,87].

The largest effects on the methylome that have been identified to date in epigenetic studies have been due to the extreme environmental influence of tobacco smoking. Using Infinium Human Methylation Bead Chip (27 and 450 K), EWAS have discovered numerous smoking-related CpG sites in whole blood samples collected through epidemiological studies. A total of 1460 smoking-related CpG sites were identified in 14 EWAS, of which 62 were reported by multiple (≥3) studies. The most frequently reported sites were cg05575921 (*AHRR*), cg03636183 (*F2RL3*), cg19859270 (*GPR15*), and other loci within the intergenic regions *2q37.1* and *6p21.33* [88]. However, in most of these CpG sites a trend to recover methylation levels typical of never smokers was observed among those who quit smoking [89]. In EWAS investigating the role of maternal smoking in newborns, differentially methylated CpG sites were also identified in several smoking-related genes, such as *AHRR*, *MYO1G*, and *GFI1*, but with less pronounced effect sizes. Interestingly, none of these studies reported differential methylation of cg03636183 (*F2RL3*) and cg19859270 (*GPR15*), two critical loci associated with adult smoking [88,90].

14.19 EPIGENETICS AND NUTRITION

A relationship between diet and stroke has been previously demonstrated. Although diet may influence stroke risk via several mechanisms, the optimal dietary habits for this purpose are not clearly established. However, the Mediterranean diet is one of the best dietary patterns for stroke prevention [91]. The high complexity of nutritional studies makes it difficult to confer a pathophysiological role to isolated dietary components. Dietary recommendations to reduce stroke risk include reduced sodium intake and increased consumption of fruits and vegetables, whole grains, cereal fiber, and fatty fish. The recommended healthy lifestyle consists of physical activity, increased fruit and vegetable intake, moderation in alcohol consumption, maintaining a healthy weight, and reduced intake of salt, saturated fat, and total fat [92–94].

Nutriepigenomics is an area of epigenomics that explores and defines the rapidly evolving field of diet-genome interactions. Nutrient-dependent epigenetic variations can significantly affect genome stability, mRNA and protein expression, and metabolic changes. Dietary bioactive compounds can affect epigenetic alterations, which are accumulated over time and are shown to be involved in the pathogenesis of age-related diseases, such as diabetes, cancer, and cardiovascular diseases [95].

Dietary factors that influence one-carbon metabolism, which is centered around folate, also modulate DNA. Folate, an important B vitamin (a family of vitamins that are coenzymes in one-carbon metabolism), is required for DNA synthesis. It supplies carbon for purine and pyrimidine synthesis and is required for DNA methylation. Therefore, folate deficiency impairs DNA synthesis and leads to expression changes in some genes due to hypomethylation of DNA. Furthermore, folate deficiencies cause homocysteine accumulation. Homocysteine is an amino acid not used in protein synthesis. Instead, its role is to serve as an intermediate in methionine metabolism. Folate acts as a cofactor for the enzyme methionine synthase, and disruption of the folate pathway leads to hyperhomocysteinemia [96,97].

14.20 TISSUE SPECIFICITY: WHICH IS THE BEST TISSUE FOR STUDYING STROKE EPIGENETICS?

DNA methylation variants can be tissue-specific or shared across tissues. It is important to identify and sample the tissue that is most relevant for the trait, recognizing that the tissue of interest is a product of a mixed cell population. However, often the most appropriate tissue may not be available or easily accessible, and for many studies only whole-blood DNA will be available. The suitability of whole blood as a surrogate for DNA methylation in other tissues is extensively debated [20,98]. Undoubtedly, the lack of the most representative tissue or cell type, that is, the brain, to study stroke pathology is one of the main limitations of stroke epigenetic studies.

Some studies have shown consistency in methylation between tissues, such as peripheral blood compared to the primary tissue for a specific disease [99]. Nevertheless, certain primary tissues are simply unavailable for sampling from live subjects (e.g., brain tissue). Thus, we may be limited to mixed or indirect signals of "true" effects. There is a growing body of literature to suggest that blood or lymphocyte epigenetic profiles may be related to tissues of interest. Interindividual differences in gene expression—and, in a limited set, DNA methylation—were consistent across blood and brain tissues; however, the utility of blood DNA methylation for nonblood-related disorders is still an open question [100].

14.21 NEUROPSYCHIATRIC ASPECTS OF CEREBROVASCULAR DISEASES

In the context of the current book, patients with stroke are well known to develop psychiatric symptoms [101]. Immediately after a stroke, there can be confusion and disorientation which may be slow to clear if cerebral damage has been extensive. Personality changes are very troublesome sequelae of stroke. Such changes include being slowed down, excessive worry, complaints of aches and pains, irritability, boredom, and frustration. Cognitive impairment is common in patients after stroke, and is known to occur in about 35% of such patients, commonly involving memory, orientation, language, and attention [102]. From 6% to 32% of patients are also known to develop dementia after stroke. This wide estimate of prevalence is thought to be due to the different classification criteria used, the ages of the cohorts studied, and the differences in the time interval between the stroke and the time of assessment [101].

It is hardly surprising that patients who survive stroke develop depression. Reasons for the development of depression in such patients include an emotional response to the sudden development of disability and its associated changes; the direct result of injury to the brain leading to changes in the biochemical balance within the brain resulting in mood changes; and a preceding tendency for depression or a history of depression. The presence of depression in patients with stroke correlates with increased mortality in such patients. A generalized anxiety disorder is also found in about 25% of stroke patients, usually in association with depression [101].

14.22 CONCLUSIONS

Stroke is a complex and multifactorial disease due to a wide variety of genetic and environmental factors, in which epigenetic mechanisms are emerging as important players (Fig. 14.1). The study of epigenetics in stroke is moving forward in recent years and, as appears evident from the studies included in this chapter, many areas of stroke epigenetics need to be further explored, both by a mechanistic and a clinical approach. It is important to point out that epigenetic events, in contrast to genetic

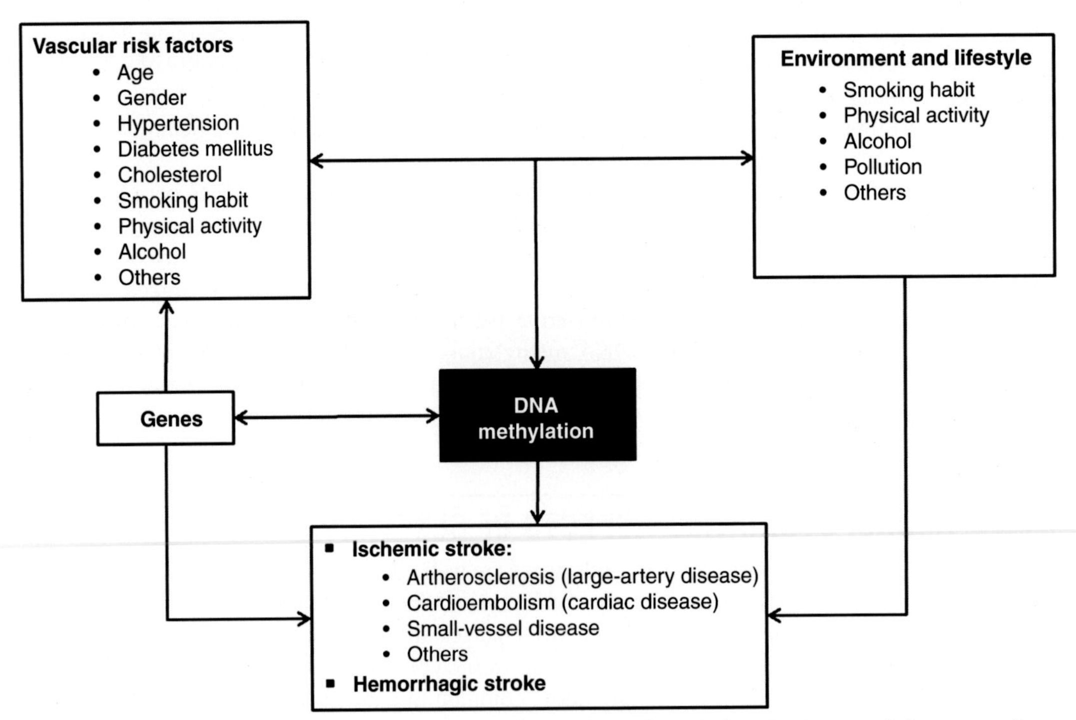

FIGURE 14.1 The Relationship Between Genetic Variation,Eepigenetic Variation, Environment Influence, and Vascular Risk Factors in Stroke

Stroke risk is the result of the complex interaction of all these factors.

sequencing, are potentially reversible, giving epigenetics an even more challenging role in finding innovative preventive and therapeutic approaches to complex, multifactorial diseases. Certainly a combined and integrative genetic-epigenetic approach can illuminate how our genome is expressed differently in a disease state. The rapidly growing field of epigenetics may indeed help to provide more insights into mechanisms involved in bringing the genotype into being and therefore explaining, at least partially, the plasticity of our genome. Three key ideas must be highlighted: first, epigenetic mechanisms are involved in the onset and development of stroke; second, epigenetic markers could be useful in the early diagnosis of stroke and the determination of prognosis; and third, nutritional or pharmacological agents might be used as novel therapeutic strategies due to their ability to modulate epigenetic processes.

ABBREVIATIONS

AF Atrial fibrillation
Bcl-2 B-cell lymphoma 2
CT Computerized tomography
DM Diabetes mellitus
EWAS Epigenome-wide association study
GWAS Genome-wide association study
5-hmc 5-Hydroxymethylcytosine
Hsp70 Heat shock protein 70
ICH Intracerebral hemorrhage
IMT Intima-media thickness
IS Ischaemic stroke
LINE-1 Long interspersed nucleotide element1
miRNA MicroRNA
MRI Magnetic resonance imaging
MTHFR Methylenetetrahydrofolate reductase

GLOSSARY

Cryptogenic stroke Cerebral ischaemia of obscure or uncertain origin
Epigenome-wide association study A microarray-based and sequencing-based approach allowing economical, high-throughput profiling of epigenetic marks, with a primary focus on DNA methylation
Exome sequencing A technique for sequencing all the expressed genes in a genome (known as the exome)
Genome-wide association study An approach that involves rapidly scanning markers across the complete genomes of many individuals to find genetic variations associated with a particular disease
Grp78 A member of the heat-shock protein-70 family of proteins
microRNA A noncoding RNAs that is 21–23 nucleotides long
Noncoding RNA An RNA that is transcribed from DNA but are not translated into protein
Penumbra area The area surrounding an ischaemic event like stroke
Stroke Also known as cerebrovascular accident, or cerebrovascular insult. Occurs when poor blood flow to the brain results in cell death

REFERENCES

[1] WHO. Stroke, cerebrovascular accident. World Health Organization; 2014. http://www.who.int/topics/cerebrovascular_accident/en/

[2] Adams HP, Bendixen BH, Kappelle LJ, Biller J, Love BB, Gordon DL, et al. Classification of subtype of acute ischemic stroke. Definitions for use in a multicenter clinical trial. TOAST. Trial of Org 10172 in Acute Stroke Treatment. Stroke 1993;24:35–41.

[3] Wolf PA, D'Agostino RB, Belanger AJ, Kannel WB. Probability of stroke: a risk profile from the Framingham Study. Stroke 1991;22:312–8.

[4] Goldstein LB, Adams R, Alberts MJ, Appel LJ, Brass LM, Bushnell DC, et al. Primary prevention of ischemic stroke: a guideline from the American Heart Association/American Stroke Association Stroke Council: cosponsored by the Atherosclerotic Peripheral Vascular Disease Interdisciplinary Working Group; Cardiovascular Nursing Council. Circulation 2006;113:e873–923.

[5] Fox CS, Polak JF, Chazaro I, Cupples A, Wolf PA, D'Agostino RA, et al. Genetic and environmental contributions to atherosclerosis phenotypes in men and women: heritability of carotid intima-media thickness in the Framingham Heart Study. Stroke 2003;34:397–401.

[6] Turner ST, Jack CR, Fornage M, Mosley TH, Boerwinkle E, de Andrade M. Heritability of leukoaraiosis in hypertensive sibships. Hypertension 2004;43:483–7.

[7] Bevan S, Traylor M, Adib-Samii P, Malik R, Paul NL, Jackson C, et al. Genetic heritability of ischemic stroke and the contribution of previously reported candidate gene and genomewide associations. Stroke 2012;43:3161–7.

[8] Holliday EG, Maguire JM, Evans T-J, Koblar SA, Jannes J, Sturm JW, et al. Common variants at 6p21.1 are associated with large artery atherosclerotic stroke. Nat Genet 2012;44:1147–51.

[9] Dichgans M. Genetics of ischaemic stroke. Lancet Neurol 2007;6:149–61.

[10] Flossmann E, Schulz UGR, Rothwell PM. Systematic review of methods and results of studies of the genetic epidemiology of ischemic stroke. Stroke 2004;35:212–27.

[11] Traylor M, Farrall M, Holliday EG, Sudlow C, Hopewell JC, Cheng YC, et al. Genetic risk factors for ischaemic stroke and its subtypes (the METASTROKE collaboration): a meta-analysis of genome-wide association studies. Lancet Neurol 2012;11:951–62.

[12] Bellenguez C, Bevan S, Gschwendtner A, Spence CC, Burgess AL, Pirinen M, et al. Genome-wide association study identifies a variant in HDAC9 associated with large vessel ischemic stroke. Nat Genet 2012;44:328–33.

[13] NINDS Stroke Genetics Network (SiGN) ISGC (ISGC). Loci associated with ischaemic stroke and its subtypes (SiGN): a genome-wide association study. Lancet Neurol 2016;15(2):174–84.

[14] Auer PL, Nalls M, Meschia JF, Worrall BB, Longstreth WT Jr, Seshadri S, et al. Rare and coding region genetic variants associated with risk of ischemic stroke: the NHLBI Exome Sequence Project. JAMA Neurol 2015;72:781–8.

[15] Eichler EE, Flint J, Gibson G, Kong A, Leal SM, Moore JH, et al. Missing heritability and strategies for finding the underlying causes of complex disease. Nat Rev Genet 2010;11:446–50.

[16] Ordovás JM, Smith CE. Epigenetics and cardiovascular disease. Nat Rev Cardiol 2010;7:510–9.

[17] Feinberg AP. Phenotypic plasticity and the epigenetics of human disease. Nature 2007;447:433–40.

[18] Kato T, Iwamoto K, Kakiuchi C, Kuratomi G, Okazaki Y. Genetic or epigenetic difference causing discordance between monozygotic twins as a clue to molecular basis of mental disorders. Mol Psychiatry 2005;10:622–30.

[19] Qureshi IA, Mehler MF. Emerging role of epigenetics in stroke: part 1: DNA methylation and chromatin modifications. Arch Neurol 2010;67:1316–22.

[20] Portela A, Esteller M. Epigenetic modifications and human disease. Nat Biotechnol 2010;28:1057–68.

[21] Sandoval J, Esteller M. Cancer epigenomics: beyond genomics. Curr Opin Genet Dev 2012;22:50–5.

[22] Petronis A. Epigenetics as a unifying principle in the aetiology of complex traits and diseases. Nature 2010;465:721–7.

[23] Baccarelli A, Wright R, Bollati V, Litonjua A, Zanobetti A, Tarantini L, et al. Ischemic heart disease and stroke in relation to blood DNA methylation. Epidemiology 2010;21:819–28.

[24] Grimaldi V, Vietri MT, Schiano C, Picascia A, De Pascale MR, Fiorito C, et al. Epigenetic reprogramming in atherosclerosis. Curr Atheroscler Rep 2015;17:476.

[25] Turunen MP, Aavik E, Yla-Herttuala S. Epigenetics and atherosclerosis. Biochim Biophys Acta 2009;1790:886–91.

[26] Kim M, Long TI, Arakawa K, Wang R, Yu MC, Laird PW. DNA methylation as a biomarker for cardiovascular disease risk. PLoS One 2010;5:e9692.

[27] Udali S, Guarini P, Moruzzi S, Choi S-W, Friso S. Cardiovascular epigenetics: from DNA methylation to microRNAs. Mol Aspects Med 2013;34:883–901.

[28] Deininger PL, Moran JV, Batzer MA, Kazazian HH. Mobile elements and mammalian genome evolution. Curr Opin Genet Dev 2003;13:651–8.

[29] Weisenberger DJ, Campan M, Long TI, Kim M, Woods C, Fiala E, et al. Analysis of repetitive element DNA methylation by MethyLight. Nucleic Acids Res 2005;33:6823–36.

[30] Schulz WA, Steinhoff C, Florl AR. Methylation of endogenous human retroelements in health and disease. Curr Top Microbiol Immunol 2006;310:211–50.

[31] Baccarelli A, Tarantini L, Wright RO, Bollati V, Litonjua AA, Zanobetti A, et al. Repetitive element DNA methylation and circulating endothelial and inflammation markers in the VA normative aging study. Epigenetics 2010;5.

[32] Karimi M, Johansson S, Ekstrom TJ. Using LUMA: a Luminometric-based assay for global DNA-methylation. Epigenetics 2006;1:45–8.

[33] Soriano-Tárraga C, Jiménez-Conde J, Giralt-Steinhauer E, Mola M, Ois A, Rodriguez-Campello A, et al. Global DNA methylation of ischemic stroke subtypes. PLoS One 2014;9:e96543.

[34] Kelly PJ, Rosand J, Kistler JP, Shih VE, Silveira S, Plomaritoglou A, et al. Homocysteine, MTHFR 677C—>T polymorphism, and risk of ischemic stroke: results of a meta-analysis. Neurology 2002;59:529–36.

[35] Casas JP, Hingorani AD, Bautista LE, Sharma P. Meta-analysis of genetic studies in ischemic stroke: thirty-two genes involving approximately 18,000 cases and 58,000 controls. Arch Neurol 2004;61:1652–61.

[36] Wei LK, Sutherland H, Au A, Camilleri E, Haupt LM, Gan SH, et al. A potential epigenetic marker mediating serum folate and vitamin B12 levels contributes to the risk of ischemic stroke. Biomed Res Int 2015;2015:167976.

[37] Zaina S, Heyn H, Carmona FJ, Varol N, Sayols S, Condom E, et al. DNA methylation map of human atherosclerosis. Circ Cardiovasc Genet 2014;7:692–700.

[38] Zaina S, Gonçalves I, Carmona FJ, Gomez A, Heyn H, Mollet IG, et al. DNA methylation dynamics in human carotid plaques after cerebrovascular events. Arterioscler Thromb Vasc Biol 2015;35:1835–42.

[38a] Gallego-Fabrega C, Carrera C, Reny J-L, Fontana P, Slowik A, Pera J, et al. PPM1A methylation is associated with vascular recurrence in aspirin-treated patients. Stroke 2016;47:1926–9.

[38b] Gallego-Fabrega C, Carrera C, Reny J-L, Fontana P, Slowik A, Pera J, et al. TRAF3 epigenetic regulation is associated with vascular recurrence in patients with is chemic stroke. Stroke 2016;47:1180–6.

[39] Ito S, D'Alessio AC, Taranova OV, Hong K, Sowers LC, Zhang Y. Role of Tet proteins in 5mC to 5hmC conversion, ES-cell self-renewal and inner cell mass specification. Nature 2010;466:1129–33.

[40] Ruzov A, Tsenkina Y, Serio A, Dudnakova T, Fletcher J, Bai Y, et al. Lineage-specific distribution of high levels of genomic 5-hydroxymethylcytosine in mammalian development. Cell Res 2011;21:1332–42.

[41] Pastor WA, Pape UJ, Huang Y, Henderson HR, Lister R, Ko M, et al. Genome-wide mapping of 5-hydroxymethylcytosine in embryonic stem cells. Nature 2011;473:394–7.

[42] Chouliaras L, Mastroeni D, Delvaux E, Grover A, Kenis G, Hof PR, et al. Consistent decrease in global DNA methylation and hydroxymethylation in the hippocampus of Alzheimer's disease patients. Neurobiol Aging 2013;34:2091–9.
[43] Wang F, Yang Y, Lin X, Wang JQ, Wu YS, Xie W, et al. Genome-wide loss of 5-hmC is a novel epigenetic feature of Huntington's disease. Hum Mol Genet 2013;22:3641–53.
[44] Miao Z, He Y, Xin N, Sun M, Chen L, Lin L, et al. Altering 5-hydroxymethylcytosine modification impacts ischemic brain injury. Hum Mol Genet 2015;24:5855–66.
[45] Fraineau S, Palii CG, Allan DS, Brand M. Epigenetic regulation of endothelial-cell-mediated vascular repair. FEBS J 2015;282:1605–29.
[46] Baltan S, Bachleda A, Morrison RS, Murphy SP. Expression of histone deacetylases in cellular compartments of the mouse brain and the effects of ischemia. Transl Stroke Res 2011;2:411–23.
[47] Picascia A, Grimaldi V, Iannone C, Soricelli A, Napoli C. Innate and adaptive immune response in stroke: focus on epigenetic regulation. J Neuroimmunol 2015;289:111–20.
[48] Kassis H, Chopp M, Liu XS, Shehadah A, Roberts C, Zhang ZG. Histone deacetylase expression in white matter oligodendrocytes after stroke. Neurochem Int 2014;77:17–23.
[49] Yan W, Fang Z, Yang Q, Dong H, Lu Y, Lei C, et al. SirT1 mediates hyperbaric oxygen preconditioning-induced ischemic tolerance in rat brain. J Cereb Blood Flow Metab 2013;33:396–406.
[50] Schweizer S, Meisel A, Märschenz S. Epigenetic mechanisms in cerebral ischemia. J Cereb Blood Flow Metab 2013;33:1335–46.
[51] Felling RJ, Song H. Epigenetic mechanisms of neuroplasticity and the implications for stroke recovery. Exp Neurol 2015;268:37–45.
[52] Tan KS, Armugam A, Sepramaniam S, Lim KY, Setyowati KD, Wang CW, et al. Expression profile of MicroRNAs in young stroke patients. PLoS One 2009;4:e7689.
[53] Jickling GC, Ander BP, Zhan X, Noblett D, Stamova B, Liu D. microRNA expression in peripheral blood cells following acute ischemic stroke and their predicted gene targets. PLoS One 2014;9:e99283.
[54] Long G, Wang F, Li H, Yin Z, Sandip C, Lou Y, et al. Circulating miR-30a, miR-126 and let-7b as biomarker for ischemic stroke in humans. BMC Neurol 2013;13:178.
[55] Menni C, Mangino M, Zhang F, Clement G, Snieder H, Padmanabhan S, et al. Heritability analyses show visit-to-visit blood pressure variability reflects different pathological phenotypes in younger and older adults: evidence from UK twins. J Hypertens 2013;31:2356–61.
[56] Lewington S, Clarke R, Qizilbash N, Peto R, Collins R. Age-specific relevance of usual blood pressure to vascular mortality: a meta-analysis of individual data for one million adults in 61 prospective studies. Lancet 2002;360:1903–13.
[57] Levy D, Ehret GB, Rice K, Verwoert GC, Launer LJ, Dehghan A, et al. Genome-wide association study of blood pressure and hypertension. Nat Genet 2009;41:677–87.
[58] Ehret GB, Munroe PB, Rice KM, Bochud M, Johnson AD, Chasman DI, et al. Genetic variants in novel pathways influence blood pressure and cardiovascular disease risk. Nature 2011;478:103–9.
[59] Friso S, Carvajal CA, Fardella CE, Olivieri O. Epigenetics and arterial hypertension: the challenge of emerging evidence. Transl Res 2015;165:154–65.
[60] Friso S, Pizzolo F, Choi S-W, Guarini P, castagna A, Ravagnani V, et al. Epigenetic control of 11 beta-hydroxysteroid dehydrogenase 2 gene promoter is related to human hypertension. Atherosclerosis 2008;199:323–7.
[61] Esler M, Eikelis N, Schlaich M, Lambert G, Alvarenga M, Kaye D, et al. Human sympathetic nerve biology: parallel influences of stress and epigenetics in essential hypertension and panic disorder. Ann NY Acad Sci 2008;1148:338–48.
[62] Kato N, Loh M, Takeuchi F, Verweij N, Wang X, Zhang W, et al. Trans-ancestry genome-wide association study identifies 12 genetic loci influencing blood pressure and implicates a role for DNA methylation. Nat Genet 2015;47:1282–93.

[63] Roquer J, Rodríguez-Campello A, Cuadrado-Godia E, Giralt-Steinhauer E, Jimenez-Conde J, Soriano C, et al. The role of HbA1c determination in detecting unknown glucose disturbances in ischemic stroke. PLoS One 2014;9:e109960.
[64] Voight BF, Scott LJ, Steinthorsdottir V, Morris AP, Dina C, Welch RP, et al. Twelve type 2 diabetes susceptibility loci identified through large-scale association analysis. Nat Genet 2010;42:579–89.
[65] Pirola L, Balcerczyk A, Okabe J, El-Osta A. Epigenetic phenomena linked to diabetic complications. Nat Rev Endocrinol 2010;6:665–75.
[66] Martín-Núñez GM, Rubio-Martín E, Cabrera-Mulero R, Rojo-Martinez G, Olveira G, Valdes S, et al. Type 2 diabetes mellitus in relation to global LINE-1 DNA methylation in peripheral blood: a cohort study. Epigenetics 2014;9:1322–8.
[67] Soriano-Tárraga C, Jiménez-Conde J, Giralt-Steinhauer E, Mola-Caminal M, Vivanco-Hidalgo RM, Ois A, et al. Epigenome-wide association study indentifies TXNIP gene associated with type 2 diabetes mellitus and sustained hyperglycemia. Hum Mol Genet 2016;25:609–19.
[68] Mozaffarian D, Benjamin EJ, Go AS, Arnett DK, Blaha MJ, Cushman M, et al. Heart disease and stroke statistics—2015 update: a report from the American Heart Association. Circulation 2015;131:e29–e322.
[69] Almgren P, Lehtovirta M, Isomaa B, Sarelin L, Taskinen MR, Lyssenko V, et al. Heritability and familiality of type 2 diabetes and related quantitative traits in the Botnia Study. Diabetologia 2011;54:2811–9.
[70] Asselbergs FW, Guo Y, van Iperen EPA, Sivapalaratnam S, Tragante V, Lanktree MB, et al. Large-scale gene-centric meta-analysis across 32 studies identifies multiple lipid loci. Am J Hum Genet 2012;91: 823–38.
[71] Kelley GA, Kelley KS, Roberts S, Haskell W. Comparison of aerobic exercise, diet or both on lipids and lipoproteins in adults: a meta-analysis of randomized controlled trials. Clin Nutr 2012;31:156–67.
[72] Teslovich TM, Musunuru K, Smith AV, Edmondson AC, Stylianou IM, Koseki M, et al. Biological, clinical and population relevance of 95 loci for blood lipids. Nature 2010;466:707–13.
[73] Abifadel M, Varret M, Rabès J-P, Allard D, Ougerram K, Devillers M, et al. Mutations in PCSK9 cause autosomal dominant hypercholesterolemia. Nat Genet 2003;34:154–6.
[74] Sabatine MS, Giugliano RP, Wiviott SD, Raal FJ, Blom DJ, Robinson J, et al. Efficacy and safety of evolocumab in reducing lipids and cardiovascular events. N Engl J Med 2015;372:1500–9.
[75] Guay S-P, Brisson D, Lamarche B, Gaudet D, Bouchard L. Epipolymorphisms within lipoprotein genes contribute independently to plasma lipid levels in familial hypercholesterolemia. Epigenetics 2014;9:718–29.
[76] Almén MS, Nilsson EK, Jacobsson JA, Kalnina I, Klovins J, Fredriksson R, et al. Genome-wide analysis reveals DNA methylation markers that vary with both age and obesity. Gene 2014;548:61–7.
[77] Dick KJ, Nelson CP, Tsaprouni L, Sandling JK, Aissi D, Wahl S, et al. DNA methylation and body-mass index: a genome-wide analysis. Lancet 2014;383:1990–8.
[78] Ollikainen M, Ismail K, Gervin K, Kyllonen A, Hakkarainen A, Lundbom J, et al. Genome-wide blood DNA methylation alterations at regulatory elements and heterochromatic regions in monozygotic twins discordant for obesity and liver fat. Clin Epigenet 2015;7:39.
[79] Olesen MS, Nielsen MW, Haunsø S, Svendsen JH. Atrial fibrillation: the role of common and rare genetic variants. Eur J Hum Genet 2014;22:297–306.
[80] Iacobas DA, Iacobas S, Haddad GG. Heart rhythm genomic fabric in hypoxia. Biochem Biophys Res Commun 2010;391:1769–74.
[81] Duygu B, Poels EM, da Costa Martins PA. Genetics and epigenetics of arrhythmia and heart failure. Front Genet 2013;4:219.
[82] Kim GH. MicroRNA regulation of cardiac conduction and arrhythmias. Transl Res 2013;161:381–92.
[83] Oliveira-Carvalho V, Carvalho VO, Bocchi EA. The emerging role of miR-208a in the heart. DNA Cell Biol 2013;32:8–12.
[84] Carter BD, Abnet CC, Feskanich D, Freedman ND, Hartge P, Lewis CE, et al. Smoking and mortality—beyond established causes. N Engl J Med 2015;372:631–40.

[85] Messner B, Bernhard D. Smoking and cardiovascular disease: mechanisms of endothelial dysfunction and early atherogenesis. Arterioscler Thromb Vasc Biol 2014;34:509–15.
[86] Shenker NS, Polidoro S, van Veldhoven K, Sacerdote C, Ricceri F, Birrell MA, et al. Epigenome-wide association study in the European Prospective Investigation Into Cancer And Nutrition (EPIC-Turin) identifies novel genetic loci associated with smoking. Hum Mol Genet 2013;22:843–51.
[87] Zeilinger S, Kühnel B, Klopp N, Baurecht H, Kleinschmidt A, Gieger C, et al. Tobacco smoking leads to extensive genome-wide changes in DNA methylation. PLoS One 2013;8:e63812.
[88] Gao X, Jia M, Zhang Y, Breitling LP, Brenner H. DNA methylation changes of whole blood cells in response to active smoking exposure in adults: a systematic review of DNA methylation studies. Clin Epigenet 2015;7:113.
[89] Sayols-Baixeras S, Lluís-Ganella C, Subirana I, Salas LA, Vilahur N, Corella D, et al. Identification of a new locus and validation of previously reported loci showing differential methylation associated with smoking. The REGICOR study. Epigenetics 2015;10:1156–65.
[90] Lee KWK, Richmond R, Hu P, French L, Shin J, Bourdon C, Reischl E, et al. Prenatal exposure to maternal cigarette smoking and DNA methylation: epigenome-wide association in a discovery sample of adolescents and replication in an independent cohort at birth through 17 years of age. Environ Health Perspect 2015;123:193–9.
[91] Estruch R, Ros E, Salas-Salvadó J, Covas MI, Corella D, Aros F, et al. Primary prevention of cardiovascular disease with a Mediterranean diet. N Engl J Med 2013;368:1279–90.
[92] Lin JS, O'Connor E, Whitlock EP, Beil TL. Behavioral counseling to promote physical activity and a healthful diet to prevent cardiovascular disease in adults: a systematic review for the U.S. Preventive Services Task Force. Ann Intern Med 2010;153:736–50.
[93] Rodríguez-Campello A, Jiménez-Conde J, Ois Á, Cuadrado-Godia E, Giralt-Steinhauer E, Schroeder H, et al. Dietary habits in patients with ischemic stroke: a Case-Control Study. PLoS One 2014;9:e114716.
[94] Ding EL, Mozaffarian D. Optimal dietary habits for the prevention of stroke. Semin Neurol 2006;26:11–23.
[95] Vahid F, Zand H, Nosrat-Mirshekarlou E, Najafi R, Hekmatdoost A. The role dietary of bioactive compounds on the regulation of histone acetylases and deacetylases: a review. Gene 2015;562:8–15.
[96] Fenech M. Folate (vitamin B9) and vitamin B12 and their function in the maintenance of nuclear and mitochondrial genome integrity. Mutat Res 2012;733:21–33.
[97] Agodi A, Barchitta M, Quattrocchi A, Maugeri A, Canto C, Marchese AE, et al. Low fruit consumption and folate deficiency are associated with LINE-1 hypomethylation in women of a cancer-free population. Genes Nutr 2015;10:480.
[98] Rakyan VK, Down TA, Balding DJ, Beck S. Epigenome-wide association studies for common human diseases. Nat Rev Genet 2011;12:529–41.
[99] Bakulski KM, Fallin MD. Epigenetic epidemiology: promises for public health research. Environ Mol Mutagen 2014;55:171–83.
[100] Davies MN, Volta M, Pidsley R, Lunnon K, Dixit A, Lovestone S, et al. Functional annotation of the human brain methylome identifies tissue-specific epigenetic variation across brain and blood. Genome Biol 2012;13:R43.
[101] Fleminger S. Cerebrovascular disorders. In: David AS, Fleminger S, Kopelman MD, Lovestone S, Mellers JDC, editors. Lishman's organic psychiatry. 4th ed. Chichester: Wiley-Blackwell; 2009. p. 473–542.
[102] Tatemichi TK, Desmond DW, Stern Y, Paik M, Sano M, Bagiella E. Cognitive impairment after stroke: frequency, patterns, and relationship to functional abilities. J Neurol Neurosurg Psychiatry 1994;57:202–7.

CHAPTER

EPIGENETICS AND EATING DISORDERS

15

H. Frieling, V. Buchholz

Department of Psychiatry, Social Psychiatry and Psychotherapy, Molecular Neuroscience Laboratory, Hannover Medical School (MHH), Hannover, Lower Saxony, Germany

CHAPTER OUTLINE

15.1 INTRODUCTION

Eating disorders (ED) are a group of psychiatric disorders characterized by disturbed eating behavior and preoccupation with eating, and weight and shape concerns. Currently, three diagnostic entities are defined: anorexia nervosa (AN), bulimia nervosa (BN), and binge-eating disorder (BED). Probably even more common than these distinct phenotypes are ED not otherwise specified (EDNOS), which are ED with some but not all classical features of AN, BN, or BED. Obesity is not yet classified as an eating disorder.

AN is described as an eating disorder characterized by a persistent restriction of energy intake, fear of gaining weight and a disturbed perceivement of body shape and weight (DSM-V). The illness has a high prevalence especially in adolescent girls and one of the highest mortality rates in mental disorders [1] and the costs it imposes to afflicted individuals, families, and societies are high [2]. Two distinct subtypes of AN exist: restrictive AN with patients maintaining a low body weight by restricting energy intake without any bingeing or purging behavior, and AN with bingeing and purging, where episodes of binging and purging (i.e., vomiting, abuse of laxatives, excessive physical activation, etc.) occur.

BN is characterized by episodes of binge eating followed by purging behavior like vomiting or abuse of laxatives. Most patients with BN have normal weight, due to effective purging and dieting strategies. Patients with BED suffer from recurrent episodes of binge eating. Contrary to BN, no purging behaviors occur, which explains why most patients with BED are overweight or obese. Subthreshold forms of all disorders are very common and in the natural history of ED, many patients experience

Neuropsychiatric Disorders and Epigenetics. http://dx.doi.org/10.1016/B978-0-12-800226-1.00015-0

AN, BN, and binge eating episodes. The etiology of the disorders is still not entirely understood, but they seem to be multifactorial disorders. Notably, many patients suffering from EDs experience other psychiatric problems, such as anxiety, affective disorders, especially depression, obsessive-compulsive disorder, emotional-instability, or poor impulse control [3,4].

Heritability rates for AN and BN range from 0% to 80%, [5] although a distinct genetic pattern has not yet been determined. Recent GWAS studies suggest that for AN several common genetic variants might lead to a small increase in risk for the disorder (GCAN) [6,7]. Genetic studies in BN have not been performed in a comparable fashion, especially no GWAS for BN has been accomplished so far.

The observed variance may be explained at least in part by environmental factors influencing gene expression and leading to a multifactorial model of the development of the disease. Several environmental risk factors for the development of ED with known effects on epigenetic regulation have been discovered through the last years which will be reviewed in this chapter. Strikingly, almost all patients developing any ED had prolonged episodes of dieting or restriction of energy intake and were trying to lose weight. Metabolic and epigenetic consequences of dieting and malnutrition may play an important role as initiating and maintaining factors of the acute disease.

15.2 DEVELOPMENTAL RISK FACTORS

A large body of evidence has linked pre- and perinatal events to the risk of developing complex disorders like obesity and cardiovascular disease as well as schizophrenia, anxiety, and depression. For EDs it has been shown that female offspring of mothers suffering from anxiety disorder during pregnancy are at an increased risk for developing EDs later in life [8]. Apart from a potential overlapping genetic risk architecture of EDs and anxiety disorders, maternal anxiety is associated with overreactivity of the hypothalamic-pituitary-adrenal (HPA) axis, lower birth weights, increase in obstetric complications, fearful temperament, hyperactivity, disturbances in catecholamine metabolism and structural and functional changes in the brain of the offspring [9]. It has also been shown that offspring of mothers with anxiety disorder differ in learning styles with a preference toward inflexible, habit-based learning strategies. Strikingly, many of these traits are overrepresented in patients with EDs. For example, Favaro and coworkers have repeatedly shown that obstetric complications, such as maternal anemia, preeclampsia, gestational diabetes, or placental infarctions increase the offspring's risk for AN and BN as well as low birth weight for gestational age, neonatal hyporeactivity, and early feeding problems [10,11]. The experience of maternal stress during pregnancy has been proven to exert epigenetic effects on the offspring. Maternal depression during pregnancy is associated with changes in DNA methylation of different genes in the offspring, among them the glucocorticoid receptor (NR3C1) [12,13]. In those children with altered epigenetic control of the HPA axis, increased stress sensitivity was shown at the age of 3 months. Maternal antepartal anxiety and depression were found to lead to alterations in several CpG sites in a recent genome-wide scan of offspring umbilical cord blood cells [14]. But also shorter but more extreme stress experiences affect the unborn child: Project IceStorm followed up children who were in the third gestational trimester when Quebec was hit by an icestorm in 1998. When girls from this project were at the age of 13 years, increased rates of ED were observed [15]. In a different study from Project IceStorm, alterations in DNA methylation patterns in comparison with unaffected children were found [16,17].

Maternal malnutrition during pregnancy might be another important factor for developing ED. Several studies imply that maternal ED during pregnancy can increase the risk for offsprings' ED as well as maternal obesity [18]. The Dutch hunger winter (1944–45) led to a large variety of epigenetically mediated physiological and psychological consequences in persons experiencing the famine in utero during the second or third gestational trimester. Notably, adults who were in the third trimester during the famine had an increased weight, increased risk for obesity, diabetes, and metabolic syndrome and depression, while earlier exposure dramatically increased the risk for neurodevelopmental alterations including an elevated prevalence of schizophrenia in this cohort [19–21]. Similarities in prenatal environmental risk factors between EDs and obesity hint at shared developmental trajectories. Many patients developing ED suffer from being overweight during childhood and especially regarding BN, weight-related teasing and mobbing experience during early puberty often precede the disorder's onset [22]. Early life stress (ELS), especially childhood abuse, but probably also broader types of attachment trauma, are independent risk factors for EDs. Recently, an additive interaction model between pre- and perinatal events and ELS was proposed, indicating that ELS dramatically increases the risk of later ED especially in those subjects with one or more perinatal risk factors [23]. Experiences of sexual abuse are especially present in a high number of patients with bingeing and purging behaviors. Howard Steiger's group showed that in bulimic patients, alterations in the DNA methylation of brain derived neurotrophic factor gene (BDNF) are mainly associated with childhood traumatization [24].

Taken together, a picture of a plethora of developmental risk factors for ED emerges, which seems to be orchestrated by alterations in the epigenome, which fits well into current concepts of "foetal programming" or the developmental origins of health and disease (DOHaD)—hypothesis. However, none of these factors alone can account for the disorder risk and conversely, patients with ED do not necessarily have any of those factors.

15.3 EFFECTS OF MALNUTRITION AND ENERGY RESTRICTION

All ED have one thing in common, they come along with a disturbed eating behavior and consecutively alterations in energy supply and availability of vitamins, amino acids, and essential fatty acids. Normal homeostasis of B-vitamins is essential for the functioning of the methionine cycle, which provides methyl groups for all methylation reactions in the organism. Lack of small amino acids like glycine and serine, and folate, or vitamin B6 or B12 deficiencies decrease the availability of methyl groups, ultimately leading to alterations in DNA and histone methylation. Hyperhomocysteinemia, an indicator for folate deficiency and impaired methionine cycle functioning, has consistently been shown to be present in acute starvation in AN [25,26]. In adolescent anorexia, homocysteine serum levels return to normal after successful therapy [27]. In recovered adults with AN, hyperhomocysteinemia is only partially reversed by successful therapy and weight rehabilitation [28]. In healthy controls and in anorectic patients, hyperhomocysteinemia is associated with a global reduction of methylated cytosines (mC) in peripheral blood cells [29].

Several neuropeptides implicated in energy homeostasis and appetite regulation show alterations in acute starvation, among them leptin, ghrelin, alpha-melanocyte stimulating hormone, agouti-related protein, and neuropeptide Y. Extensive epigenetic control of these hormonal networks has been shown in different disease contexts and are probably also relevant in the context of EDs [30]. To date, candidate gene studies into these peptides did not find alterations in DNA methylation of the leptin gene (although the pilot study involved was probably underpowered) [31] or the proopiomelanocortin gene (POMC) which codes for the melanocyte stimulating hormones [32]. Even though experimental evidence is still lacking, it seems reasonable to assume that mal- or undernutrition due to ED symptomatology affects proper functioning of the "epigenetic machinery." Given that under normal conditions, the brain's energy supply is tightly controlled and even small disturbances or short periods of fasting provoke a multitude of physiological, hormonal, and psychological compensatory measures, it is likely that an impaired epigenetic control can become a maintaining factor of the disordered behavior [33].

15.4 EPIGENETIC ALTERATIONS IN ANOREXIA AND BULIMIA NERVOSA

Our group was the first to publish changes in DNA methylation in AN and BN. In 2007 we were able to show that adult women with AN had a reduced content of mC in peripheral blood cells [29], a finding that was independently replicated in a cohort of adolescent girls with restrictive AN [34]. In the same study, we found that patients with AN or BN had significant hypermethylation of the gene promoter of alpha synuclein that corresponded to reduced mRNA expression levels of alpha synuclein, while the nonchaperone homocysteine-induced endoplasmatic reticulum protein did not show alterations in expression or methylation. In subsequent studies we showed alterations in the DNA methylation of atrial natriuretic peptide [35], the dopamine receptor 2 (DRD2) and the dopamine transporter gene (DAT) [36]. Further candidate gene studies included the oxytocin receptor (OXTR), BDNF, NR3C1, leptin, vasopressin, and the dopamine receptor 4 gene (DRD4) as well as cannabinoid receptor type 1 (Cnr1) with mixed results [31,35–39]. An overview of current candidate gene studies is provided in Table 15.1.

To date, only a small number of genome-wide methylation studies have been published, all of them facing serious power issues. In the first study, 10 women with AN were compared to 10 women without AN. The authors investigated possible differences between patients and controls in methylation of LINE-1 repetitive elements and in the H19 imprinting control region [40]. They did not find any significant alterations in AN when compared to controls. So far only one study obtained genome-wide methylation data using Illumina's Infinium 450K HumanMethylation Beadchip from lymphocytes of 13 women with restricting type AN, 16 with AN with bingeing and purgeing and 15 normal-weight healthy control women. Restricting the number of probes analyzed to CGs in genes known to be expressed in neurons and applying a false discovery rate correction of *p*-values, Booij et al. found 14 CpG probes with significant between-group differences [41]. Two of the identified genes (NR1H3, PXDNL) are implicated in histone acetylation and RNA modifications and most interestingly, in lipid transport and cholesterol storage, as well as dopamine and glutamate signaling. Malfunction of all of these systems has been associated with ED in different studies. In a broader analysis looking for associations with disease duration, 142 CpG probes were identified. All findings are still awaiting replication in large and independent samples.

Apart from AN, no methylomic studies have yet been conducted in the field of ED, reflecting the situation in GWAS. BN and BED have not reached the attention of the "omics" field so far. Given that in some studies prominent differences in biological measures occurred between rAN and bpAN, an approach not focusing on classical disorder concepts, but more on the occurrence of binging and purging behavior will likely be successful and lead to new insights into the epigenetic underpinnings of different cerebral, hormonal, and behavioral networks altered in this large group of disorders with altered eating behavior.

Table 15.1 Overview of Candidate Gene Studies in Eating Disorders

Gene	Methylation	Condition	Remarks	References
Alpha-synuclein	↑ ↔	AN BN		[29]
ANP	↔ ↑	AN BN		[35]
BDNF	↔ ↑	AN BN		[31] [24]
CNR1	↔	AN, BN		[38]
DAT	↑	AN, BN		[36]
DRD2	↑ ↔ ↔	AN AN BN	Hypermethylation in BN associated with comorbid BPD	[36] [31] [42]
DRD4	↔	AN; BN		[36]
Leptin	↔	AN		[31]
NR3C1	↔	BN	Hypermethylation in patients with comorbid BPD, suicidality	[43]
OXTR	↑	AN		[37]
POMC	↔	AN	Association with smoking	[32,44]
SERT	↔	AN		[31]
Vasopressin	↔	AN; BN		[36]

AN, Anorexia nervosa; BN, bulimia nervosa; BPD, borderline personality disorder.

15.5 CONCLUSIONS: AN EPIGENETICALLY INFORMED MODEL OF EATING DISORDERS

Like for most psychiatric disorders, ED have been attributed to a broad variety of causes, many bearing the potential of blaming the patient herself, her parents or society to be responsible for the disorder. An epigenetic model can integrate most replicated environmental risk factors, genetic risk architecture and, most importantly, the role of disturbed nutrition and energy homeostasis in one dynamic and adaptive framework. An epigenetically informed model can help therapists by making a biological model more adaptable to the individual patient's situation thereby leading a path away from models including "character weakness" of the patient or "the bad society," to real persons with a biological susceptibility toward an eating disorder. Especially the role of disordered eating in the pathophysiology of the disorder can be made transparent to the patient stressing the need to reestablish a normal eating pattern. Such a model incorporates (1) vulnerability, and (2) a first prolonged diet initiating a cascade of metabolic and epigenetic changes. If these changes are strengthening certain unhealthy perceptions and behaviors, such as occupation with food, eating and body shape ultimately leading to further restriction of food and energy intake and increasing energy expenditure, a vicious circle can be closed in which

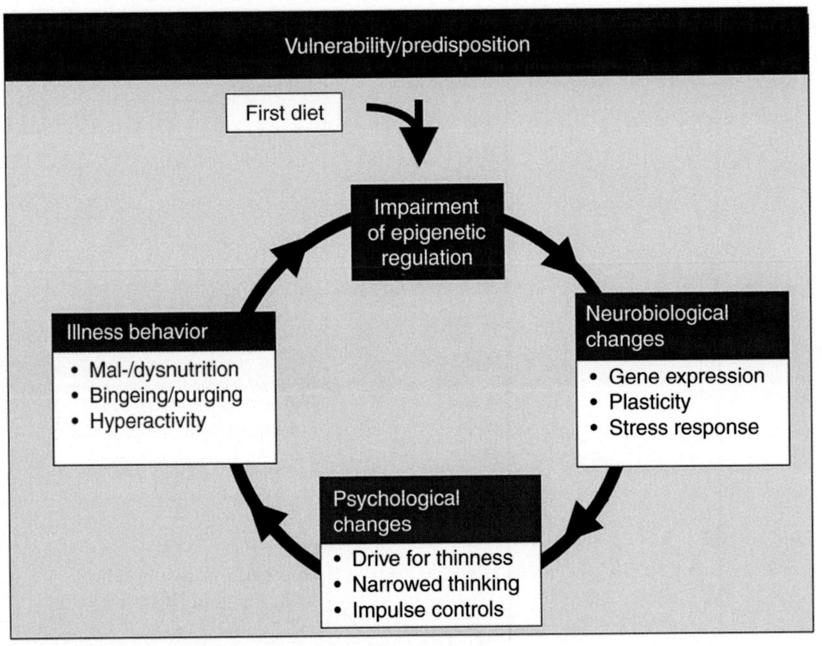

FIGURE 15.1 An Epigenetically Informed Model of Eating Disorders

Genetic and environmental risk factors interact to determine the individual's vulnerability or predisposition toward an eating disorder. A first diet impairs the regular function of the epigenetic machinery potentially leading to persistent disturbances in eating behavior.

epigenetic alterations maintain core features of the acute illness. A simplified diagram of this model that we already use in daily work with patients is provided as Fig. 15.1.

Even though this "epigenetic model of ED" is promising, many questions remain unanswered. We currently do not know which epigenetic alterations are induced by malnutrition or starvation and which are part of the underlying vulnerability and formed by the history of environmental influences of the individual patient. Fasting studies in healthy persons might help answer this question as well as studying those adolescents developing an ED in the large ongoing cohort studies like ALSPAC or IMAGEN. Long term follow-up studies of ED patients are needed to find out which alterations persist after successful weight rehabilitation or remission of bingeing behavior. Are there certain epigenetic alterations predicting response to a certain treatment or might there even be an epigenetic signature that can inform therapists about the best fitting therapy? Another large field of work to be done has not yet been addressed in the ED field at all—namely epigenetic mechanisms apart from DNA methylation. To date no studies into histone modifications, or noncoding RNAs has been performed for AN or BN. Studies investigating specific gene x environment interactions are also still sparse. However, as large networks studying the genetics of AN and BED were successfully set up during the last years, it is likely that more biological and epigenetic research in the field of EDs will come up in the near future potentially revolutionizing our current understanding of EDs and ultimately leading to better therapies for these disorders.

ABBREVIATIONS

ALSPAC Avon longitudinal study of parents and children
AN Anorexia nervosa
BED Binge eating disorder
BMI Body mass index
BN Bulimia nervosa
ED Eating disorders
EDNOS Eating disorder not otherwise specified
ELS Early life stress
GCAN Genetic consortium to study anorexia nervosa

GLOSSARY

Anorexia nervosa (AN) Patients suffering from AN maintain a very low body weight (BMI < 17.5 kg/m^2), fear to be fat and suffer from a disturbed body image. AN has a high morbidity and mortality, as ~20% of all patients die due to the disorder

Binge eating disorder (BED) Patients with binge eating disorder suffer from recurrent binge eating episodes but do not apply purging behavior. Most patients with BED are overweight or obese

Binge eating or binging Episodes of overeating without being hungry. During a binge-eating attack patients sometimes eat 5000 kcal or more. In many cases bingeing is followed by feelings of shame and disgust

Bulimia nervosa (BN) Patients suffering from BN have normal to slightly increased body weight. They suffer from recurrent episodes of binge-eating and utilize purging behavior to control their weight

IMAGEN A European research study investigating mental health and risk taking behavior in teenagers

Purging Purging behavior is behavior, such as vomiting or laxative abuse that is applied to control weight especially after bingeing

REFERENCES

[1] Nagl M, Jacobi C, Paul M, Beesdo-Baum K, Höfler M, Lieb R, et al. Prevalence, incidence, and natural course of anorexia and bulimia nervosa among adolescents and young adults. Eur Child Adolesc Psychiatry 2016;25(8):903–18.

[2] Stuhldreher N, Wild B, König H-H, Konnopka A, Zipfel S, Herzog W. Determinants of direct and indirect costs in anorexia nervosa. Int J Eat Disord 2015;48:139–46.

[3] Herpertz-Dahlmann B. Adolescent eating disorders: update on definitions, symptomatology, epidemiology, and comorbidity. Child Adolesc Psychiatr Clin N Am 2015;24:177–96.

[4] Friborg O, Martinussen M, Kaiser S, Øvergård KT, Martinsen EW, Schmierer P, et al. Personality disorders in eating disorder not otherwise specified and binge eating disorder: a meta-analysis of comorbidity studies. J Nerv Ment Dis 2014;202:119–25.

[5] Hinney A, Volckmar AL. Genetics of eating disorders. Curr Psychiatry Rep 2013;15:423.

[6] Huckins LM, Boraska V, Franklin CS, Floyd JAB, Southam L, GCAN., et al. Using ancestry-informative markers to identify fine structure across 15 populations of European origin. Eur J Hum Genet 2014;22:1190–200.

[7] Boraska V, Floyd JAB, Thornton LM, Huckins LM, Southam L, Rayner NW, Tachmazidou I, Klump KL, Treasure J, Lewis CM, Schmidt U, Tozzi F, Kiezebrink K, Hebebrand J, Gorwood P, Adan RAH, Kas MJH,

Favaro A, Santonastaso P, Fernández-Aranda F, Gratacos M, Ryba FCS. A genome-wide association study of anorexia nervosa. Mol Psychiatry 2014;19:1085–94.
[8] Strober M, Freeman R, Lampert C, Diamond J. The association of anxiety disorders and obsessive compulsive personality disorder with anorexia nervosa: evidence from a family study with discussion of nosological and neurodevelopmental implications. Int J Eat Disord 2007;40:S46–51.
[9] Strober M, Peris T, Steiger H. The plasticity of development: how knowledge of epigenetics may advance understanding of eating disorders. Int J Eat Disord 2014;47:696–704.
[10] Favaro A, Tenconi E, Santonastaso P. Perinatal factors and the risk of developing anorexia nervosa and bulimia nervosa. Arch Gen Psychiatry 2006;63:82–8.
[11] Tenconi E, Santonastaso P, Monaco F, Favaro A. Obstetric complications and eating disorders: a replication study. Int J Eat Disord 2014;1–7.
[12] Oberlander TF, Weinberg J, Papsdorf M, Grunau R, Misri S, Devlin AM. Prenatal exposure to maternal depression, neonatal methylation of human glucocorticoid receptor gene (NR3C1) and infant cortisol stress responses. Epigenetics 2008;3:97–106.
[13] Braithwaite EC, Kundakovic M, Ramchandani PG, Murphy SE, Champagne FA. Maternal prenatal depressive symptoms predict infant NR3C1 1F and BDNF IV DNA methylation. Epigenetics 2015;10:408–17.
[14] Non AL, Binder AM, Kubzansky LD, Michels KB. Genome-wide DNA methylation in neonates exposed to maternal depression, anxiety, or SSRI medication during pregnancy. Epigenetics 2014;9:964–72.
[15] St-Hilaire A, Steiger H, Liu A, Laplante DP, Thaler L, Magill T, et al. A prospective study of effects of prenatal maternal stress on later eating-disorder manifestations in affected offspring: preliminary indications based on the Project Ice Storm cohort. Int J Eat Disord 2015;48:512–6.
[16] Cao-Lei L, Dancause KN, Elgbeili G, Massart R, Szyf M, Liu A, et al. DNA methylation mediates the impact of exposure to prenatal maternal stress on BMI and central adiposity in children at age 13½ years: Project Ice Storm. Epigenetics 2015;10:749–61.
[17] Cao-Lei L, Massart R, Suderman MJ, Machnes Z, Elgbeili G, Laplante DP, et al. DNA methylation signatures triggered by prenatal maternal stress exposure to a natural disaster: Project Ice Storm. PLoS One 2014;9:e107653.
[18] Micali N, Treasure J. Biological effects of a maternal ED on pregnancy and foetal development: a review. Eur Eat Disord Rev 2009;17:448–54.
[19] Susser ES, Lin SP. Schizophrenia after prenatal exposure to the Dutch Hunger Winter of 1944–1945. Arch Gen Psychiatry 1992;49:983–8.
[20] Schulz LC. The Dutch Hunger Winter and the developmental origins of health and disease. Proc Natl Acad Sci USA 2010;107:16757–8.
[21] Tobi EW, Goeman JJ, Monajemi R, Gu H, Putter H, Zhang Y, et al. DNA methylation signatures link prenatal famine exposure to growth and metabolism. Nat Commun 2014;5:5592.
[22] Milos G, Spindler A, Schnyder U, Fairburn CG. Instability of eating disorder diagnoses: prospective study. Br J Psychiatry 2005;187:573–8.
[23] Favaro A, Tenconi E, Santonastaso P. The interaction between perinatal factors and childhood abuse in the risk of developing anorexia nervosa. Psychol Med 2010;40:657–65.
[24] Thaler L, Gauvin L, Joober R, Groleau P, de Guzman R, Ambalavanan A, et al. Methylation of BDNF in women with bulimic eating syndromes: associations with childhood abuse and borderline personality disorder. Prog Neuropsychopharmacol Biol Psychiatry 2014;54:43–9.
[25] Frieling H, Romer K, Roschke B, Bonsch D, Wilhelm J, Fiszer R, et al. Homocysteine plasma levels are elevated in females with anorexia nervosa. J Neural Transm 2005;112:979–85.
[26] Frieling H, Röschke B, Kornhuber J, Wilhelm J, Römer KD, Gruss B, et al. Cognitive impairment and its association with homocysteine plasma levels in females with eating disorders—findings from the HEaD-study. J Neural Transm 2005;112:1591–8.

[27] Moyano D, Vilaseca MA, Artuch R, Valls C, Lambruschini N. Plasma total-homocysteine in anorexia nervosa. Eur J Clin Nutr 1998;52:172–5.
[28] Wilhelm J, Müller E, de Zwaan M, Fischer J, Hillemacher T, Kornhuber J, et al. Elevation of homocysteine levels is only partially reversed after therapy in females with eating disorders. J Neural Transm 2010;117:521–7.
[29] Frieling H, Gozner A, Romer KD, Lenz B, Bonsch D, Wilhelm J, et al. Global DNA hypomethylation and DNA hypermethylation of the alpha synuclein promoter in females with anorexia nervosa. Mol Psychiatry 2007;12:229–30.
[30] Tortorella A, Brambilla F, Fabrazzo M, Volpe U, Monteleone AM, Mastromo D, et al. Central and peripheral peptides regulating eating behaviour and energy homeostasis in anorexia nervosa and bulimia nervosa: a literature review. Eur Eat Disord Rev 2014;22:307–20.
[31] Pjetri E, Dempster E, Collier DA, Treasure J, Kas MJ, Mill J, et al. Quantitative promoter DNA methylation analysis of four candidate genes in anorexia nervosa: a pilot study. J Psychiatr Res 2013;47:280–2.
[32] Ehrlich S, Weiss D, Burghardt R, Infante-Duarte C, Brockhaus S, Muschler MA, et al. Promoter specific DNA methylation and gene expression of POMC in acutely underweight and recovered patients with anorexia nervosa. J Psychiatr Res 2010;44:827–33.
[33] Peters a, Schweiger U, Pellerin L, Hubold C, Oltmanns KM, Conrad M, et al. The selfish brain: competition for energy resources. Neurosci Biobehav Rev 2004;28:143–80.
[34] Tremolizzo L, Conti E, Bomba M, Uccellini O, Rossi MS, Marfone M, et al. Decreased whole-blood global DNA methylation is related to serum hormones in anorexia nervosa adolescents. World J Biol Psychiatry 2014;15:327–33.
[35] Frieling H, Bleich S, Otten J, Romer KD, Kornhuber J, de Zwaan M, et al. Epigenetic downregulation of atrial natriuretic peptide but not vasopressin mRNA expression in females with eating disorders is related to impulsivity. Neuropsychopharmacology 2008;33:2605–9.
[36] Frieling H, Romer KD, Scholz S, Mittelbach F, Wilhelm J, De Zwaan M, et al. Epigenetic dysregulation of dopaminergic genes in eating disorders. Int J Eat Disord 2010;43:577–83.
[37] Kim Y-R, Kim J-H, Kim MJ, Treasure J. Differential methylation of the oxytocin receptor gene in patients with anorexia nervosa: a pilot study. PLoS One 2014;9:e88673.
[38] Frieling H, Albrecht H, Jedtberg S, Gozner A, Lenz B, Wilhelm J, et al. Elevated cannabinoid 1 receptor mRNA is linked to eating disorder related behavior and attitudes in females with eating disorders. Psychoneuroendocrinology 2009;34:620–4.
[39] Schroeder M, Eberlein C, de Zwaan M, Kornhuber J, Bleich S, Frieling H. Lower levels of cannabinoid 1 receptor mRNA in female eating disorder patients: association with wrist cutting as impulsive self-injurious behavior. Psychoneuroendocrinology 2012;37:2032–6.
[40] Saffrey R, Novakovic B, Wade TD. Assessing global and gene specific DNA methylation in anorexia nervosa: a pilot study. Int J Eat Disord 2014;47:206–10.
[41] Booij L, Casey KF, Antunes JM, Szyf M, Joober R, Israël M, et al. DNA methylation in individuals with anorexia nervosa and in matched normal-eater controls: a genome-wide study. Int J Eat Disord 2015;48:874–82.
[42] Groleau P, Joober R, Israel M, Zeramdini N, DeGuzman R, Steiger H. Methylation of the dopamine D2 receptor (DRD2) gene promoter in women with a bulimia-spectrum disorder: associations with borderline personality disorder and exposure to childhood abuse. J Psychiatr Res 2014;48:121–7.
[43] Steiger H, Labonté B, Groleau P, Turecki G, Israel M. Methylation of the glucocorticoid receptor gene promoter in bulimic women: associations with borderline personality disorder, suicidality, and exposure to childhood abuse. Int J Eat Disord 2013;46:246–55.
[44] Ehrlich S, Walton E, Roffman JL, Weiss D, Puls I, Doehler N, et al. Smoking, but not malnutrition, influences promoter-specific DNA methylation of the proopiomelanocortin gene in patients with and without anorexia nervosa. Can J Psychiatry 2012;57:168–76.

CHAPTER

EPIGENETICS AND OBESITY

16

B.M. Shewchuk

Department of Biochemistry and Molecular Biology, Brody School of Medicine at East Carolina University, Greenville, NC, United States

CHAPTER OUTLINE

16.1 INTRODUCTION

Obesity is defined as abnormal or excessive fat accumulation with the potential to impair health, which the World Health Organization (WHO) classifies as having a body mass index (BMI) of 30 kg/m^2 or greater. The WHO places current global estimates of individuals with obesity at about 13% and those overweight (BMI > 25 kg/m^2) at about 39%, which represents a doubling since 1980 [1]. The increasing prevalence of obesity has become a significant international public health concern that has driven increased healthcare costs and a diminishing quality of life for those afflicted. A notable feature of

Neuropsychiatric Disorders and Epigenetics. http://dx.doi.org/10.1016/B978-0-12-800226-1.00016-2

obesity as a disease is the breadth of its impact across multiple physiological systems, resulting in deleterious effects on a range of normal functions. This range of systems associated with obesity is consistent with its classification as a marker for an underlying spectrum of common comorbidities referred to metabolic syndrome (MetS), which also includes elevated triglycerides, reduced high-density lipoprotein (HDL) cholesterol levels, elevated fasting glucose concentrations, and hypertension [2]. The worsening epidemic of obesity, in contrast to the improving scenario for many other chronic diseases, has provided a major impetus for research onto the etiology of obesity. One aspect of obesity that has garnered significant attention since early in the history of this question is evidence for a developmental origin of adult obesity that appears to be associated with; the environment, the health, and the metabolic status of child bearing women during pregnancy, and also the periconceptional metabolic health and nutrition of both parents. More recently, evidence has indicated that metabolic health is susceptible to environmental inputs throughout life, which may continually modulate the propensity for obesity. Observations of the developmental origins of obesity were among the earliest to indicate the existence of an active epigenetic mechanism, one that is highly sensitive to multiple environmental and physiological inputs and can functionally program gene expression networks. This process can thereby have significant long term effects on physiological outcomes. These observations provided the initial evidence for the concept of epigenetics, and the development of underlying theories, initiating a new era in the investigation of gene regulation during development, physiological homeostasis, and disease.

The etiology of obesity involves mechanisms that occur in multiple physiological systems, encompassing both the central nervous system (CNS) and peripheral metabolic tissues. This fact illustrates the complexity of obesity as a disease process, and the potential challenges this complexity may present with regard to treatment. It also indicates that energy homeostasis is itself a complex process involving levels of central and peripheral control, and that there are multiple points at which homeostasis can be perturbed to result in a positive energy balance leading to obesity. Of particular scientific and public health significance is the potential for stable alterations in phenotype through epigenetic mechanisms sensitive to an obesogenic environment that result in the persistence of this state, and a resistance to metabolic normalization. In this chapter, the observations and biochemical underpinnings of the epigenetics of obesity will be discussed, with a particular focus on the potential for significant nutritional and environmental impacts on the obesity-related epigenome throughout development. In this context, the predominant role of the hypothalamic feeding center in the control of food intake and energy expenditure, and the epigenetic perturbation of this system in obesity, will be highlighted.

16.2 EPIGENETIC MECHANISMS

The initial concept of epigenetics did not refer to any specific molecular mechanism, but rather to all of the processes that govern the nexus of genotype and phenotype in the context of mammalian development [3]. In practical terms, this refers to the ensemble of spatial and temporal gene regulation events that program discrete cell lineages and maintain diverse cell phenotypes through the selective expression of specific subsets of genes. Operationally, epigenetics encompasses molecular events at multiple levels of interaction that integrate extracellular chemical and environmental signals with chromatin-mediated gene regulation. These signals can be differentiation factors, nutrients, hormones, or environmental variations, and are referred to as "epigenators" in the epigenetics schema. These epigenators elicit the action of "epigenetic initiators", which are DNA sequence-specific transcription factors and noncoding RNAs that target and establish the epigenetic chromatin state at the gene level through the recruitment of cofactor enzyme complexes. "Epigenetic maintainers", such as DNA and histone modifying enzymes (DNA methyltransferases and histone acetyltransferases, or histone variants) are then recruited to the chromatin sites established by the initiators to stably maintain the chromatin state through cell division [4]. Importantly, while an epigenetic initiator may be sufficient to program an epigenetic phenotype in the absence of an upstream signal, the epigenetic maintainers are not, since they lack DNA sequence specificity and are instead recruited through affinity for specific histone marks established by the action of the initiators. Despite this complexity, the predominant conventional utilization of epigenetics in the majority of published studies, including those related to obesity, refers specifically to the covalent chemical modifications of chromatin (DNA and histones) that are primarily the product of epigenetic maintainers. While the functional consequences of chromatin modifications on the establishment and maintenance of gene expression is the ultimate biochemical goal of the epigenetic process, it should be recognized that this is the endpoint of multiple levels of interacting mechanisms, any of which can be the potential target of an agent or process that is reported to have an epigenetic effect. This is an important consideration when identifying the causal agents of an epigenetic effect, the biochemical pathway that elicits the effect, or potential targets for therapeutic intervention.

The biochemical processes that are encompassed by the conventional rubric of epigenetics include DNA methylation, residue-specific covalent histone modifications (acetylation, methylation, phosphorylation, ubiquitination, biotinylation, and sumoylation), and noncoding RNA-based gene modulation. This roster of modifications will likely expand, as additional covalent histone modifications, such as O-GlcNAcylation [5], continue to be identified. These processes are variably associated with both gene activation and repression, and hundreds of published studies over the last 20 years have described the associations of each of these specific epigenetic marks to gene activity states at both the whole genomic and individual gene levels. In addition, these studies have also focused heavily on whether and how agents or disease states can alter gene expression by specifically perturbing the levels of these epigenetic marks. Mitotic persistence is an essential operational property of a regulatory determinant for it to be considered a truly epigenetic mechanism and to date most evidence suggests that this is limited to DNA methylation, and a subset of covalent histone modifications [4]. It should be noted that other histone modifications or components of the full epigenetic cascade have essential functions in the initial establishment of an epigenetic phenotype, or in more dynamic gene regulatory responses in postmitotic somatic tissues.

16.3 DNA METHYLATION

DNA methylation is a major epigenetic mechanism, and primarily occurs in mammals through the addition of a methyl group to C-5 of cytosine at CpG dinucleotides mediated by several isoforms of DNA methyltransferase (DNMT), which utilize S-adenosyl methionine (SAM) as the methyl donor. These DNMT isoforms have context-specific activities: DNMT3a and DNMT3b are de novo methyltransferases responsible for global embryonic methylation after implantation. DNMT3a can also methylate cytosine in a CH context where H is A, C, or T particularly in brain [6]. DNMT1 is a somatic maintenance methyltransferase with a substrate preference for hemimethylated CpG DNA, and is thereby associated with the mitotic maintenance of DNA methylation patterns by replicating the pattern of the methylated DNA template provided by semiconservative DNA replication. DNMT1o is an oocyte specific maintenance methyltransferase important for maintaining maternal gene imprints.

Genome wide, methylation usually occurs outside of CpG islands (CGI). CGI are genomic sequences that average 1 Kb in length, have GC content above 50%, and are enriched in CpG dinucleotides. CGI are associated with 60–70% of all gene promoters, many of which are widely expressed housekeeping genes (reviewed in Ref. [7]). These regions contain a higher density of CpG precisely because they are maintained as constitutively unmethylated, regardless of gene activity. Spontaneous deamination of 5-methylcytosine to thymine results in a mutation of CpG to TpG, such that, over evolutionary time the CpG dinucleotide frequency has been reduced to about 20% of the expected frequency in mammals based on genomic nucleotide composition (CpG/GpC ratio of ~0.2) [8]. In contrast, CGI maintain a CpG/GpC ratio of ~1, consistent with the historical lack of methylation in these regions. Protection from de novo methylation is thought to be conferred by transcription factors, such as Sp1, for which there are frequent binding elements in CpG-rich regions [9]. Thus, tissue-specific non-CGI gene promoters will typically be unmethylated (or less methylated) in expressing cells, but more heavily methylated in nonexpressing cells. At CGI promoters, transcriptional repression can occur in the absence of CpG methylation, but aberrant CGI methylation is observed in certain contexts, such as cancer. Methylation-mediated repression can occur through the direct steric blocking of transcription factor binding to cognate elements overlapping the methylated site, or through the binding of methyl-binding domain proteins, such as MeCP2 and the subsequent recruitment of corepressor complexes containing histone deacetylase (HDAC) activity.

16.4 HISTONE MODIFICATION

Histones are a component of the fundamental repeating unit of chromatin, the nucleosome, which consists of 147 bp of DNA wrapped in a right-handed helix around an octamer of histones, comprised of two molecules each of histone 2A (H2A), H2B, H3, and H4. In the majority of chromatin, nucleosomes are separated by roughly 60 bp of linker DNA, which is associated with the linker histones H1 or H5 in higher order nucleosomal arrays and condensed chromatin. In the structure of the nucleosome, the N-terminal tails of histones, particularly H3 and H4, extend from the cylindrical nucleosome core in random coils, allowing them to mediate higher-order chromatin compaction through primarily electrostatic intra- and internucleosomal interactions between the histone tails and DNA. Some covalent modifications, such as the neutralization of basic lysine residues by acetylation, disrupt these interactions and reduce higher-order chromatin condensation, thereby directly facilitating gene transcription. Other modifications, such as lysine and arginine methylation, are more predominantly involved in the subsequent recruitment of coactivator, corepressor, and nucleosome remodeling complexes through the binding of chromodomain-containing subunits to the methyl groups. These nucleosome-remodeling complex interactions are associated with both activation and repression, depending on the modification site and context. Acetylation can also play a role at this level through the analogous recruitment of bromodomain-containing coactivator complexes that recognize acetylated histones.

One class of coactivators recruited by modified histones are nucleosome remodeling complexes that utilize resident ATP-dependent helicase-like activities to loosen protein-DNA contacts in the nucleosome, allowing the removal or repositioning of entire nucleosomes. These complexes are classified based on the ATPase subunit, and include the SWI/SNF, ISWI, CHD and INO80 complexes. Significantly, these complexes have been shown to be essential for mammalian development from an early embryonic stage, illustrating their role in epigenetic programming during zygotic genome activation and development [10]. The recognition of acetylated and methylated histones by bromodomains and chromodomains, respectively, also provides a mechanism for the mitotic propagation of the histone modifications. In the currently dominant model for histone segregation, fully conservative random partitioning of intact modified nucleosomes between chromatids during mitosis allows for the reading and reestablishment of domains of histone modification when nucleosomes are synthesized de novo [11].

16.5 NORMAL DEVELOPMENTAL EPIGENETIC PROCESSES

The molecular mechanisms that are encompassed by the epigenetics rubric are all involved in the normal modulation of gene expression, at multiple time scales and degrees of dynamism. DNA methylation, for which the correlation with gene function and role in developmental programming is best understood, can be considered in broad contiguous domains (hundreds to thousands of base pairs) encompassing genes, with very low levels of CpG methylation at the 5' ends and gene bodies with elevated methylation levels that positively correlate with transcription [12]. Large domains of heterochromatin, comprised largely of low-complexity DNA sequences and other repetitive elements are also maintained in a highly methylated state (80–90%) to enforce transcriptional repression. DNA methylation serves to maintain this repression to limit expression to only a subset of the genomic housekeeping and tissue-specific genes in terminally differentiated somatic cells, in a modification pattern referred to as the methylome. This process begins in the zygote, where the genomic methylation patterns of the male and female gametes are largely erased, and zygotic methylation patterns are established to begin the process of embryogenesis. During prenatal development, a reformatting of methylation patterns occurs again in the germ cell lineage, in which the primordial germ line methylome is erased, and germ cell-specific methylation patterns are established; this process extends postnatally in oocytes. In addition to DNA methylation, coordinated phases of specific histone methylation (at H3 lysine 4 and lysine 27), also occurs during both embryonic development and germ line ontogeny [13]. It is this process of reformatting the zygotic epigenome, and the role that cell lineage-specific chromatin modification has in programming phenotype during prenatal development, that the maternal environment plays a critical role in metabolic programming and obesity.

16.6 DEVELOPMENTAL ORIGINS OF OBESITY

The hypothesis that proposed an epigenetic mechanism contributing to the development of obesity has its origins in observations from several historical natural experiments. The elevated prevalence of persistent obesity and other MetS-associated disorders in human populations associated with episodes of extreme malnutrition provided some of the first evidence for a susceptibility of human phenotype to the developmental environment, and led to the modern mechanism-focused conceptualization of epigenetics. A primary example of these scientifically valuable episodes was the Dutch famine of 1944–45 (Hunger Winter), caused by a German embargo of food shipments and the institution of highly restricted food rations in the densely populated western provinces of the Netherlands during World War II. Because food intake levels and nutrition were not limited prior to the embargo, this allowed subsequent researchers to study the effects of maternal nutrition during specific developmental windows on the subsequent long term phenotype and health of offspring by comparison to unexposed siblings. The results of these Dutch Famine Cohort Studies found that offspring of women exposed to the famine had an increased propensity to develop diabetes, obesity, cardiovascular disease, and several additional health problems [14]. Thus, while the underlying biochemical mechanisms were not yet known, these studies provided considerable evidence for a labile epigenetic process that can program developmental outcomes and phenotypes, and supported an emerging concept of the developmental origins of disease.

The first formalization of this concept was the "Thrifty Phenotype" hypothesis of Hales and Barker. Observing a correlation between low birth weight and the incidence of Type 2 diabetes (T2D) in a British male cohort, these investigators hypothesized that prenatal malnutrition functionally programs the fetus for metabolic efficiency in order to accommodate a perceived nutrient-poor environment, such that a subsequent nutrient-rich environment and caloric excess can more readily lead to obesity and T2D [15]. Another example of a natural cohort demonstrating a link between the prenatal environment and adult metabolic health is the Pima indigenous people of the American southwest. Until the past few generations, this population subsisted by farming the banks of the Gila and Salt rivers in what is now southern Arizona, during which time they encountered protracted periods of famine due to regular drought cycles. A transition to modern lifestyles and processed food following the move of the majority of the population to the Gila River Indian Community resulted in widespread obesity and the emergence of an abnormally high incidence of T2D, as predicted by the Thrifty Phenotype hypothesis. The levels of T2D among the Pima population reaches 35% for males (50% for females) by age 40, and nearly 70% for females (40% for males) by age 60, providing a relatively isogenic population with which to study the potential for epigenetic programming during development in the context of obesity and MetS [16]. Subsequent analysis of this population showed that offspring of mothers with T2D had a higher prevalence of T2D at adulthood (45%) compared to offspring of nondiabetic (1.4%) or prediabetic (8.6%) mothers, suggesting the importance of the intrauterine environment in determining the development of T2D, beyond the contribution of genetic factors [17].

16.7 GENETIC IMPRINTING DISORDERS ASSOCIATED WITH OBESITY

An indication of a correlation between epigenetic processes and obesity came from resolution of the mechanisms underlying several types of early onset syndromes that present with obesity and are caused by defects in genomic imprinting. This relatively rare gene regulatory process (only about 50 loci have been identified as imprinted in humans [18]) involves the origin-specific transcriptional repression of one of the two parental alleles of imprinted genes, such that only one of the allelic pairs is expressed. Many imprinted genes are involved in the regulation of somatic growth, and therefore it is thought that this mechanism has origins as a means to regulate nutritional resource partitioning from the mother, where the nutrients benefit the survival of the mother and thus the offspring of any father, to the fetus, which represents the survival of a single paternal genetic lineage and thus benefits from increased resource extraction [19]. Consistent with this hypothesis, the placenta, which mediates nutrient flux between the fetus and mother, is a predominant location of tissue-specific imprinting [20,21]. This parental conflict theory posits that imprinting essentially represents the competition between maternal and paternal genomes for reproductive propagation. The selective transcriptional repression is conferred by parent-specific DNA methylation of the imprinted allele, resulting in an identifiable differentially methylated region (DMR) that mediates the allele-selective expression. Imprinted genes are typically localized in clustered domains, where allele-specific expression is controlled by regulatory imprinting control regions (ICR) that target a complex mechanism of noncoding RNA and DNA methylation that establish and delimit the imprinted domains [20,22].

Imprinting disorders that display an obesity phenotype include Prader–Willi syndrome, Maternal Uniparental Disomy 14, Pseudohypoparathyroidism, and Angelman syndrome [23,24]. The commonality of obesity among these disorders suggests that defects in resource partitioning during development may play a role in epigenetic causes of obesity [21]. In addition, the obese phenotype of these syndromes illuminates the nexus of a role for imprinting in brain development with the important role of the CNS, particularly the feeding center of the hypothalamus, in regulating energy homeostasis [25]. In the case of Prader–Willi syndrome, obesity is the outcome of a neurological impairment that results in the apparent lack of a functioning satiety signaling mechanism, causing uncontrolled appetite and food seeking behavior [21,26]. This points to a likely disruption of the balanced interplay between the orexigenic neuropeptide Y (NPY)/Agouti-related peptide (AgRP) hypothalamic neurons and the countervailing proopiomelanocortin (POMC)/cocaine and amphetamine responsive transcript (CART) neurons, such that orexigenic signaling exceeds the stimulation of the anorexigenic pathway [27]. The observed increase in ghrelin, a gut-derived hormone that functions as a nutrient-activated orexigenic agent, in Prader–Willi subjects is consistent with this shift [28]. Ghrelin is a natural ligand of the growth hormone secretagogue receptor (GHSR), and modulates pituitary somatotroph GH release in addition to stimulating NPY/AgRP neurons. Although the causal role of ghrelin in the Prader–Willi obese phenotype has not been resolved, it appears to be different from its association with nonsyndromic common obesity [28].

Another example of an imprinting disorder with an obesity phenotype is Pseudohypoparathyroidism (PHP), which is caused by variants in the *GNAS* (guanine nucleotide binding protein alpha subunit) gene locus and takes multiple forms depending on which parental allele is affected, and the site of the nucleotide sequence variant [29]. *GNAS* encodes isoforms of the stimulatory Gα_s subunit of the heterotrimeric G-proteins responsible for transducing G-protein-coupled receptor signaling. Biallelic Gα_s expression is observed in most tissues except in cells of the renal proximal tubule, thyroid, pituitary, and gonads where the paternal allele is imprinted, conferring maternal allele-specific expression. Thus, the phenotype is only observed when the disease allele is inherited maternally [30]. PHP type 1a derives from a loss of function mutation of the *GNAS* gene on the maternal allele, such that Gα_s is expressed only from the paternal allele. The syndrome of multiple clinical features associated with PHP 1a is referred to as Albright hereditary osteodystrophy (AHO), and includes obesity, short stature, several skeletal abnormalities, and sporadic mental retardation [31]. As would be predicted from a defect in a G-protein subunit, this disorder is associated with target organ resistance to multiple trophic hormones including parathyroid hormone (PTH), thyroid-stimulating hormone (TSH), and multiple gonadotropic hormones. A distinct form of this syndrome, PHP 1b, results from deletions of the DMR of the *GNAS* locus, a region which is also encompassed by a regulatory *GNAS* locus antisense transcript [32]. PHP 1b can also arise from a more remote *cis* deletion in the *STX16* gene, which functions as a distal regulatory element that contributes to the control of DNA methylation in the *GNAS* locus [33]. The loss of methylation control of the *GNAS* locus by these lesions causes the loss of expression of the maternal Gα_s isoform predominantly in renal tissue, thereby manifesting the attendant renal hormone resistance but more rarely presenting with the full AHO phenotype associated with PHP 1a [34].

16.8 GENETIC ASSOCIATIONS WITH OBESITY

In contrast to a few limited examples of monogenic and syndromic obesity, research into the potential genetic basis of common obesity has reinforced a general conclusion that obesity cannot be fully accounted for by specific genetic variants, but instead reflects the dynamic interaction of the environment with this genetic framework through epigenetic processes. It is estimated from studies of twins that about 40–70% of obesity is due to heritable genetic causes [35], and the outcomes of multiple analyses of genome-wide association study (GWAS) data have identified more than 50 genetic alleles associated with obesity and lipid storage distribution. The genetic propensity for obesity appears to be due to the incremental input of alleles of multiple genes resulting in a cumulative effect, rather than the predominant input of single alleles [36]. Common alleles of only two loci, *FTO* and *MCR4*, have reproducibly been associated with BMI in humans. However, the genes with known functions that have been identified are consistent with the metabolic dysfunction one would expect with obesity and MetS, and include genes involved in adipogenesis, glucose transport, lipolysis and lipid transport, neural development, insulin signaling, vascular development, feeding behavior, and energy balance [36].

The most well-known example of a gene identified in association with BMI heritability is *FTO* (fat mass and obesity associated), which encodes a nuclear localized member of the 2-oxoglutarate-dependent oxygenase superfamily related to AlkB and was the first gene locus reproducibly associated with nonsyndromic obesity [37]. This gene emerged previously in a GWAS of T2D, and the association with obesity, which accounts for about 1% of BMI heritability, is abrogated when adjusted for T2D incidence [37], illustrating common metabolic and physiological origins of MetS phenotypes. Interestingly, FTO was shown to be a nucleic acid demethylase, which can utilize both methyl-DNA (single and double-stranded) and methyl-mRNA as substrates. It is expressed abundantly in the brain, particularly in the hypothalamic arcuate nucleus where feeding behavior is governed. *FTO* mRNA levels in the hypothalamus are regulated by feeding and fasting, consistent with its potential involvement in the control of energy balance [38]. FTO is also expressed in adipose tissue, where it affects lipolytic flux in adipocytes, indicating that *FTO* risk alleles may contribute to obesity through actions in peripheral metabolic tissues as well [39]. Another gene locus that has been repeatedly linked to obesity is *MCR4* (melanocortin receptor 4) also expressed in the hypothalamus. A haplo-insufficiency of *MCR4* is associated with morbid obesity in humans, and MCR4-deficient mice display marked hyperphagia leading to obesity, consistent with a central role in the regulation of energy balance through the modulation of feeding behavior [40].

16.9 EPIGENETIC CONTROL OF METABOLISM AND OBESITY ASSOCIATED GENES

Despite clear evidence for a cumulative genetic component to obesity, it is also apparent that these variants cannot fully explain the heritability of obesity, illustrating the significant contribution of developmental and environmental epigenetic interactions. While any discussion of genetic alleles associated with obesity may appear to be orthogonal to a focus on epigenetics, it is important to recognize that gene loci displaying an association with obesity are also potential targets of epigenetic processes that affect expression levels, thereby leading to similar outcomes. For example, long-term exposure to a high fat diet in mice can lead to altered methylation of the *Mcr4* gene, demonstrating the potential for environmental impacts on *Mcr4* expression [41]. Several additional examples of obesity-associated genes have been identified as being subject to environmentally sensitive epigenetic modulation, including *INS* (insulin) [42], *INSR* (insulin receptor) [43], *IGF2* (insulin-like growth factor 2) [44], *GLUT4* (glucose transporter 4) [45], *ObLep* (leptin), and *TNF* (tumor necrosis factor) [46]. The epigenetic regulation of these genes involves multiple biochemical mechanisms, including DNA methylation, histone methylation, and histone acetylation, depending on the locus. The significance of the distinctions in the types of epigenetic marks found at different loci, as well as the identity of the upstream pathways and ultimate causal determinants of these marks remain to be elucidated.

Comparison of individuals in the aforementioned Dutch famine cohort with sibling controls identified differentially methylated regions (DMR) proximal to multiple genes involved in growth and metabolism that correlated with exposure to periconceptual famine. These loci include *INSIGF* (an insulin-insulin like growth factor 2 read through transcript), *IL10* (interleukin 10), *LEP* (leptin), *ABCA1* (ATP-binding cassette A1), *GNASAS* (a *GNAS1* antisense transcript), and *MEG3* (maternally expressed 3) [47,48]. DMRs near the *INSR* and *CPT1A* genes were also identified, and functional analysis showed that these sequences possessed enhancer activity that was abrogated by CpG methylation, consistent with the potential physiological significance of the differential methylation [49].

Genes at loci identified by GWAS in association with obesity traits may themselves encode components of epigenetic mechanisms. As indicated earlier, the *FTO* gene possessing the strongest association with obesity is a nucleic acid demethylase, suggesting that the modulation of DNA methylation is central to energy homeostasis. Additional examples can be found, at least those for which there is a known function. Genes encoding several transcription factors have been identified in obesity GWAS screens, including *MAF*, a regulator of adipogenesis and insulin/glucagon interplay [50], *HOXC13*, involved in embryonic morphological development, and *TBX15*, involved in adipose tissue development [51]. Considering the essential role of transcription factors in recruiting histone modifying enzyme complexes to specific genomic loci, these genes likely contribute to epigenetic regulation. As the identification of the target gene determinants of SNP associations are resolved, additional components of epigenetic processes will likely be elucidated, which will help to clarify the biochemical basis for the connection between obesity and epigenetics.

16.10 METABOLIC REGULATION BY HISTONE MODIFYING ENZYMES

The concept of a general role for epigenetic processes in the regulation of energy balance has been reinforced by targeted null mutations of chromatin modifying enzymes in mice. These studies serendipitously revealed the possibility of a global role of specific epigenetic mechanisms in the regulation of metabolism and energy homeostasis. Histone 3 lysine 9 (H3K9) trimethylation is a predominant mark of heterochromatin and transcriptionally repressed genes. This modification is reversed by the activity of histone demethylases, allowing for countervailing transcriptional activation to occur. It was shown that H3K9 demethylation is mediated by members of the jumonji C (JmjC) domain family of proteins, such as JHDM2a (JmjC-domain-containing histone demethylase 2a) through iron and α-ketoglutarate-dependent oxidation [52]. Studies into the physiological role of JHDM2a in mouse knockout models revealed a predominant and broad role for this H3K9 demethylase in energy homeostasis. Mice lacking JHDM2a develop obesity, hyperlipidemia, hyperinsulinemia, and hyperleptinemia, representing a classic presentation of MetS. They also display reduced energy expenditure and lipid oxidation, consistent with metabolic dysregulation [53]. Additional studies revealed the broad impact of this deletion, which directly affects the expression levels of genes, namely *Ppara* and *Ucp1*, involved in metabolic regulation in multiple peripheral tissues (brown and white fat, skeletal muscle, and liver) consistent with the physiological effects of impaired energy balance. This system is under the control of ββ-adrenergic signaling, as stimulation of this pathway upregulates JHDM2a, PPARα, and UCP1 expression [54].

A role for histone deacetylation in global metabolic regulation has also been indicated by recent studies. The nuclear receptor corepressor 1 (NCOR1)-histone deacetylase 3 (HDAC3) complex is recruited to genes by DNA-binding silencers to elicit regional histone deacetylation, leading to a closed chromatin structure refractory to transcription factor binding, thereby repressing transcription. NCOR1-HDAC3 is involved in the regulation of energy expenditure, adiposity, and insulin signaling through a circadian mechanism, reinforcing a general role of epigenetic processes, in this case histone modification, in energy homeostasis [55].

16.11 METABOLIC PATHWAY CONNECTIONS TO EPIGENETIC MECHANISMS

The apparent sensitivity of the epigenome to the environmental influence raises the question of the identity of the causal agents and the biochemical mechanisms by which they alter DNA methylation and histone modification. A predominant factor is the role that dietary nutrients play in providing substrates for epigenetic mechanisms. For example, the multiple methyltransferases that methylate DNA and histones as components of epigenetic regulatory mechanisms utilize S-adenosylmethionine (SAM) as the methyl group donor. SAM is a product of one-carbon metabolism, which utilizes several essential dietary micronutrients as carriers of a methyl group, and the level of SAM available for DNA and histone methylation is thus linked to the levels of these substrates in the diet [56]. The nutrients involved in one-carbon metabolism include folate, vitamin B2, vitamin B6, vitamin B12, choline, betaine, and methionine. Studies in rodent models of have shown that dietary supplementation with methyl donor nutrients can affect the methylation level and transcriptional activities of what are termed "metastable epialleles." Epialleles are gene loci that possess a unique sensitivity to epigenetic variability during early embryonic development that becomes stabilized in diverse tissues, enforcing a programmed phenotype [57]. In the seminal demonstration of this phenomenon, the expression pattern of the murine viable yellow (A^{vy}/a) allele of the Agouti coat color pigment gene was shown to be sensitive to maternal dietary methyl donor supplementation, resulting in a correlation between maternal supplementation and Agouti expression. This was due to the methylation-dependent silencing of a cryptic constitutively active promoter conferred by an intracisternal A particle (IAP) retrotransposon insertion upstream of the normally regulated Agouti promoter in the A^{vy} locus. Interestingly, the phenotype of the A^{vy} mice includes obesity, due to the ability of the Agouti protein to mimic the orexigenic action of AgRP in the hypothalamus, and this outcome is also suppressed by methyl donor supplement-dependent methylation of the cryptic IAP promoter [58]. While a homologous locus in humans does not exist, these data provided direct evidence for the susceptibility of certain alleles to the perturbation of epigenetic mechanisms during development. Subsequent studies have expanded the observations of this phenomenon at other loci, reinforcing the potential sensitivity of the methylome to dietary and environmental influence in the context of metabolic health in both animal models and studies of human cohorts [57,59].

While the role of one-carbon metabolism at the nexus of diet and epigenetic methylation mechanisms is well established, additional links between metabolism and epigenetic processes have been identified that may impact the propensity for obesity. One example is the production of acetyl-CoA for use as a substrate for histone acetylation. The cytoplasmic enzyme ATP-citrate lyase (ACL) synthesizes acetyl-CoA from citrate that is transported from mitochondria to the cytoplasm of mammalian cells under carbohydrate-replete conditions. This acetyl-CoA is then used to synthesize lipids for storage as triglycerides. Citrate also enters the nucleus, where a newly discovered nuclear pool of ACL can synthesize acetyl-CoA exclusively for histone acetylation, which was found to result in the upregulation of multiple genes involved in glucose uptake and metabolism, thereby linking epigenetic gene regulation with glucose availability and metabolism [60]. Another notable example of a metabolism-epigenetics link is the multiple NAD^+-dependent sirtuins (SIRT), which represent Class III histone deacetylases (HDAC) and have the potential to mediate an epigenetic response to changes in metabolic and redox states. Sirtuins have been shown to be involved in several aspects of metabolic health, and have significant correlations with obesity and T2D [61].

16.12 HYPOTHALAMIC REGULATION OF ENERGY BALANCE

While the precise etiology of obesity is a complex process involving central and peripheral inputs, the development of obesity ultimately requires an imbalance of caloric intake and net energy expenditure. Feeding behavior is a primary governor of this balance as the sole source of caloric intake, and thus the CNS plays a major role in the etiology of obesity, a fact supported by the CNS expression of many of the candidate loci associated with obesity identified by GWAS. Many additional genes encode peripheral signaling molecules that interact with the feeding center and regulate energy balance, and the epigenetic modulation of these signaling molecules is an additional factor in the development of obesity.

The hypothalamus is the predominant central governor of energy balance, through the function of neurons principally within the arcuate nucleus (ARC), but also involving the paraventricular nucleus (PVN) and dorsomedial hypothalamus (DMH), a system often referred to as the feeding center [62]. The ARC contains counteracting orexigenic AgRP/NPY/γ-butyric acid (GABA) neurons and anorexigenic POMC/CART neurons that release their cognate neuropeptides and neurotransmitters in response to peripheral signaling molecules to regulate feeding behavior and energy expenditure, thereby contributing to the regulation of body mass (Fig. 16.1) [62]. The peripheral hormones that can regulate ARC neuron function include glucocorticoids, leptin, insulin, and ghrelin, and dysregulation of these signaling circuits is appreciated to be a likely contributor to the pathophysiology of obesity [63]. In addition, the epigenetic programming of ARC neurons by the embryonic environment and beyond is widely thought to be a potential source of metabolic dysregulation, and a causal source of the association between periconceptual and gestational malnutrition, and obesity and T2D observed in human cohorts [62]. Controlled studies in animal models have supported this hypothesis, showing that maternally undernourished animals have impaired glycemic control and insulin signaling, and altered hypothalamic signaling pathways [64,65].

The anorexigenic POMC/CART neurons in the ARC synthesize POMC and CART peptides, which decrease feeding behavior when released. Axons of these neurons project primarily to the PVN, lateral hypothalamus (LH) and brainstem, which are involved in energy homeostasis. POMC is a precursor peptide for additional signaling molecules that are derived by protease cleavage of POMC, including

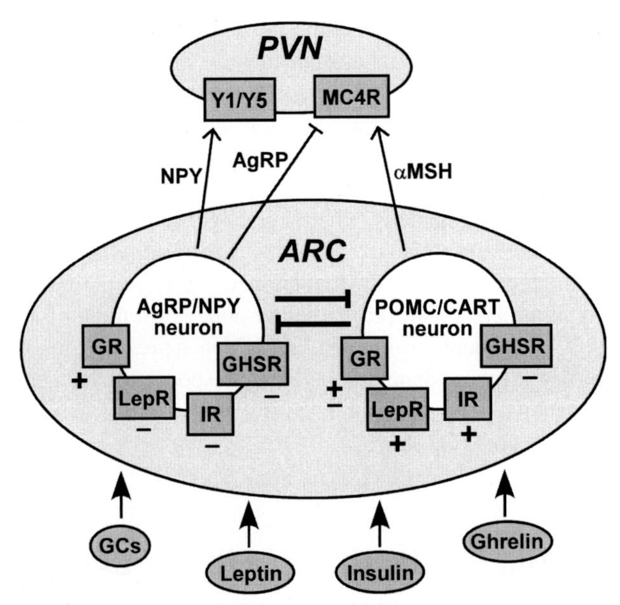

FIGURE 16.1 Hypothalamic Signaling in the Control of Energy Balance

AgRP/NPY and POMC/CART neurons project from the arcuate nucleus *(ARC)* to the paraventricular nucleus *(PVN)* to regulate food intake and energy expenditure. Multiple peripheral hormones regulate this process through countervailing postreceptor effects in orexigenic and anorexigenic neurons, indicated by the plus and minus symbols. *GCs*, Glucocorticoids; *GHSR*, growth hormone secretagogue receptor; *GR*, glucocorticoid receptor; *IR*, insulin receptor; *LepR*, leptin receptor; *αMSH*, α-melanocyte stimulating factor.

adrenocorticotropic hormone (ACTH), α-melanocyte stimulating factor (α-MSH), and β-endorphin. The most well defined anorexigenic agent produced by the ARC is α-MSH, which is an agonist for the melanocortin 3 and 4 receptors (MC3R and MC4R) that are highly expressed in PVN neurons. Activation of these receptors in the PVN results in blunted feeding behavior, as well as an increase in energy expenditure. A knockout mutation of the *Pomc* gene in mice was shown to cause hyperphagia culminating in obesity, illustrating this function [66]. Null mutations of *Mc3r* or *Mc4r* genes in mice also lead to obesity due to increased fat mass [67], and mutations in the *POMC* gene have been associated with obesity in humans [68], reinforcing the role of hypothalamic POMC and its derivatives in energy homeostasis. The ARC also secretes CART from anorexigenic neurons, but this system is less studied largely due to the lack of characterization of the cognate receptors for CART, which has an effect on feeding behavior and energy expenditure parallel to that of POMC. A knockout of CART in mice resulted in increased food intake and body mass when challenged with a high caloric diet, confirming the anorexigenic activity of CART and its role in metabolic regulation [69].

To counter the anorexigenic actions of POMC/CART neurons, the orexigenic NPY/AgRP neurons release NPY and AgRP, as well as the neurotransmitter GABA to increase food intake. Axon projections from these neurons terminate in the PVN, DMH, LH, and medial preoptic region (MPO) where these peptides are released to elicit feeding behavior. A dramatic demonstration of the mandatory role of these neurons and their cognate signaling peptides in the regulation of food intake came from the selective ablation of NPY/AgRP neurons in adult mice, which resulted in an acute reduction in feeding and parallel loss of body mass [70,71]. Subsequent studies in mice resolved temporally and dynamically distinct roles of AgRP, NPY and GABA in the synergistic modulation of feeding behavior, with AgRP being uniquely involved in prolonged feeding [72]. In the PVN, AgRP acts as an antagonist of MC4R, thereby countering the action of α-MSH and increasing food intake. NPY functions as an agonist of the Y1 and Y5 receptors in the PVN and other areas to increase feeding behavior and promote a positive energy balance and increased adiposity [73]. GABA has a function similar to that of NPY in mediating the initial acute phase of feeding, complementing the longer-acting orexigenic function of AgRP [72].

The region of the brain encompassing the ARC is near the third ventricle and median eminence where the blood brain barrier is characteristically leaky, allowing the ingress of circulating peripheral hormones. This allows multiple signaling molecules to access the orexigenic and anorexigenic ARC neurons, and elicit the release of their cognate peptides and transmitters. A significant example of peripheral signals that control the feeding center are the glucocorticoids (GCs) secreted by the adrenal gland in response to ACTH, which is released primarily from pituitary corticotrophs under the control of hypothalamic corticotrophin-releasing hormone (CRH). GCs function as ligands for a specific nuclear GC receptor (GR or NR3C1), which acts as ligand-dependent transcription factor to upregulate AgRP and NPY expression through promoter glucocorticoid response elements (GRE) [74]. Highlighting their orexigenic action, GCs have been demonstrated to increase food intake and body mass in mice [75]. Conversely, ablation of endogenous GC by adrenalectomy in genetically obese *ob/ob* (leptin deficient) mice resulted in reduced body mass [76]. GCs have a role in the regulation of both anorexigenic and orexigenic peptide expression. AgRP expression has been reproducibly shown to be positively regulated by GCs [74]. However, the role of GCs in NPY expression is less clear, with only the synthetic GC dexamethasone having a positive effect on NPY expression in mice [77], while the endogenous rodent GC did not [66]. The regulation of POMC expression by GCs is also less clear, with both positive [78] and negative [79] effects observed.

In addition to GCs, several peripheral peptide hormones are significant regulators of NPY/AgRP and POMC/CART neuron function, thereby modulating energy balance. Insulin released in the fed state functions as a satiety factor, binding to specific receptors (INSR) on ARC neuron bodies to elicit an anorexigenic effect and body weight reduction. INSR is located on both NPY/AgRP and POMC/CART neurons, indicating the likelihood of countervailing actions in each neuron. Consistent with this, activation of insulin signaling in the ARC decreases NPY and AgRP expression, while increasing POMC expression [80,81].

Leptin, produced by adipocytes at circulating levels proportional to adipose mass, is also an anorexigenic satiety factor that functions as a marker for positive energy balance. It acts in part by increasing αMSH production by POMC neurons, and also by reducing the expression of AgRP and NPY. Intraventricular administration of leptin in rodents resulted in decreased food intake and body mass consistent with an anorexigenic role [82]. However, leptin does not have potential as a therapeutic agent for obesity, as central leptin resistance, as well as central insulin resistance, appears to be a significant mediator of obesity, and may be associated with neuroinflammation [63].

Another peripheral peptide hormone that is significantly involved in the regulation of energy balance is the stomach-produced orexigenic hormone ghrelin. Ghrelin functions uniquely as a nutrient activated hormone, and is modified in the gut to its active acylated form by exposure to diet-derived fatty acids, thereby acting as a lipid content sensor [83]. Ghrelin is the natural ligand for the growth hormone secretagogue receptor (GHSR), which is highly expressed in the hypothalamic feeding center. The orexigenic action of ghrelin requires the function of NPY/AgRP neurons, as their ablation blocks the effects of ghrelin [84], and ghrelin increases NPY and AgRP expression [85].

16.13 EPIGENETIC PROGRAMMING OF HYPOTHALAMIC DYSFUNCTION BY PERINATAL MALNUTRITION

The development of the hypothalamus occurs in distinct phases, which involve both initial neurogenesis and the subsequent formation of functional neuronal circuits. In humans, the development of hypothalamic axonal projections and functional synapses occurs in the fetus, while in rodents this process continues during early postnatal life. Because of this critical developmental window, any perturbation or extremes in the perinatal environment that can affect hypothalamic gene expression through epigenetic mechanisms has the potential to impact the functional development of the hypothalamus, thereby programming the regulation of energy balance in offspring. This potential has been demonstrated in several rodent models of maternal undernutrition, in which offspring exposed to undernutrition display low birth weight and postnatal hyperphagia, and ultimately develop obesity and insulin resistance in adulthood, particularly when challenged with a high caloric diet [86]. Impairment of hypothalamic cell proliferation and axonal elongation has been observed in mice exposed to perinatal malnourishment, which alters the density of NPY/AgRP and POMC/CART neurons [87]. These structural alterations could also have direct effects on feeding behavior and energy expenditure, in addition to the effects mediated by the alteration of neuropeptide expression levels or postreceptor signaling by epigenetic mechanisms.

In addition to the structural effects of maternal undernutrition on the hypothalamus, rodent models have also revealed that the expression of the key ARC neuropeptides is altered by this exposure, including the anorexigenic POMC, and the orexigenic AgRP and NPY. These studies have consistently shown that maternal undernourishment results in increased hypothalamic NPY and AgRP expression, and decreased POMC expression in offspring, consistent with the associated hyperphagia and obesity [88]. In addition, investigators have taken advantage of the extended postnatal period of hypothalamic development in rodents to determine if undernourishment during this period can reprogram the metabolic phenotype. By adjusting litter size, and thus the maternal milk allotment of each pup, it was found that undernourished pups raised in large litters displayed reduced preweaning growth. This effect was associated with increased NPY neurons and NPY concentrations in the ARC [89], demonstrating that postnatal nutrition during the critical window of hypothalamus development can modify the ARC neural networks involved in energy homeostasis.

While these results are significant, answering the question of the long-term stability of early metabolic programming is the ultimate goal with respect to the implications for adult health. Addressing this question requires longitudinal comparison of programmed individuals with unaffected controls, an approach for which there are relatively few published studies. However, the results of these longitudinal analyses in rodents have shown that the association of maternal undernutrition with elevated GR expression levels is persistent throughout adult life. A reduction in POMC expression in response to perinatal undernutrition was also sustained into adulthood [64,90]. These observations are in accordance with those from the aforementioned studies of human famine cohorts, which indicated a persistent alteration of metabolic phenotype following perinatal exposure to malnutrition, and provide insight into the possible biochemical basis of these phenotypes and the connection to early epigenetic programming.

A major caveat to the findings of perinatal malnutrition studies in rodents is the discordance in the timing and dynamics of the morphological and functional development of the hypothalamus with that of humans. Thus, it is unclear whether the observations of maternal metabolic programming of offspring in rodent models can be extended to an analogous scenario in humans, despite the support of this possibility from population-level analyses. This limitation has been addressed by maternal malnutrition studies in animals for which the process of hypothalamus ontogeny is congruent with that in humans, such as sheep and other nonhuman primates. Studies in sheep have also shown that maternal undernutrition during early gestation could increase glucocorticoid receptor (GR) expression, but no changes in POMC or NPY expression were observed [64,90]. However, observations in baboon indicated reduced POMC expression and increased NPY expression in the fetus following gestational maternal undernutrition, consistent with an orexigenic outcome [91]. Taken together, these observation warrant caution in considering the transferability of findings in animal models to the situation in human, but confirm a general sensitivity of hypothalamic development and function to the perinatal environment that can result in obesogenic metabolic programming of offspring that likely extends to humans.

The predominant focus of research into the developmental programming of metabolic dysregulation leading to obesity has been the effects of perinatal undernutrition, largely due to the initial observations in human cohorts exposed to maternal undernutrition during historical famines. However, the rapidly compounding frequency of obesity worldwide, particularly in developed Western countries where caloric excess is widespread at all socioeconomic levels, indicates some aspect of maternal over nutrition as an alternative mediator of fetal metabolic programming leading to obesity. Thus, more recently the focus of research has transitioned to studying the effects of maternal over nutrition. Extensive epidemiological studies have established a clear link between increased maternal BMI and the susceptibility of offspring to adult metabolic and cardiovascular disease hallmarks of MetS, reinforcing the general validity of this hypothesis [92]. The breadth of the effects is striking, affecting the development and function of diverse tissues including adipose, pancreas, muscle, vasculature, and brain. In addition, cohorts of patients who undergo bariatric surgery provide the opportunity to directly test the role of the maternal over nourished environment in a somewhat longitudinal fashion. In one such study, children born to obese mothers who underwent bariatric surgery had a decreased propensity to develop obesity compared to those born prior to surgery, illustrating the deleterious effects of maternal over nutrition on the metabolic health of offspring [93].

Experiments designed in rodent models have described the effects of maternal high caloric content and high fat diets on the metabolic programming of offspring. Significantly, these studies have revealed changes that occur during fetal prenatal development, but also those that may occur postnatally through nursing [94]. These analyses reinforced a common outcome of obesity in the offspring accompanied by impaired insulin and glucose control, as well as other MetS markers, such as cardiovascular disease [95]. However, the underlying causal mechanisms for these phenotypes are not well understood due to conflicting results, in contrast with the more decided outcome of maternal undernutrition. Some studies reproducibly demonstrate impaired anorexigenic and orexigenic axon projections from the ARC to the PVN, as was observed in the case of maternal undernutrition [96]. However, the effects of over nutrition on the levels of neuropeptide expression were highly variable between studies. In one study, mothers fed a high fat diet throughout gestation and lactation had offspring with elevated POMC expression and reduced NPY expression [97]. In other studies, either decreases in both POMC and NPY expression [98], or increases in CART, NPY, and AgRP expression were observed [99]. Thus, the precise outcomes of maternal over nutrition in rodent models appears to be sensitive to experimental design, indicating a degree of complexity that differentially affects components of the feeding center.

While these animal models of maternal over nutrition gave rise to adverse outcomes in offspring, it was unclear to which period of exposure the effects of over nutrition were attributable. Along these lines, the importance in the timing of the maternal diet on the phenotype of the programmed offspring was elucidated in a study which found that a maternal high fat diet during lactation elicited changes in hypothalamic POMC and AgRP nerve fiber projections. This was accompanied by hyperinsulinemia in the programmed offspring due to the significantly elevated glucose content in the milk of high fat diet-fed mothers, suggesting that the hypothalamic structural effects may be due to aberrant insulin signaling. Consistent with this, selective deletion of the insulin receptor gene in POMC neurons rescued the aberrant ARC neural projections in the programmed offspring, and the accompanying glucose intolerance. These findings suggest that elevated neuronal insulin signaling in offspring in response to maternal over nutrition can predispose offspring for impaired glucose tolerance [100]. Since ARC neuronal projections form during the third trimester of gestation in humans, these findings indicate that maternal glucose control during this period of human gestation may be a critical determinant of epigenetic metabolic programming of offspring. This phenomenon is not dependent on a maternal high fat diet, as general overfeeding in preweaned mice due to litter size restriction resulted in hyperphagia and obesity, and was accompanied by increased numbers of orexigenic galanin-producing neurons in the hypothalamus [89]. Studies by other groups have identified a similar potential linkage between aberrant leptin signaling in the ARC and metabolic dysfunction in rats programmed by maternal over nutrition [96], indicating that multiple metabolic signaling mechanisms may be acting in this context of a hypercaloric maternal milieu to alter hypothalamic architecture and function.

16.14 HYPOTHALAMIC EPIGENETIC TARGETS ASSOCIATED WITH OBESITY

The lack of strong genetic associations with obesity from individual loci limit the utility of this information in identifying potential genomic targets of epigenetic mechanisms associated with metabolic programming of obesity. While the loci identified in these studies have in limited instances been shown to be perturbed epigenetically in obesity, a candidate gene approach informed by established functional networks in the hypothalamus and peripheral metabolic tissues has identified several epigenetic targets that may underlie programmed metabolic dysregulation. The anorexigenic POMC is a potential target for epigenetic modifications associated with obesity considering its regulatory role, and the *POMC* gene is regulated by promoter methylation [101]. *POMC* expression in the hypothalamus is also regulated by two conserved distal neuronal enhancers, nPE1 and nPE2 located 10–12 Kb upstream of the transcriptional start site [102]. Perinatal nutrition can affect epigenetic marks in these distal regulatory elements of the *POMC* gene, thereby affecting feeding behavior and eliciting an obese phenotype. For example, maternal periconceptual undernutrition in sheep results in hypomethylation of the nPE1 and nPE2 *POMC* enhancers in the fetal hypothalamus [103]. Neonatal over nutrition in rats was associated with hypermethylation of CpG sites within a promoter Sp1 binding site required for leptin and insulin mediated activation of *POMC* in the hypothalamus. This was accompanied by obesity and increased leptin, insulin, and glucose levels [104]. Further studies have shown hypermethylation of the *POMC* promoter in 28-day-old postnatal mice reared by dams with dietary supplementation of conjugated linoleic acids, which suppressed *POMC* expression. This resulted in hyperphagia, hyperglycemia and insulin resistance, confirming the physiological significance of this mechanism [105]. While more data is needed to establish the generality *POMC* as an epigenetic target in obesity, these studies support its contention as a viable candidate.

GR functions in the hypothalamic control of energy homeostasis by regulating the expression of POMC, NPY, and AgRP in response to circulating GCs. As such, GR also represents a potential epigenetic target for fetal programming that affects energy balance, predisposing offspring to metabolic disease. Evidence in mice has shown a strong link between a maternal care-dependent stress response and the epigenetic status of the *Nr3c1* gene in the offspring, consistent with the potential for the *Nr3c1* gene as a target for metabolic programming [106]. It was also shown in sheep that maternal undernutrition resulted in decreased *Nr3c1* promoter methylation and altered histone modifications associated with increased *GR* expression in the fetal hypothalamic regions involved in energy balance [90]. A significant finding of these studies was that the epigenetic modifications of the *Nr3c1* gene persisted for up to 5 years and were accompanied by decreased *Pomc* expression, confirming the stability of epigenetic effects beyond the initial insult [64]. Epigenetic changes in the *Nr3c1* gene as a result of altered maternal nutrition have also been observed in other tissues, such as liver and hippocampus, and folic acid supplementation was shown to abrogate epigenetic changes to *NR3C1* in liver induced by a protein-restricted diet [107].

In addition to *POMC* and *NR3C1*, some additional hypothalamic genes involved in the regulation of feeding behavior and energy expenditure have been identified as epigenetic targets. For example, newborn rats fed a high carbohydrate formula showed increased *Npy* expression correlated with epigenetic changes in the proximal promoter region of *Npy* that were associated with later adult onset obesity. These changes included reduced DNA methylation and increased H3 lysine 9 acetylation [108]. Neonatal over nutrition also leads to DNA hypermethylation in the promoter of the *INSR* gene in the hypothalamus, consistent with the hypothalamic insulin resistance associated with obesity [43]. Offspring of mice fed a high fat diet during pregnancy and lactation displayed promoter DNA hypomethylation of hypothalamic dopamine, and opioid-related genes, resulting in increased expression. These offspring also displayed a preference for energy-dense food rich in sucrose and fat [109]. This suggests that dietary modulation during critical stages of the growth and development can affect the epigenetic regulation of multiple hypothalamic genes involved in food intake, body weight, and energy homeostasis, thereby programming the subsequent risk of developing obesity and related metabolic and cardiovascular disorders.

16.15 HYPOTHALAMIC-PITUITARY AXIS DYSFUNCTION IN OBESITY

While the role of the hypothalamic feeding center in metabolic homeostasis is clear, there are additional CNS functions that are dysregulated in association with obesity and are a significant component of its pathophysiology. For example, some functions of the hypothalamic-pituitary (H-P) axis are aberrant in obesity, leading to dysfunction in the peripheral target tissues of pituitary trophic hormones [110]. The pituitary gland plays a predominant role in homeostasis, responding to regulatory signals from the hypothalamus by modulating the synthesis and secretion of multiple trophic hormones that control the function of target glands and peripheral tissues. In this way, the H-P axis integrates multiple feedback mechanisms to control essentially all physiological systems. Significantly, the neurons that regulate pituitary function are found in the same regions where orexigenic and anorexigenic peptides are expressed, indicating the potential for overlapping epigenetic effects. Consistent with this possibility, obesity is associated with the dysregulation of multiple pituitary hormones. For example, growth hormone (GH) has a primary physiological role to promote cell proliferation and tissue expansion, and as such is a predominant mediator of postnatal growth. GH also regulates metabolic flux, promoting muscle protein anabolism and adipose lipolysis, and inhibiting lipogenesis. Obesity is correlated with reduced circulating GH levels, which may exacerbate obesity-related metabolic derangements in a reinforcing cycle by promoting adipose lipogenesis and reducing muscle protein anabolism.

In addition to GH, other pituitary functions are dysregulated in obesity. The production of gonadotropins may also be epigenetically affected by obesity, evidenced by the increased prevalence of precocious puberty associated with childhood obesity that is linked to increased luteinizing hormone (LH). This increase appears to be through the elevation of hypothalamic LH releasing hormone (LHRH) secretion, which may be in response to inflammatory eicosanoids. Obesity may be a risk factor for developing polycystic ovary syndrome, which is also correlated with elevated LH levels and gonadal steroidogenesis. Thyrotropes in the anterior pituitary produce TSH in response to hypothalamic thyrotropin-releasing hormone (TRH). The function of TSH is to stimulate the thyroid production of thyroxine (T4) and triiodothyronine (T3), which in turn regulate metabolic flux, tissue growth, and responsiveness to other endocrine factors. Elevated TSH levels have likewise been correlated with obesity and insulin resistance, and obese human subjects respond to TRH administration with a greater level of TSH secretion compared to lean subjects.

The H-P axis function for which the most evidence for an obesity connection exists is the regulation of ACTH production by corticotropes. Secretion of ACTH is activated by hypothalamic corticotropin-releasing hormone (CRH) and arginine vasopressin (AVP), and functions primarily to stimulate the synthesis and release of adrenal corticosteroids in response to physiological stress. Modulation of adrenal activity is associated with a broad range of physiological processes, including metabolic flux, electrolyte and water balance, and inflammation. Significantly, hyperactivation of the adrenal axis appears to be tightly associated with the pathophysiology of obesity due to its responsiveness to multiple inflammation factors that are elevated in obesity. These observations illustrate the central and far-reaching role of the hypothalamus in the etiology and pathophysiology of obesity.

16.16 NEUROPSYCHIATRIC ASPECTS OF OBESITY

In the context of this book, psychosocial factors are known to be crucial to the development of obesity [111]. One such psychosocial factor is known to be stress [112]. However, how psychosocial factors result in obesity is at present unclear [111]. Obesity is also known to be associated with the presence of mood disorders, such as major depression and bipolar disorder, as well as anxiety disorders, such as generalized anxiety and panic disorder [113]. In addition, psychotropic drugs are known to cause obesity as an adverse effect [114]. It is also thought that individuals with obesity can develop psychiatric problems, such as depression as a psychological reaction to their obesity [113].

16.17 CONCLUSIONS

A considerable history of observations in human populations and experimental animal models have established and reinforced a connection between the developmental environment and the propensity for obesity. The overall conclusion form these findings is that the optimal outcome for metabolic health through adulthood is attained when the developmental environment, in particular maternal perinatal metabolic and nutritional status, is maintained to avoid extremes of under- or overnutrition. The correlation between environmental factors and obesity is mediated by epigenetic mechanisms that program the function of genes through chemical modifications to DNA and histones in chromatin and these biochemical mechanisms are sensitive to specific nutrients and metabolic signals. In the context of obesity, these epigenetic programming mechanisms largely affect the hypothalamic feeding center in the ARC and PVN where food intake and energy expenditure are regulated, and where other neuroendocrine functions occur, such as the control of pituitary hormone synthesis and release. While changes in epigenetic marks associated with obesity in other tissues also likely contribute to obesity due to the programming of oxidative metabolism and other pathways in peripheral tissues, it is the central control of energy homeostasis in the hypothalamus that is the key for understanding the etiology of obesity.

Despite the wealth of observations to date, there are several areas in which future expanded research will benefit. The causal agents of the correlation between extremes of perinatal nutrition and obesity in offspring remain to be fully characterized, although several possibilities, such as elevated glucose have been proposed. The genome-scale epigenetic effects of an obesogenic developmental environment are also not well understood, and will benefit from advances in technology for epigenome-wide association studies, analogous to GWAS. This would allow the unbiased identification of novel target loci for epigenetic mechanisms, which would also expand the knowledge of risk loci for obesity. Despite these limitations, candidate gene approaches informed by knowledge of the functional roles of various proteins in metabolic regulation will continue to be a useful approach. Finally, the field needs a better understanding of the long-term susceptibility of the energy homeostasis-related epigenome to environmental factors, beyond the effects of the perinatal environment. Seminal studies in monozygotic twins have demonstrated that the whole epigenome with respect to DNA methylation diverges between twins to a degree proportional to their age, and to the degree of difference in their environments and lifestyles [115], reinforcing this potential. As the mechanistic underpinnings of the epigenetic modulation of obesity are fully resolved, the ability to understand and counteract the growing problem of obesity will progress.

ABBREVIATIONS

AgRP Agouti-related peptide
ARC Arcute nucleus
BMI Body mass index
CNS Central nervous system
CART Cocaine and amphetamine responsive transcript
DMN Dorsomedial hypothalamus

DNMT	DNA methyltransferase
GABA	γ-Amino butyric acid
GC	Glucocorticoid
GHSR	Growth hormone secretagogue receptor
GR	Glucocorticoid receptor
HDAC	Histone deacetylase
HDL	High-density lipoprotein
IR	Insulin receptor
LepR	Leptin receptor
LH	Lateral hypothalamus or luteinizing hormone
MetS	Metabolic syndrome
NPY	Neuropeptide Y
POMC	Proopiomelanocortin
PVN	Paraventricular nucleus
T2D	Type 2 diabetes

GLOSSARY

Anorexigenic Conferring a decrease in food intake

Body mass index (BMI) A measurement of the ratio of weight to height as an objective indicator of obesity and overweight. Defined as body mass in kilograms divided by the square of body height in meters (kg/m^2)

Epigenators Signals such as differentiation factors, nutrients, hormones, or environmental variations that trigger an epigenetic response

Epigenetic initiators DNA sequence-specific transcription factors and noncoding RNAs that target and establish the epigenetic chromatin state at the gene level through the recruitment of cofactor enzyme complexes

Epigenetic maintainers DNA and histone modifying enzymes recruited chromatin sites established by the initiators to stably maintain the chromatin state through cell division

Epigenome The whole genomic patterns of specific markers of functional chromatin modifications, including DNA methylation, histone methylation and histone acetylation, among others

Feeding center Hypothalamic nuclei where satiety, feeding behavior, and energy expenditure are controlled

Histone A globular protein subunit of nucleosomes in chromatin

Hypothalamic-pituitary axis The functional integration between the hypothalamus and pituitary that regulates pituitary trophic hormone synthesis and release

Imprinting The mechanism for parental origin-specific monoallelic expression of a gene

Metabolic syndrome (MetS) A spectrum of common comorbidities that includes obesity, elevated triglycerides, reduced high-density lipoprotein (HDL) cholesterol levels, elevated fasting glucose concentrations, and hypertension

Metastable epialleles Gene loci that possess a unique sensitivity to epigenetic variability during early embryonic development that becomes stabilized in diverse tissues, enforcing a programmed phenotype

Nucleosome An octamer of histone subtypes wrapped by ~150 basepairs of DNA to form the fundamental repeating unit of chromatin

Obesity Abnormal or excessive fat accumulation with the potential to impair health

Orexigenic Conferring an increase in food intake

Periconceptual The period of time immediately surrounding oocyte fertilization and zygote implantation

Perinatal The period of time immediately surrounding birth

REFERENCES

[1] Obesity and overweight fact sheet: World Health Organization; 2015. Available from: http://www.who.int/mediacentre/factsheets/fs311/en/.

[2] Grundy SM. Obesity, metabolic syndrome, and cardiovascular disease. J Clin Endocrinol Metab 2004;89(6):2595–600.

[3] Waddington CH. The epigenotype. Endeavor 1942;1:18–21.

[4] Berger SL, Kouzarides T, Shiekhattar R, Shilatifard A. An operational definition of epigenetics. Genes Dev 2009;23(7):781–3.

[5] Sakabe K, Wang Z, Hart GW. Beta-*N*-acetylglucosamine (O-GlcNAc) is part of the histone code. Proc Natl Acad Sci USA 2010;107(46):19915–20.

[6] Guo JU, Su Y, Shin JH, Shin J, Li H, Xie B, et al. Distribution, recognition and regulation of non-CpG methylation in the adult mammalian brain. Nat Neurosci 2014;17(2):215–22.

[7] Deaton AM, Bird A. CpG islands and the regulation of transcription. Genes Dev 2011;25(10):1010–22.

[8] Holliday R, Grigg GW. DNA methylation and mutation. Mutat Res 1993;285(1):61–7.

[9] Brandeis M, Frank D, Keshet I, Siegfried Z, Mendelsohn M, Nemes A, et al. Sp1 elements protect a CpG island from de novo methylation. Nature 1994;371(6496):435–8.

[10] Ho L, Crabtree GR. Chromatin remodelling during development. Nature 2010;463(7280):474–84.

[11] Martin C, Zhang Y. Mechanisms of epigenetic inheritance. Curr Opin Cell Biol 2007;19(3):266–72.

[12] Zemach A, Gaspan O, Grafi G. The three methyl-CpG-binding domains of AtMBD7 control its subnuclear localization and mobility. J Biol Chem 2008;283(13):8406–11.

[13] Reik W. Stability and flexibility of epigenetic gene regulation in mammalian development. Nature 2007;447(7143):425–32.

[14] Lumey LH, Stein AD, Kahn HS, van der Pal-de Bruin KM, Blauw GJ, Zybert PA, et al. Cohort profile: the Dutch Hunger Winter families study. Int J Epidemiol 2007;36(6):1196–204.

[15] Hales CN, Barker DJ. Type 2 (non-insulin-dependent) diabetes mellitus: the thrifty phenotype hypothesis. Diabetologia 1992;35(7):595–601.

[16] Bennett PH, Burch TA, Miller M. Diabetes mellitus in American (Pima) Indians. Lancet 1971;2(7716):125–8.

[17] Pettitt DJ, Aleck KA, Baird HR, Carraher MJ, Bennett PH, Knowler WC. Congenital susceptibility to NIDDM. Role of intrauterine environment. Diabetes 1988;37(5):622–8.

[18] Morison IM, Ramsay JP, Spencer HG. A census of mammalian imprinting. Trends Genet 2005;21(8):457–65.

[19] Moore T, Haig D. Genomic imprinting in mammalian development: a parental tug-of-war. Trends Genet 1991;7(2):45–9.

[20] Ferguson-Smith AC. Genomic imprinting: the emergence of an epigenetic paradigm. Nat Rev Genet 2011;12(8):565–75.

[21] Monk D. Genomic imprinting in the human placenta. Am J Obstet Gynecol 2015;213(4 Suppl.):S152–62.

[22] Reik W, Walter J. Genomic imprinting: parental influence on the genome. Nat Rev Genet 2001;2(1):21–32.

[23] Butler MG. Genomic imprinting disorders in humans: a mini-review. J Assist Reprod Genet 2009;26(9–10):477–86.

[24] Delrue MA, Michaud JL. Fat chance: genetic syndromes with obesity. Clin Genet 2004;66(2):83–93.

[25] Chen M, Wang J, Dickerson KE, Kelleher J, Xie T, Gupta D, et al. Central nervous system imprinting of the G protein G(s)alpha and its role in metabolic regulation. Cell Metab 2009;9(6):548–55.

[26] Tauber M, Diene G, Mimoun E, Cabal-Berthoumieu S, Mantoulan C, Molinas C, et al. Prader-Willi syndrome as a model of human hyperphagia. Front Horm Res 2014;42:93–106.

[27] Holland A, Whittington J, Hinton E. The paradox of Prader–Willi syndrome: a genetic model of starvation. Lancet 2003;362(9388):989–91.

[28] Cummings DE, Clement K, Purnell JQ, Vaisse C, Foster KE, Frayo RS, et al. Elevated plasma ghrelin levels in Prader–Willi syndrome. Nat Med 2002;8(7):643–4.

[29] Wilson LC, Oude Luttikhuis ME, Clayton PT, Fraser WD, Trembath RC. Parental origin of Gs alpha gene mutations in Albright's hereditary osteodystrophy. J Med Genet 1994;31(11):835–9.
[30] Mantovani G, Spada A. Mutations in the Gs alpha gene causing hormone resistance. Best Pract Res Clin Endocrinol Metab 2006;20(4):501–13.
[31] Mantovani G, Romoli R, Weber G, Brunelli V, De Menis E, Beccio S, et al. Mutational analysis of GNAS1 in patients with pseudohypoparathyroidism: identification of two novel mutations. J Clin Endocrinol Metab 2000;85(11):4243–8.
[32] Liu J, Litman D, Rosenberg MJ, Yu S, Biesecker LG, Weinstein LS. A GNAS1 imprinting defect in pseudohypoparathyroidism type IB. J Clin Invest 2000;106(9):1167–74.
[33] Bastepe M, Frohlich LF, Hendy GN, Indridason OS, Josse RG, Koshiyama H, et al. Autosomal dominant pseudohypoparathyroidism type Ib is associated with a heterozygous microdeletion that likely disrupts a putative imprinting control element of GNAS. J Clin Invest 2003;112(8):1255–63.
[34] Mariot V, Maupetit-Mehouas S, Sinding C, Kottler ML, Linglart A. A maternal epimutation of GNAS leads to Albright osteodystrophy and parathyroid hormone resistance. J Clin Endocrinol Metab 2008;93(3):661–5.
[35] Wardle J, Carnell S, Haworth CM, Plomin R. Evidence for a strong genetic influence on childhood adiposity despite the force of the obesogenic environment. Am J Clin Nutr 2008;87(2):398–404.
[36] Herrera BM, Keildson S, Lindgren CM. Genetics and epigenetics of obesity. Maturitas 2011;69(1):41–9.
[37] Frayling TM, Timpson NJ, Weedon MN, Zeggini E, Freathy RM, Lindgren CM, et al. A common variant in the FTO gene is associated with body mass index and predisposes to childhood and adult obesity. Science 2007;316(5826):889–94.
[38] Jia G, Fu Y, Zhao X, Dai Q, Zheng G, Yang Y, et al. N6-methyladenosine in nuclear RNA is a major substrate of the obesity-associated FTO. Nat Chem Biol 2011;7(12):885–7.
[39] Wahlen K, Sjolin E, Hoffstedt J. The common rs9939609 gene variant of the fat mass- and obesity-associated gene FTO is related to fat cell lipolysis. J Lipid Res 2008;49(3):607–11.
[40] Chambers JC, Elliott P, Zabaneh D, Zhang W, Li Y, Froguel P, et al. Common genetic variation near MC4R is associated with waist circumference and insulin resistance. Nat Genet 2008;40(6):716–8.
[41] Widiker S, Karst S, Wagener A, Brockmann GA. High-fat diet leads to a decreased methylation of the Mc4r gene in the obese BFMI and the lean B6 mouse lines. J Appl Genet 2010;51(2):193–7.
[42] Kuroda A, Rauch TA, Todorov I, Ku HT, Al-Abdullah IH, Kandeel F, et al. Insulin gene expression is regulated by DNA methylation. PLoS One 2009;4(9):e6953.
[43] Plagemann A, Roepke K, Harder T, Brunn M, Harder A, Wittrock-Staar M, et al. Epigenetic malprogramming of the insulin receptor promoter due to developmental overfeeding. J Perinat Med 2010;38(4):393–400.
[44] Zhou D, Pan YX. Gestational low protein diet selectively induces the amino acid response pathway target genes in the liver of offspring rats through transcription factor binding and histone modifications. Biochim Biophys Acta 2011;1809(10):549–56.
[45] Zheng S, Rollet M, Pan YX. Maternal protein restriction during pregnancy induces CCAAT/enhancer-binding protein (C/EBPbeta) expression through the regulation of histone modification at its promoter region in female offspring rat skeletal muscle. Epigenetics 2011;6(2):161–70.
[46] Cordero P, Campion J, Milagro FI, Goyenechea E, Steemburgo T, Javierre BM, et al. Leptin and TNF-alpha promoter methylation levels measured by MSP could predict the response to a low-calorie diet. J Physiol Biochem 2011;67(3):463–70.
[47] Heijmans BT, Tobi EW, Stein AD, Putter H, Blauw GJ, Susser ES, et al. Persistent epigenetic differences associated with prenatal exposure to famine in humans. Proc Natl Acad Sci USA 2008;105(44):17046–9.
[48] Tobi EW, Lumey LH, Talens RP, Kremer D, Putter H, Stein AD, et al. DNA methylation differences after exposure to prenatal famine are common and timing- and sex-specific. Hum Mol Genet 2009;18(21):4046–53.
[49] Tobi EW, Goeman JJ, Monajemi R, Gu H, Putter H, Zhang Y, et al. DNA methylation signatures link prenatal famine exposure to growth and metabolism. Nat Commun 2014;5:5592.

[50] Willer CJ, Speliotes EK, Loos RJ, Li S, Lindgren CM, Heid IM, et al. Six new loci associated with body mass index highlight a neuronal influence on body weight regulation. Nat Genet 2009;41(1):25–34.
[51] Heid IM, Jackson AU, Randall JC, Winkler TW, Qi L, Steinthorsdottir V, et al. Meta-analysis identifies 13 new loci associated with waist-hip ratio and reveals sexual dimorphism in the genetic basis of fat distribution. Nat Genet 2010;42(11):949–60.
[52] Shi Y, Whetstine JR. Dynamic regulation of histone lysine methylation by demethylases. Mol Cell 2007;25(1):1–14.
[53] Inagaki T, Tachibana M, Magoori K, Kudo H, Tanaka T, Okamura M, et al. Obesity and metabolic syndrome in histone demethylase JHDM2a-deficient mice. Genes Cells 2009;14(8):991–1001.
[54] Tateishi K, Okada Y, Kallin EM, Zhang Y. Role of Jhdm2a in regulating metabolic gene expression and obesity resistance. Nature 2009;458(7239):757–61.
[55] Alenghat T, Meyers K, Mullican SE, Leitner K, Adeniji-Adele A, Avila J, et al. Nuclear receptor corepressor and histone deacetylase 3 govern circadian metabolic physiology. Nature 2008;456(7224):997–1000.
[56] Anderson OS, Sant KE, Dolinoy DC. Nutrition and epigenetics: an interplay of dietary methyl donors, one-carbon metabolism and DNA methylation. J Nutr Biochem 2012;23(8):853–9.
[57] Waterland RA, Kellermayer R, Laritsky E, Rayco-Solon P, Harris RA, Travisano M, et al. Season of conception in rural gambia affects DNA methylation at putative human metastable epialleles. PLoS genet 2010;6(12):e1001252.
[58] Waterland RA, Jirtle RL. Transposable elements: targets for early nutritional effects on epigenetic gene regulation. Mol Cell Biol 2003;23(15):5293–300.
[59] Pannia E, Cho CE, Kubant R, Sanchez-Hernandez D, Huot PS, Harvey Anderson G. Role of maternal vitamins in programming health and chronic disease. Nutr Rev 2016;74:166–80.
[60] Wellen KE, Hatzivassiliou G, Sachdeva UM, Bui TV, Cross JR, Thompson CB. ATP-citrate lyase links cellular metabolism to histone acetylation. Science 2009;324(5930):1076–80.
[61] Covington JD, Bajpeyi S. The sirtuins: Markers of metabolic health. Mol Nutr Food Res 2016;60(1):79–91.
[62] Breton C. The hypothalamus-adipose axis is a key target of developmental programming by maternal nutritional manipulation. J Endocrinol 2013;216(2):R19–31.
[63] Thaler JP, Schwartz MW. Minireview: inflammation and obesity pathogenesis: the hypothalamus heats up. Endocrinology 2010;151(9):4109–15.
[64] Begum G, Davies A, Stevens A, Oliver M, Jaquiery A, Challis J, et al. Maternal undernutrition programs tissue-specific epigenetic changes in the glucocorticoid receptor in adult offspring. Endocrinology 2013;154(12):4560–9.
[65] Rumball CW, Oliver MH, Thorstensen EB, Jaquiery AL, Husted SM, Harding JE, et al. Effects of twinning and periconceptional undernutrition on late-gestation hypothalamic-pituitary-adrenal axis function in ovine pregnancy. Endocrinology 2008;149(3):1163–72.
[66] Coll AP, Challis BG, Lopez M, Piper S, Yeo GS, O'Rahilly S. Proopiomelanocortin-deficient mice are hypersensitive to the adverse metabolic effects of glucocorticoids. Diabetes 2005;54(8):2269–76.
[67] Butler AA, Marks DL, Fan W, Kuhn CM, Bartolome M, Cone RD. Melanocortin-4 receptor is required for acute homeostatic responses to increased dietary fat. Nat Neurosci 2001;4(6):605–11.
[68] Creemers JW, Lee YS, Oliver RL, Bahceci M, Tuzcu A, Gokalp D, et al. Mutations in the amino-terminal region of proopiomelanocortin (POMC) in patients with early-onset obesity impair POMC sorting to the regulated secretory pathway. J Clin Endocrinol Metabol 2008;93(11):4494–9.
[69] Moffett M, Stanek L, Harley J, Rogge G, Asnicar M, Hsiung H, et al. Studies of cocaine-and amphetamine-regulated transcript (CART) knockout mice. Peptides 2006;27(8):2037–45.
[70] Gropp E, Shanabrough M, Borok E, Xu AW, Janoschek R, Buch T, et al. Agouti-related peptide-expressing neurons are mandatory for feeding. Nat Neurosci 2005;8(10):1289–91.
[71] Luquet S, Perez FA, Hnasko TS, Palmiter RD. NPY/AgRP neurons are essential for feeding in adult mice but can be ablated in neonates. Science 2005;310(5748):683–5.

[72] Krashes MJ, Shah BP, Koda S, Lowell BB. Rapid versus delayed stimulation of feeding by the endogenously released AgRP neuron mediators GABA, NPY, and AgRP. Cell Metab 2013;18(4):588–95.

[73] Stanley BG, Kyrkouli SE, Lampert S, Leibowitz SF, Neuropeptide Y. chronically injected into the hypothalamus: a powerful neurochemical inducer of hyperphagia and obesity. Peptides 1986;7(6):1189–92.

[74] Lee B, Kim SG, Kim J, Choi KY, Lee S, Lee SK, et al. Brain-specific homeobox factor as a target selector for glucocorticoid receptor in energy balance. Mol Cell Biol 2013;33(14):2650–8.

[75] Karatsoreos IN, Bhagat SM, Bowles NP, Weil ZM, Pfaff DW, McEwen BS. Endocrine and physiological changes in response to chronic corticosterone: a potential model of the metabolic syndrome in mouse. Endocrinology 2010;151(5):2117–27.

[76] Makimura H, Mizuno TM, Roberts J, Silverstein J, Beasley J, Mobbs CV. Adrenalectomy reverses obese phenotype and restores hypothalamic melanocortin tone in leptin-deficient ob/ob mice. Diabetes 2000;49(11):1917–23.

[77] Yi CX, Foppen E, Abplanalp W, Gao Y, Alkemade A, la Fleur SE, et al. Glucocorticoid signaling in the arcuate nucleus modulates hepatic insulin sensitivity. Diabetes 2012;61(2):339–45.

[78] Uchoa ET, Silva LE, de Castro M, Antunes-Rodrigues J, Elias LL. Glucocorticoids are required for meal-induced changes in the expression of hypothalamic neuropeptides. Neuropeptides 2012;46(3):119–24.

[79] Gyengesi E, Liu ZW, D'Agostino G, Gan G, Horvath TL, Gao XB, et al. Corticosterone regulates synaptic input organization of POMC and NPY/AgRP neurons in adult mice. Endocrinology 2010;151(11):5395–402.

[80] Schwartz MW, Sipols AJ, Marks JL, Sanacora G, White JD, Scheurink A, et al. Inhibition of hypothalamic neuropeptide Y gene expression by insulin. Endocrinology 1992;130(6):3608–16.

[81] Sipols AJ, Baskin DG, Schwartz MW. Effect of intracerebroventricular insulin infusion on diabetic hyperphagia and hypothalamic neuropeptide gene expression. Diabetes 1995;44(2):147–51.

[82] Campfield LA, Smith FJ, Guisez Y, Devos R, Burn P. Recombinant mouse OB protein: evidence for a peripheral signal linking adiposity and central neural networks. Science 1995;269(5223):546–9.

[83] Kirchner H, Gutierrez JA, Solenberg PJ, Pfluger PT, Czyzyk TA, Willency JA, et al. GOAT links dietary lipids with the endocrine control of energy balance. Nat Med 2009;15(7):741–5.

[84] Chen HY, Trumbauer ME, Chen AS, Weingarth DT, Adams JR, Frazier EG, et al. Orexigenic action of peripheral ghrelin is mediated by neuropeptide Y and agouti-related protein. Endocrinology 2004;145(6):2607–12.

[85] Kamegai J, Tamura H, Shimizu T, Ishii S, Sugihara H, Wakabayashi I. Chronic central infusion of ghrelin increases hypothalamic neuropeptide Y and Agouti-related protein mRNA levels and body weight in rats. Diabetes 2001;50(11):2438–43.

[86] Yura S, Itoh H, Sagawa N, Yamamoto H, Masuzaki H, Nakao K, et al. Role of premature leptin surge in obesity resulting from intrauterine undernutrition. Cell Metab 2005;1(6):371–8.

[87] Garcia AP, Palou M, Priego T, Sanchez J, Palou A, Pico C. Moderate caloric restriction during gestation results in lower arcuate nucleus NPY- and alphaMSH-neurons and impairs hypothalamic response to fed/fasting conditions in weaned rats. Diabetes Obes Metab 2010;12(5):403–13.

[88] Shin BC, Dai Y, Thamotharan M, Gibson LC, Devaskar SU. Pre- and postnatal calorie restriction perturbs early hypothalamic neuropeptide and energy balance. J Neurosci Res 2012;90(6):1169–82.

[89] Plagemann A, Harder T, Rake A, Waas T, Melchior K, Ziska T, et al. Observations on the orexigenic hypothalamic neuropeptide Y-system in neonatally overfed weanling rats. J Neuroendocrinol 1999;11(7):541–6.

[90] Stevens A, Begum G, Cook A, Connor K, Rumball C, Oliver M, et al. Epigenetic changes in the hypothalamic proopiomelanocortin and glucocorticoid receptor genes in the ovine fetus after periconceptional undernutrition. Endocrinology 2010;151(8):3652–64.

[91] Li C, McDonald TJ, Wu G, Nijland MJ, Nathanielsz PW. Intrauterine growth restriction alters term fetal baboon hypothalamic appetitive peptide balance. J Endocrinol 2013;217(3):275–82.

[92] Reynolds RM, Osmond C, Phillips DI, Godfrey KM, Maternal BMI. parity, and pregnancy weight gain: influences on offspring adiposity in young adulthood. J Clin Endocrinol Metab 2010;95(12):5365–9.

[93] Patti ME. Intergenerational programming of metabolic disease: evidence from human populations and experimental animal models. Cell Mol Life Sci 2013;70(9):1597–608.

[94] Guo F, Jen KL. High-fat feeding during pregnancy and lactation affects offspring metabolism in rats. Physiol Behav 1995;57(4):681–6.

[95] Williams L, Seki Y, Vuguin PM, Charron MJ. Animal models of in utero exposure to a high fat diet: a review. Biochim Biophys Acta 2014;1842(3):507–19.

[96] Kirk SL, Samuelsson AM, Argenton M, Dhonye H, Kalamatianos T, Poston L, et al. Maternal obesity induced by diet in rats permanently influences central processes regulating food intake in offspring. PLoS One 2009;4(6):e5870.

[97] Chen H, Simar D, Lambert K, Mercier J, Morris MJ. Maternal and postnatal overnutrition differentially impact appetite regulators and fuel metabolism. Endocrinology 2008;149(11):5348–56.

[98] Morris MJ, Chen H. Established maternal obesity in the rat reprograms hypothalamic appetite regulators and leptin signaling at birth. Int J Obes 2009;33(1):115–22.

[99] Lopez M, Seoane LM, Tovar S, Garcia MC, Nogueiras R, Dieguez C, et al. A possible role of neuropeptide Y, agouti-related protein and leptin receptor isoforms in hypothalamic programming by perinatal feeding in the rat. Diabetologia 2005;48(1):140–8.

[100] Vogt MC, Paeger L, Hess S, Steculorum SM, Awazawa M, Hampel B, et al. Neonatal insulin action impairs hypothalamic neurocircuit formation in response to maternal high-fat feeding. Cell 2014;156(3):495–509.

[101] Newell-Price J. Proopiomelanocortin gene expression and DNA methylation: implications for Cushing's syndrome and beyond. J Endocrinol 2003;177(3):365–72.

[102] de Souza FS, Santangelo AM, Bumaschny V, Avale ME, Smart JL, Low MJ, et al. Identification of neuronal enhancers of the proopiomelanocortin gene by transgenic mouse analysis and phylogenetic footprinting. Mol Cell Biol 2005;25(8):3076–86.

[103] Stevens A, Begum G, White A. Epigenetic changes in the hypothalamic pro-opiomelanocortin gene: a mechanism linking maternal undernutrition to obesity in the offspring? Eur J Pharmacol 2011;660(1):194–201.

[104] Plagemann A, Harder T, Brunn M, Harder A, Roepke K, Wittrock-Staar M, et al. Hypothalamic proopiomelanocortin promoter methylation becomes altered by early overfeeding: an epigenetic model of obesity and the metabolic syndrome. J Physiol 2009;587(Pt 20):4963–76.

[105] Zhang X, Yang R, Jia Y, Cai D, Zhou B, Qu X, et al. Hypermethylation of Sp1 binding site suppresses hypothalamic POMC in neonates and may contribute to metabolic disorders in adults: impact of maternal dietary CLAs. Diabetes 2014;63(5):1475–87.

[106] Zhang TY, Labonte B, Wen XL, Turecki G, Meaney MJ. Epigenetic mechanisms for the early environmental regulation of hippocampal glucocorticoid receptor gene expression in rodents and humans. Neuropsychopharmacology 2013;38(1):111–23.

[107] Lillycrop KA, Phillips ES, Jackson AA, Hanson MA, Burdge GC. Dietary protein restriction of pregnant rats induces and folic acid supplementation prevents epigenetic modification of hepatic gene expression in the offspring. eJ Nutr 2005;135(6):1382–6.

[108] Mahmood S, Smiraglia DJ, Srinivasan M, Patel MS. Epigenetic changes in hypothalamic appetite regulatory genes may underlie the developmental programming for obesity in rat neonates subjected to a high-carbohydrate dietary modification. J Dev Orig Health Dis 2013;4(6):479–90.

[109] Vucetic Z, Kimmel J, Totoki K, Hollenbeck E, Reyes TM. Maternal high-fat diet alters methylation and gene expression of dopamine and opioid-related genes. Endocrinology 2010;151(10):4756–64.

[110] Shewchuk BM. Prostaglandins and n-3 polyunsaturated fatty acids in the regulation of the hypothalamic-pituitary axis. Prostaglandins Leukot Essent Fatty Acids 2014;91(6):277–87.

[111] Sadock BJ, Sadock VA, Ruiz P. Kaplan & Sadock's synopsis of psychiatry. Philadelphia, PA: Wolters Kluwer; 2015.

[112] Sinha R, Jastreboff AM. Stress as a common risk factor for obesity and addiction. Biol Psychiatry 2013;73(9):827–35.

[113] Simon GE, Von Korff M, Saunders K, Miglioretti DL, Crane PK, van Belle G, et al. Association between obesity and psychiatric disorders in the US adult population. Arch Gen Psychiatry 2006;63(7):824–30.
[114] Devlin MJ, Yanovski SZ, Wilson GT. Obesity: what mental health professionals need to know. Am J Psychiatry 2000;157(6):854–66.
[115] Fraga MF, Ballestar E, Paz MF, Ropero S, Setien F, Ballestar ML, et al. Epigenetic differences arise during the lifetime of monozygotic twins. Proc Natl Acad Sci USA 2005;102(30):10604–9.

CHAPTER

EPIGENETICS AND DRUG ADDICTION: TRANSLATIONAL ASPECTS

17

J. Feng

Department of Biological Science, Neuroscience PhD Program, Florida State University, Tallahassee, FL, United States

CHAPTER OUTLINE

17.1 INTRODUCTION: DRUG ADDICTION, REWARD PATHWAY, AND EPIGENETICS

Addiction can be defined as the loss of control over drug use and the compulsive seeking and taking of a drug despite adverse effects [1]. It is a serious leading cause of disease and public health concern across the world. For instance, over 20 million people in the United States have a substance use disorder. Among them, over 1 million are addicted to cocaine and more than 350,000 to heroin. The number of prescription opiate abusers has also risen to over one million [2]. Though addiction continues to exact enormous costs on society, available treatments remain insufficient, and demands an improved understanding of the biological basis of addiction that will lead to more effective prevention, diagnosis, and treatments.

Drug use does not always result in addiction. For example, only ~20% of cocaine users eventually become addicted and the rate is much lower for alcohol [3]. Features of drug addiction include the development of dependence after gradual use, such that addicts require the drug to function

Neuropsychiatric Disorders and Epigenetics. http://dx.doi.org/10.1016/B978-0-12-800226-1.00017-4

properly. Ultimately, larger doses of the drug are needed (tolerance) and symptoms will occur once the drug is not available (withdrawal). There is therefore a great interest in understanding the transition from drug use to the addiction state. It is known that addiction is influenced by the convergence of genetic and environmental factors. Although the specific genes that comprise this risk remain largely unknown, twin studies have revealed common heritable genetic factors that may contribute up to 50% of the risk for addiction [4]. However, the other ~50% is thought to reflect a range of environmental influences. Recently, studies have elucidated that environmental experience can influence addiction through epigenetic regulation of gene expression changes in key brain structures.

The word "epigenetics" was originally defined as "the study of mitotically and/or meiotically heritable changes in gene function that cannot be explained by changes in DNA sequence" [5,6]. Epigenetics mainly refers to histone modifications, DNA modifications (such as DNA methylation), and noncoding RNAs. Epigenetic mechanisms adjust gene activity by altering accessibility of DNA to the transcriptional machinery. Given the fact that many chromatin marks are short-lived and may not be transmissible between generations and that DNA methylation patterns can also be rapidly removed during development, the emphasis on heritability in epigenetics may not be necessary [5]. It is generally accepted that "epigenetics" is the study of changes in the regulation of gene activity and expression that are not dependent on gene sequence (NIH Roadmap Epigenomics Mapping Consortium). In the past few decades, aberrant epigenetic regulation has been increasingly linked to human diseases and can be influenced by a multitude of environmental inputs. In fact, it has been demonstrated that epigenetic mechanisms play pivotal roles in translating environmental stimuli into long-lasting gene expression changes in the nervous system, which is key to both physiological (learning and memory), as well as pathological conditions (psychiatric disorders). Drug addiction is generally deemed as an aberrant form of neural plasticity [7]. As we will elaborate throughout this chapter, epigenetic regulation of gene expression provides an attractive regulatory mechanism that leads to relatively stable neural changes and thus contributes to the long-lasting behavioral abnormalities that characterize drug addiction [1,8]. This chapter will mainly focus on histone and DNA modifications in drug addiction. Although there are promising data for the involvement of noncoding RNAs, they are relatively few and have been explored in previous publications [8–12].

Repeated drug use stimulates dopamine transmission in the brain circuitry that normally regulates adaptive behavioral responses to a changing environment. It is believed that the drug-induced adaptations impair synaptic plasticity in neuronal projections and thereby deregulate the ability of addicts to control their drug-taking habits. In the past decades, great effort has been made to identify key brain regions involved in addiction. The circuit that has received the most attention is referred to as the mesolimbic dopamine system [1,13], which comprises dopamine neurons in the ventral tegmental area (VTA) of the midbrain projecting to medium spiny neurons in the nucleus accumbens (NAc, a part of the ventral striatum) [14] (Fig. 17.1). This VTA–NAc circuit is crucial for the recognition of rewards in the environment and for initiating their consumption, but they also respond to aversive stimuli. The VTA neurons also innervate many other forebrain regions, including hippocampus, amygdala, prefrontal cortex (PFC), and others. This brain reward pathway is in fact very much interconnected, as the NAc also receives glutamatergic innervation from other brain regions. Together, these various interconnected circuits are referred to as the brain's reward pathway. While these brain structures are crucial for mediating responses to natural rewards, they are also the sites where drugs of abuse produce the long-lasting changes that underlie addiction.

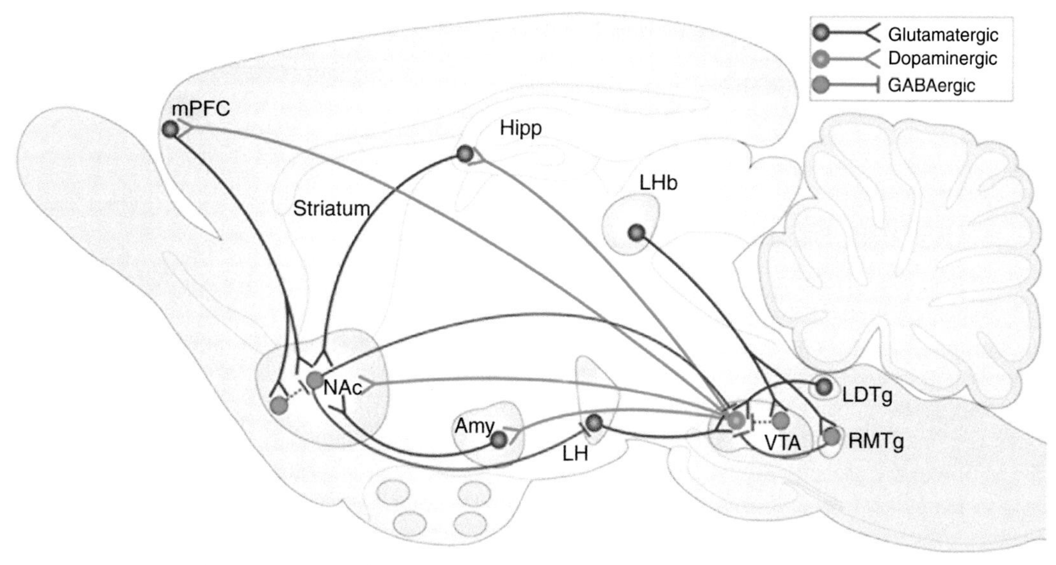

FIGURE 17.1 The Brain's Reward Circuitry

A schematic of the major dopaminergic, glutamatergic, and GABAergic connections to and from the ventral tegmental area (*VTA*) and nucleus accumbens (*NAc*) (dopaminergic = green; glutamatergic = red; and GABAergic = blue) in the rodent brain. The primary reward circuit includes dopaminergic projections from the VTA to the NAc. There are also GABAergic projections from the NAc to the VTA; projections through the direct pathway [mediated primarily by D1-type medium spiny neurons (MSNs)] directly innervate the VTA, whereas projections through the indirect pathway (mediated primarily by D2-type MSNs) innervate the VTA via intervening GABAergic neurons in ventral pallidum (not shown). The NAc also contains numerous types of interneurons (not shown). The NAc receives innervation from glutamatergic monosynaptic circuits from the medial prefrontal cortex (*mPFC*) and other PFC regions (not shown), hippocampus (*Hipp*), and amygdala (*Amy*), among other regions. The VTA receives such inputs from amygdala, lateral dorsal tegmentum (*LDTg*), lateral habenula (*LHb*), and lateral hypothalamus (*LH*), among others. These various glutamatergic inputs control aspects of reward-related perception and memory. The glutamatergic circuit from LH to VTA is also mediated by orexin (not shown).

Reprinted with permission from Nature Publishing Group. Russo SJ, Nestler EJ. The brain reward circuitry in mood disorders. Nat Rev Neurosci 2014;14(9):609–625 [14]

Rodent models have been widely utilized to mimic drug addiction in order to successfully recapitulate the key features of addiction seen in humans. In one model, the operant self-administration paradigm, an animal performs a behavior reinforced by the delivery of a drug (e.g., intravenous for cocaine self-administration) paired with a sensory cue. The animals thus produce behavioral patterns that are reminiscent of human addicts, such as escalation of drug administration. In this model, depending on the experimental conditions, a subset of animals become compulsive drug users and show high rates of relapse during abstinence. However, drug self-administration paradigms are very labor intensive and technically challenging in mice, which are a better species for molecular/genomic research compared to rats. Many studies continue to utilize passive drug administration paradigms (e.g., repeated intraperitoneal injection (IP) injections of cocaine). Although this cannot capture the consequences of volitional control over drug use, it is rewarding and can produce sensitized drug responses that provide essential insights into drug addiction. In the meantime, findings from passive drug administration paradigms should also be validated in rodent self-administration models as well as human patients whenever possible.

17.2 HISTONE MODIFICATIONS IN ADDICTION

In the nuclei of all eukaryotic cells, genomic DNA is highly folded and compacted by bound histone and nonhistone proteins in a dynamic polymer called chromatin. The founding unit of chromatin is the nucleosome, which is composed of 147 base pairs of double stranded DNA wrapped around a histone octamer that consist of two copies each of H2A, H2B, H3, and H4. The distinct levels of chromatin organization are dependent on the dynamic higher order structuring of nucleosomes and DNA. Histones are small basic proteins consisting of a more flexible and charged N-terminus (histone "tail") in addition to the globular domain. Histones undergo posttranslational modification (PTM) that alters their structure and interactions with neighboring DNA. The long histone tails protruding away from the nucleosome can be covalently modified at several places by acetylation, methylation, phosphorylation, ubiquitination, SUMOylation, citrullination, and ADP-ribosylation [15], thus altering their charge and regulating chromatin structure and accessibility to the underlying DNA. The enzymes which establish these histone tail modifications "writers" are highly specific for particular amino acid positions. In parallel, the modifications can also be removed with corresponding modifying enzymes "erasers" [16]. Addition and removal of these posttranslational modifications of histone tails leads to the addition and/or removal of other marks in a highly complicated histone code. These make histone modifications a reversible, labile epigenetic mark, thereby extending the informational content of the genome beyond the genetic (DNA) code [17].

Being an aberrant form of neural plasticity [7], drug addiction has been shown to be mediated in part by altering the levels of histone epigenetic modifications [2,8,10,18–28]. By far, histone acetylation and methylation are the most highly studied histone modifications in drug addiction and will be discussed further herein.

17.2.1 HISTONE ACETYLATION IN DRUG ADDICTION

Histone acetylation negates the positive charge of lysine residues in the histone tail to open up chromatin structure and, is therefore associated with transcriptional activation. It is regulated by histone acetyltransferases (HATs), which usually facilitate transcription, and histone deacetylases (HDACs), which usually repress transcription [29]. Cocaine and other drugs of abuse induce changes in histone acetylation in key brain regions, which underlie some of the functional abnormalities seen in addiction. Acute and chronic cocaine administration induces different histone acetylation changes in the NAc, which appear to be gene specific [30]. A robust induction of H4 acetylation, but not H3 acetylation, was observed at the *c-Fos* promoter within 30 min of acute cocaine administration and remained elevated until 3 h following drug exposure. However, no acetylation change was seen at the same locus after chronic cocaine administration, which is consistent with cocaine's effect to only induce *c-Fos* acutely. In contrast, at the promoters of genes that are induced by chronic but not acute cocaine exposure, such as *Bdnf* (brain-derived neurotrophic factor) and *Cdk5* (cyclin-dependent kinase 5), H3 acetylation was observed. This enrichment was long-lasting and even stably accumulated over the first week of drug withdrawal [30]. These results provide the first evidence that different histone modifications can specifically encode acute and chronic drug exposure in a gene-specific manner. Under a rat cocaine self-administration paradigm, it was also found that among the genes activated by chronic cocaine, the expression of *CaMKIIα* correlated positively with motivation for the drug. Viral mediated shRNA knockdown of *CaMKIIα* in the NAc shell indicated that chronic drug-use-induced H3 acetylation modulated transcriptional activation of genes such as *CaMKIIα*, and is essential for the maintenance of motivation to self-administer cocaine [31].

This cocaine-mediated histone acetylation change also participates in gene expression regulation in other brain regions involved in addiction. For example, increased *Bdnf* expression in the VTA coincides with incubation of cocaine-seeking behavior. In rats with 7 days of forced drug abstinence after self-administration of cocaine, BDNF protein and exon I-containing transcripts were both significantly increased in cocaine-experienced rats compared to yoked saline controls [32]. Interestingly, cocaine-induced changes in *Bdnf* mRNA were associated with increased H3 acetylation as well as CREB-binding protein enrichment at exon I-containing promoters in the VTA. These changes suggest that drug abstinence following cocaine self-administration utilizes histone acetylation to modulate *Bdnf* expression in the VTA, which may contribute to neuroadaptations underlying cocaine craving and relapse [32].

Accumulating evidence also suggests that histone acetylation is implicated in the action of other substances of abuse as well. It was revealed that on day 7 of morphine abstinence there was a significant increase in H3 acetylation (H3K9/K14) at the *Bdnf* promoter II in the locus coeruleus of the rat brain [33]. It was also recognized that the systemic administration of methamphetamine (METH) caused decreases in histone H3 acetylated at lysine 9 (H3K9ac) and lysine 18 (H3K18ac) in the rat NAc [34]. In contrast, METH injection caused time-dependent increases in acetylated H4K5 and H4K8. The histone acetyltransferase ATF2 also showed significant METH-induced increases in protein expression. These results suggest that METH-induced alterations in the transcriptome might be related to METH-induced dynamic changes in histone acetylation. Indeed, in a follow-up study, it was revealed that METH led to a decrease in acetylation of histone H4 on glutamate receptor *GluA1*, *GluA2*, and *GluN1* promoters. METH exposure also increased enrichment of repressor element-1 silencing transcription factor corepressor 1, methylated CpG binding protein 2 (MeCP2), and histone deacetylase 2 (HDAC2) at the *GluA1* and *GluA2* genes [35]. As we mentioned earlier, *c-Fos* is rapidly induced in the striatum after acute psychostimulant exposure, such as cocaine [30] and amphetamine (AMPH) [36]. *c-Fos* is also a notable downstream target that is repressed chronically by ΔFosB, a transcription factor that accumulates in the striatum after repeated drug exposure and mediates sensitized behavioral responses to psychostimulants and other drugs of abuse. It was shown that accumulation of ΔFosB in the striatum after chronic AMPH treatment desensitizes *c-Fos* mRNA induction to a subsequent drug dose. ΔFosB desensitizes *c-Fos* expression by recruiting HDAC1 to the *c-Fos* gene promoter, leading to deacetylation of nearby histones and attenuation of gene transcription [36]. This study demonstrated an epigenetic pathway through which ΔFosB mediates distinct transcriptional programs that may ultimately alter behavioral plasticity to chronic drug exposure.

While HATs attach an acetyl group onto a lysine amino acid of a histone, HDACs are a class of enzymes that remove acetyl groups from an acetyl-lysine, allowing the DNA to wrap more tightly around the nucleosomes. Moreover, the availability of various HDAC inhibitors (HDACi) provides an additional pharmacological pathway to further elucidate the role of histone acetylation in drug addiction. As HDAC4 is highly expressed within the striatum, its function in drug addiction was first probed among all HDACs [30]. It was found that overexpression of HDAC4 in the striatum dramatically decreased the rewarding effects of cocaine in the conditioned place preference (CPP) assay, coupled with reduced levels of H3 acetylation. Subsequently, HDAC5 was found to be a central integrator in cocaine action [37]. Chronic, but not acute, exposure to cocaine or stress decreased HDAC5 function in the NAc, which allowed for increased histone acetylation and transcription of HDAC5 target genes. This regulation is behaviorally important, as loss of HDAC5 caused hypersensitive responses to chronic, not acute, cocaine or stress. These findings suggest that proper balance of histone acetylation is a crucial factor in the transition from an acute adaptive response to a chronic psychiatric illness. It was also found that the role of HDAC5 in cocaine reward may be regulated by cyclic adenosine monophosphate (cAMP) signaling [38]. Specifically it was shown that cAMP-stimulated nuclear import of HDAC5 requires transient, protein phosphatase 2A (PP2A)-dependent dephosphorylation of a Cdk5 site (S279) found within the HDAC5 nuclear localization sequence. Dephosphorylation of HDAC5 increased its nuclear accumulation by accelerating its nuclear import rate and reducing its nuclear export rate. Importantly, dephosphorylation of HDAC5 S279 in the NAc suppressed the development of cocaine reward behavior in vivo. This provided a mechanism by which cocaine regulates HDAC5 function to antagonize the rewarding impact of cocaine. Interestingly, as HDAC3 functions in vitro with HDAC4 or HDAC5 in multiprotein transcriptional repressor complexes, it was speculated HDAC3 may work in concert with HDAC4 and/or HDAC5 in the NAc in drug addiction. To address the hypothesis that HDAC3 is a negative regulator of cocaine-context-associated memory formation in mice, researchers examined the role of HDAC3 during the conditioning phase of CPP, when the mouse has the opportunity to form an associative memory between the cocaine-paired context and the subjective effects of cocaine. To address this hypothesis, viral cre-mediated HDAC3 deletion in NAc was achieved in *Hdac3* (flox/flox) mice, which exhibited significantly enhanced CPP acquisition [39]. Moreover, increased gene expression of *c-Fos* and *Nr4a2* was correlated with decreased HDAC3 occupancy and increased histone H4 lysine 8 acetylation at their promoters. The results from this study demonstrated that HDAC3 negatively regulates cocaine-induced CPP acquisition. A similar conclusion was also drawn by applying selective HDAC3 inhibitor RGFP966 to study cocaine behavior [40]. It was found that systemic treatment with RGFP966 facilitates extinction in mice in a manner resistant to reinstatement.

HDAC inhibitors have also been applied in other drug addiction research to further elucidate the role of histone acetylation. For example, to study the behavioral effects of HDAC inhibition and the underlying molecular alterations to morphine, rats were pretreated with the class I and II HDACi, trichostatin A (TSA), in the basolateral amygdala (BLA), a brain structure critically involved in cue-associated memory [41]. It was found that intra-BLA pretreatment with TSA significantly enhanced morphine-induced CPP acquisition and expression, facilitated extinction, and reduced reinstatement of morphine-induced CPP. These behavioral changes were associated with a general increase in histone H3 lysine14 acetylation in the BLA, together with upregulation of BDNF and ΔFosB, as well as increased CREB activation. Collectively, these results imply that HDAC inhibition in the BLA promotes some aspects of the memory that develops during conditioning and extinction training. Therefore, H3 acetylation may play a role in learning and memory for morphine addiction in the BLA. In another approach [42], HDACi butyric acid (BA) and valproic acid (VPA) treatment were shown to have no locomotor effects, but significantly potentiated AMPH-induced behavioral sensitization in mice together with an increase of histone H4 acetylation in the striatum. However, repeated BA and VPA administration after the induction of AMPH sensitization inhibited the expression of the sensitized response to AMPH upon a challenge dose. Thus, HDACi differentially modulate the induction and expression of AMPH-induced effects, which suggest that dynamic changes in histone acetylation may be an important mechanism underlying AMPH-induced neuronal plasticity and associative learning. Furthermore, HDACi may interact additively with psychostimulants at both histone acetylation and CREB phosphorylation through the CREB:HDAC protein complex in the striatum to modulate ΔFosB and psychomotor behavioral sensitization [43].

It should be noted that studies investigating the effects of HDAC inhibition on psychostimulant-induced behavioral plasticity have yielded conflicting results, with some reporting that HDAC inhibition suppresses the behavioral effects of cocaine or AMPH [44–46]. These discrepancies suggest different affinities of various HDAC inhibitors for different HDAC isoforms mediate highly specific biological actions of different HDAC isoforms in regulating psychostimulant responses and time-dependent effects of HDAC inhibition in the brain. To address these possibilities, local knockout of HDAC1, HDAC2, or HDAC3 in NAc of adult mice having a loxP-flanked gene by viral expression of *Cre* recombinase, revealed that only prolonged knockdown of HDAC1 substantially suppressed cocaine-induced behavioral plasticity [47]. Although acute HDAC inhibition enhances the behavioral effects of cocaine or AMPH, prolonged HDAC inhibition paired with repeated cocaine selectively induced a form of repressive histone methylation. Specifically dimethylation of Lys9 of histone H3 (H3K9me2) in the NAc, which, through the repression of GABA-A receptor subunits and inhibitory tone on NAc medium spiny neurons, contributed to the blockade of cocaine-induced plasticity. Together, these findings demonstrate cross-talk among different types of histone modifications in the adult brain.

17.2.2 HISTONE METHYLATION AND OTHER HISTONE MODIFICATIONS IN ADDICTION

In contrast to the general activation role of histone acetylation, methylation at different histone lysine sites can have repressive or activating effects on transcriptional regulation. A recent study has investigated the role of histone methylation modifications in METH-associated memory [48]. In this study, the impact of histone methylation (H3K4me2/3) on the formation and expression of METH-associated memory was determined by intra-NAc knockdown of methyltransferase mixed-lineage leukemia 1 (*Mll1*) expression, and an eraser of H3K4me2/3, the histone lysine specific demethylase 5C (KDM5C). It was revealed *Mll1* knockdown reduced H3K4me3, *c-Fos* and *Oxtr* gene transcription levels and disrupted METH-associated memory, whereas deficiency of KDM5C resulted in hypermethylation of H3K4 and prevented the expression of METH-associated memory. These data indicate that permissive histone methylation marks may serve as potential targets for the treatment of substance abuse relapse.

Recently, repressive histone methylation has also been found to be involved in cocaine action. As H3K9me2 and H3K27me2 are regulated at many gene promoters after repeated cocaine exposure [49], numerous lysine methyltransferases (KMTs) and demethylases (KDMs) that are known to control H3K9 or H3K27 methylation were therefore profiled for changes in gene expression. It turned out that only two enzymes, G9a and G9a-like protein (GLP), displayed persistent downregulation after repeated cocaine administration. Since G9a and GLP specifically catalyze the dimethylation of H3K9 (H3K9me2), their downregulation by cocaine is consistent with decreased global levels of euchromatic H3K9me2 observed at this time point [50]. Moreover, this G9a downregulation increased the dendritic spine plasticity in NAc, which was correlated with increased behavioral sensitivity to the drug, thereby establishing a crucial role for histone methylation in the long-term action of cocaine. Interestingly, there appears to be a negative feedback of accumulation of ΔFosB after chronic cocaine to repress *G9a* and hence reduce global levels of H3K9me2 [50]. By using a ribosomal affinity purification approach [51], it was found that *G9a* repression by cocaine occurred in both *Drd1* expressing and *Drd2* expressing medium spiny neurons, the two major neuronal cell types in NAc [52]. Conditional knockout and overexpression of *G9a* within these two cell types displayed divergent behavioral phenotypes in response to cocaine. A developmental deletion of *G9a* selectively in *Drd2* neurons resulted in the unsilencing of transcriptional programs specific to *Drd1* neurons and in the acquisition of Drd1-associated circuit formation. This suggests a critical function for cell type-specific histone methylation patterns in the regulation of behavioral responses to environmental stimuli [51]. Similarly, chronic morphine also decreased *G9a* expression and global levels of H3K9me2 in mouse NAc. Viral-mediated gene transfer and conditional mutagenesis demonstrated that overexpression of *G9a* in NAc opposes morphine reward and locomotor sensitization and concomitantly promotes analgesic tolerance and naloxone-precipitated withdrawal, whereas downregulation of *G9a* in NAc enhances locomotor sensitization and delays the development of analgesic tolerance. This indicates that G9a/H3K9me2 may be a universal key factor regulating drug reward [53].

Though histone acetylation and methylation are the most frequently studied epigenetic mechanisms in drug addiction, the effect of other histone modifications remains poorly understood. Recently, it was found that repeated cocaine administration increased global levels of poly(ADP-ribose) polymerase-1 (PARP-1) and its histone modification PAR in mouse NAc. PARP-1 is a nuclear protein catalyzing the synthesis of a negatively charged polymer called poly (ADP-ribose) or PAR on histones and other substrate proteins and forms transcriptional regulatory complexes with other chromatin proteins. Using PARP-1 inhibitors and viral-mediated gene transfer, it was established that *Parp-1* induction in NAc mediates enhanced behavioral responses to cocaine, including increased self-administration of the drug. It was further identified that *Sidekick-1* (*Sdk1*), important for synaptic connections during development, is a critical PARP-1 target gene involved in cocaine's behavioral effects. It is foreseeable, with increasing forms of histone modification recognized in response to drugs of abuse, their novel role in drug addiction will need be understood in the years to come.

17.3 GENOME-WIDE MAPPING OF HISTONE MODIFICATIONS IN ADDICTION

Alterations in gene expression, potentially regulated by histone and/or other epigenetic modifications, contribute importantly to the long-lasting changes that drugs of abuse induce in the brain's reward circuitry. Numerous studies to date have utilized candidate gene/locus approaches to explore epigenetic alterations and associated transcriptional changes implicated in drug addiction. However, it is crucial to obtain a global view to gain insights of other gene/locus alterations across the genome.

Chromatin immunoprecipitation (ChIP) against acetylated H3 or H4 followed by gene promoter microarray, revealed that more genes demonstrated hyperacetylation on H3, in comparison to H4, in mouse NAc after repetitive cocaine administration [49]. Numerous gene promoters were also shown to be hyperacetylated on H4 following chronic cocaine treatment in NAc. In addition, the H4 hyperacetylated genes have only a minimal overlap with genes that displayed hyperacetylation on H3 [49]. This indicates that H4 acetylation not only is involved in both acute and chronic cocaine responses, but also functions independently from H3 acetylation in regulating repeated cocaine-induced transcription. Though combinatory patterns of histone PTM are speculated to constitute the "histone code" that is thought to control the transcriptional machinery [15], this study suggests the interpretation of the code can be more sophisticated. Thus, the same histone modification may exert its function in a context dependent manner, [varied drug administration paradigms (acute or chronic) at varied gene promoters], which is less likely to be universally translated. To address this complexity, it is important to carry out further studies to: (1) analyze more genomic regions beyond gene promoters, (2) utilize a more sophisticated technique which queries multiple histone modifications at the same time, and (3) employ a more sensitive transcriptome analysis that can identify expression of multiple mRNA splicing isoforms. To achieve this, investigators carried out next generation sequencing (NGS) to profile the transcriptome (mRNAseq) and epigenome (ChIP-seq), including several histone marks such as H3K4me1, H3K-4me3, K3K9me2, H3K9me3, H3K27me3, H3K36me3, and total RNA polymerase II [54]. Compared with microarray technology, NGS provides several unsurpassed advantages that include whole genome coverage, single base resolution, high sensitivity, and less bias. NAc tissue was collected 24 h after 7 daily cocaine IP injections, the classical chronic cocaine administration paradigm used in the previous publications. To account for interanimal variations, three biological replicates were obtained for each treatment group, with each replicate containing NAc pooled from five animals. Hundreds to thousands of differential sites (with size ranging from 200 to 1200 bp) for various ChIP-seq H3 and RNA pol II marks were identified, which represent a massive amount of epigenomic change induced in the mouse NAc by chronic cocaine. Although only a moderate number of genes were observed to be differentially expressed after cocaine, comprising more than 1% of the total 22,000 genes, it was also revealed that splicing changes are much more widespread, as 4,129 isoforms were shown to be differentially expressed, comprising more than 5% of the transcriptome. These results suggest that differentially spliced transcripts play substantial roles in the transcriptional perturbations induced by chronic cocaine [54]. Alternative splicing is a highly regulated process by which pre-mRNAs are differentially spliced, leading to expression of several different mRNAs from a single gene. This is especially common for genes expressed in the brain, where production of a particular protein isoform determines different aspects of neuron development and function. In recent years, emerging evidence indicates that some histone marks can act as beacons in exon definition and in recruiting splicing factors to pre-mRNA [55]. Therefore, a systematic approach was utilized called "chromatin signature" that allowed for the profiling of the epigenomic changes of each splicing transcript in a unified fashion. In total, researchers defined 335,779 and 441,648 unique exonic and intronic regions on the genome and calculated each mark's enrichment difference between cocaine and saline at each region. By comparing only the epigenomic alterations coupled with differential splicing changes, K-means clustering identified 29 clusters of chromatin signatures that shared similar epigenetic patterns and were coexpressed together [54] (Fig. 17.2). Similar mRNAseq transcriptome profilings were also carried out in postmortem hippocampi of cocaine addicts [56] and mouse NAc after seven daily cocaine IP injections [57], with or without a coupled H3K4me3 ChIP-seq [56]. Variable differential lists were generated from each dataset, which suggests the NGS results need to be validated through independent approaches and the data analysis pipelines are better to be standardized in the future to minimize cross-experimental variations.

One promising feature of current genomic approaches is that they are high throughput. A sophisticated bioinformatics data analysis can predict promising candidate genes that may play key roles in addiction. For the 29 signature clusters identified in mouse NAc after chronic cocaine [54], each cluster may be coregulated by a few common protein regulators which interact with chromatin during transcription. By predicting transcription factors and splicing factors that may be involved in this process, researchers indeed recognized enrichment of numerous splicing factor and transcription factor binding

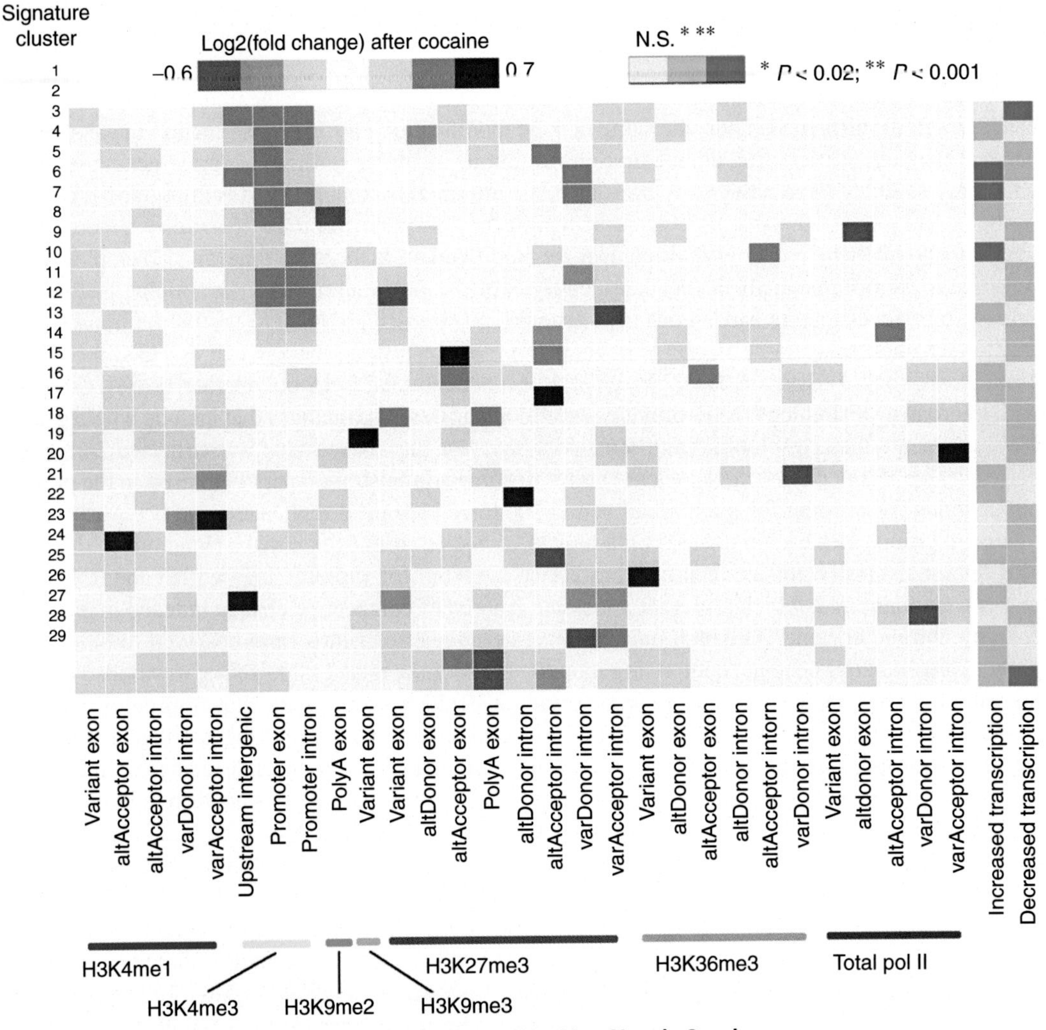

FIGURE 17.2 Chromatin Signature Clustering in Mouse NAc After Chronic Cocaine

In the left heat map, each row represents a cluster and each column represents an epigenomic mark at a specific genomic region; each square represents the averaged chromatin modification change in log2 scale; purple and brown colors indicate increased or decreased binding and the darkness indicates the magnitude of change. The two column heat map in the right panel illustrates the statistical significance of each cluster's association with transcriptional change. Each cluster represents a combination of enrichment differences of the seven marks which lead to increased or decreased transcript levels. For example, the H3K4me3 generally shows increased binding around TSSs for clusters 1–12. However, depending on the combination with other histone marks, clusters 1–12 show either increased or decreased transcription H3K4me1 overall shows decreased binding at variant and alternative acceptor exons including neighboring intronic regions across all 29 clusters. H3K27me3 displays the most dynamics among the seven marks for the 29 clusters, while H3K9me2 and H3K9me3 only regulate polyA and variant exons, respectively. Notably, the majority of the marks show regulation at the intronic regions, indicating the importance of introns in determining splicing.

Reprinted with permission from BioMed Central. Feng J, Wilkinson M, Liu X, Purushothaman I, Ferguson D, Vialou V, et al. Chronic cocaine-regulated epigenomic changes in mouse nucleus accumbens. Genome Biol 2014;15(4):R65 [54] motifs across these clusters. It was also confirmed that one splicing factor, *Rbfox1* (or *A2bp1*), plays a previously unrecognized role in drug reward behavior and related transcriptional regulation [54]. Likewise, the ChIP-chip study [49] also provided comprehensive insight into the molecular pathways regulated by cocaine, including a novel role for sirtuins (*Sirt1* and *Sirt2*)-which are induced in the NAc by cocaine and dramatically enhance the behavioral effects of the drug. Sirtuins (SIRT) are class III histone deacetylases that were first characterized for their control of cellular physiology in peripheral tissues, but their influence in brain under normal and pathological conditions remains poorly understood. Ferguson et al. showed that chronic cocaine administration increased *Sirt1* and *Sirt2* expression in the mouse NAc, while chronic morphine administration induced *Sirt1* expression alone, with no regulation of other sirtuin family members observed [58]. Drug induction of SIRT1 and SIRT2 is mediated in part via ΔFosB and is associated with robust histone modifications at the *Sirt1* and *Sirt2* genes. Viral-mediated overexpression of SIRT1 or SIRT2 in the NAc enhances the rewarding effects of both cocaine and morphine, which established Sirtuins as key mediators of the molecular and cellular plasticity induced by drugs of abuse in NAc. To further determine the mechanisms by which SIRT1 mediates cocaine-induced plasticity in the NAc ChIP-seq was used to map SIRT1 binding genome-wide in mouse NAc [59]. The results revealed two modes of SIRT1 action: (1) chronic cocaine caused depletion of SIRT1 from most affected gene promoters in concert with enrichment of H4K16ac, and (2) the Forkhead transcription factor (FOXO) family is a downstream mechanism through which SIRT1 regulates cocaine action. It was found that SIRT1 induction caused the deacetylation and activation of FOXO3a, which alone was sufficient to promote cocaine-elicited behavioral responses. The validated functions of A2BP1, SIRT, and FOXO3a in drug addiction, all predicted from genomic approaches, further indicate the need for more exploration of novel candidates identified from the growing number of genomic approaches in the field.

One other advantage of the NGS genomic approach is that it truly covers whole genomes. As the majority (>90%) of the mammalian genome has recently been proven to be transcribed and have regulatory function [60], it is essential to look at these previously ignored intergenic regions in future research. Indeed, ChIP-seq of the mouse NAc identified close to 3000 cocaine-induced differential H3K9me3 sites and more than 9000 morphine-induced differential H3K9me2 sites, most of which are located at repetitive genomic sequences [53,61]. Concomitantly, cocaine-mediated decrease of H3K-9me3 at specific repeats, such as *LINE-1*, was further correlated with increased expression. This likely reflects a global pattern of genomic destabilization after repeated cocaine administration, which may affect related neural plasticity [62].

Genomic research also provides a means to study DNA–nucleosome interactions at a global level. An investigation into the DNA–nucleosome interactions within promoter regions of 858 genes in human neuroblastoma cells in response to nicotine or cocaine was recently carried out [63]. Widespread, drug- and time-resolved repositioning of nucleosomes was identified at the transcription start sites and promoter regions of multiple genes. Nicotine and cocaine produced unique and shared changes in terms of the numbers and types of genes affected, as well as repositioning of nucleosomes at sites which could increase or decrease the probability of gene expression based on DNA accessibility. These findings suggest that nucleosome repositioning represents an initial dynamic genome-wide alteration of the transcriptional landscape preceding more selective downstream transcriptional reprogramming, which ultimately characterizes the cell- and tissue-specific responses to drugs of abuse. Lately, the ACF (ATP-utilizing chromatin assembly and remodeling factor) ATP-dependent chromatin-remodeling complex was also shown to be necessary for stress-induced depressive-like behaviors in the NAc [64].

Altered ACF binding after chronic stress was also correlated with altered nucleosome positioning, particularly around the transcription start sites of affected genes. Based on these cumulative studies, it would be interesting to continue to explore the role of nucleosome remodeling in drug addiction in vivo in the future.

17.4 DNA MODIFICATIONS IN ADDICTION

DNA methylation is an epigenetic mechanism in which a methyl group is covalently coupled to the C5 position of a cytosine residue, predominantly at CpG dinucleotides [6]. This reaction is catalyzed by a group of enzymes called DNA methyltransferases (DNMTs), which include the maintenance enzyme DNMT1, responsible for methylating the unmethylated DNA strand during DNA replication, and two DNA methyltransferases, DNMT3a and DNMT3b, which establish de novo methylation patterns on unmethylated DNA. During development, the methylome undergoes distinct waves of methylation changes. As a result, there are cell-type and tissue-specific DNA methylation patterns. The importance of DNA methylation has also been widely demonstrated in genomic imprinting (silencing of specific genes depends on parental origin), retroviral silencing, X-chromosome inactivation and embryonic development [6]. In general, methylated DNA at gene promoters leads to gene silencing. This is possibly because: (1) methylated DNA blocks transcriptional activator binding, (2) DNMT protein recruits transcriptional repressors and plausibly mediates nonenzymatic roles in transcriptional silencing, and (3) methyl-CpG-binding proteins directly recognize methylated DNA and recruit corepressor molecules to silence transcription and further modify the surrounding chromatin [65]. The family of methyl-CpG-binding proteins includes six members designated MBD1–MBD4, Kaiso, and MeCP2. They are characterized by the presence of a methyl-CpG-binding domain (MBD), the protein motif responsible for binding methylated CpG dinucleotides.

Compared to most other epigenetic mechanisms, DNA methylation is more static and was initially believed to mediate long-term gene silencing. This view was also supported by the fact that no definitive DNA demethylation machinery has been identified in the nervous system [66]. Moreover, as DNA methylation patterns are largely established and maintained during DNA replication, it is unclear how dynamic is DNA methylation in nondividing neurons. Despite these earlier views, recent evidence indicates that DNA methylation turnover does occur in the mature brain [67–69], which suggests the existence of some previously uncharacterized DNA demethylation machinery. The major breakthrough came in the end of 2009, when two groups independently demonstrated that ten-eleven translocation protein 1 (TET1) oxidizes 5-methylcytosine (5mC) into 5-hydroxymethylcytosine (5hmC) [70,71]. Subsequently, two other members of the same family, TET2 and TET3, were also shown to possess similar enzymatic activities. Additionally, TET enzymes can further oxidize 5hmC into 5-formylcytosine (5fC) and 5-carboxylcytosine (5caC) successively [72,73]. As these newer forms of DNA epigenetic modifications were derived directly from methylated cytosine, it was speculated that 5mC oxidation may lead to DNA demethylation. Indeed, conversion of 5mC to 5hmC can facilitate both passive and active DNA demethylation through several pathways such as base excision repair [74–76]. Given the fact that TET-mediated 5mC oxidation is independent of DNA replication; these discoveries provide mechanistic details for DNA demethylation in postmitotic neurons.

Though still elusive, 5hmC also appears to serve transcriptional regulatory roles by either promoting or repressing transcription [76]. The potential pathways include, but are not limited to: (1) 5hmC inhibits the binding of MBD proteins and therefore is thought to induce transcriptional activation, (2) 5hmC prevents DNMT1 recruitment, (3) 5hmC impacts accessibility of transcription factors, and/or (4) 5hmC reduces the melting temperature of DNA duplexes to promote transcriptional elongation.

Thus far, accumulating evidence has implicated DNA methylation in neural plasticity, learning and memory, and cognition [77–82]. However, studies of DNA methylation in addiction are still relatively few, with a number of studies focused on the role of DNMTs or the methyl-cytosine binding protein MeCP2. Here we review the available literature on both DNA methylation and novel forms of DNA epigenetic modifications in addiction, with a focus on cocaine research.

As noted above, MeCP2 encoded by *MECP2/Mecp2* is an X-linked methyl-DNA binding protein that is highly expressed in mature neurons. *MECP2* mutation is recognized to cause the vast majority of Rett syndrome cases. More recently, MeCP2 has also been implicated in neural and behavioral responses to psychostimulants. For example, *Mecp2* was induced in the dorsal striatum of rats with extended access to intravenous cocaine self-administration, a process that mimics the increasingly uncontrolled cocaine use seen in addicts. Viral-mediated MeCP2 knockdown in this region decreased the rats' cocaine intake [83]. Interestingly, it was found that MeCP2 regulates this effect through an interaction with microRNA-212 (*miR-212*) to control the cocaine response on striatal BDNF levels, which brought up a plausible interplay between epigenetic modifications in addiction. Concomitantly, it was shown that manipulation of MeCP2 expression in the NAc bidirectionally modulates AMPH-induced CPP. Viral-mediated knockdown of MeCP2 in this region increased AMPH place conditioning, whereas local MeCP2 overexpression had the opposite effect [84]. *Mecp2* hypomorphic mutant mice showed deficient AMPH-induced structural plasticity of NAc dendritic spines, as well as deficient plasticity of striatal immediate early gene inducibility after repeated AMPH administration. Notably, psychostimulants induce phosphorylation of MeCP2 at Ser421, a site that regulates MeCP2s function as a repressor. This phosphorylation strongly predicts the degree of behavioral sensitization. Importantly, research has shown more recently that mice with a Ser421Ala mutation in MeCP2 display greater locomotor sensitization to experimenter-administered cocaine as well as greater self-administration of the drug [85]. The mutant *Mecp2* mice also display reduced neuronal electrical excitability of NAc and altered transcriptional responses to cocaine. These studies together link MeCP2 function in the NAc and dorsal striatum with psychostimulant addiction.

The expression of the maintenance DNA methylation catalyzing enzyme DNMT1 and the de novo DNA methyltransferases DNMT3a and DNMT3b have been characterized in the nervous system [86,87]. Almost all mature neurons in the adult mouse brain express DNMT1 at substantially high levels as compared with other organs, whereas DNMT3a is predominantly expressed in later embryonic stages through adulthood within neural precursor cells, maturing neurons, oligodendrocytes, and a subset of astrocytes, indicating their functional importance in the nervous system. LaPlant et al. demonstrated that repeated cocaine exposure regulates *Dnmt3a* transcription in mouse NAc [88]. DNMT3a, but no other DNMTs, were upregulated at an early time point of withdrawal (4 h after the last cocaine dose), followed by downregulation after 24 h. Whether this complicated pattern of DNMT3a regulation is associated with fluctuations in DNA methylation awaits further investigation. However, DNMT3a was upregulated in the NAc after an extended 28 days of withdrawal, following either cocaine IP injections or cocaine self-administration [88]. This long-lasting induction of *Dnmt3a* may be important in addiction, given its potential influence on downstream regulation of target genes, a possibility which requires further examination. Furthermore, pharmacological and viral gene transfer approaches have been used to examine the behavioral influence of DNMTs on drug addiction. Viral-mediated overexpression of DNMT3a in the NAc attenuated the rewarding effects of cocaine [88]. At the cellular level, *Dnmt3a* overexpression increased thin dendritic spines of NAc neurons to comparable levels seen in response to chronic cocaine administration. Meanwhile, viral-mediated DNMT3a knockdown, or inhibition of DNMTs via local infusion of the DNMT inhibitor RG108, had the opposite effect. These cellular and behavioral findings establish DNMT3a as having an important role in addiction. In contrast, different effects were reported by another group [89], which found induction of both *Dnmt3a* and *Dnmt3b* in mouse NAc but only after acute (not chronic) cocaine administration [89]. The reasons for these discrepancies are unknown, but could be due to the different experimental paradigms used. Furthermore, the same group also observed that cocaine treatment resulted in DNA hypermethylation and increased binding of MeCP2 at the *Pp1c* gene promoter, which is associated with transcriptional downregulation of *Pp1c*. In contrast, acute and repeated cocaine administrations induced hypomethylation and decreased binding of MeCP2 at the *FosB* promoter, and these are associated with transcriptional upregulation of *FosB* in NAc. DNMT inhibitor zebularine also decreased cocaine-induced DNA hypermethylation at the *Pp1c* promoter and associated transcriptional downregulation, which further delayed the development of cocaine-induced behavioral sensitization. Together, these results suggest a role for DNMT as well as DNA methylation in cocaine-induced behavioral sensitization. Interestingly, it has been shown that drug-induced regulation of *Dnmt1* expression in the NAc and other regions is dependent on the genetic background. It was reported that METH treatment induced opposing patterns of *Dnmt1* mRNA expression in the nucleus caudatus and NAc of two inbred rat strains, Fischer 344/N (increased *Dnmt1*) and Lewis/N (decreased *Dnmt1*). As Fischer rats have a hyperresponsive negative feedback in their hypothalamic-pituitary-adrenocortical (HPA) axis and are resistant to sensitizing effects of METH, Lewis rats have a hyporesponsive feedback in their HPA axis and are prone to METH sensitization. These data suggest METH-induced differences in *Dnmt1* expression and DNA methylation might be related to the contrasting susceptibilities of these two rat strains to behavioral and neurochemical effects of METH [90].

Though the roles of DNMT3a and MeCP2 in drug reward are interesting, direct evidence of differential DNA methylation is needed to further establish the function of DNA methylation in addiction. As is observed for many epigenetic modifications, global DNA methylation change is not always indicative of an alteration at particular genetic loci. Mass spectrometry-based measurements showed that chronic cocaine decreased total levels of methylated DNA in the PFC [91]. However, there was no such change in the NAc [92] or in response to other drugs of abuse, such as morphine [91]. A major requirement in the field is to obtain genome-wide maps of DNA methylation in the NAc and other brain reward regions following chronic drug administration (see subsequent section). In the absence of such genome-wide studies, a small number of candidate genes have been shown to exhibit altered methylation in addiction models, as previously indicated for PP1c (Table 17.1). For example, chronic cocaine was shown to induce *c-Fos* expression in the NAc, which was associated with reduced methylation at CpG dinucleotides in the *c-Fos* gene promoter [93]. In contrast, cocaine administration induced hypomethylation and decreased binding of MeCP2 at the *FosB* promoter in the NAc, associated with induction of *FosB* [89]. Beyond brain reward pathways, significant hypomethylation at multiple CpG sites of the *Sox10* promoter region was observed in the corpus callosum of rats at 30 days of forced abstinence from cocaine self-administration [94]. As *Sox10* expression is enriched in oligodendrocytes, the major cell type in corpus callosum, this indicates that cocaine regulation of DNA methylation in some nonneuronal cell types must be considered.

Table 17.1 Examples of Candidate Genes Exhibiting Altered DNA Methylation in Addiction

Gene Name	Drug of Abuse	Differential Methylation Region	Direction of Change	Associated mRNA/Protein Change	Species/ Tissue or Cell Type	DNA Methylation Methodology	References
c-Fos	Cocaine	Promoter	Hypomethylation	mRNA increase	Rat/nucleus accumbens	Bisulfite sequencing	[93]
FosB	Cocaine	Promoter	Hypomethylation	mRNA increase	Mouse/ nucleus accumbens	MeDIP/ methylation-specific qPCR	[89]
GluA1	Methamphetamine	Promoter	Hypomethylation	Both decrease	Rat/striatum	MeDIP	[35]
GluA2	Methamphetamine	Promoter	Hypomethylation	Both decrease	Rat/striatum	MeDIP	[35]
OPRM1	Opioids	Promoter	Hypermethylation	N/A	Human/ blood, sperm	Pyrosequencing; bisulfite sequencing	[129,130]
PP1c	Cocaine	Promoter	Hypermethylation	mRNA decrease	Mouse/ nucleus accumbens	MeDIP/ methylation-specific qPCR	[89]
Sox10	Cocaine	Promoter	Hypomethylation	N/A	Rat/white matter	Bisulfite sequencing	[94]

N/A, Data not available.

When 5hmC was first identified, it was instantly recognized as the sixth DNA base in its own right, rather than an intermediate of methyl cytosine (5mC) on its way to demethylation. One notable feature of 5hmC is that it is most enriched in various brain regions compared with other organs [71,95,96]. Indeed, TET enzymes and 5hmC have been shown to play important roles in active DNA demethylation in the hippocampus, where they have been implicated in neural development, aging, and learning and memory [97–102]. The role of TET1 and 5hmC in drug addiction was also recently identified [92]. Researchers found that chronic cocaine administration selectively decreased TET1, but not TET2 or TET3, in the mouse NAc as well as in the NAc of cocaine addicts examined postmortem. To understand the functional effect of TET1 on cocaine action, TET1 was virally knocked down or overexpressed in the adult NAc and showed that TET1 negatively regulates drug reward behavior. These results suggest that the cocaine-induced downregulation of TET1 in the NAc contributes to increased drug reward behavior. Since TET1 oxidizes 5mC into 5hmC, global measurements of both modifications in NAc were performed using a mass spectroscopy-based approach and did not detect any global regulation by cocaine administration. This indicates that any 5hmC changes induced by cocaine in this brain region may be locus specific, indicating the need for genome-wide mapping of 5hmC. Additionally, there have also been reports for other mechanisms of active DNA demethylation. In particular, a role has been suggested for growth arrest and DNA damage 45 (GADD45) protein family members, which have been shown to mediate DNA demethylation during cell differentiation and cellular stress responses [69,103]. Of note, *Gadd45b* was reported as a neural activity-induced immediate early gene in mouse hippocampal neurons. Mice with *Gadd45b* deletion exhibit specific deficits in neural activity-induced neural progenitor cell proliferation and dendritic arborization in adult hippocampus. GADD45b was then recognized to be required for activity-induced DNA demethylation of specific gene promoters, including *Bdnf* and fibroblast growth factor. From in vitro neuronal culture, it was also reported that a knockdown of GADD45b blocks the ability of NMDA receptor activation to reduce methylation of the *Bdnf* gene promoter and induce *Bdnf* mRNA levels in cultured neurons [69,104]. Thus, GADD45b links neuronal activity to epigenetic DNA modification and expression of key factors in mature neurons for extrinsic modulation in the adult brain. Given the importance of BDNF signaling in drug addiction, work is now needed to study a possible role for GADD45 in addiction-related abnormalities and a link to potential DNA methylation alterations. In fact, Koo et al. [105] demonstrated that chronic morphine administration decreases GADD45g in the NAc and that overexpression of GADD45g promoted rewarding responses to the drug.

17.5 GENOME-WIDE MAPPING OF DNA MODIFICATIONS IN ADDICTION

In the past, most studies have taken a candidate gene approach to identify genes regulated by DNA methylation in addiction. However, with the availability and affordability of tools to investigate genomic DNA modifications, it is time to gather a genome-wide view of such regulation.

A recent study demonstrated dynamic DNA methylation in the NAc during incubation of cocaine craving in rat [106]. By using methyl-DNA antibody-based immunoprecipitation followed by microarray, the study examined DNA methylation changes at promoters of all coding genes, as well as the full gene length of a custom panel of 47 candidate genes previously implicated in drug addiction. They identified broad and time-dependent increases in DNA methylation in the NAc after cocaine withdrawal and cue-induced cocaine seeking. Interestingly, there were quite dynamic DNA methylation changes in the rat brain between 1 day and 30 days after cocaine self-administration. This indicates that the DNA methylome is not static in the brain during the withdrawal period and certain alterations may possibly serve as prognostic biomarkers, pending further research. Unexpectedly, cue-induced cocaine seeking after withdrawal triggered widespread DNA methylation changes within 1 h following exposure to the cue, and they generally changed in the reverse direction as methylation changes following withdrawal. These results demonstrate a role for DNA methylation in incubation of cocaine craving [106].

Cocaine-induced 5hmC alterations were recently profiled genome-wide in the NAc, given the role of TET1 in cocaine action, as described above. This technique used the β-glucosyltransferase to transfer an engineered glucose moiety containing an azide group onto the hydroxyl group of 5hmC. The azide group is then chemically modified with biotin for downstream avidin-biotin based affinity enrichment and deep sequencing of 5hmC containing DNA fragments [107]. In total, over 20,000 differential regions were recognized, the majority of which existed in both coding regions (such as gene body) and intergenic regions [92]. To better understand the potential function of 5hmC regulation in these intergenic regions, researchers focused on 5hmC dynamics at putative distal enhancer regions, which are regulatory elements that exist long distances away from transcription start sites. ChIP-seq maps were generated for two histone marks of enhancers, H3K4me1 and H3K27ac [108]. The "chromatin state" approach was then applied [109] to explore combinatorial patterns of these two histone marks, 5hmC, and H3K4me3 (a mark typically enriched at promoters). After excluding promoter regions that have high enrichment of H3K4me3 from the analysis, dynamic regulation of chromatin states at putative distal enhancers was observed in response to cocaine [92]. Interestingly, these enhancers also displayed dynamic alterations with gain or loss of 5hmC. By using a preliminary screen to identify plausible targets of these enhancers, they were assigned to the nearest neighboring genes. These enhancer-associated genes were narrowly enriched in a few meaningful categories, such as immune genes and domains binding to methylated DNA. With respect to the coding regions of genes having altered 5hmC content, cocaine regulation of 5hmC was particularly concentrated around exon boundaries. These changes also positively correlated with pre-mRNA alternative splicing as detected by RNA-seq. Overlay of RNA-seq data with 5hmC data in coding regions further revealed that increased levels of 5hmC at gene bodies is associated positively with both increased steady-state transcription of that gene and elevated inducibility in response to a subsequent cocaine challenge [92]. Again, the genes that displayed 5hmC regulation were highly enriched in addiction-related gene categories. While 5hmC was first recognized as an intermediate between methylated and unmethylated DNA, it was unclear how stable or dynamic this novel epigenetic mark was. When loci specific for 5hmC changes were examined 30 days after cocaine administration it was found that, at least for some loci, the cocaine-induced changes in 5hmC were sustained, which further suggests a role of 5hmC as an independent epigenetic mark [92].

17.6 TRANSLATIONAL FUTURE OF EPIGENETIC STUDIES IN ADDICTION RESEARCH

Recent years have seen remarkable progress in understanding of human genetics, enabled by the availability of the human genome sequence and increasingly high-throughput technologies for DNA analysis. However, DNA sequence level investigations do not shed light on a crucial component of biology. For example, how an identical genome sequence gives rise to hundreds of different cell types during differentiation and even more diversified transcriptional programs during various diseases. Epigenetics is now known to help regulate these processes. Aberrant regulation of such phenomena has been linked to human diseases, such as drug addiction, and has been shown to be influenced by various environmental inputs.

While epigenetics was initially coined as the regulation of the heritable changes in gene expression that occur without alterations of the underlying DNA sequence, the heritability of DNA epigenetics may be better reflected in fast dividing cells during DNA replication compared to nondividing neurons. Nevertheless, recent studies indicate a plausible role of epigenetic modifications on the regulation of trans-generational inheritance of addiction response. The inheritance of psychiatric experiences has been demonstrated on numerous occasions [110–112]. For example, a cocaine addiction experience can affect an offspring's cocaine acquisition behavior [113]. It appears that DNA methylation can mediate such inheritance of experience-influenced behavior. For example, sperm DNA from parental odor fear conditioned F0 males and F1 naive offspring revealed CpG hypomethylation in the *Olfr151* gene, which was associated with increased behavioral sensitivity. In addition, in vitro fertilization, F2 inheritance and cross-fostering revealed that these transgenerational effects are inherited via parental gametes [114]. Although precisely how addiction can also impact transgenerational response to drugs of abuse through DNA epigenetic modification is not fully understood, one early study to probe this question indeed demonstrated dynamic DNA methylation [115]. By using a genome-wide DNA methylation profiling technology, the authors compared the NAc methylome in animals with and without parental THC (the main psychoactive component of marijuana) exposure. They identified 1027 differentially methylated regions associated with parental THC exposure in F1 adults. Intriguingly, many of the regions were related to genes involved in glutamatergic synaptic regulation. A long-standing question in neuro-epigenetics is whether epigenetic alterations constitute a response to drugs of abuse or predispose to addiction. Studies from transgenerational epigenetics indicate that, under certain circumstances, epigenetics may contribute to an individual's vulnerability to addiction.

Another obstacle to epigenetic research in drug addiction is the in vivo studies' deficiency of cellular resolution. Epigenetic changes are thought to participate in physiological memory mechanisms and to be critical for long-term behavioral alterations associated with addiction. However, the brain is composed of multiple cell types and little is known concerning the cell-type specificity of epigenetic modifications. To address this question, research has used bacterial artificial chromosome transgenic mice which express EGFP fused to the N-terminus of the large subunit ribosomal protein L10a driven by the D1 or D2 dopamine receptor (D1R, D2R) promoter, respectively. Fluorescence in nucleoli was used to sort nuclei from D1R- or D2R-expressing neurons and to quantify by flow cytometry the cocaine-induced changes in histone acetylation and methylation specifically in these two types of nuclei [116]. Differential epigenetic responses to cocaine in D1R- and D2R-positive neurons and their potential regulation were identified. These cell-type specific changes may be involved in the persistent effects of cocaine in these. The method described should have general utility for studying nuclear modifications in different types of neuronal or nonneuronal cell types in the future [116].

Another challenge for epigenetic studies in addiction is how to seamlessly integrate the various methodologies employed [117]. For the study of histone modifications, there are locus specific approaches using ChIP-microarray, or more recently ChIP-seq, whereas there are numerous approaches developed to measure DNA epigenetic modifications. With novel forms of DNA modification discovered in the past few years (e.g., 5hmC, 5fC, and 5caC), it would not be surprising to see still additional approaches being introduced [118–121]. Each currently used approach has its own pros and cons, which makes the choice of the right methodology particularly difficult. For instance, investigators have to decide between cost and coverage, coding regions and intergenic regions, base resolution and fragmented resolution, quantitative and relative, small amounts and large amounts of starting DNA, and so on. Though NGS based studies are widely applied and promise unprecedented advantages, the relative cost is still high and demands extensive bioinformatic support. It is predicted that the sequencing approach will become more feasible with time, as sequencing costs continue to decline. In the meantime, it is essential to cross-compare datasets derived not only from different brain regions, peripheral tissues, and addiction paradigms, but also derived from various bioinformatic platforms and technical approaches [122].

The influence of epigenetics in addiction models raises the possibility that epigenetic manipulations might provide a plausible path for addiction therapy. Various HDACi have previously been applied to the treatment of neuropsychiatric disorders. Though as noted earlier, the detailed mechanisms, specifically in drug addiction, still require further investigation. In the meantime, methyl supplementation through administration of the methyl donor methionine was found to inhibit cocaine reward in mice [21,91]. More recently, rats receiving methionine underwent either a sensitization regimen of intermittent cocaine injections or intravenous cocaine self-administration, followed by cue-induced and drug-primed reinstatement. It was found that methionine not only blocked locomotor sensitization, but also attenuated drug-primed reinstatement [93]. Systemic methionine administration was also associated with reversal of DNA hypomethylation at the *c-Fos* gene in the NAc [93]. In contrast, a similar cocaine self-administration approach, intra-NAc injection of a methyl donor promoted cue-induced cocaine seeking after prolonged withdrawal, whereas injection of the DNMT inhibitor RG108 had the opposite effect [106]. These seemingly opposite effects of methyl donor supplementation on drug behavior may be due to differences in route of administration (systemic vs. intra-NAc). Further research is needed to investigate these possibilities, as well as to test the effects of methyl supplementation in humans. As more epigenetic changes are identified, it will become crucial to manipulate these epigenetic states at selective loci to obtain causal insight into their role in gene regulation. A recent study, using engineered zinc finger proteins or transcription activator-like effectors to target single types of histone modifications to single genes within a single brain region of interest in vivo, provides the means of obtaining such causal data [123]. The increasing availability of genome-editing tools [124] offers additional technical approaches to achieve this important goal.

In the future, we also need to better define if epigenetic modifications are specific to a given drug of abuse or specific brain region, and which common modifications are most relevant in the progression of addiction, so to better define the role of epigenetics in addiction. The existing literature, while still limited, suggests some common actions as well as many distinct ones. For example, one recent study demonstrated that DNA methylation at selected genes in VTA is required for the formation of reward-related memories, effects not seen for the NAc [125]. It also appears that epigenetic regulation may prime the genome to respond to drugs of abuse. In human populations, cigarettes generally serve as a gateway drug, which is used first before progressing to other illicit substances, such as cocaine. To understand the biological basis of this gateway sequence of drug use, it was found that pretreatment of mice with nicotine increased the response to cocaine, as assessed by behaviors and synaptic plasticity in the striatum [126]. Nicotine primed the response to cocaine by enhancing its induction of the *FosB* gene through inhibition of histone deacetylase. Additionally, an HDAC inhibitor simulated the actions of nicotine by priming the cocaine response and enhancing *FosB* gene expression in the NAc. Similarly, this histone acetylation-dependent metaplastic effect of nicotine on cocaine also exists in the amygdala and hippocampal dentate gyrus [127,128]. This suggests that a decrease in smoking rates may lead to a decrease in cocaine addiction and the priming effect of nicotine may potentially be achieved by the inhibition of histone acetylation in human populations.

ABBREVIATIONS

AMPH Amphetamine
BA Butyric acid
BDNF Brain-derived neurotrophic factor
BLA Basolateral amygdala
Cdk5 Cyclin-dependent kinse 5
ChIP Chromatin immunoprecipitation
CPP Conditioned place preference
DNA Deoxyribonucleic acid
DNMT DNA methyltransferase
GABA Gamma-aminobutyric acid
GADD45 Growth arrest and DNA damage 45

H3	Histone 3
H4	Histone 4
HAT	Histone acetyltransferase
HDAC	Histone deacetyltransferase
HDACi	Histone deacetyltransferase inhibitor
IP	Intraperitoneal injection
KDM	Lysine demethylase
KMT	Lysine methyltransferase
LC	Locus coeruleus
MeCP2	Methylated CpG binding protein 2
METH	Methamphetamine
Mll1	Methyltransferase mixed-lineage leukemia 1
MSN	Medium spiny neuron
NAc	Nucleus accumbens
NGS	Next generation sequencing
PARP	Poly (ADP-ribose) polymerase
PFC	Prefrontal cortex
PTM	Posttranslational modification
SIRT	Sirtuin
TSA	Trichostatin A
TET	Ten-eleven translocation protein
VPA	Valproic acid
VTA	Ventral tegmental area

GLOSSARY

5mC 5-Methylcytosine is a methylated form of the DNA base cytosine

5hmC 5-Hydroxymethylcytosine is a DNA pyrimidine nitrogen base that is formed from the DNA base cytosine by adding a methyl group and then a hydroxyl group

ChIP-chip A technology that combines chromatin immunoprecipitation (ChIP) with DNA microarray

ChIP-seq A technology that combines chromatin immunoprecipitation (ChIP) with massively parallel DNA sequencing to analyze protein interactions with DNA

RNA-seq A technology that uses next-generation sequencing to reveal the presence and quantity of RNA in a biological sample

Methylome Refers to the set of nucleic acid methylation modifications in an organisms genome

Self-administration A form of operant conditioning in which a animal learns to administer a rewarding drug

Transcriptome The entire set of all messenger RNA molecules

Epigenome Usually refers to a record of the chemical changes to the DNA and histone proteins of an organism

ACKNOWLEDGMENT

The author apologizes for work not cited in this review due to space limitations.

REFERENCES

[1] Nestler EJ. Molecular basis of long-term plasticity underlying addiction. Nat Rev Neurosci 2001;2(2):119–28.

[2] Nielsen DA, Utrankar A, Reyes JA, Simons DD, Kosten TR. Epigenetics of drug abuse: predisposition or response. Pharmacogenomics 2012;13(10):1149–60.

[3] Anthony JC, Warner LA, Kessler RC. Comparative epidemiology of dependence on tobacco, alcohol, controlled substances, and inhalants: basic findings from the National Comorbidity Survey. Exp Clin Psychopharmacol 1994;2(3):244–68.

[4] Kendler KS, Prescott CA, Myers J, Neale MC. The structure of genetic and environmental risk factors for common psychiatric and substance use disorders in men and women. Arch Gen Psychiatry 2003;60(9):929–37.

[5] Bird A. Perceptions of epigenetics. Nature 2007;447(7143):396–8.

[6] Jaenisch R, Bird A. Epigenetic regulation of gene expression: how the genome integrates intrinsic and environmental signals. Nat Genet 2003;33(Suppl.):245–54.

[7] Hyman SE, Malenka RC, Nestler EJ. Neural mechanisms of addiction: the role of reward-related learning and memory. Annu Rev Neurosci 2006;29:565–98.

[8] Feng J, Nestler EJ. Epigenetic mechanisms of drug addiction. Curr Opin Neurobiol 2013;23(4):521–8.

[9] Heyer MP, Kenny PJ. Corticostriatal microRNAs in addiction. Brain Res 2015;1628(Pt A):2–16.

[10] Schmidt HD, McGinty JF, West AE, Sadri-Vakili G. Epigenetics and psychostimulant addiction. Cold Spring Harb Perspect Med 2013;3(3):a012047.

[11] Sartor GC, St Laurent G III, Wahlestedt C. The emerging role of noncoding RNAs in drug addiction. Front Genet 2012;3:106.

[12] Most D, Workman E, Harris RA. Synaptic adaptations by alcohol and drugs of abuse: changes in microRNA expression and mRNA regulation. Front Mol Neurosci 2014;7:85.

[13] Kalivas PW, Volkow ND. New medications for drug addiction hiding in glutamatergic neuroplasticity. Mol Psychiatry 2011;16(10):974–86.

[14] Russo SJ, Nestler EJ. The brain reward circuitry in mood disorders. Nat Rev Neurosci 2014;14(9):609–25.

[15] Jenuwein T, Allis CD. Translating the histone code. Science 2001;293(5532):1074–80.

[16] Allis CD, Berger SL, Cote J, Dent S, Jenuwien T, Kouzarides T, et al. New nomenclature for chromatin-modifying enzymes. Cell 2007;131(4):633–6.

[17] Strahl BD, Allis CD. The language of covalent histone modifications. Nature 2000;403(6765):41–5.

[18] Kenny PJ. Epigenetics, microRNA, and addiction. Dialog Clin Neurosci 2014;16(3):335–44.

[19] Starkman BG, Sakharkar AJ, Pandey SC. Epigenetics-beyond the genome in alcoholism. Alcohol Res 2012;34(3):293–305.

[20] Ponomarev I. Epigenetic control of gene expression in the alcoholic brain. Alcohol Res 2013;35(1):69–76.

[21] LaPlant Q, Nestler EJ. CRACKing the histone code: cocaine's effects on chromatin structure and function. Horm Behav 2011;59(3):321–30.

[22] Maze I, Nestler EJ. The epigenetic landscape of addiction. Ann NY Acad Sci 2011;1216:99–113.

[23] Robison AJ, Nestler EJ. Transcriptional and epigenetic mechanisms of addiction. Nat Rev Neurosci 2011;12(11):623–37.

[24] Walker DM, Cates HM, Heller EA, Nestler EJ. Regulation of chromatin states by drugs of abuse. Curr Opin Neurobiol 2015;30:112–21.

[25] Rogge GA, Wood MA. The role of histone acetylation in cocaine-induced neural plasticity and behavior. Neuropsychopharmacology 2013;38(1):94–110.

[26] Kyzar EJ, Pandey SC. Molecular mechanisms of synaptic remodeling in alcoholism. Neurosci Lett 2015;11–9.

[27] Godino A, Jayanthi S, Cadet JL. Epigenetic landscape of amphetamine and methamphetamine addiction in rodents. Epigenetics 2015;10(7):574–80.

[28] Grayson DR, Kundakovic M, Sharma RP. Is there a future for histone deacetylase inhibitors in the pharmacotherapy of psychiatric disorders? Mol Pharmacol 2010;77(2):126–35.
[29] Borrelli E, Nestler EJ, Allis CD, Sassone-Corsi P. Decoding the epigenetic language of neuronal plasticity. Neuron 2008;60(6):961–74.
[30] Kumar A, Choi KH, Renthal W, Tsankova NM, Theobald DE, Truong HT, et al. Chromatin remodeling is a key mechanism underlying cocaine-induced plasticity in striatum. Neuron 2005;48(2):303–14.
[31] Wang L, Lv Z, Hu Z, Sheng J, Hui B, Sun J, et al. Chronic cocaine-induced H3 acetylation and transcriptional activation of CaMKIIalpha in the nucleus accumbens is critical for motivation for drug reinforcement. Neuropsychopharmacology 2010;35(4):913–28.
[32] Schmidt HD, Sangrey GR, Darnell SB, Schassburger RL, Cha JH, Pierce RC, et al. Increased brain-derived neurotrophic factor (BDNF) expression in the ventral tegmental area during cocaine abstinence is associated with increased histone acetylation at BDNF exon I-containing promoters. J Neurochem 2012;120(2):202–9.
[33] Mashayekhi FJ, Rasti M, Rahvar M, Mokarram P, Namavar MR, Owji AA. Expression levels of the BDNF gene and histone modifications around its promoters in the ventral tegmental area and locus ceruleus of rats during forced abstinence from morphine. Neurochem Res 2012;37(7):1517–23.
[34] Martin TA, Jayanthi S, McCoy MT, Brannock C, Ladenheim B, Garrett T, et al. Methamphetamine causes differential alterations in gene expression and patterns of histone acetylation/hypoacetylation in the rat nucleus accumbens. PLoS One 2012;7(3):e34236.
[35] Jayanthi S, McCoy MT, Chen B, Britt JP, Kourrich S, Yau HJ, et al. Methamphetamine downregulates striatal glutamate receptors via diverse epigenetic mechanisms. Biol Psychiatry 2014;76(1):47–56.
[36] Renthal W, Carle TL, Maze I, Covington HE III, Truong HT, Alibhai I, et al. Delta FosB mediates epigenetic desensitization of the c-fos gene after chronic amphetamine exposure. J Neurosci 2008;28(29):7344–9.
[37] Renthal W, Maze I, Krishnan V, Covington HE III, Xiao G, Kumar A, et al. Histone deacetylase 5 epigenetically controls behavioral adaptations to chronic emotional stimuli. Neuron 2007;56(3):517–29.
[38] Taniguchi M, Carreira MB, Smith LN, Zirlin BC, Neve RL, Cowan CW. Histone deacetylase 5 limits cocaine reward through cAMP-induced nuclear import. Neuron 2012;73(1):108–20.
[39] Rogge GA, Singh H, Dang R, Wood MA. HDAC3 is a negative regulator of cocaine-context-associated memory formation. J Neurosci 2013;33(15):6623–32.
[40] Malvaez M, McQuown SC, Rogge GA, Astarabadi M, Jacques V, Carreiro S, et al. HDAC3-selective inhibitor enhances extinction of cocaine-seeking behavior in a persistent manner. Proc Natl Acad Sci USA 2013;110(7):2647–52.
[41] Wang Y, Lai J, Cui H, Zhu Y, Zhao B, Wang W, et al. Inhibition of histone deacetylase in the basolateral amygdala facilitates morphine context-associated memory formation in rats. J Mol Neurosci 2015;55(1):269–78.
[42] Kalda A, Heidmets LT, Shen HY, Zharkovsky A, Chen JF. Histone deacetylase inhibitors modulates the induction and expression of amphetamine-induced behavioral sensitization partially through an associated learning of the environment in mice. Behav Brain Res 2007;181(1):76–84.
[43] Shen HY, Kalda A, Yu L, Ferrara J, Zhu J, Chen JF. Additive effects of histone deacetylase inhibitors and amphetamine on histone H4 acetylation, cAMP responsive element binding protein phosphorylation and DeltaFosB expression in the striatum and locomotor sensitization in mice. Neuroscience 2008;157(3):644–55.
[44] Romieu P, Host L, Gobaille S, Sandner G, Aunis D, Zwiller J. Histone deacetylase inhibitors decrease cocaine but not sucrose self-administration in rats. J Neurosci 2008;28(38):9342–8.
[45] Kim WY, Kim S, Kim JH. Chronic microinjection of valproic acid into the nucleus accumbens attenuates amphetamine-induced locomotor activity. Neurosci Lett 2008;432(1):54–7.
[46] Schroeder FA, Penta KL, Matevossian A, Jones SR, Konradi C, Tapper AR, et al. Drug-induced activation of dopamine D(1) receptor signaling and inhibition of class I/II histone deacetylase induce chromatin remodeling in reward circuitry and modulate cocaine-related behaviors. Neuropsychopharmacology 2008;33(12):2981–92.

[47] Kennedy PJ, Feng J, Robison AJ, Maze I, Badimon A, Mouzon E, et al. Class I HDAC inhibition blocks cocaine-induced plasticity by targeted changes in histone methylation. Nat Neurosci 2013;16(4):434–40.

[48] Aguilar-Valles A, Vaissiere T, Griggs EM, Mikaelsson MA, Takacs IF, Young EJ, et al. Methamphetamine-associated memory is regulated by a writer and an eraser of permissive histone methylation. Biol Psychiatry 2014;76(1):57–65.

[49] Renthal W, Kumar A, Xiao G, Wilkinson M, Covington HE III, Maze I, et al. Genome-wide analysis of chromatin regulation by cocaine reveals a role for sirtuins. Neuron 2009;62(3):335–48.

[50] Maze I, Covington HE III, Dietz DM, LaPlant Q, Renthal W, Russo SJ, et al. Essential role of the histone methyltransferase G9a in cocaine-induced plasticity. Science 2010;327(5962):213–6.

[51] Maze I, Chaudhury D, Dietz DM, Von Schimmelmann M, Kennedy PJ, Lobo MK, et al. G9a influences neuronal subtype specification in striatum. Nat Neurosci 2014;17(4):533–9.

[52] Lobo MK, Covington HE III, Chaudhury D, Friedman AK, Sun H, Damez-Werno D, et al. Cell type-specific loss of BDNF signaling mimics optogenetic control of cocaine reward. Science 2010;330(6002):385–90.

[53] Sun H, Maze I, Dietz DM, Scobie KN, Kennedy PJ, Damez-Werno D, et al. Morphine epigenomically regulates behavior through alterations in histone H3 lysine 9 dimethylation in the nucleus accumbens. J Neurosci 2012;32(48):17454–64.

[54] Feng J, Wilkinson M, Liu X, Purushothaman I, Ferguson D, Vialou V, et al. Chronic cocaine-regulated epigenomic changes in mouse nucleus accumbens. Genome Biol 2014;15(4). R65.

[55] Luco RF, Allo M, Schor IE, Kornblihtt AR, Misteli T. Epigenetics in alternative pre-mRNA splicing. Cell 2011;144(1):16–26.

[56] Zhou Z, Yuan Q, Mash DC, Goldman D. Substance-specific and shared transcription and epigenetic changes in the human hippocampus chronically exposed to cocaine and alcohol. Proc Natl Sci USA 2011;108(16):6626–31.

[57] Eipper-Mains JE, Kiraly DD, Duff MO, Horowitz MJ, McManus CJ, Eipper BA, et al. Effects of cocaine and withdrawal on the mouse nucleus accumbens transcriptome. Genes Brain Behav 2013;12(1):21–33.

[58] Ferguson D, Koo JW, Feng J, Heller E, Rabkin J, Heshmati M, et al. Essential role of SIRT1 signaling in the nucleus accumbens in cocaine and morphine action. J Neurosci 2013;33(41):16088–98.

[59] Ferguson D, Shao N, Heller E, Feng J, Neve R, Kim HD, et al. SIRT1-FOXO3a regulate cocaine actions in the nucleus accumbens. J Neurosci 2015;35(7):3100–11.

[60] Dunham I, Kundaje A, Aldred SF, Collins PJ, Davis CA, Doyle F, et al. An integrated encyclopedia of DNA elements in the human genome. Nature 2012;489(7414):57–74.

[61] Maze I, Feng J, Wilkinson MB, Sun H, Shen L, Nestler EJ. Cocaine dynamically regulates heterochromatin and repetitive element unsilencing in nucleus accumbens. Proc Natl Acad Sci USA 2011;108(7):3035–40.

[62] Muotri AR, Chu VT, Marchetto MC, Deng W, Moran JV, Gage FH. Somatic mosaicism in neuronal precursor cells mediated by L1 retrotransposition. Nature 2005;435(7044):903–10.

[63] Brown AN, Vied C, Dennis JH, Bhide PG. Nucleosome Repositioning: a novel mechanism for nicotine- and cocaine-induced epigenetic changes. PLoS One 2015;10(9):e0139103.

[64] Sun H, Damez-Werno DM, Scobie KN, Shao NY, Dias C, Rabkin J, et al. ACF chromatin-remodeling complex mediates stress-induced depressive-like behavior. Nat Med 2015;21(10):1146–53.

[65] Feng J, Fan G. The role of DNA methylation in the central nervous system and neuropsychiatric disorders. Int Rev Neurobio 2009;89:67–84.

[66] Ooi SK, Bestor TH. The colorful history of active DNA demethylation. Cell 2008;133(7):1145–8.

[67] Miller CA, Sweatt JD. Covalent modification of DNA regulates memory formation. Neuron 2007;53(6):857–69.

[68] Feng J, Zhou Y, Campbell SL, Le T, Li E, Sweatt JD, et al. Dnmt1 and Dnmt3a maintain DNA methylation and regulate synaptic function in adult forebrain neurons. Nat Neurosci 2010;13(4):423–30.

[69] Ma DK, Jang MH, Guo JU, Kitabatake Y, Chang ML, Pow-Anpongkul N, et al. Neuronal activity-induced Gadd45b promotes epigenetic DNA demethylation and adult neurogenesis. Science 2009;323(5917):1074–7.

[70] Tahiliani M, Koh KP, Shen Y, Pastor WA, Bandukwala H, Brudno Y, et al. Conversion of 5-methylcytosine to 5-hydroxymethylcytosine in mammalian DNA by MLL partner TET1. Science 2009;324(5929):930–5.
[71] Kriaucionis S, Heintz N. The nuclear DNA base 5-hydroxymethylcytosine is present in Purkinje neurons and the brain. Science 2009;324(5929):929–30.
[72] He YF, Li BZ, Li Z, Liu P, Wang Y, Tang Q, et al. Tet-mediated formation of 5-carboxylcytosine and its excision by TDG in mammalian DNA. Science 2011;333(6047):1303–7.
[73] Ito S, Shen L, Dai Q, Wu SC, Collins LB, Swenberg JA, et al. Tet proteins can convert 5-methylcytosine to 5-formylcytosine and 5-carboxylcytosine. Science 2011;333(6047):1300–3.
[74] Wu H, Zhang Y. Mechanisms and functions of Tet protein-mediated 5-methylcytosine oxidation. Genes Dev 2011;25(23):2436–52.
[75] Branco MR, Ficz G, Reik W. Uncovering the role of 5-hydroxymethylcytosine in the epigenome. Nat Rev Genet 2012;13(1):7–13.
[76] Pastor WA, Aravind L, Rao A. TETonic shift: biological roles of TET proteins in DNA demethylation and transcription. Nat Rev Mol Cell Biol 2013;14(6):341–56.
[77] Day JJ, Sweatt JD. Epigenetic mechanisms in cognition. Neuron 2011;70(5):813–29.
[78] Nelson ED, Monteggia LM. Epigenetics in the mature mammalian brain: effects on behavior and synaptic transmission. Neurobiol Learn Mem 2011;96(1):53–60.
[79] Shin J, Ming GL, Song H. DNA modifications in the mammalian brain. Philos Trans R Soc Lond B Biol Sci 2014;369.(1652).
[80] Moore LD, Le T, Fan G. DNA methylation and its basic function. Neuropsychopharmacology 2013;38(1):23–38.
[81] Lubin FD, Gupta S, Parrish RR, Grissom NM, Davis RL. Epigenetic mechanisms: critical contributors to long-term memory formation. Neuroscientist 2011;17(6):616–32.
[82] Mikaelsson MA, Miller CA. The path to epigenetic treatment of memory disorders. Neurobiol Learn Mem 2011;96(1):13–8.
[83] Im HI, Hollander JA, Bali P, Kenny PJ. MeCP2 controls BDNF expression and cocaine intake through homeostatic interactions with microRNA-212. Nat Neurosci 2010;13(9):1120–7.
[84] Deng JV, Rodriguiz RM, Hutchinson AN, Kim IH, Wetsel WC, West AE. MeCP2 in the nucleus accumbens contributes to neural and behavioral responses to psychostimulants. Nat Neurosci 2010;13(9):1128–36.
[85] Deng JV, Wan Y, Wang X, Cohen S, Wetsel WC, Greenberg ME, et al. MeCP2 phosphorylation limits psychostimulant-induced behavioral and neuronal plasticity. J Neurosci 2014;34(13):4519–27.
[86] Feng J, Chang H, Li E, Fan G. Dynamic expression of de novo DNA methyltransferases Dnmt3a and Dnmt3b in the central nervous system. J Neurosci Res 2005;79(6):734–46.
[87] Goto K, Numata M, Komura JI, Ono T, Bestor TH, Kondo H. Expression of DNA methyltransferase gene in mature and immature neurons as well as proliferating cells in mice. Differentiation 1994;56(1–2):39–44.
[88] LaPlant Q, Vialou V, Covington HE III, Dumitriu D, Feng J, Warren BL, et al. Dnmt3a regulates emotional behavior and spine plasticity in the nucleus accumbens. Nat Neurosci 2010;13(9):1137–43.
[89] Anier K, Malinovskaja K, Aonurm-Helm A, Zharkovsky A, Kalda A. DNA methylation regulates cocaine-induced behavioral sensitization in mice. Neuropsychopharmacology 2010;35(12):2450–61.
[90] Numachi Y, Shen H, Yoshida S, Fujiyama K, Toda S, Matsuoka H, et al. Methamphetamine alters expression of DNA methyltransferase 1 mRNA in rat brain. Neurosci Lett 2007;414(3):213–7.
[91] Tian W, Zhao M, Li M, Song T, Zhang M, Quan L, et al. Reversal of cocaine-conditioned place preference through methyl supplementation in mice: altering global DNA methylation in the prefrontal cortex. PLoS One 2012;7(3):e33435.
[92] Feng J, Shao N, Szulwach KE, Vialou V, Huynh J, Zhong C, et al. Role of Tet1 and 5-hydroxymethylcytosine in cocaine action. Nat Neurosci 2015;18(4):536–44.
[93] Wright KN, Hollis F, Duclot F, Dossat AM, Strong CE, Francis TC, et al. Methyl supplementation attenuates cocaine-seeking behaviors and cocaine-induced c-Fos activation in a DNA methylation-dependent manner. J Neurosci 2015;35(23):8948–58.

[94] Nielsen DA, Huang W, Hamon SC, Maili L, Witkin BM, Fox RG, et al. Forced abstinence from cocaine self-administration is associated with DNA methylation changes in myelin genes in the corpus callosum: a preliminary study. Front Psychiatry 2012;3:60.
[95] Globisch D, Munzel M, Muller M, Michalakis S, Wagner M, Koch S, et al. Tissue distribution of 5-hydroxymethylcytosine and search for active demethylation intermediates. PLoS One 2010;5(12):e15367.
[96] Szwagierczak A, Bultmann S, Schmidt CS, Spada F, Leonhardt H. Sensitive enzymatic quantification of 5-hydroxymethylcytosine in genomic DNA. Nucleic Acids Res 2010;38(19):e181.
[97] Guo JU, Su Y, Zhong C, Ming GL, Song H. Hydroxylation of 5-methylcytosine by TET1 promotes active DNA demethylation in the adult brain. Cell 2011;145(3):423–34.
[98] Rudenko A, Dawlaty MM, Seo J, Cheng AW, Meng J, Le T, et al. Tet1 is critical for neuronal activity-regulated gene expression and memory extinction. Neuron 2013;79(6):1109–22.
[99] Kaas GA, Zhong C, Eason DE, Ross DL, Vachhani RV, Ming GL, et al. TET1 controls CNS 5-methylcytosine hydroxylation, active DNA demethylation, gene transcription, and memory formation. Neuron 2013;79(6):1086–93.
[100] Szulwach KE, Li X, Li Y, Song CX, Wu H, Dai Q, et al. 5-hmC-mediated epigenetic dynamics during post-natal neurodevelopment and aging. Nat Neurosci 2011;14(12):1607–16.
[101] Li X, Wei W, Zhao QY, Widagdo J, Baker-Andresen D, Flavell CR, et al. Neocortical Tet3-mediated accumulation of 5-hydroxymethylcytosine promotes rapid behavioral adaptation. Proc Natl Acad Sci USA 2014;111(19):7120–5.
[102] Zhang RR, Cui QY, Murai K, Lim YC, Smith ZD, Jin S, et al. Tet1 regulates adult hippocampal neurogenesis and cognition. Cell Stem Cell 2013;13(2):237–45.
[103] Niehrs C, Schafer A. Active DNA demethylation by Gadd45 and DNA repair. Trends Cell Biol 2012;22(4):220–7.
[104] Gavin DP, Kusumo H, Sharma RP, Guizzetti M, Guidotti A, Pandey SC. Gadd45b and N-methyl-d-aspartate induced DNA demethylation in postmitotic neurons. Epigenomics 2015;7(4):567–79.
[105] Koo JW, Mazei-Robison MS, Chaudhury D, Juarez B, LaPlant Q, Ferguson D, et al. BDNF is a negative modulator of morphine action. Science 2012;338(6103):124–8.
[106] Massart R, Barnea R, Dikshtein Y, Suderman M, Meir O, Hallett M, et al. Role of DNA methylation in the nucleus accumbens in incubation of cocaine craving. J Neurosci 2015;35(21):8042–58.
[107] Song CX, Szulwach KE, Fu Y, Dai Q, Yi C, Li X, et al. Selective chemical labeling reveals the genome-wide distribution of 5-hydroxymethylcytosine. Nat Biotechnol 2011;29(1):68–72.
[108] Creyghton MP, Cheng AW, Welstead GG, Kooistra T, Carey BW, Steine EJ, et al. Histone H3K27ac separates active from poised enhancers and predicts developmental state. Proc Natl Acad Sci USA 2010;107(50):21931–6.
[109] Ernst J, Kheradpour P, Mikkelsen TS, Shoresh N, Ward LD, Epstein CB, et al. Mapping and analysis of chromatin state dynamics in nine human cell types. Nature 2011;473(7345):43–9.
[110] Dietz DM, Laplant Q, Watts EL, Hodes GE, Russo SJ, Feng J, et al. Paternal transmission of stress-induced pathologies. Biol Psychiatry 2011;70(5):408–14.
[111] Gapp K, Jawaid A, Sarkies P, Bohacek J, Pelczar P, Prados J, et al. Implication of sperm RNAs in transgenerational inheritance of the effects of early trauma in mice. Nat Neurosci 2014;17(5):667–9.
[112] Bale TL. Epigenetic and transgenerational reprogramming of brain development. Nat Rev Neurosci 2015;16(6):332–44.
[113] Vassoler FM, White SL, Schmidt HD, Sadri-Vakili G, Pierce RC. Epigenetic inheritance of a cocaine-resistance phenotype. Nat Neurosci 2013;16(1):42–7.
[114] Dias BG, Ressler KJ. Parental olfactory experience influences behavior and neural structure in subsequent generations. Nat Neurosci 2014;17(1):89–96.
[115] Watson CT, Szutorisz H, Garg P, Martin Q, Landry JA, Sharp AJ, et al. Genome-wide DNA methylation profiling reveals epigenetic changes in the rat nucleus Aaccumbens associated with cross-generational effects of adolescent THC exposure. Neuropsychopharmacology 2015;40(13):2993–3005.

[116] Jordi E, Heiman M, Marion-Poll L, Guermonprez P, Cheng SK, Nairn AC, et al. Differential effects of cocaine on histone posttranslational modifications in identified populations of striatal neurons. Proc Natl Acad Sci USA 2013;110(23):9511–6.

[117] Bock C, Tomazou EM, Brinkman AB, Muller F, Simmer F, Gu H, et al. Quantitative comparison of genome-wide DNA methylation mapping technologies. Nat Biotechnol 2010;28(10):1106–14.

[118] Wu H, Wu X, Shen L, Zhang Y. Single-base resolution analysis of active DNA demethylation using methylase-assisted bisulfite sequencing. Nat Biotechnol 2014;32(12):1231–40.

[119] Yu M, Hon GC, Szulwach KE, Song CX, Zhang L, Kim A, et al. Base-resolution analysis of 5-hydroxymethylcytosine in the mammalian genome. Cell 2012;149(6):1368–80.

[120] Booth MJ, Branco MR, Ficz G, Oxley D, Krueger F, Reik W, et al. Quantitative sequencing of 5-methylcytosine and 5-hydroxymethylcytosine at single-base resolution. Science 2012;336(6083):934–7.

[121] Song CX, Szulwach KE, Dai Q, Fu Y, Mao SQ, Lin L, et al. Genome-wide profiling of 5-formylcytosine reveals its roles in epigenetic priming. Cell 2013;153(3):678–91.

[122] Maze I, Shen L, Zhang B, Garcia BA, Shao N, Mitchell A, et al. Analytical tools and current challenges in the modern era of neuroepigenomics. Nat Neurosci 2014;17(11):1476–90.

[123] Heller EA, Cates HM, Pena CJ, Sun H, Shao N, Feng J, et al. Locus-specific epigenetic remodeling controls addiction- and depression-related behaviors. Nat Neurosci 2014;17(12):1720–7.

[124] Tuesta LM, Zhang Y. Mechanisms of epigenetic memory and addiction. EMBO J 2014;33(10):1091–103.

[125] Day JJ, Childs D, Guzman-Karlsson MC, Kibe M, Moulden J, Song E, et al. DNA methylation regulates associative reward learning. Nat Neurosci 2013;16(10):1445–52.

[126] Levine A, Huang Y, Drisaldi B, Griffin EA Jr, Pollak DD, Xu S, et al. Molecular mechanism for a gateway drug: epigenetic changes initiated by nicotine prime gene expression by cocaine. Sci Transl Med 2011;3(107). 107ra109.

[127] Huang YY, Kandel DB, Kandel ER, Levine A. Nicotine primes the effect of cocaine on the induction of LTP in the amygdala. Neuropharmacology 2013;74:126–34.

[128] Huang YY, Levine A, Kandel DB, Yin D, Colnaghi L, Drisaldi B, et al. D1/D5 receptors and histone deacetylation mediate the Gateway Effect of LTP in hippocampal dentate gyrus. Learn Mem 2014;21(3):153–60.

[129] Nielsen DA, Yuferov V, Hamon S, Jackson C, Ho A, Ott J, et al. Increased OPRM1 DNA methylation in lymphocytes of methadone-maintained former heroin addicts. Neuropsychopharmacology 2009;34(4):867–73.

[130] Chorbov VM, Todorov AA, Lynskey MT, Cicero TJ. Elevated levels of DNA methylation at the OPRM1 promoter in blood and sperm from male opioid addicts. J Opioid Manag 2011;7(4):258–64.

CHAPTER

EPIGENETICS AND ALCOHOL USE DISORDERS

18

S. Sagarkar, A. Sakharkar

Department of Biotechnology, Savitribai Phule Pune University, Ganeshkhind, Pune, Maharashtra, India

CHAPTER OUTLINE

18.1 INTRODUCTION

Drug addiction exerts an enormous health burden worldwide with serious economic and societal consequences. It incurs huge loss to the society in terms of family disturbance, loss of work force, and medical burden. Clinically, drinking over a long period of time is associated with alcohol use disorders (AUD) and other health issues including increased risk of certain cancers. In 2012, the 3.3 million deaths, which make 5.9% of all global deaths, were attributed to alcohol consumption (from the NIAAA website). In recent years, the sharp decline in the average age of drug addicts is especially alarming as growing populations of teenagers are diagnosed with drug abuse tendencies. Although the currently available medical and counseling services could provide some help in the treatment of addiction, the high rate of relapse substantiate the urgency in putting more efforts into research on finding brain mechanisms of addiction. Alcohol is one of the most detrimental addictive drugs with no known specific receptors found to date and only a few AUD medications are available with limited efficacy. The central nervous system (CNS) is one of the most vulnerable target organs for the deleterious actions of

Neuropsychiatric Disorders and Epigenetics. http://dx.doi.org/10.1016/B978-0-12-800226-1.00018-6

alcohol. According to a recent study by Collins et al. [1], AUD emerged as the second most detrimental neuropsychiatric disorder which traverses through different phases of drinking patterns from initiation and binging through dependence and craving. Consequently, the withdrawal from alcohol exposure manifests into the negative affective states of alcoholism explained by comorbid behaviors, such as depression and anxiety [2]. These negative affective states of alcoholism lead to an incessant drinking pattern as a phenomenon of self-medication to relieve the symptoms of withdrawal leading the patient into relapse [3,4]. One of the most prominent of these comorbid behaviors is expression of anxiety following alcohol withdrawal [5]. It has been found that the anxiety disorder is often associated with high alcohol consumption in clinical population [6,7].

The multifactorial nature of alcohol addiction includes crosstalk between ones' genetic constitution and physiological and environmental factors in conjunction with other prevalent psychiatric disorders [8]. The central mechanisms that underlie the different etiopathologies of alcohol use and abuse due to environmental perturbations are complex and are difficult to interpret. It is by now understood that environmental stimuli determine an individual's response to drugs [9] and alcohol is no exception. Repeated alcohol abuse causes maladaptations in various neural circuitries involved in a range of psychological phenotypes. For example, emotion and cognition, which drives the transition to the addictive state [10]. These alterations in gene expression due to environmental impacts are one of the core cellular mechanisms that occur during the neuroadaptation. Perhaps, one may not be overstating by saying that the altered gene expression to large extent is at the core of the central mechanisms of drug addiction.

During the last couple of decades, it has become evident that the processes that regulate gene expression during differentiation and development of normal cells continue to impact the functioning of almost all cell types during adulthood. This offers the flexibility to adapt to environmental perturbations [11]. Broadly, these processes are indicated as epigenetic mechanisms, which include chromatin modifications, for example, histone posttranslational modifications (PTMs), DNA methylation, and noncoding RNA-mediated gene transcription. For example, the core histone tails undergo a variety of covalent modifications, such as acetylation, phosphorylation, methylation, and ubiquitination. These modifications that occur on different amino acids in the histone tails have been known for decades, but their function have only recently begun to be revealed. These mechanisms are substantially involved in mediating the effects of drugs of abuse to induce highly stable, but reversible changes in the neuronal functions that precipitate into addictive phenotypes [12]. Chromatin structure is mainly referred to as open or closed which allows or blocks the accessibility of transcription factors to gene regulatory elements, such as promoter regions [13]. The field of chromatin biology received momentum when it was first demonstrated that chromatin is composed of a unit structure, called the nucleosome [14]. A nucleosome consists of approximately 147 base pairs of DNA wrapped around an octamer made up of heterodimers of four histone proteins, that is, H2A, H2B, H3, and H4 [15–17]. However, the field of chromatin research experienced some obscurity until it was found that transcription factors recruit the histone modifying enzymes and chromatin remodeling complexes, which essentially shape chromatin architecture to regulate gene transcription [18–20]. The covalent modifications of histone proteins, such as acetylation, methylation, phosphorylation and ubiquitination, either helps to relax or condense the chromatin structure, whereas methylation of cytosine residues in DNA is believed to condense chromatin. Multiple families of enzymes, for example, histone deacetylases (HDACs), histone acetyltransferases (HATs), histone demethylases (HDMs), and histone methyltransferases regulate histone modifications at lysine residues in histone tails. DNA methylation at the 5th carbon of cytosine residues in DNA is induced by DNA methyltransferases (DNMTs), whereas an opposing mechanism of DNA demethylation operates via the base excision repair (BER) pathway to maintain the methylation status of DNA. Chromatin remodeling via these mechanisms have been shown to alter the course of gene expression in health and disease conditions. In addition to these mechanisms, recent evidence supports the hypothesis that noncoding RNA-induced transcriptional regulation is affected by the action of alcohol on the brain which directly or indirectly causes neuroadaptive changes underlying the alcohol addictive phenotype [8,21,22]. Here we review the progress made during the last decade in understanding different epigenetic mechanisms involved in the pathophysiology of alcohol use disorders and comorbid behaviors.

18.2 HISTONE MODIFICATIONS

Histone covalent modifications at different amino acids, such as acetylation (lysine), methylation (arginine and lysine), phosphorylation (serine and threonine), ubiquitination (lysine), sumoylation (lysine), ADP ribosylation (lysine), and proline isomerization are characterized as potential transcription regulatory marks [23]. Histone H3 and H4 acetylation and phosphorylation have been widely reported to activate transcription, while sumoylation is associated with repression. Histone methylation is situation specific; it may activate or represses gene transcription based on which lysine residues are modified [24]. Not only does the type of the histone modification affect the regulatory outcome, but also the residue at which the modification is taking place. For example, various modifications at lysine (K) residues in histones H3 and H4 at different positions define variable cellular outcomes, for example, methylation at H3-K4, H3-K36, and H3-K79 causes transcriptional activation. However, H3-K9, H3-K27, and H4-K20 methylation cause repression [25]. The epigenetic machinery consists of various modifiers, for example, HATs and HDACs, enzymes with opposing effects, which maintain the histone code. Although acetylation and deacetylation work in an opposite fashion, the equilibrium between the two aids in the maintenance of nuclear histone acetylation status. Increased HDAC activity results in repressive chromatin domains that restrict access for RNA polymerase or other promoter activity regulatory factors. On the other hand, the promoters of stably expressed genes harbor hyperacetylation of histones H3 and H4 and also H3-K4 trimethylation [23]. Acetylation and methylation are among the modifications of histones that are studied in quite some detail with respect to alcohol actions on the brain.

18.2.1 HISTONE ACETYLATION AND DEACETYLATION

The acetylation status of histones is governed by two enzymes working in opposing manners, that is, HATs and HDACs. Acetylation by HATs is predominantly a lysine residue specific modification which acts by neutralizing the basic charge of lysine, alleviating DNA and nucleosomal association thereby creating space for nuclear protein–DNA interaction. HATs are associated with various basic cellular processes, such as transcriptional initiation and elongation, genome stability, and cell cycle–regulated DNA repair [26,27]. Transcriptional activation takes place through acetylation of specific lysine tails predominantly located on the amino-termini of core histones (H2A, H2B, H3, and H4). Various lysine residues undergo acetylation including K5, K8, K12, and K16 in histone H4 and K9, K14, K18, and K56 in histone H3 [25]. Gcn5-associated *N*-acetyltransferase (GNAT), MYST and the global coactivators cAMP response element-binding protein (CREB)-binding protein/p300 (CBP/p300) [28] are three major families of HATs. The CBP/p300 family of HATs functions via directly binding to phosphorylated CREB (p-CREB) at the 133rd serine residue and are also associated with different multimeric

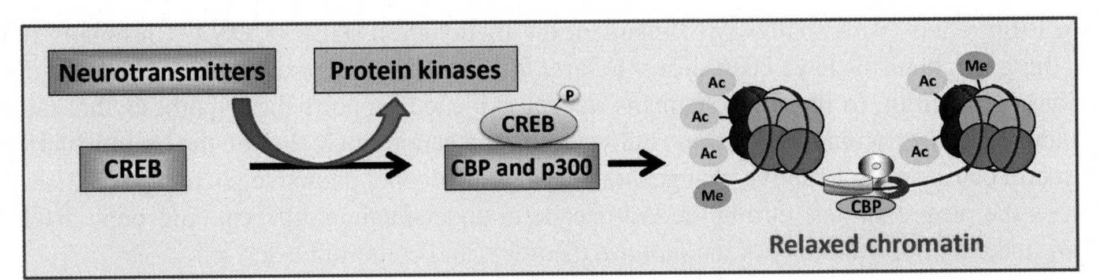

FIGURE 18.1 A Hypothetical Schematic Representation of a Possible Mechanism of Neurotransmitter Regulated Signaling and Its Role in Maintenance of Chromatin Architecture

Neurotransmitters bind to G-protein coupled receptors at the cell membrane activating downstream adenylyl cyclase-dependent signaling mechanisms, which in response elevates cyclic AMP (cAMP) levels. Activity of several protein kinases depends on cellular cAMP levels. For example, CREB (cAMP response element-binding protein) is a cellular transcription factor known to regulate the gene promoter activity upon phosphorylation by protein kinases. Protein kinases phosphorylate CREB converting it into P-CREB, which further directly interacts with CREB-binding protein (*CBP*) and p300. CBP is known to exhibit intrinsic HAT activity and has been shown to increase histone acetylation, which relaxes chromatin and allow transcription factor to access promoters.

complexes thereby remodeling chromatin architecture [29–31]. CREB is a neuronal transcription factor and several studies have shown its involvement in modulation of the epigenome during alcoholism [9,32,33]. Some of the neuropeptides, for example, neuropeptide Y (NPY) and brain-derived neurotrophic factor (BDNF), which have been previously implicated in alcoholic phenotypes, are CREB-target genes [5,34]. Recent studies link these previous findings in CREB-related mechanisms to epigenetic events via histone acetylation [35,36] (Figs. 18.1 and 18.2).

As we have discussed earlier, HDACs represses transcription by removing acetyl groups from histones resulting in compact chromatin. HDACs are also called lysine deacetylases, and are grouped into four classes depending on sequence homology to the originally described yeast enzymes and domain organization [37]. Class I HDACs [1–3,8], class II HDACs [4–7,9], and class IV HDACs are classical HDACs which require zinc for their activity, whereas class III HDACs, also referred as Sirtuins, require NAD+ for enzymatic activity [38,39]. HDACs have been implicated in isoform-specific regulation of learning and memory in both vertebrate and invertebrate models [40–45]. HDAC2 is associated with the maintenance of neuronal cognitive function and hippocampal memory regulation, especially remote fear memory, which may have profound effects in treatment of posttraumatic stress disorders [46,47]. Sirtuins (Class III HDACs) play a critical role in the regulation of aging and neurodegeneration as they are involved in the maintenance of metabolic activity, cell cycle, and circadian rhythms [48–51]. SIRT1 deficiency has been shown to attenuate cyclic AMP-CREB expression via *microRNA-134* impairing synaptic plasticity [52]. These studies have been leading the way in developing HDAC or sirtuin inhibitors as promising new therapeutic reagents to treat neurological disorders [53]. It should be noted that the HDAC inhibitor suberoyl hydroxamic acid (SAHA) marketed as Vorinostat has been approved for treatment of cancers [54].

Cyclic AMP response element-binding (CREB) protein signaling is among the most studied pathway in alcohol addiction and other drugs of abuse in both vertebrates and invertebrate models [5,27,55,56]. In vertebrate models, the anxiolytic effects of ethanol and related behaviors were shown to be associated with CREB signaling in the amygdaloid circuitry [5,55,57,58]. Alcohol exposure

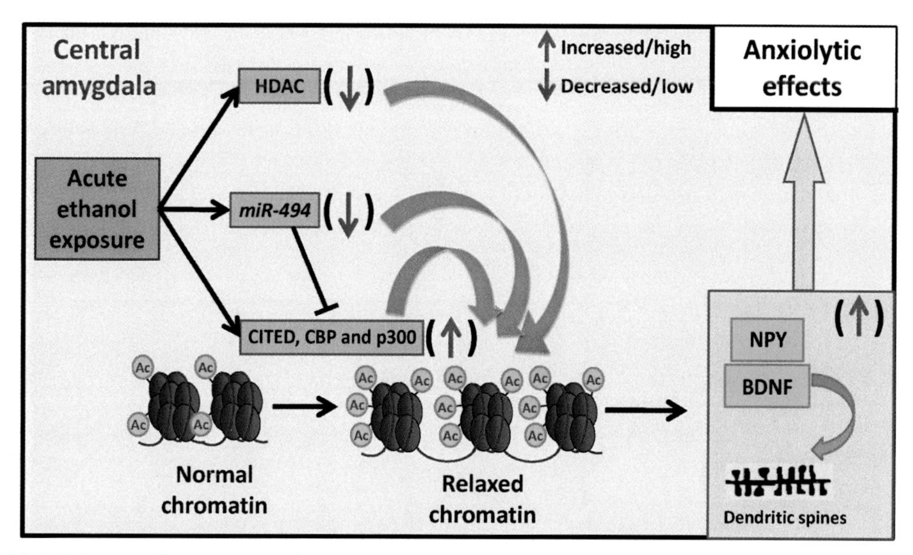

FIGURE 18.2 A Schematic Representation of the Histone Acetylation and Downstream Mechanisms Within the Framework of Amygdaloid Neurocircuitry After Acute Ethanol Exposure in Rodents

Ethanol decreases histone deacetylases (HDAC) function with concomitant increase in human CREB-binding protein/p300-interacting transactivator with ED-rich tail (CITED), CREB-binding protein (CBP), and p300 levels, thereby increasing histone acetylation. Acute ethanol exposure also downregulates the levels of *microRNA-494 (miR-494)*, which in turn results into the increase in CBP levels. CBP is a putative target of the *miR-494*. Therefore, the increase in CBP levels post ethanol exposure might be the result of downregulation of *miR-494*. These mechanisms may work in concert to hyperacetylate the chromatin at the promoters of genes involved in neural transmission and synaptic plasticity, such as neuropeptide Y (*NPY*), brain derived neurotrophic factor (*BDNF*), and activity-regulated cytoskeleton-associated protein (*Arc*). This could result in the upregulation of NPY, BDNF, and Arc levels leading to sprouting of dendritic spines. This hyperactivation of gene promoter activity due to hyperacetylation and synaptic plasticity in the amygdala following acute ethanol exposure might be resulting into anxiolytic response.

during development decreased CBP levels and histone acetylation, which was associated with motor deficits, one of the features of fetal alcohol spectrum disorders (FASD) [59]. In their seminal paper, Pandey et al. [35] have shown an upregulation in CREB-binding protein (CBP), coinciding with HDAC inhibition resulting in hyperacetylation of histones H3-K9 and H4-K8 possibly increasing neuropeptide Y (NPY) levels in the central and medial nucleus of the amygdala with acute ethanol administration. These changes were associated with the anxiolytic effects of acute ethanol exposure (Fig. 18.2). Most importantly, two opposing mechanisms of histone acetylation via CBP, a protein with intrinsic HAT activity and deacetylation via HDACs are modulated by a single dose of ethanol. However, another dose of ethanol 24 h later failed to inhibit HDAC activity in the amygdala, resulting in alcohol tolerant chromatin which correlated with the development of a rapid tolerance to anxiolytic effects of ethanol via NPY [60]. Similarly, chronic exposure to ethanol was also found to produce a chronic tolerance to altering these epigenetic events with no changes in histone acetylation status in the amygdala [35]. Furthermore, withdrawal after chronic ethanol exposure (chronic intermittent ethanol, CIE) induced anxiety-like behaviors and alleviated CBP levels with hypoacetylation of H3-K9. These changes were associated with a decrease in NPY, BDNF, and Arc levels [35,61,62]. While changes in the levels of NPY, BDNF, and Arc were not observed while ethanol was present during chronic exposure, the withdrawal from ethanol reduced the levels of these proteins in the amygdaloid neurocircuitry (Fig. 18.3). These changes were associated with alterations in histone acetylation levels and concomitant changes in CBP and HDAC functions.

Among the many different animal models, alcohol preferring (P) and nonpreferring (NP) rats have been widely used as genetically inbred models to study the mechanisms of alcohol preference.

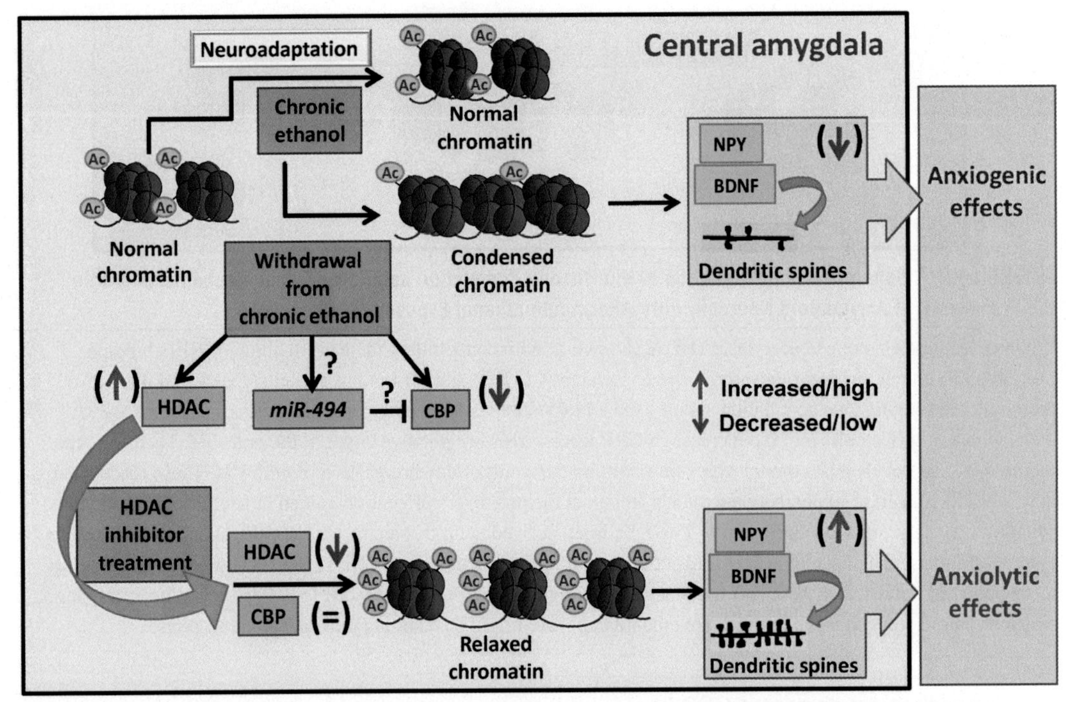

FIGURE 18.3 A Schematic Representation of the Possible Histone Acetylation Mechanism Operative in the Amygdala During the Process of Alcohol Addiction and Comorbid Anxiety Behaviors in Rodents

Unlike acute exposure (Fig. 18.2), the chronic exposure to ethanol may cause neuroadaptations via development of epigenetic tolerance. However, withdrawal from ethanol has been associated with a rise in histone deacetylase (HDAC) activity and lower levels of CREB binding protein (CBP), which could be further downregulating histone acetylation causing condensed chromatin via hypoacetylation. This results in the decline in neuropeptide Y (NPY), brain derived neurotrophic factor (BDNF) and activity-regulated cytoskeleton-associated protein (Arc) levels causing aberrant neural transmission and deficits in synaptic plasticity. Therefore, withdrawal is found to be associated with increased anxiety behaviors and ethanol preference as a means of self-medication to relieve anxiety. These comorbid behaviors can be reversed by the treatment with HDAC inhibitors, for example, trichostatin A (TSA) which mimic the action of acute ethanol exposure by blocking histone deacetylases, leading to hyperacetylation of histones, resulting in anxiolytic effects. As shown in terms of acute ethanol exposure, the role of *microRNA-494 (miR-494)* and its association with CITED and CBP is not known after chronic ethanol exposure and its withdrawal, which warrants further investigation.

Interestingly, P rats display high levels of anxiety and consume high amounts of alcohol during a single episode of exposure in a two-bottle choice model of alcohol preference [63,64]. As mentioned earlier, anxiety-like behavior is one of the major phenotypic outcomes of ethanol withdrawal observed in Sprague-Dawley rats [35,62]. HDAC2-mediated histone acetylation in the amygdala could be an underlying reason for the comorbid behavior of anxiety and alcohol preference in P rats [34,63–65]. Specifically, HDAC2 expression in the amygdala of P rats is comparatively higher than in NP rats, which is correlated with higher HDAC activity. However, ethanol exposure was found to simultaneously reduce *Hdac2* expression and inhibit HDAC activity, thereby increasing the acetylation of H3-K9 (acH3-K9) in the amygdala. *Bdnf*, *Arc*, and *Npy* expression is known to be lower in the amygdala of P rats and this reduction has been shown to be involved in comorbidity of anxiety and alcohol preference [34,63,66,67]. Interestingly, the acetylation levels of histone H3-K9 at the *Arc* promoter and *Bdnf* exon 4 were found to be lower in P rats as compared to NP rats. These deficits in histone acetylation were reversed by acute ethanol exposure by virtue of a decrease in HDAC2 levels and inhibition of HDAC activity [64]. Interestingly, ethanol exposure also normalized BDNF, Arc and NPY levels [64,65] (Fig. 18.4). Guan et al. [40] implicated the HDAC2 isoform in synaptic plasticity and memory formation. It is interesting to note here that epigenetic events might be involved in genetic predispositions to some of the neurological diseases.

Chronic exposure in adult rodents is a heavily used animal model to simulate the clinical situation in adult alcoholic populations. However, adolescents mostly indulge in a binge pattern of alcohol drinking in which heavy amounts of alcohol are consumed during a single episode [68]. Adolescence is a crucial phase of neurodevelopment during which new memories are formed with each new experience via synaptic plasticity driven by axonal and dendritic sprouting [69]. Therefore, binge drinking during adolescence may cause neuropsychiatric disorders later in life [70]. Recent studies in adolescents have employed the CIE exposure paradigm partly modeling teenage drinking to investigate the epigenetic mechanisms that could be affected during this vulnerable age of neurodevelopment. In the last few years, attempts have been made to understand the epigenetic events that may underlie alcohol's damaging and long-lasting effects during this crucial phase of neurodevelopment [33,71,72]. Like adults, the acute ethanol exposure also inhibited HDAC activity in the amygdala of adolescent rats [71]. But, adolescent rats required higher amounts of ethanol for a similar degree of anxiolytic effects which was also associated with a lack of development of tolerance [60,73]. Similar to acute exposure, CIE exposure inhibited HDAC activity. However, the withdrawal from CIE exposure increased the expression of HDAC2 and HDAC4 levels with a simultaneous increase in HDAC function. These changes were associated with reduced histone acetylation in the amygdala and anxiety-like behaviors during adolescence. Most interestingly, these epigenetic changes prevailed through development until adulthood with high HDAC activity, HDAC expression, and lower histone acetylation in the amygdala, consistent with a reduction in *Bdnf* and *Arc* expression and reduced synaptic plasticity. These deficits in synaptic plasticity related events are presumably precipitating high levels of anxiety and therefore alcohol consuming behaviors displayed by adolescent intermittent ethanol (AIE)-exposed adult rats (Fig. 18.4). In line with this, another study by Pascual et al. [74] also found that AIE-upregulated HAT activity with concomitant increases in histone acetylation (H3 and H4) in the prefrontal cortex. These findings are important in the light of the clinical findings that binge alcohol drinking during adolescence poses vulnerability to negative affective states and alcohol use disorders [70,75,76].

As discussed aforementioned, P rats also happen to harbor a similar histone acetylation-induced deficit in *Bdnf* and *Arc* expression which coincidently displays comorbid anxiety and alcohol preference.

It should be noted, that in both of these animal models of comorbid high anxiety and alcohol drinking behaviors, identical histone acetylation–induced molecular events starting from gene expression to synaptic plasticity deficits were found. These are interesting findings in that the HDAC-mediated histone acetylation mechanisms in the amygdala are converging as a cause as well as a consequence underlying the alcohol addictive phenotype wherein anxiety is displayed as a comorbid behavior (Fig. 18.4).

Previous studies have shown that neurogenesis in the hippocampus is affected by adolescent alcohol exposure [77–79]. In a recent study, histone acetylation was observed to be involved in the AIE-induced reduction of *Bdnf* expression and markers of neurogenesis in the hippocampus at adulthood [33]. Similar to amygdaloid changes, binge-like exposure of adolescent rats caused upregulation of HDAC activity with reciprocal changes in CBP and global histone acetylation (H3-K9) levels in hippocampal areas. These events culminated in the reduction in acetylated H3-K9 levels at the *Bdnf* exon IV promoter causing reduction in BDNF levels. One possibility was raised that these deficits in BDNF might be the underlying reason for the observed reductions in Ki-67 and Doublecortin (DCX)-positive cells which marks the ongoing processes of differentiation and proliferation during neurogenesis [33]. These comparable observations between amygdala and hippocampus are especially important because it is thought that adults expresses stronger hippocampal-amygdaloid connectivity compared to adolescents which is likely indicative of stages of emotional maturation [80]. AIE exposure is found to affect both the neurocircuitries via histone acetylation-mediated BDNF function, which is involved in the maintenance of synaptic plasticity. It is highly possible that AIE might be weakening connectivity by affecting chromatin architecture via histone acetylation.

In addition to synaptic plasticity, opioid signaling pathways in the amygdala have also been identified as an epigenetic target of ethanol [81,82]. Nor-binaltorphimine (nor-BNI), a selective κ-opioid receptor antagonist, decreased ethanol self-administration in dependent animals indicating a dysregulation of dynorphin/κ-opioid systems which could be targeted to alleviate the negative emotional states associated with ethanol withdrawal (Walker and Koob,2008). Additionally, the nociceptin (NOC)/nociceptin opioid receptor (NOP) system is associated with the regulation of anxiolytic effects of ethanol administration [82]. Increased levels of the pronociceptin (*Pnoc*) and prodynorphin (*Pdyn*) genes in the amygdala after alcohol administration were linked to reduced levels of H3-K27 trimethylation (repressive mark) and elevated H3-K9 acetylation at the corresponding gene promoter (activating mark) [83]. These results underline the importance of histone acetylation and methylation, working in concert to regulate the process of addiction.

Alcohol is a teratogen and its consumption by child-bearing women increases the possibility of children born with debilitating neurological outcomes categorically referred as FASD. FASD are the nondiagnostic umbrella term used to refer to the full range of effects that can occur following prenatal alcohol exposure. Such exposure can produce a variety of effects, including physical birth defects, growth retardation, and facial dysmorphism [84,85]. The disabilities associated with prenatal alcohol exposure are variable, influenced by numerous factors, and can have a life-long impact. Therefore, early diagnosis and intervention are essential for improved clinical outcomes [86]. Epigenetic mechanisms, which include histone modifications and DNA methylation, are linked to alcohol-induced fetal programming via physiological and morphological abnormalities in the developing brain [87,88]. Guo et al. reported a decrease in CBP expression and histone acetylation in the naïve rat cerebellum with ethanol exposure which may be responsible for the motor coordination deficits that characterize FASD [89]. In yet another study, ethanol treatment at postnatal day 7 (PND7), which is presumed to coincide with the third trimester of human pregnancy, impaired spatial and social recognition memory in adult rats. These behavioral deficits were reversed by pharmacological manipulation or depletion

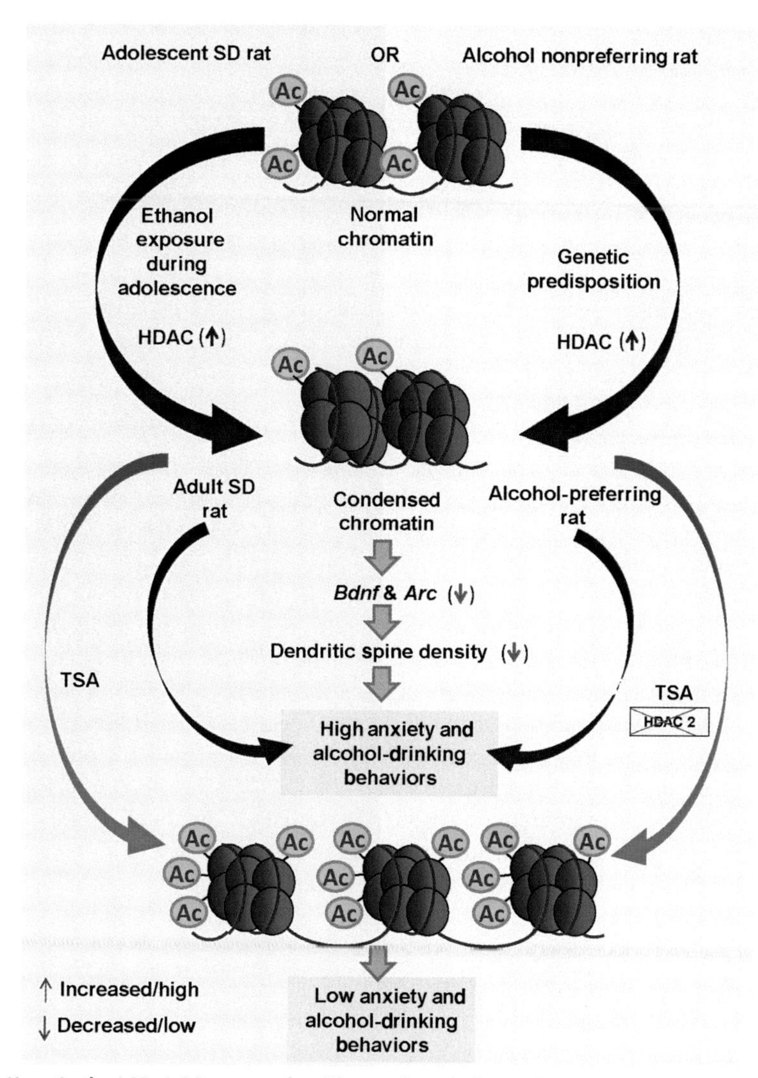

FIGURE 18.4 A Hypothetical Model Representing Histone Acetylation-Induced Molecular Mechanisms and Synaptic Plasticity Operative in the Amygdala of Adult Sprague-Dawley Rats After Adolescent Intermittent Ethanol (AIE) Exposure and of Alcohol-Preferring (P) and Nonpreferring (NP) Rats

Compared to NP rats, P rats exhibit higher histone deacetylases (HDAC) activity and HDAC2 expression resulting in comparatively less acetylation of histone proteins (H3-K9) which might be the underlying reason for condensed chromatin at the promoters of genes, for example, activity-regulated cytoskeleton-associated protein *(Arc)* and brain derived neurotrophic factor (*BDNF*) in amygdaloid areas. The resultant down regulation of these genes due to lower promoter activity might be causing P rats to consume high amounts of alcohol and display comorbid anxiety-like behaviors via reductions in dendritic spine density (DSD). AIE exposure during adolescence also causes deficits in HDAC-induced histone acetylation along with BDNF and Arc expression resulting in a decline in DSD similar to what has been observed in the amygdala of P rats as compared to that in the NP rats. Both adult SD rats (AIE exposed) and P rats display heightened anxiety levels and alcohol drinking behaviors. Treatment with HDAC inhibitor, trichostatin A (TSA) or knock-down of HDAC2 expression by a specific siRNA strategy in the amygdala were able to correct the deficits in histone acetylation causing concomitant declines in comorbid anxiety-like and alcohol-preferring behaviors. These observations emphasize the importance of HDAC-mediated histone acetylation in reversing both the alcohol-exposure-induced or genetically predisposed alcohol drinking behaviors with comorbid anxiety.

of cannabinoid receptor type 1 (CB1R) at PND7 [90]. It was observed that ethanol exposure at PND7 hyperacetylated histone H4 at lysine 8 (acH4-K8) at the promoter of the *Cnr1* gene encoding CB1R in the hippocampus and neocortex. Interestingly, CREB phosphorylation and *Arc* expression was also increased following ethanol treatment at PND7. It is likely that histone acetylation might be playing an important role in prenatal ethanol-induced behavioral deficits. This notion can be further strengthened by the observation that ethanol exposure to pups at PND7 increased histone H3-K9 acetylation at *G9a* exon 1 with enhancement of G9a protein levels followed by neurodegeneration. These observations in turn prompt us to believe that the histone acetylation and methylation processes, specifically at H3-K9, are intricately associated in FASD phenotypes [90]. Therefore, more mechanistic studies examining the role of HDACs and HATs in FASD need urgent attention. Moreover, refining and optimizing HDAC inhibitory therapies may provide an attractive option to treat alcoholism and its underlying causes including symptoms of FASD.

The most exciting aspect of epigenetics is in its potential to absorb external impacts and reestablish the normal program of gene function, which is reversible in nature. Some of the recent findings defining epigenetic pathways of neurological disorders highlight the importance of HDAC inhibitors as a major therapeutic target despite some caution in generalizing and over-estimating their possible application [8,53,91,92]. It seems more promising for the fact that SAHA is available on the shelves as "Vorinostat" for cancer treatment. In fact, the notion that epigenetic mechanisms may be involved in addictive behaviors came from the earliest studies showing that the pharmacological or genetic manipulation of HDAC activity can alter cocaine-induced behavioral outcomes [93]. Similarly, the pharmacological inhibition or genetic manipulation of specific HDAC isoforms in the brain has also been shown to modulate some of the alcoholic phenotypes [35,60,64,65,71,72,94,95]. In one of the early studies in rats, systemic administration of trichostatin A (TSA), a pan HDAC inhibitor, was shown to reverse the anxiety-like behaviors displayed by the ethanol-withdrawn rats exposed to chronic ethanol exposure [35,62]. Similarly, TSA could also prevent the development of a rapid tolerance to anxiolytic effects expressed by acute ethanol exposure in rats [60]. These behavioral corrections were associated with normalization in global histone acetylation, NPY and BDNF levels, and dendritic spine density in the amygdala of SD rats [35,60–62]. One key question here that still remains to be addressed is whether other downstream targets that might be altered during alcohol exposure could be functioning in various neuropsychopathological aspects of alcoholism. Not only does chronic drinking by adults, but also ethanol exposure in binge-like patterns during adolescence cause upregulation of HDAC activity in the amygdala at adulthood [72], which was normalized following HDAC inhibition by TSA treatment. Surprisingly, HDAC inhibition by sodium butyrate also upregulated HAT activity in prefrontal cortex of the adolescent rats, but not in the adult rats [74]. Warnault et al. have also shown reduced histone H4 acetylation in NAc with ethanol exposure in rodents and SAHA reduced ethanol-seeking behaviors in rats and binge-like alcohol drinking in mice [94]. However, conflicting observations were made in another mice study [96]. In disagreement with Warnault et al. [94], TSA treatment was found to facilitate CIE-induced ethanol preference [96]. The underlying reasons for these discrepancies could be related to ethanol exposure paradigms and the timing of inhibitor (TSA) treatment. Nonetheless, these studies clearly show that HDAC inhibition alters ethanol drinking behaviors in mice. As mentioned previously, P rats are genetically predisposed to alcohol preference and also display comorbid anxiety behaviors [58,63,97,98]. TSA treatment as well as HDAC2 knock-down by siRNA attenuated both of these behaviors and also corrected the deficits in histone H3 acetylation (H3-K9) at the *Bdnf* promoter which normalized BDNF function as well as dendritic spine density [60,64]. Chromatin immunoprecipitation studies revealed normalization in histone acetylation (H3-K9) levels in promoters of *Bdnf* exon IV and *Arc* in the amygdala of P rats following amygdaloid infusion of HDAC2 siRNA. P rats are also known to express low levels of NPY in the amygdala, one of the possible underlying reasons for their high anxiety and alcohol preference [63]. In addition, TSA treatment also upregulated NPY levels in the amygdaloid region of P rats [60]. On the other hand, H4 acetylation in the NAc was increased with ethanol exposure, which corresponds to a decrease in HDAC activity in the striatum of mice corresponding to ethanol-induced behavioral sensitization [99]. However, surprisingly the treatment with the HDAC inhibitor, sodium butyrate prevented ethanol induced sensitization and further modulations in *Bdnf* expression in the striatum and prefrontal cortex [100]. Tolerance to inhaled benzyl alcohol, which regulates potassium channels, was found operative through H4 acetylation not only in rodents but also in the insect model *Drosophila melanogaster* [101]. All these findings suggest that chromatin modulation via HDAC plays a critical role in expression of genes and concomitant synaptic plasticity implicated in alcohol preference, tolerance, and dependence. HDAC2-specific inhibitors/regulators might essentially prove beneficial to coordinate negative and positive affective states of alcoholism. However, despite these studies potentially implicating HDAC-mediated histone acetylation in the brain to alcoholic phenotypes, in-depth analysis in a region- and cell-type specific manner is required to identify the histone acetylation-mediated pathways in different neurocircuits. Further research into designing specific subtypes of HDAC isoform inhibitors and identifying specific signaling pathways that could be mediating epigenetic events in different areas of the brain linked to AUD need to be further investigated.

18.2.2 HISTONE ACETYLATION IN PERIPHERAL TISSUES

Some efforts have been made in the last decade to understand the actions on epigenetic events that alcohol has in peripheral tissues for clinical relevance. Peripheral tissues, such as the digestive track and liver are the primary targets of alcohol administration and chromatin remodeling via histone modifications have been associated with alcohol consumption in peripheral tissues [102]. Like the amygdala, binge drinking also caused changes in H3-K9 acetylation in liver, lung, and spleen of rats, with a maximum of a ~6 fold up regulation in liver [102]. These effects occurred in a dose- and time-dependent manner in primary cultures of hepatocytes and inhibition of deacetylation by TSA treatment was able to mimic the effects of acute ethanol exposure, supporting the notion that ethanol can cause H3-K9 acetylation [103]. In contrast, chronic ethanol exposure in rodents did not alter global H3-K9 acetylation, but alcohol dehydrogenase-1 gene (*Adh1*) specific H3-K9 acetylation increased in both promoter and coding regions [104]. High levels of ethanol and its metabolites were reported to induce H3-K9 acetylation in genes related to heart development in cardiac progenitor cells with increased expression of *Gata4* and *Mef2c*, which could be possible mediators of ethanol induced congenital heart disease [105]. These studies prompt us to assume that histone H3-K9 acetylation is a prominent histone modification based on observations from brain and peripheral studies. This finding needs to be validated as a common epigenetic mark that may be involved in ethanol's hazards to the peripheral and central nervous system.

18.2.3 HISTONE METHYLATION MECHANISMS

Methylation occurs at lysine and arginine residues in histone tails and is one of the major modifications that is associated with chromatin remodeling and gene transcription. However, methylation is believed to regulate gene transcription in many different ways. First, the single lysine/arginine residue in histone tails can be mono and dimethylated or lysine can be even present in trimethylated form. Second, the varying degree of methylation at single lysine/arginine residues in a specific histone tail can offer conducive or repressive chromatin structure. The third layer of regulation by methylation is dependent on the position of lysine and arginine in the given histone tails which results in different readouts. For example, Grewal and coworkers reported that euchromatin and heterochromatin have distinctive site-specific histone H3 methylation patterns [106]. Whereas H3-K9 methylation is localized to a 20 kb silent heterochromatic region, H3-K4 methylation was strictly found in surrounding euchromatic regions. In higher vertebrates including humans, dimethylation of histone H3-K9 has been shown to be associated with transcriptional silencing, while trimethylation of histone H3 at lysine 4 (H3-K4me3) is correlated with active transcription [107,108]. Similarly, trimethylation of H3-K4 (H3-K4me3) is associated with active promoters in humans [109]. Histone methylation is dynamically governed by the histone methyltransferases, for example, G9a and G9a-related protein (GLP) [110,111] and histone demethylases, for example, lysine-specific demethylases (KDMs). Histone methylation and histone modifiers have also been studied in the nervous system using preclinical animal models [112] and alterations can be found to be associated with brain pathologies. Of note, dysregulation of H3-K9 was seen in postmortem brains of subjects diagnosed with Huntingtons disease [113] and Friedreich's ataxia [114]. Similarly, drug-induced neuronal plasticity and subsequent mental health issues including memory formation and cognition were found to be affected by specific alterations in histone methylation [115–118]. Acetylation of H3-K9 and H3-K14 potentiates transcription factor interaction with H3-K4me3, which reveals the effect of histone modifications on the assembly of transcriptional apparatus [119]. It was observed that H3-K4me3 (open chromatin mark) was elevated in the promoters of the *Bdnf* and *Zif-268* genes within the hippocampus in a contextual fear conditioning model. Also the HDAC inhibitor sodium butyrate induced elevated trimethylation at H3-K4 with a simultaneous decrease in dimethylation at H3-K9 (repressive mark), suggesting the role of histone methylation in facilitating long-term memory formation [116]. Enrichment in H3-K4me3 levels in the hippocampus induced expression of the solute carrier family 17 member 6 (*Slc17a6*) gene, which encodes for vesicular glutamate transporter 2 (VGLUT2) in the hippocampus of prenatally ethanol-exposed adult. However, on the contrary, VGLUT2 protein levels were significantly decreased, suggesting an additional level of posttranscriptional control [120]. Neurodegeneration caused by ethanol exposure in the developing brain (postnatal day 7) was associated with increased G9a levels followed by a concomitant rise in H3-K9me2 and H3-K27me2 levels. G9a is potentially involved in this neurodegenerative action of alcohol since treatment with a G9a inhibitor prevented neurodegeneration along with restoration of H3-K9me2 and H3-K27me2 levels [121].

Glutamate is the primary excitatory neurotransmitter in the mammalian brain, especially in all cortical pyramidal neurons and thalamic relay neurons [122] and aberrations in *N*-methyl-D-aspartate (NMDA) function is known to translate into neurodegeneration. Excitatory synaptic transmission in the CNS is predominantly mediated via NMDA receptors (NMDAR). Through a series of in vivo and in vitro studies, chronic ethanol exposure and withdrawal was shown to alter the levels of NMDAR [123,124]. Studies over the last few years have provided strong evidence offering epigenetic dimensions to the process of alcohol-mediated neuroadaptations via changes in NMDAR expression [125]. As mentioned previously, in cultured primary cortical neurons, ethanol treatment induced *Nr2b* expression, which could be related to the increase in H3-K9 acetylation and the reciprocal decline in H3-K9me2 at its promoter as revealed by chromatin immunoprecipitation assay [126]. In the same way, levels of H3-K9me2 decreased in the promoter of cytochrome p450 subunit and lactate dehydrogenase in cultured rat hepatocytes exposed to ethanol. However, H3-K4me2 levels were increased in a promoter of alcohol dehydrogenase in the similar study [127].

One of the brain areas, which have not been explored for epigenetic actions of alcohol is the hypothalamus, in spite of the fact that the neural stress axis plays an important role in alcoholism [128]. Clinically, abnormalities in the proopiomelanocortin (POMC) system were associated with alcohol preference and a family history of alcoholism [129,130]. Along these lines, Sarkar and coworkers have explored the effects of fetal alcohol exposure (FAE) on the POMC system in the hypothalamic region in a rodent model. Prenatal alcohol treatment has been shown to downregulate the number of POMC neurons in the hypothalamus, possibly due to decreased *Pomc* gene expression with a consistent decline in β-endorphin peptide [131,132]. Epigenetic mechanisms were also found to regulate the fetal programming of the POMC system resulting in changes in the neural-stress axis [131]. Furthermore, FAE decreased mRNA levels of Set domain histone lysine methyltransferase (*Set7/9*) that catalyzes the methylation of H3-K4, and reciprocally increased the mRNA level of *G9a* and histone-lysine methyltransferase Set domain bifurcated 1 (*Setdb1*) that regulate methylation of H3-K9. In accordance with these findings, FAE reduced the activation mark H3-K4me2, 3 in β-endorphin positive neurons, while on the other hand FAE also increased the level of the repressive mark. Gestational choline supplementation not only reversed the alcohol-induced changes in the mRNA levels of these histone-methylating enzymes, but also normalized the methylation levels of H3-K4me2, 3 and H3-K9me2 [133]. These studies clearly highlight another mechanism underlying how choline supplementation might help prevent developmental pathologies due to FAE.

Like any other epigenetic event, histone methylation is also a reversible process; dynamic equilibrium between the KMTs and KDMs maintains histone methylation thereby contributing to chromatin architecture modulation [134]. Clinically two studies are highly relevant in terms of alcoholism and histone methylation. A genome-wide association study has revealed that three SNP's of *KDM4C* (a lysine demethylase of the Jumonji domain 2 family) gene at loci 9p24.1 are associated with alcohol withdrawal [135] and prodynorphin (*Pdyn*) SNPs have exhibited differential methylation patterns in the postmortem brain of alcoholic patients [136]. Having said that, histone methylation seems to record the harms caused due to FAE to induce faulty transcriptional schemes. The role of methionine precursors in reversing the teratogenic and neurodevelopmental abnormalities of alcohol exposure is shown using fetal supplementation studies, for example, folate and choline [137–139]. Interestingly, choline deficiency upregulated DNA methylation in genes controlling histone methylation, (e.g., *G9a* and *Suv39h1*) thereby reducing expression of the genes subsequently modulating histone methylation marks which could be manipulated with choline supplementation [140]. These studies directly implicate the interplay between methylation of both DNA and histones in driving underlying gene expression. In case of FAE, such possibilities are highly likely to occur and should instil urgent intervention.

18.2.4 HISTONE PHOSPHORYLATION

Phosphorylation of histones occurs at serine (S), threonine (T), and tyrosine (Y) residues and its presence may differ from miniscule to the majority of nucleosomes depending on the phases of cell cycle [141–143]. Histone phosphorylation is associated with chromatin remodeling and is mediated by protein kinases and phosphatases, hence is critical to transcriptional activation, DNA repair, and cell cycle regulation [143–146]. In brain, H3 phosphorylation in the hippocampus was induced by the ERK/MAPK pathway in a contextual fear-conditioning model and MEK (the kinase upstream of ERK) inhibitor treatment blocked this phosphorylation [147]. As described aforementioned, various signaling pathways converge at transcription factor CREB [148], the phosphorylation of which to p-CREB modulates the binding of CBP at CRE elements. In the absence of mitogens, kinase activity of ribosomal S6 kinase 2 (RSK2) and the HAT activity of CBP was downregulated as they associate in a complex. Within a few minutes of mitogenic stimulation both of the activities are upregulated with the dissociation of the complex. RSK2 further phosphorylates histone H3 and CREB thereby mediating chromatin remodeling and transcriptional activation [149]. These studies highlight the convergence of the CREB mediated signaling mechanisms in histone phosphorylation. As mentioned earlier in the section on histone acetylation, the CREB signaling pathway is substantially involved in alcohol's addictive processes [57]. A few studies also link it with histone phosphorylation in mediating alcohol's mode of action. In primary rat hepatocytes, ethanol and its metabolites are shown to activate MAP kinases, specifically p38 MAP kinase, which phosphorylates histone H3 at serine 10 and serine 28 [150], which could be potentially involved in various pathological states contributing to alcoholic liver disease. Similarly, increased phosphorylation of histone H3 was observed with psychostimulant drugs, such as amphetamines, cocaine, and morphine due to sequestration of dopamine- and cyclic AMP-regulated phosphoprotein 32 (DARPP-32) in the nucleus which is a protein phosphatase-1 inhibitor [151]. Recently, McClain and Nixon [152] demonstrated that alcohol in low doses (1 g/kg) increases the number of histone H3-S10 positive cells in the hippocampus. On the contrary, high doses (2.5 and 5 g/kg) and chronic exposure reduces the histone H3-S10 phosphorylation. However, withdrawal from chronic exposure augmented the H3-S10 phosphorylated cells, which is thought to be involved in activation of *c-Fos* expression. A plethora of information is available on the posttranslational modifications carried on histone tails associated with CNS diseases. However, few studies could show PTMs in the core region of histones, such as phosphorylation [15,153]. Phosphorylation of H3 could be offering more complex mechanisms of chromatin remodeling than other histones, because of the presence of several potential threonine and serine phosphorylation sites. There is growing evidence that the mechanisms of chromatin modifications closely interplay in neuropsychiatric illnesses. Given the importance of protein kinases and CREB in alcohol abuse disorders, more efforts are required to put into research on histone phosphorylation and its crosstalk with other regulatory mechanisms.

18.3 DNA METHYLATION

Among all the epigenetic mechanisms, DNA methylation is the most stable and well-studied modification in almost all aspects of gene regulation during development and in health and disease conditions. Recent development in high throughput sequencing technologies and bioinformatics are providing a deeper insight into epigenetic profiling of DNA methylation across the genome in relation to biological and pathological processes. The most prominent biological outcomes driven by complex epigenetic mechanisms include gene silencing [154,155], genomic imprinting [156], X-chromosome inactivation [157], and various mechanisms of cancer induction [158]. Methylation mainly occurs at sites rich in cytosine and guanosine dinucleotides commonly referred as CpG islands. Hypermethylation is mostly associated with transcriptional repression, whereas hypomethylation leads to activation of gene transcription. The process occurs in two ways: directly hindering the interaction of DNA-binding proteins and indirectly through binding of methyl-CpG-binding proteins (MBD) and recruitment of HDACs, corepressors, and other heterochromatin-associated proteins [159–162]. If these CpG dinucleotides are located within regulatory sequences, such as promoter regions, their methylation induces chromatin condensation and can block the binding of transcription factors [91].

The dynamic equilibrium in DNA methylation is maintained by two opposing mechanisms, DNA methylation and DNA demethylation, which are among the best-studied epigenetic pathways. DNA methyltransferases are a class of methyltransferase enzymes comprising DNMT1, DNMT3a, and DNMT3b that mediate DNA methylation by catalyzing the addition of a methyl group to the C5 position on cytosine residues, which is alternatively denoted as 5-methylcytosine (5-mc) using S-adenosylmethionine (SAM) as a methyl group donor [162,163]. DNMT3a and 3b carry out de novo CpG methylation, whereas DNMT1 helps in the maintenance of the methylation pattern during DNA replication and repair. However, DNMT1 is also known to exhibit catalytic activity for de novo methylation at non-CpG cytosines [164]. In some instances, it has also been observed that DNMT1 also methylates cytosines in a de novo pattern at CpG islands [165,166]. On the contrary, inability of DNMT1 to maintain the methylation on daughter strands during replication is technically termed as a nonenzymatic or passive method of DNA demethylation. However, in principle, this notion does not stand as mechanistically true given the postmitotic nature of neurons. Therefore, the mechanism of active or enzymatic DNA demethylation holds special importance in DNA methylation studies of the brain. In this sense, DNA demethylation is the active removal of methyl group from cytosine through a series of well-governed mechanism essentially requiring a battery of factors, for example, ten–eleven translocation methylcytosine dioxygenase 1 (Tet1) and growth arrest and DNA damage factors (GADD45a and b). One of the first steps involves the conversion of 5-methylcytosine to 5-hydroxymethylcytosine (5-hmc) via hydroxylation. TET proteins have recently been shown to function as 5-mc hydroxylases, which involves the oxidation of 5-mc to 5-hmc [167,168].

The CNS is the central controlling unit of the body hence any dysfunction in the maintenance of the mechanism of DNA methylation in CNS may have long term irreversible effects. Dysfunction of the pathways that regulate DNA methylation have been implicated in a variety of brain disorders, such as fragile-X syndrome [169], Rett syndrome (an X-linked autism spectrum disorder caused by mutations in the *MECP2* gene encoding methyl-CpG-binding protein 2) [170], synaptic plasticity [171,172], abnormal learning and memory [173,174], schizophrenia, bipolar disorder [175], and autism spectrum disorder [176]. Mutations or duplications in *MECP2* are known to cause Rett and Rett-like syndromes with autistic behaviors, motor dysfunction, and mental retardation [177]. The increase in *Bdnf* expression after neuronal depolarization is associated with decreased methylation in the *Bdnf* promoter region and also dissociation of the MeCP2–histone deacetylase–mSin3A repressor complex from its promoter further regulating neuronal plasticity [171,178]. Genome-wide mapping of 5-hmC in embryonic stem cells, brain, and other tissues revealed a strong enrichment within exons and near transcriptional start sites specifically of genes whose promoters bear dual H3K27me3 and H3K4me3 marks potentially indicating its role as a transcriptional activator [59,179–183]. Methylcytosine dioxygenase (TET1) has been shown to promote DNA demethylation in the nervous system regulating various memory-associated genes and synaptic plasticity [184,185]. Various studies have also revealed a role for the Gadd45 family of proteins in DNA demethylation including hippocampal synaptic plasticity associated with learning and memory [186–188]. Although a reasonable number of studies have been carried out to understand the molecular basis of alcohol use disorder stemming from aberrant DNA methylation, its role in controlling specific downstream pathways remains elusive, which could be attributed to multifactorial nature of AUD. For example, low levels of BDNF in the amygdala have been associated with anxiety-like and alcohol-drinking behaviors [64,67,189]. However, the role of DNMTs or MeCP2 in the regulation of BDNF/*Bdnf* gene expression during the comorbidity of anxiety and alcoholism is largely unknown. We will review here the recent hypothesis-driven gene-oriented and whole genome-wide studies that shed some light on the role of DNA methylation as one of the major epigenetic mechanisms operative in driving the alcohol actions in the brain.

18.3.1 ALCOHOL AND DNA METHYLATION

The idea that DNA methylation could be involved in alcoholism can be traced back to 1940s and 1950s, to the works of Dr. Roger J. Williams, a biochemistry professor at the University of Texas at Austin. He showed for the first time that dietary changes could affect beverage alcohol (i.e., ethanol) consumption in rodents. Specifically, diets deficient in B vitamins (e.g., folic acid and choline) increased consumption of solutions containing 10% ethanol in some rats, whereas vitamin rich diets decreased it [190]. It is now well established that folates and several other B vitamins are critical for one-carbon metabolism and the synthesis of a compound S-adenosyl-methionine (SAM), which serves as the primary methyl group donor in most trans-methylation reactions, including DNA methylation [191]. This suggested that alcohol consumption could be controlled via changes in DNA methylation and downstream gene expression. Chronic alcohol exposure is associated with widespread alterations in neuronal gene expression leading to behavioral tolerance and alcohol dependence. These gene expression changes contribute to the brain pathology and brain plasticity associated with alcohol abuse and dependence [192]. In a variety of alcoholic phenotypes in central and peripheral tissues, DNA methylation as well as demethylation has been implicated. In a rodent model of chronic alcohol withdrawal followed by dependence, alcohol exposure increased DNA methylation in the medial prefrontal cortex (mPFC) via elevated levels of DNMT1, specifically in neurons. Whole transcriptome sequencing of the mPFC revealed down regulation of genes encoding synaptic proteins involved in neurotransmitter release specifically Synaptotagmin 2 (*Syt2*). Treatment with the DNMT inhibitor RG108 prevented alcohol consumption, with down regulation of 4 out of the 7 transcripts. Hypermethylation on CpG#5 of first exon of *Syt2* was reversed by RG108 treatment and *Syt2* knockout induced compulsivity-like behavior in mice implicating role of DNA methylation in chronic ethanol dependence-induced neuroadaptation specifically of *Syt2* gene [192].

As discussed previously, glutamate is the primary neurotransmitter and increased glutamate transmission via elevated levels of NMDA receptor was observed in response to ethanol exposure. Chronic ethanol exposure leads to demethylation of cytosine residues in the CpG island at the NMDA receptor 2B subunit (*Nr2b*) gene promoter, which in turn upregulated *Nr2b* gene expression in adult mouse cortex and cultured fetal cortical neurons [193]. Moreover, the DNMT inhibitor 5-azacytidine mimicked these effects independently and also in combination with ethanol [193,194]. As shown in Fig. 18.5, CIE exposure decreases *Dnmt* expression, implicating alleviated levels of DNA methylation in the 5' regulatory region of the *Nr2b* gene promoter, specifically at a site proximal to the transcription factor, AP-1 and CBP binding sites. This allows transcription factor and RNA polymerase binding at the promoter, which activates transcription of *Nr2b* gene encoding the NR2B protein. Treatment with DNMT inhibitor 5-azacytidine mimicked these effects independent of ethanol [195,196], confirming the role of DNMT in regulation of NR2B gene expression. Therefore, these findings clearly identify ethanol-induced chromatin remodeling at the promoter of the *Nr2b* gene, allowing transcription factor binding thereby activating transcription of the *Nr2b* gene. Consequently, elevating demand of corresponding ligand glutamate, leads to neuroadaptation and alcohol dependency.

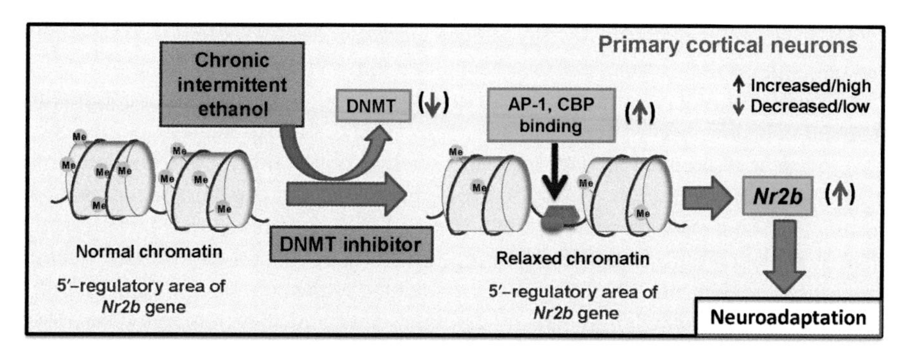

FIGURE 18.5 A Schematic Representation Showing Chronic Intermittent Ethanol-Induced Chromatin Remodeling at the 5′-Regulatory Area of *N*-Methyl-D-Aspartate (NMDA) Receptor 2B Subunit (*Nr2b*) Promoter Region

Chronic intermittent ethanol exposure decreases DNA methyltranseferase 1 (*DNMT1*) gene expression culminating into specific demethylation of *NR2B* gene promoter region. The demethylation might be leading to increased accessibility of the promoter region to transcription factors, AP-1 and CREB thus activating the *NR2B* promoter and increasing transcription. Similar effects were observed with DNMT inhibitor 5-azacytidine (5-AzaC) in the absence of ethanol in the internal milieu via blocking promoter DNA methylation. These observations substantiate the notion that chronic intermittent ethanol exposure might be implicated in promoter DNA methylation thereby increasing the levels of NR2B protein in cortical neurons leading to neuroadaptation with the subsequent addiction phenotype.

Recent clinical findings implicate DNA methylation in alcoholic phenotypes. Prodynorphin is a precursor for endorphins, the chemical messengers in the brain that are associated with the anticipation of pain and the formation of deep emotional bonds. Three single-nucleotide polymorphisms have been detected in postmortem studies of human alcoholics in the prodynorphin (*PDYN*) gene that overlaps with CpG sites that are differentially methylated [136]. In a study on psychotic patients with and without a history of alcohol abuse compared with nonpsychotic patients, elevated levels of DNMT1, GADD45B, and TET1 were found in prefrontal cortex in psychotic patients when compared with nonpsychotic patients [197]. However, this increase in *DNMT1* expression was not observed in psychotic patients with a history of alcohol abuse. On the contrary, the increase in Tet1 in psychosis was more pronounced in patients who had abused alcohol. This may lead to genome-wide chromatin demethylation activating genes critical for psychic maintenance, suggesting that psychotic patients may take ethanol to self medicate themselves.

BDNF has been best characterized for its role in many neuropsychiatric and neurodegenerative disorders through its potential activity in maintaining synaptic plasticity and dendritic spine morphology [198]. It is a complex gene consisted of nine exons with four promoters, and its epigenetic dysregulation has been very well documented in alcoholism [61,66,199–203]. C57BL/6J (C57) and DBA/2J (DBA) mice are reported to consume high and low amounts of alcohol respectively [201,203]. In the shell of the nucleus accumbens, lower levels of *Gadd45b*, *g* and *Bdnf9a* mRNA along with GADD45b protein, were found [204]. 9a is the only protein coding exon of *Bdnf*; lower levels of H3-K9 and H3-K14 acetylation of the *Bdnf9a* promoter along with higher 5-hmc and 5-mc, may hinder transcription factor binding to the promoter consequently inhibiting transcription of *Bdnf9a*. Exactly opposite results were reported with acute ethanol exposure, suggesting that epigenetic-modifiers regulate neuroadaptation, leading to increase in tolerance to ethanol. *Gadd45b* (+/−) haplodeficient mice, exhibited significantly higher alcohol drinking, than control wild-type mice *Gadd45b* (+/+), whereas alcohol-drinking behavior of knockout animals didn't change. These results established the role of *Gadd45b* toward modulation of *Bdnf* promoter chromatin architecture, regulating alcohol dependence phenotype [204].

As discussed in histone acetylation, FASD is a term used to refer to the range of psychosomatic effects that can occur with prenatal alcohol exposure, such as physical birth defects, growth retardation, and facial dysmorphism [205]. The implications of prenatal alcohol exposure are variable and can last lifelong, hence treatment strategies need to be identified [86]. Several studies have linked dysregulation of DNA methylation with FASD [88]. An in vitro study of alcohol exposure of neural stem cells revealed inhibition of DNA methylation of genes associated with neuronal development, further affecting cellular differentiation, which may lead to abnormal embryonic development [206]. Hypothalamic POMC neurons regulate immune functions and activity of the hypothalamic-pituitary-adrenal (HPA) axis [132]. A prenatal ethanol exposure study in a rodent model demonstrated a reduction in POMC neurons, with a decrease in *Pomc* mRNA expression and corresponding peptide product ß-endorphin levels [131,132]. The *Pomc* gene is mainly involved in the regulation of stress responses. Increased promoter methylation status and reduced expression of hypothalamic *Pomc* with prenatal ethanol exposure could be correlated to the modulation of epigenetic modifiers, such as suppressed histone activation marks (H3K4me3, Set7/9, acetylated H3K9, phosphorylated H3S10) and elevated histone repressive marks (H3K9me2, G9a, Setdb1) along with *Dnmt1* and *Mecp2* expression. Gestational choline attenuated these effects normalizing the stress response of adult offspring [133].

The availability of tissue samples is the main limitation of clinical studies on brain disorders; blood is the primary available source of live cells from the human system. Hence biomarkers corresponding to altered methylation levels as well as gene transcripts levels, which will aid in the prediction of consequential predisposition or diseased phenotype, are warranted in blood cells [207–210]. In chronic ethanol exposure, elevated levels of homocysteine and associated brain atrophies were reported [211]. The reduced mRNA expression of homocysteine-induced endoplasmic reticulum protein (HERP) and hypermethylation of its promoter in blood cells have been associated with elevated homocysteine levels in the blood of alcoholic patients [212]. Also, hypermethylation of the α-synuclein gene promoter (*Snca*) and elevated mRNA as well as protein levels were also been correlated to elevated homocysteine in alcoholics and alcohol craving behavior [213,214]. Modulation of the dopaminergic system is a crucial factor for the development of alcohol dependence (AD) [215]. Hypermethylation of gene promoters involved in dopaminergic neurotransmission, such as the dopamine transporter gene (*Dat*) and *Snca* has been observed in peripheral blood cells of alcohol dependent patients [208,214]. This may increase dopamine levels in the synaptic gap reflecting in a lower craving. In contrast, Jasiewicz et al. [215] reported, hypomethylation of position 12 out of 23 CpG islands in the *Dat* gene promoter of alcohol dependent patients but further studies are needed to confirm relevance of this position specific methylation in the AD-related phenotype.

In a step toward biomarker development, Ruggeri et al. [216] analyzed genome-wide DNA methylation of 18 monozygotic twin pairs discordant for AUD. They have found an association between hypermethylation of the 3'-protein-phosphatase-1G (*Ppm1g*) gene locus with AUD and early escalation of alcohol use and increased impulsiveness. The elevated blood-oxygen-level-dependent response during impulsiveness with *Ppm1g* hypermethylation in the right subthalamic nucleus was also reported. In a recent study, a genome-wide approach was utilized to examine the DNA methylation signatures of lymphocytes from subjects with heavy alcohol consumption [217]. Alcohol drinking has been shown to cause widespread changes in DNA methylation, which were reversible with abstinence.

Further studies may help develop relevant biomarkers for analyzing the vulnerability of patients to alcohol use disorders.

DNMT inhibitors, zebularine and 5-azacytidine were reported to regulate learning and memory behaviors, synaptic plasticity, and long-term potentiation induction in the hippocampus [172,218]. They have also been used to study other drugs of abuse, such as cocaine regulated phenotypes [219,220]. Prevention of excessive alcohol drinking behavior in mice has been achieved by systemic administration of 5-azacitidine (5-AzaC) via repression of DNA methylation by inhibiting DNA methyltransferase activity [94]. Brueckner et al. [221] has characterized a small molecule known as RG108 as a potential DNA methyl-transfersae inhibitor. It was able to activate a tumor suppresser gene which was under the regulation of DNA methyltransferase, without affecting the methylation status of centromeric satellite sequences. It was also employed to study the epigenetic regulation of *Syt2* [192].

It is important to take note of some relevant studies in another drug of abuse, that is, cocaine which has been shown to cause closely similar epigenetic changes in different parts of the brain to establish relevance to alcohol's epigenetic impacts. Cocaine has shown to cause widespread epigenetic changes including histone covalent modifications and DNA methylation [92]. Acute cocaine administration inhibited *Dnmt3a* in the nucleus accumbens, 24 h after the last exposure, whereas chronic exposure followed by 28 days withdrawal, significantly elevated *Dnmt3a* expression which could be linked to increased dendritic spines on NAc neurons (Fig. 18.6). Inhibition of *Dnmt3a*, increased cocaine reward behavioral effects and also prevented the increase of neuronal dendritic spines, on the other hand *Dnmt* overexpression attenuated these effects [222]. Inhibition of spinogenesis is known to elevate cocaine reward [223]. These results suggest that the cocaine reward is regulated by NAc dendritic spine density. However, the functional circuitry leading to this synaptic plasticity and their epigenetic regulation needs to be studied. Cocaine self-administration induces *Mecp2* expression in the dorsal striatum, which further potentiates *Bdnf*, thereby elevating cocaine-induced synaptic plasticity. Homeostatic interaction between micro RNA 212 (*mir212*) and *Mecp2* regulates cocaine intake, suggesting a multifactorial mechanism of regulation (Fig. 18.6) [224]. An increase in *Bdnf* expression in the dorsal striatum was also observed with self-ethanol administration [202], but its epigenetic relevance and behavioral consequences needs to be elucidated. It could be possible that signaling networks for ethanol and other drugs of abuse functioning in drug-seeking behaviors may overlap in specific brain areas. More research is necessitated in addressing the convergent epigenetic pathways of drug addiction in specific neurocircuits.

18.3.2 ADOLESCENT DRINKING AND ALCOHOL USE DISORDERS AT ADULTHOOD

Neuroadaptations due to adolescent alcohol drinking may induce progression of alcoholism in adulthood [79,225]. The role of epigenetics in regulating neuroplasticity has been evident in some recent literature. Binge-like alcohol drinking is the most common pattern of drinking among adolescents. Elevated levels of H3-K9 andH4-K12 acetylation were reported in the frontal cortex and nucleus accumbens alone with a concomitant decrease in the striatum of adolescent rats exposed to binge-like alcohol drinking. However, these methylation marks were absent in adult rats suggesting chromatin remodeling, confirming the vulnerability of the adolescent brain to binge-like alcohol drinking [226]. Acute ethanol exposure has been reported to cause rapid tolerance to the anxiolytic effects of ethanol in adult rats [60]. On the contrary, reduced sensitivity to anxiolysis and the lack of rapid tolerance to the anxiolytic effects of ethanol was observed in adolescent rats. This was predictably attributed to the epigenetic players controlling the part of emotional circuitry of the brain, that is, amygdala and bed nucleus of stria terminalis (BNST). Acute ethanol exposure attenuated DNMT activity with concomitant alterations in *Dnmt* expression in amygdala and BNST [71].

Epigenetic mechanisms of psychiatric disorders are immensely complex and neuropsychological outcomes cannot be fully attributed to a single or even to a well coordinated epigenetic network. Therefore, it is also advisable to find the potential of existing pharmacological agents that could potentially normalize the symptoms on one hand, but also offer epigenetic advantages for long-term prognosis. For example, dibenzepine derivatives (clozapine, quetiapine, and olanzapine) alter chromatin remodeling and activate DNA-demethylation of GABAergic and glutamatergic promoters [227]. Similarly, the elucidation of different mechanisms and possible drug targets which could interfere with alcohol's deleterious effects via DNA methylation and demethylation may pave the way for better development of diagnostic and pharmacotherapeutic strategies.

18.3.3 ALCOHOL AND DNA METHYLATION MECHANISMS IN PERIPHERAL TISSUES

The liver is the main organ involved in the detoxification of chemicals, such as alcohol from the body and hence is susceptible to chemical induced DNA methylation, which may alter physiological functions. In case of chronic ethanol exposure, inhibition of methionine adenosyltransferase, which converts methionine to SAM leads to global DNA hypomethylation [228,229]. Low levels of DNMT3b were observed in alcoholic liver and metabolites of alcohol, such as acetaldehyde have been shown to inhibit DNMT activity [230]. The *Dnmt1* hypomorphic mice were resistant to hepatic steatosis suggesting a protective role of the $Dnmt1^{Ni}$ allele in alcohol-induced hepatosteatosis [231]. Alcohol exposure modulates epigenetic modifiers, revealing the cascade of alcohol induced epigenetics and it's the functional implications will be beneficial for understanding of advanced diseases and may provide new therapeutic avenues [229].

18.4 MicroRNAs

MicroRNAs (miRNAs) are small noncoding RNAs, which controls the translation of mRNA by binding to 3′ untranslated region (3′ UTR) which usually occurs either by mRNA degradation or by translational repression [232]. A wide variety of miRNAs have been implicated in a range of disease conditions including developmental disorders to cancer. Circulating miRNAs in the plasma have gained a special interest in clinical field for its potential in the development of the biomarkers. MicroRNAs are involved in neuronal migration, maturation, synaptic plasticity, and adult neurogenesis in addition to regulating modifiers of other key epigenetic processes [233,234]. A number of studies have implicated altered miRNA expression in a majority of neuropsychiatric disorders in preclinical models as well as in clinical populations [235]. Not only are the miRNAs, but the proteins involved in the miRNA processing and production are also linked to psychiatric disorders. For example, DICER1, which helps miRNA processing, is affected in schizophrenia, while DGCR8, another protein which is involved in miRNA processing is deleted in DiGeorge syndrome, a rare neurodevelopmental disorder [236]. In a study by Im et al. [224] an interesting phenomena of regulation of *miRNA-212* via MeCP2 mediated DNA methylation was observed which resulted in upregulation of *Bdnf* in the dorsal striatum (Fig. 18.6). This mechanism was possibly involved in cocaine-induced alterations in synaptic plasticity and cocaine addictive behaviors.

Ethanol consumption modulates expression of miRNAs that can regulate alcohol-induced organ dysfunctions or neurological complications [237–240]. MicroRNAs have been implicated in alcohol

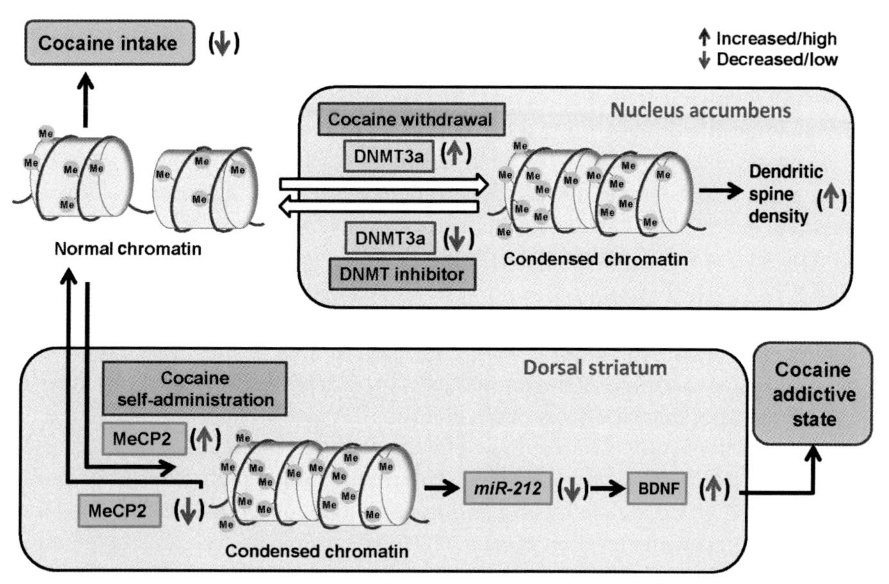

FIGURE 18.6 A Schematic Representation of DNA Methylation Mechanisms in the Nucleus Accumbens (NAc) and Dorsal Striatum (DS) of the Rodent Animal Models, Involved in Cocaine-Seeking Behaviors

Withdrawal from prolonged exposure to cocaine increased DNA methyltransferase (DNMT) 3a expression, which may be lead to condensed chromatin by virtue of hypermethylation and increase of the dendritic spine density (DSD) in neurons of the nucleus accumbens, helping to maintain the cocaine addictive state. Nucleus accumbens-specific inhibition of DNA methylation by DNMT inhibitors normalized these effects. On the other hand, similar condensation of chromatin by DNA hypermethylation was speculated to be triggered by increased methyl CpG-binding protein 2 (MeCP2) expression in the dorsal striatum after chronic cocaine self-administration. Methyl CpG-binding protein 2-induced condensed chromatin might upregulate BDNF levels by downregulating the expression of microRNA-212 (*miR-212*), which is known to target brain derived neurotrophic factor (BDNF) mRNA and possibly maintain cocaine addiction. Methyl CpG-binding protein 2 knockdown in the dorsal striatum reversed the cascade of these events thereby normalizing cocaine-seeking behaviors in rats. These mechanisms highlight the importance of different DNA methylation mechanisms in drug addiction and possibly might also be involved in alcohol use disorders similar to cocaine-seeking behaviors.

addiction, toxicity, and teratology. Two classes of miRNAs are known, one encoded by a long intergenic noncoding RNA (lincRNA) transcript and another encoded by intronic regions of the protein-coding parent gene locus. Some of the miRNAs are present in clusters at particular loci. Ethanol exposure induces teratogenesis in fetal neuronal stem cells via suppression of four miRNAs, *miR-9*, *miR-21*, *miR-153*, and *miR-335*, whereas suppression of the individual miRNAs like *mir-21* resulted in apoptosis and suppression of *miR-335* increased NSC proliferation [241]. This study in concert with other findings suggests that the miRNA profiles of development and ethanol exposure have a similar pattern [242–246]. Altered microRNA expression was also reported in acute ethanol tolerance, intermittent ethanol exposure, chronic exposure in humans, and ethanol-induced neurotoxicity [36,247–251]. MicroRNAs are known to regulate complex interactions in a cooperative and combinational fashion. In a rodent model study of alcohol dependence and relapse drinking, significantly altered levels of microRNAs were observed in the cortex and mid brain of chronic intermittent ethanol-exposed dependent mice compared with their nondependent controls. Several modules of coexpressed microRNAs were also observed. In an integrative analysis, different miRNA and protein clusters were identified at all the conditional transitions, suggesting a behavioral transition from alcohol consumption to dependence [252]. Downregulation of *mir-140*, which targets dynamin-1, in the cortex and upregulation of the same in midbrain, suggest midbrain specific adaptations [253]. Predictive analysis by Vlachos et al. [254], also reported *mir-140-3p* with the alcohol-associated pathway. Similarly, EtOH-sensitive (*miR-9*, *miR-21*, *miR-153*, and *miR-335*) miRNAs levels along with *mir-140* were also attenuated in neuronal stem cells with ethanol exposure [255]. A specific set of miRNAs with their cogent targets may regulate behavioral outcome of alcoholism. *Mir-9* is critical for neuronal development [256], also high levels of *mir-9* were maintained in the adult brain as well [257]. Suppression of *mir-9* was observed with ethanol exposure using both in vivo as well as in vitro models [241,255]. This suggested that *mir-9* can be used as a biomarker for developmental effects of ethanol [255]. Apart from development, *mir-9* has tumor suppressor function in nonneuronal tissue. In several nonneuronal carcinomas, hypermethylation of the *mir-9* locus was associated with decreased *mir-9* levels. In the medial prefrontal cortex (mPFC), persistently elevated *mir-206* was correlated with a history of alcohol dependence. The 3′ UTR of *Bdnf* has three *mir-206* target sites, with elevation of *mir-206*, *Bdnf* expression decreases in the mPFC, thereby enforcing ethanol self-administration through decreased synaptic plasticity. These effects were also observed with viral mediated overexpression of *mir-206*, independent of ethanol, confirming association of *mir-206* with alcohol-induced dependence [258].

A recent study by Pandey and coworkers [36] investigated the role of microRNA's underlying the mechanisms of ethanol-induced anxiolysis specifically in the amygdala. With acute ethanol exposure, decreased levels of *miR-494* were associated with increased expression of CBP/p300-interacting transactivator 2 (CITED) and CBP as they have *miR-494* binding sites, the effects of which further potentiated level of histone acetylation in the amygdaloid circuitry. As discussed previously, acute ethanol exposure in the same animal model is known to increase NPY levels in correlation with increased histone acetylation [35,60]. The phenomena of *miRNA-494* regulation during acute exposure can be found intertwined with histone acetylation-induced changes in the amygdala leading to the anxiolytic effects of ethanol. The dynamic changes in histone acetylation in the amygdala within 1 h of alcohol exposure might be via *miR-494* regulation of CBP levels, in addition to the inhibition of HDAC activity. This notion gained support due to anxiolytic action of amygdaloid infusion of an antagomir of *miR-494*, which was associated with increase in CITED, CBP and acetylated H3-K9 levels in amygdala.

MicroRNAs control gene expression not only by directly binding to the cognate mRNA, but also by regulation of miRNA expression and its interaction with other epigenetic modifiers could be critical for downstream gene expression. Elevated levels of 5-hmc mediated by TET1/TET3 in the *miR-365-3p* promoter in the spinal cord increases *mir-365* levels, that in turn inhibit potassium channel, *Kcnh2* gene expression, further modulating the nociceptive pathway involving in pain process [259]. *MicroRNA-140-3p* regulates target protein levels either directly or indirectly through epigenetic modifiers, such as Sirt1, which is a class III HDAC [260]. A study conducted on peripubertal mice, supported this hypothesis with elevated expression of *Sirt1* in the dorsal hippocampus upon chronic alcohol exposure [261].

One intriguing observation is that the miRNA regulation of gene function seems tightly entangled with other epigenetic mechanisms, such as histone acetylation and DNA methylation in neuropathologies due to drug addiction. Therefore, the epigenetic drugs that maintain histone acetylation or DNA methylation may also offer a remedial strategy wherein miRNA might be involved in causing dysregulation of gene functions.

18.5 CONCLUSIONS

Alcohol use disorder is a chronic condition, which is multifactorial and often progresses through positive reinforcing steps due to loss in control over consumption. The physiological and psychological consequences of the AUD is a result of complex interactions of an individual's genetic makeup and environmental interactions which includes societal factors, exposures to stress (prenatal and postnatal) and also to other drugs of abuse [8,262,263]. We here attempted to summarize the accumulating evidence implicating regulatory potential of myriad epigenetic mechanisms including DNA methylation, histone posttranscriptional modifications, and miRNAs each providing diverse levels of gene expression culminating in neuroadaptations to alcohol abuse. We have also provided the relevant findings from other neuropsychiatric disorders including addiction to other drugs of abuse, such as cocaine.

The observations discussed previously clearly emphasize that the various epigenetic mechanisms are highly complex and intertwined. For example, DNA methylation or histone acetylation can regulate the expression of different miRNAs and therefore the correct cataloguing of the vast network of regulatory events in different neurocircuits are needed to precisely match the underlying cellular and psychological outcomes. In these endeavors, one of the most important caveats is the use of different alcohol exposure paradigms resulting in a multitude of results, some even conflicting at times. It further adds an undue burden in understanding the already enormous complexities of epigenetic regulation. Although some remarkable progress has been made in certain areas of alcoholism, we still lack the in-depth understanding as to how the recruitment of different epigenetic factors and enzymes occurs to appropriately modulate the transcriptional machinery at a specific gene locus.

Methodological advances may provide a great help in this direction. We have observed that our current understanding is based on the manipulation of epigenetic factors by drugs resulting in a variety of transcriptional outcomes. For example, trichostatin A (TSA) or sodium butyrate are nonspecific HADC inhibitors; whereas our understanding about the CNS disorders by now implicate specific HDAC isoforms in different neurological outcomes. Therefore, the use of more specific drugs, such as inhibitors against specific HDAC and DNMT isoforms could provide better insights. Most of the time, these drugs have been administered peripherally with an unknown amount of drug reaching to the brain. Thus, these drug manipulation strategies, although helpful, are the product of a series of assumptions. Being able to manipulate the specific epigenetic drug to the particular cell type (neuron, astrocyte, microglia) in discrete brain regions would provide a great help in advancement of the field. The rationale for this stems from the fact that genome-wide studies have shown distinctive patterns of epigenetic changes in neuronal and nonneuronal cell types within a unique brain region [264].

Current advances in genome-wide studies in the field of alcoholism do not match to the progress made in other fields, such as cancer to develop the networks between signaling mechanisms and epigenetic factors to further downstream gene expression outcomes. These studies are limited to genome-wide DNA sequencing with specific epigenetic marks coupled with transcriptomics. However, additional bioinformatic tools such as proteomics, metabolomics, and protein-docking analysis are underutilized in revealing the complex epigenetic interactions in alcoholism. For example, the data obtained from preclinical models needs to be tested with the observations from postmortem brain studies from other psychiatric disorders to test possible mechanistic links.

ABBREVIATIONS

5-AzaC	5-Azacytidine
Arc	Activity-regulated cytoskeleton-associated protein
5-hmc	5-Hydroxy methylcytosine
5-mc	5-Methylcytosine
AIE	Aldolescent intermittent ethanol
AMP	Adenosine monophosphate
AUD	Alcohol use disorders
BDNF	Brain-derived neurotrophic factor
BER	Base excision repair
BNST	Bed nucleus of stria terminalis
CNS	Central nervous system
CBP	CREB-binding protein
CIE	Chronic intermittent ethanol
CITED2	CBP/p300-interacting transactivator 2
CpG	Cytosine and guanosine dinucleotides
CREB	cAMP response element binding protein
DARPP-32	Dopamine and cyclic AMP regulated phosphoprotein 32
DAT	Dopamine transporter
DNMT	DNA methyltransferase
FAE	Fetal alcohol exposure
FASD	Fetal alcohol spectrum disorders
GLP	G9a-like protein
GNAT	Gcn5-associated N-acetyltransferase
GWAS	Genome-wide association study
HAT	Histone acetyltransferase
HDAC	Histone deacetylase
HDM	Histone demethylase
HERP	Homocysteine-induced endoplasmic reticulum protein
HPA	Hypothalamic-pituitary-adrenal axis
KDM	Lysine-specific demethylase
KMT	Lysine methyltransferase
LincRNA	Long intergenic noncoding RNA
MBD	Methyl-CpG-binding domain
MeCP2	Methyl-CpG-binding protein 2
NAc	Nucleus accumbens
NMDA	*N*-methyl-D-aspartate
NP rat	Alcohol nonpreferring rat
NOC	Nociceptine
NOP	Nociceptine Opioid receptor
Nor-BNI	Norbinaltorphimine
NPY	Neuropeptide Y
NR2B	NMDA receptor 2b subunit
NSC	Neuronal stem cell
P rat	Alcohol preferring rat
PDNY	Prodynorphin
PFC	Prefrontal cortex

PND	Postnatal day
PNOC	Pronociceptin
POMC	Proopiomelanocortin
PPM1G	3′-protein phosphatase-1G
PTM	Posttranslational Modification
SAHA	Suberoyl hydroxamic acid
SAM	S-adenosylmethionine
SD	Sprague-Dawley
Setdb1	Set domain bifurcated 1
SIRT	Sirtuin
SNCA	α-Synuclein gene
SNP	Single nucleotide polymorphism
SYT2	Synaptotagmin 2
TET	Ten–eleven translocation methylcytosine dioxygenase
TSA	Trichostatin A
UTR	Untranslated region
VGLUT2	Vesicular glutamate transporter 2

ACKNOWLEDGMENTS

The preparation of the manuscript is supported by the Department of Biotechnology, Savitribai Phule Pune University, Pune, India (SS and AS) and University Grant Commission, Government of India, New Delhi, India through UGC-FRP program to AS.

REFERENCES

[1] Collins PY, Patel V, Joestl SS, March D, Insel TR, Daar AS, et al. Grand challenges in global mental health. Nature 2011;475:27–30.
[2] Koob GF, Le Moal M. Drug abuse: hedonic homeostatic dysregulation. Science 1997;278:52–8.
[3] Koob GF. Alcoholism: allostasis and beyond. Alcohol Clin Exp Res 2003;27:232–43.
[4] Robinson J, Sareen J, Cox BJ, Bolton JM. Role of self-medication in the development of comorbid anxiety and substance use disorders: a longitudinal investigation. Arch Gen Psychiatry 2011;68:800–7.
[5] Pandey SC. Anxiety and alcohol abuse disorders: a common role for CREB and its target, the neuropeptide Y gene. Trends Pharmacol Sci 2003;24:456–60.
[6] Schuckit MA, Hesselbrock V. Alcohol dependence and anxiety disorders: what is the relationship? Am J Psychiatry 1994;151:1723–34.
[7] Grant BF, Stinson FS. Prevalence and co-occurrence of substance use disorders and independent mood and anxiety disorders: results from the national epidemiologic survey on alcohol and related conditions. Arch Gen Psychiatry 2004;61:807–16.
[8] Krishnan HR, Sakharkar AJ, Teppen TL, Berkel TD, Pandey SC. The epigenetic landscape of alcoholism. Int Rev Neurobiol 2014;115:75–116.
[9] Starkman BG, Sakharkar AJ, Pandey SC. Epigenetics-beyond the genome in alcoholism. Alcohol Res 2012;34:293–305.
[10] Koob GF, Le Moal M. Drug addiction, dysregulation of reward, and allostasis. Neuropsychopharmacology 2001;24:97–129.

[11] Sutherland JE, Costa MAX. Epigenetics and the environment. Ann NY Acad Sci 1992;983:151–60. 2003.
[12] Robison AJ, Nestler EJ. Transcriptional and epigenetic mechanisms of addiction. Nat Rev Neurosci 2011;12:623–37.
[13] Maze I, Nestler EJ. The epigenetic landscape of addiction. Ann NY Acad Sci 2011;1216:99–113.
[14] Hewish DR, Burgoyne LA. Chromatin sub-structure. The digestion of chromatin DNA at regularly spaced sites by a nuclear deoxyribonuclease. Biochem Biophy Res Commun 1973;52:504–10.
[15] Jenuwein T, Allis CD. Translating the histone code. Science 2001;293:1074–80.
[16] Smith MM. Histone structure and function. Curr Opin Cell Biol 1991;3:429–37.
[17] Luger K, Mäder AW, Richmond RK, Sargent DF, Richmond TJ. Crystal structure of the nucleosome core particle at 2.8 Å resolution. Nature 1997;389:251–60.
[18] Hirschhorn JN, Brown SA, Clark CD, Winston F. Evidence that SNF2/SWI2 and SNF5 activate transcription in yeast by altering chromatin structure. Genes Dev 1992;6:2288–98.
[19] Kwon H, Imbalzano AN, Khavari PA, Kingston RE, Green MR. Nucleosome disruption and enhancement of activator binding by a human SW1/SNF complex. Nature 1994;370:477–81.
[20] Brownell JE, Zhou J, Ranalli T, Kobayashi R, Edmondson DG, Roth SY, et al. Tetrahymena histone acetyltransferase A: a homolog to yeast Gcn5p linking histone acetylation to gene activation. Cell 1996;84: 843–51.
[21] Nunez YO, Mayfield RD. Understanding alcoholism through microRNA signatures in brains of human alcoholics. Front Genet 2012;3:43.
[22] Miranda RC. MicroRNAs and ethanol toxicity. Int Rev Neurobiol 2014;115:245–84.
[23] Harr JC, Gonzalez-Sandoval A, Gasser SM. Histones and histone modifications in perinuclear chromatin anchoring: from yeast to man. EMBO Rep 2016;17:139–55.
[24] Lachner M, O'Sullivan RJ, Jenuwein T. An epigenetic road map for histone lysine methylation. J Cell Sci 2003;116:2117–24.
[25] Allis CD, Berger SL, Cote J, Dent S, Jenuwien T, Kouzarides T, et al. New nomenclature for chromatin-modifying enzymes. Cell 2007;131:633–6.
[26] Masumoto H, Hawke D, Kobayashi R, Verreault A. A role for cell-cycle regulated histone H3 lysine 56 acetylation in the DNA damage response. Nature 2005;436:294–8.
[27] Wang Z, Zang C, Cui K, Schones DE, Barski A, Peng W, et al. Genome wide mapping of HATs and HDACs reveals distinct functions in active and inactive genes. Cell 2009;138:1019–31.
[28] Sterner DE, Berger SL. Acetylation of histones and transcription-related factors. Microbiol Mol Biol Rev 2000;64:435–59.
[29] Chrivia JC, Kwok RPS, Lamb N, Hagiwara M, Montminy MR, Goodman RH. Phosphorylated CREB binds specifically to the nuclear protein CBP. Nature 1993;365:855–9.
[30] Arany Z, Newsome D, Oldread E, Livingston DM, Eckner R. A family of transcriptional adaptor proteins targeted by the E1A oncoprotein. Nature 1995;374:81–4.
[31] Parker D, Ferreri K, Nakajima T, LaMorte VJ, Evans R, Koerber SC, et al. Phosphorylation of CREB at Ser-133 induces complex formation with CREB-binding protein via a direct mechanism. Mol Cell Biol 1996;16:694–703.
[32] Mayr B, Montminy M. Transcriptional regulation by the phosphorylation dependent factor CREB. Nat Rev Mol Cell Biol 2001;2:599–609.
[33] Sakharkar AJ, Vetreno RP, Zhang H, Kokare DM, Crews FT, Pandey SC. A role for histone acetylation mechanisms in adolescent alcohol exposure-induced deficits in hippocampal brain-derived neurotrophic factor expression and neurogenesis markers in adulthood. Brain Struct Funct 2016;.
[34] Moonat S, Starkman BG, Sakharkar A, Pandey SC. Neuroscience of alcoholism: molecular and cellular mechanisms. Cell Mol Life Sci 2010;67:73–88.
[35] Pandey SC, Ugale R, Zhang H, Tang L, Prakash A. Brain chromatin remodeling: a novel mechanism of alcoholism. J Neurosci 2008;28:3729–37.

[36] Teppen TL, Krishnan HR, Zhang H, Sakharkar AJ, Pandey SC. The potential role of amygdaloid microRNA-494 in alcohol-induced anxiolysis. Biol Psychiatry 2015;.
[37] Dokmanovic M, Clarke C, Marks PA. Histone deacetylase inhibitors: overview and perspectives. Mol Cancer Res 2007;5:981–9.
[38] Blander G, Guarente L. The Sir2 family of protein deacetylases. Annu Rev Biochem 2004;73:417–35.
[39] Haberland M, Montgomery RL, Olson EN. The many roles of histone deacetylases in development and physiology: Implications for disease and therapy. Nat Rev Genet 2009;10:32–42.
[40] Guan JS, Haggarty SJ, Giacometti E, Dannenberg JH, Joseph N, Gao J, et al. HDAC2 negatively regulates memory formation and synaptic plasticity. Nature 2009;459:55–60.
[41] Fitzsimons HL, Scott MJ. Genetic modulation of Rpd3 expression impairs long-term courtship memory in Drosophila. PLoS One 2011;6:e29171.
[42] McQuown SC, Barrett RM, Matheos DP, Post RJ, Rogge GA, Alenghat T. HDAC3 is a critical negative regulator of long-term memory formation. J Neurosci 2011;31:764–74.
[43] Wang WH, Cheng LC, Pan FY, Xue B, Wang DY, Chen Z, et al. Intracellular trafficking of histone deacetylase 4 regulates long-term memory formation. Anat Rec (Hoboken) 2011;294:1025–34.
[44] Kim MS, Akhtar MW, Adachi M, Mahgoub M, Bassel-Duby R, Kavalali ET, et al. An essential role for histone deacetylase 4 in synaptic plasticity and memory formation. J Neurosci 2012;32:10879–86. 2012.
[45] Fitzsimons HL, Schwartz S, Given FM, Scott MJ. The histone deacetylases HDAC4 regulates long-term memory in Drosophila. PLoS One 2013;8:e83903.
[46] Graff J, Rei D, Guan JS, Wang WY, Seo J, Hennig KM, et al. An epigenetic blockade of cognitive functions in the neurodegenerating brain. Nature 2012;483:222–6.
[47] Graff J, Joseph NF, Horn ME, Samiei A, Meng J, Seo J, et al. Epigenetic priming of memory updating during reconsolidation to attenuate remote fear memories. Cell 2014;156(1–2):261–76.
[48] Cohen HY, Miller C, Bitterman KJ, Wall NR, Hekking B, Kessler B, et al. Calorie restriction promotes mammalian cell survival by inducing the SIRT1deacetylase. Science 2004;305(5682):390–2.
[49] Hirayama J, Sahar S, Grimaldi B, Tamaru T, Takamatsu K, Nakahata Y, et al. CLOCK-mediated acetylation of BMAL1 controls circadian function. Nature 2007;450:1086–90.
[50] Kim D, Nguyen MD, Dobbin MM, Fischer A, Sananbenesi F, Rodgers JT, et al. SIRT1 deacetylase protects against neurodegeneration in models for Alzheimer's disease and amyotrophic lateral sclerosis. EMBO J 2007;26:3169–79.
[51] Nakahata Y, Kaluzova M, Grimaldi B, Sahar S, Hirayama J, Chen D, et al. The NAD+−dependent deacetylase SIRT1 modulates CLOCK-mediated chromatin remodeling and circadian control. Cell 2008;134:329–40.
[52] Gao J, Wang WY, Mao YW, Graff J, Guan JS, Pan L, et al. A novel pathway regulates memory and plasticity via SIRT1 and miR-134. Nature 2010;466:1105–9.
[53] Kazantsev AG, Thompson LM. Therapeutic application of histone deacetylase inhibitors for central nervous system disorders. Nat Rev Drug Discov 2008;7:854–68.
[54] Richon VM. Cancer biology: Mechanism of antitumour action of vorinostat (suberoylanilidehydroxamic acid), a novel histone deacetylase inhibitor. Br J Cancer 2006;95:S2–6.
[55] Pandey SC, Roy A, Zhang H. The decreased phosphorylation of cyclic adenosinemonophosphate (cAMP) response element binding (CREB) protein in the central amygdala acts as a molecular substrate for anxiety related to ethanol withdrawal in rats. Alcohol Clin Exp Res 2003;27:396–409.
[56] Levine AA, Guan Z, Barco A, Xu S, Kandel ER, Schwartz JH. CREB binding protein controls response to cocaine by acetylating histones at the fosB promoter in the mouse striatum. Proc Natl Acad Sci USA 2005;102:19186–91.
[57] Pandey SC. The gene transcription factor cyclic AMP-responsive element binding protein: role in positive and negative affective states of alcohol addiction. Pharmacol Ther 2004;104:47–58.
[58] Pandey SC, Zhang H, Roy A, Xu T. Deficits in amygdaloid cAMP responsive element-binding protein signaling play a role in genetic predisposition to anxiety and alcoholism. J Clin Investig 2005;115:2762–73.

[59] Guo JU, Su Y, Zhong C, Ming GL, Song H. Hydroxylation of 5-methylcytosine by TET1 promotes active DNA demethylation in the adult brain. Cell 2011;145:423–34.
[60] Sakharkar AJ, Zhang H, Tang L, Shi G, Pandey SC. Histone deacetylases (HDAC)-induced histone modifications in the amygdala: A role in rapid tolerance to the anxiolytic effects of ethanol. Alcohol Clin Exp Res 2012;36:61–71.
[61] Pandey SC, Zhang H, Ugale R, Prakash A, Xu T, Misra K. Effector immediate-early gene Arc in the amygdala plays a critical role in alcoholism. J Neurosci 2008;28:2589–600.
[62] You C, Zhang H, Sakharkar AJ, Teppen T, Pandey SC. Reversal of deficits in dendritic spines BDNF and Arc expression in the amygdala during alcohol dependence by HDAC inhibitor treatment. Int J Neuropsychopharmacol 2014;17:313–22.
[63] Zhang H, Sakharkar AJ, Shi G, Ugale R, Prakash A, Pandey SC. Neuropeptide Y signaling in the central nucleus of amygdala regulates alcohol-drinking and anxiety-like behaviors of alcohol-preferring rats. Alcohol Clin Exp Res 2010;34:451–61.
[64] Moonat S, Sakharkar AJ, Zhang H, Tang L, Pandey SC. Aberrant histone deacetylase 2-mediated histone modifications and synaptic plasticity in the amygdala predisposes to anxiety and alcoholism. Biol Psychiatry 2013;73:763–73.
[65] Sakharkar AJ, Zhang H, Tang L, Baxstrom K, Shi G, Moonat S, et al. Effects of histone deacetylase inhibitors on amygdaloid histone acetylation and neuropeptide Y expression: a role in anxiety-like and alcohol-drinking behaviours. Int J Neuropsychopharmacol 2014;17(8):1207–20.
[66] Prakash A, Zhang H, Pandey SC. Innate differences in the expression of brain-derived neurotrophic factor in the regions within the extended amygdala between alcohol preferring and nonpreferring rats. Alcohol Clin Exp Res 2008;32:909–20.
[67] Pandey SC, Zhang H, Roy A, Misra K. Central and medial amygdaloid brain-derived neurotrophic factor signaling plays a critical role in alcohol-drinking and anxiety-like behaviors. J Neurosci 2006;26:8320–31.
[68] Courtney KE, Polich J. Binge drinking in young adults: data, definitions, and determinants. Psychol Bull 2009;135:142–56.
[69] Szyf M, McGowan P, Meaney MJ. The social environment and the epigenome. Environ Mol Mutagen 2008;49:46–60.
[70] DeWit DJ, Adlaf EM, Offord DR, Ogborne AC. Age at first alcohol use: a risk factor for the development of alcohol disorders. Am J Psychiatry 2000;157:745–50.
[71] Sakharkar AJ, Tang L, Zhang H, Chen Y, Grayson DR, Pandey SC. Effects of acute ethanol exposure on anxiety measures and epigenetic modifiers in the extended amygdala of adolescent rats. Int J Neuropsychopharmacol 2014;17:2057–67.
[72] Pandey SC, Sakharkar AJ, Tang L, Zhang H. Potential role of adolescent alcohol exposure-induced amygdaloid histone modifications in anxiety and alcohol intake during adulthood. Neurobio Dis 2015;82:607–19.
[73] Varlinskaya EI, Spear LP. Acute effects of ethanol on social behavior of adolescent and adult rats: role of familiarity of the test situation. Alcohol Clin Exp Res 2002;26(10):1502–11.
[74] Pascual M, Do Couto BR, Alfonso-Loeches S, Aguilar MA, Rodriguez-Arias M, Guerri C. Changes in histone acetylation in the prefrontal cortex of ethanol exposed adolescent rats are associated with ethanol-induced place conditioning. Neuropharmacology 2012;62:2309–19.
[75] Grant BF, Dawson DA. Age at onset of alcohol use and its association with DSM-IV alcohol abuse and dependence: results from the National Longitudinal Alcohol Epidemiologic Survey. J Subst Abuse 1997;9:103–10.
[76] Winward JL, Hanson KL, Tapert SF, Brown SA. Heavy alcohol use, marijuana use, and concomitant use by adolescents are associated with unique and shared cognitive decrements. J Int Neuropsychol Soc 2014;20:784–95.
[77] Crews FT, Braun CJ, Hoplight B, Switzer RC, Knapp DJ. Binge ethanol consumption causes differential brain damage in young adolescent rats compared with adult rats. Alcohol Clin Exp Res 2000;24:1712–23.

[78] Crews FT, Mdzinarishvili A, Kim D, He J, Nixon K. Neurogenesis in adolescent brain is potently inhibited by ethanol. Neuroscience 2006;137(2):437–45.
[79] Guerri C, Pascual M. Mechanisms involved in the neurotoxic, cognitive, and neurobehavioral effects of alcohol consumption during adolescence. Alcohol 2010;44(1):15–26.
[80] Guyer AE, Monk CS, McClure-Tone EB, Nelson EE, Roberson-Nay R, Adler AD, et al. A developmental examination of amygdala response to facial expressions. J Cogn Neurosci 2008;20:1565–82.
[81] Walker BM, Koob GF. Pharmacological evidence for a motivational role of kappa-opioid systems in ethanol dependence. Neuropsychopharmacology 2008;33:643–52.
[82] Kuzmin A, Kreek MJ, Bakalkin G, Liljequist S. Thenociceptin/orphanin FQ receptor agonist Ro 64-6198 reduces alcohol self-administration and prevents relapse-like alcohol drinking. Neuropsychopharmacology 2007;32:902–10.
[83] D'Addario C, Caputi FF, Ekstrom TJ, Di Benedetto M, Maccarrone M, Romualdi P, et al. Ethanol induces epigenetic modulation of prodynorphin and pronociceptin gene expression in the rat amygdala complex. J Mol Neurosci 2013;49(2):312–9.
[84] Drew PD, Kane CJ. Fetal alcohol spectrum disorders and neuroimmune changes. Int Rev Neurobiol 2014;118:41–80.
[85] Dörrie N, Föcker M, Freunscht I, Hebebrand J. Fetal alcohol spectrum disorders. Eur Child Adolesc Psychiatry 2014;28:863–75.
[86] Streissguth AP, Bookstein FL, Barr HM, Sampson PD, O'Malley K, Young JK. Risk factors for adverse life outcomes in fetal alcohol syndrome and fetal alcohol effects. J Dev Behav Pediatr 2004;25:228–38.
[87] Haycock PC, Ramsay M. Exposure of mouse embryos to ethanol during preimplantation development: Effect on DNA methylation in the h19 imprinting control region. Biol Reprod 2009;81:618–27.
[88] Liu Y, Balaraman Y, Wang G, Nephew KP, Zhou FC. Alcohol exposurealters DNA methylation profiles in mouse embryos at early neurulation. Epigenetics 2009;4:500–11.
[89] Guo W, Crossey EL, Zhang L, Zucca S, George OL, Valenzuela CF, et al. Alcohol exposure decreases CREB binding protein expression and histone acetylation in the developing cerebellum. PLoS One 2011;6:e19351.
[90] Subbanna S, Nagre NN, Umapathy NS, Pace BS, Basavarajappa BS. Ethanol exposure induces neonatal neurodegeneration by enhancing CB1R Exon1 histone H4K8 acetylation and upregulating CB1R function causing neurobehavioral abnormalities in adult mice. Int J Neuropsychopharmacol 2015;18:pyu028.
[91] Renthal W, Nestler EJ. Chromatin regulation in drug addiction and depression. Dialogues Clin Neurosci 2009;11:257–68.
[92] Nestler E. Epigenetic mechanisms of drug addiction. Neuropharmacology 2014;76:259–68.
[93] Kumar A, Choi KH, Renthal W, Tsankova NM, Theobald DE, Truong HT, et al. Chromatin remodeling is a key mechanism underlying cocaine-induced plasticity in striatum. Neuron 2005;48:303–14.
[94] Warnault V, Darcq E, Levine A, Barak S, Ron D. Chromatin remodelling—a novel strategy to control excessive alcohol drinking. Transl Psychiatry 2013;19:e231.
[95] Arora DS, Nimitvilai S, Teppen TL, McElvain MA, Sakharkar AJ, You C, et al. Hyposensitivity to gamma-aminobutyric acid in the ventral tegmental area during alcohol withdrawal: reversal by histone deacetylase inhibitors. Neuropsychopharmacology 2013;38:1674–84.
[96] Qiang M, Li JG, Denny AD, Yao JM, Lieu M, Zhang K, et al. Epigenetic mechanisms are involved in the regulation of ethanol consumption in mice. Int J Neuropsychopharmacol 2014;18.
[97] Li T-K, Lumeng L, Doolittle DP. Selective breeding for alcohol preference and associated responses. Behav Genet 1993;23:163–70.
[98] Murphy JM, Stewart RB, Bell RL, Badia-Elder NE, Carr LG, McBride WJ, et al. Phenotypic and genotypic characterization of the Indiana University rat lines selectively bred for high and low alcohol preference. Behav Genet 2002;32:363–88.
[99] Botia B, Legastelois R, Alaux-Cantin S, Naassila M. Expression of ethanol induced behavioral sensitization is associated with alteration of chromatin remodeling in mice. PLoS One 2012;7:e47527.

[100] Legastelois R, Botia B, Naassila M. Blockade of ethanol-induced behavioral sensitization by sodium butyrate: Descriptive analysis of gene regulations in the striatum. Alcohol Clin Exp Res 2013;37:1143–53.
[101] Wang Y, Krishnan HR, Ghezzi A, Yin JCP, Atkinson NS. Drug induced epigenetic changes produce drug tolerance. PLoS Biol 2007;5:e265.
[102] Kim JS, Shukla SD. Acute in vivo effect of ethanol (binge drinking) on histone H3 modifications in rat tissues. Alcohol Alcohol 2006;41:126–32.
[103] Park PH, Miller R, Shukla SD. Acetylation of histone H3 at lysine 9 by ethanol in rat hepatocytes. Biochem Biophy Res Comm 2003;306:501–4.
[104] Park PH, Lim RW, Shukla SD. Gene-selective histone H3 acetylation in the absence of increase in global histone acetylation in liver of rats chronically fed alcohol. Alcohol Alcohol 2012;47:233–9.
[105] Zhong L, Zhu J, Lv T, Chen G, Sun H, Yang X, Huang X, et al. Ethanol and its metabolites induce histone lysine 9 acetylation and an alteration of the expression of heart development-related genes in cardiac progenitor cells. Cardiovasc Toxicol 2010;10:268–74.
[106] Noma K, Allis CD, Grewal SI. Transitions in distinct histone H3 methylation patterns at the heterochromatin domain boundaries. Science 2001;293:1150–5.
[107] Schneider R, Bannister AJ, Myers FA, Thorne AW, Crane-Robinson C, Kouzarides T. Histone H3 lysine 4 methylation patterns in higher eukaryotic genes. Nat Cell Biol 2004;6:73–7.
[108] Barski A, Cuddapah S, Cui K, Roh TY, Schones DE, Wang Z, et al. High-resolution profiling of histone methylations in the human genome. Cell 2007;129:823–37.
[109] Bernstein BE, Kamal M, Lindblad-Toh K, Bekiranov S, Bailey DK, Huebert DJ, et al. Genomic maps and comparative analysis of histone modifications in human and mouse. Cell 2005;120:169–81.
[110] Tachibana M, Sugimoto K, Nozaki M, Ueda J, Ohta T, Ohki M, et al. G9a histone methyltransferase plays a dominant role in euchromatic histone H3 lysine 9 methylation and is essential for early embryogenesis. Genes Dev 2002;16:1779–91.
[111] Ogawa H, Ishiguro KI, Gaubatz S, Livingston DM, Nakatani Y. A complex with chromatin modifiers that occupies E2F- and Myc responsive genes in G0 cells. Science 2002;296:1132–6.
[112] Jiang Y, Langley B, Lubin FD, Renthal W, Wood MA, Yasui DH, et al. Epigenetics in the nervous system. J Neurosci 2008;28:11753–9.
[113] Ryu H, Lee J, Hagerty SW, Soh BY, McAlpin SE, Cormier KA, et al. ESET/SETDB1 gene expression and histone H3 (K9) trimethylation in Huntington's disease. Pro Natl Acad Sci USA 2006;103:19176–81.
[114] Al-Mahdawi S, Pinto RM, Ismail O, Varshney D, Lymperi S, Sandi C, et al. The Friedreich ataxia GAA repeat expansion mutation induces comparable epigenetic changes in human and transgenic mouse brain and heart tissues. Hum Mol Genet 2008;17:735–46.
[115] Schaefer A, Sampath SC, Intrator A, Min A, Gertler TS, Surmeier DJ, et al. Control of cognition and adaptive behavior by the GLP/G9a epigenetic suppressor complex. Neuron 2009;64:678–91.
[116] Gupta S, Kim SY, Artis S, Molfese DL, Schumacher A, Sweatt JD, et al. Histone methylation regulates memory formation. J Neurosci 2010;30:3589–99.
[117] Maze I, Covington HE, Dietz DM, LaPlant Q, Renthal W, Russo SJ, et al. Essential role of the histone methyltransferase G9a in cocaine-induced plasticity. Science 2010;327:213–6.
[118] Shinkai Y, Tachibana M. H3K9 methyltransferase G9a and the related molecule GLP. Genes Dev 2011;25:781–8.
[119] Vermeulen M, Mulder KW, Denissov S, Pijnappel WP, van Schaik FM, Varier RA, et al. Selective anchoring of TFIID to nucleosomes by trimethylation of histone H3 lysine 4. Cell 2007;131:58–69.
[120] Zhang CR, Ho MF, Vega MCS, Burne TH, Chong S. Prenatal ethanol exposure alters adult hippocampal VGLUT2 expression with concomitant changes in promoter DNA methylation, H3K4 trimethylation and miR-467b-5p levels. Epigenetics Chromatin 2015;8:40.
[121] Subbanna S, Shivakumar M, Umapathy NS, Saito M, Mohan PS, Kumar A, et al. G9a-mediated histone methylation regulates ethanol-induced neurodegeneration in the neonatal mouse brain. Neurobiol Dis 2013;54:475–85.

[122] Nieuwenhuys R. The neocortex. An overview of its evolutionary development, structural organization and synaptology. Anat Embryol (Berl) 1994;190:307–37.
[123] Carpenter-Hyland EP, Woodward JJ, Chandler LJ. Chronic ethanol induces synaptic but not extrasynaptic targeting of NMDA receptors. J Neurosci 2004;24:7859–68.
[124] Follesa P, Ticku MK. NMDA receptor upregulation: molecular studies in cultured mouse cortical neurons after chronic antagonist exposure. J Neurosci 1996;16:2172–8.
[125] Chandrasekar R. Alcohol and NMDA receptor: current research and future direction. Front Mol Neurosci 2013;28:6–14.
[126] Qiang M, Denny A, Lieu M, Carreon S, Li J. Histone H3K9 modifications are a local chromatin event involved in ethanol- induced neuroadaptation of the NR2B gene. Epigenetics 2011;6:1095–104.
[127] Pal-Bhadra M, Bhadra U, Jackson DE, Mamatha L, Park PH, Shukla SD. Distinct methylation patterns in histone H3 at Lys-4 and Lys-9 correlate with up- and downregulation of genes by ethanol in hepatocytes. Life Sci 2007;81:979–87.
[128] Moonat S, Pandey SC. Stress, epigenetics, and alcoholism. Alcohol Res 2012;34:495–505.
[129] Xuei X, Flury-Wetherill L, Bierut L, Dick D, Nurnberger J Jr, Foroud T, et al. The opioid system in alcohol and drug dependence: family-based association study. Am J Med Genet B Neuropsychiatr Genet 2007;144:877–84.
[130] Hernandez-Avila CA, Oncken C, Van Kirk J, Wand G, Kranzler HR. Adrenocorticotropin and cortisol responses to a naloxone challenge and risk of alcoholism. Biol Psychiatry 2002;51:652–8.
[131] Govorko D, Bekdash RA, Zhang C, Sarkar DK. Male germline transmits fetal alcohol adverse effect on hypothalamic proopiomelanocortin gene across generations. Biol Psychiatry 2012;72:378–88.
[132] Sarkar DK, Kuhn P, Marano J, Chen C, Boyadjieva N. Alcohol exposure during the developmental period induces beta-endorphin neuronal death and causes alteration in the opioid control of stress axis function. Endocrinology 2007;148:2828–34.
[133] Bekdash RA, Zhang C, Sarkar DK. Gestational choline supplementation normalized fetal alcohol-induced alterations in histone modifications, DNA methylation, and proopiomelanocortin (POMC) gene expression in beta-endorphin-producing POMC neurons of the hypothalamus. Alcohol Clin Exp Res 2013;37:1133–42.
[134] Kouzarides T. Chromatin modifications and their function. Cell 2007;128:693–705.
[135] Wang KS, Liu X, Zhang Q, Wu LY, Zeng M. Genome-wide association study identifies 5q21 and 9p24.1 (KDM4C) loci associated with alcohol withdrawal symptoms. J Neural Trans (Vienna) 2012;119:425–33.
[136] Taqi MM, Bazov I, Watanabe H, Sheedy D, Harper C, Alkass K, et al. Prodynorphin CpG-SNPs associated with alcohol dependence: Elevated methylation in the brain of human alcoholics. Addict Biol 2011;16: 499–509.
[137] Gundogan F, Elwood G, Mark P, Feijoo A, Longato L, Tong M, et al. Ethanol-induced oxidative stress and mitochondrial dysfunction in rat placenta. Relevance to pregnancy loss. Alcohol Clin Exp Res 2010;34: 415–23.
[138] Yanaguita M, Gutierrez C, Ribeiro C, Lima G, Machado H, Peres L. Pregnancy outcome in ethanol-treated mice with folic acid supplementation in saccharose. Childs Nerv Syst 2008;24:99–104.
[139] Monk BR, Leslie FM, Thomas JD. The effects of perinatal choline supplementation on hippocampal cholinergic development in rats exposed to alcohol during the brain growth spurt. Hippocampus 2012;22:1750–7.
[140] Davison JM, Mellott TJ, Kovacheva VP, Blusztajn JK. Gestational choline supply regulates methylation of histone H3, expression of histone methyltransferases G9a (Kmt1c) and Suv39h1 (Kmt1a), and DNA methylation of their genes in rat fetal liver and brain. J Biol Chem 2009;284:1982–9.
[141] Barratt MJ, Hazzalin CA, Cano E, Mahadevan LC. Mitogen-stimulated phosphorylation of histone H3 is targeted to a small hyperacetylation-sensitive fraction. Proc Natl Acad Sci USA 1994;91:4781–5.
[142] Fischle W, Tseng BS, Dormann HL, Ueberheide BM, Garcia BA, Shabanowitz J, et al. Regulation of HP1–chromatin binding by histone H3 methylation and phosphorylation. Nature 2005;438:1116–22.
[143] Sawicka A, Seiser C. Histone H3 phosphorylation–a versatile chromatin modification for different occasions. Biochimie 2012;94:2193–201.

[144] Banerjee T, Chakravarti D. A peek into the complex realm of histone phosphorylation. Mol Cell Biol 2011;31:4858–73.
[145] Bannister AJ, Kouzarides T. Regulation of chromatin by histone modifications. Cell Res 2011;21:381–95.
[146] Sawicka A, Seiser C. Sensing core histone phosphorylation—a matter of perfect timing. Biochim Biophys Acta 2014;1839:711–8.
[147] Chwang WB, O'Riordan KJ, Levenson JM, Sweatt JD. ERK/MAPK regulates hippocampal histone phosphorylation following contextual fear conditioning. Learn Mem 2006;13:322–8.
[148] Shaywitz AJ, Greenberg ME. CREB: a stimulus-induced transcription factor activated by a diverse array of extracellular signals. Annu Rev Biochem 1999;68:821–61.
[149] Merienne K, Pannetier S, Harel-Bellan A, Sassone-Corsi P. Mitogen regulated RSK2-CBP interaction controls their kinase and acetylase activities. Mol Cell Biol 2001;21:7089–96.
[150] Lee YJ, Shukla SD. Histone H3 phosphorylation at serine 10 and serine 28 is mediated by p38 MAPK in rat hepatocytes exposed to ethanol and acetaldehyde. Eur J Pharmacol 2007;573:29–38.
[151] Stipanovich A, Valjent E, Matamales M, Nishi A, Ahn JH, Maroteaux M, et al. A phosphatase cascade by which rewarding stimuli control nucleosomal response. Nature 2008;453:879–84.
[152] McClain JA, Nixon K. Alcohol induces parallel changes in hippocampal histone H3 phosphorylation and c-Fos protein expression in male rats. Alcohol Clin Exp Res 2016;40:102–12.
[153] Cheung P, Allis CD, Sassone-Corsi P. Signaling to chromatin through histone modifications. Cell 2000;103:263–71.
[154] Miranda TB, Jones PA. DNA methylation: the nuts and bolts of repression. J Cell Physiol 2007;213:384–90.
[155] Lande-Diner L, Zhang J, Ben-Porath I, Amariglio N, Keshet I, Hecht M, et al. Role of DNA methylation in stable gene repression. J Biol Chem 2007;282:12194–200.
[156] Hore TA, Rapkins RW, Graves JA. Construction and evolution of imprinted loci in mammals. Trends Genet 2007;23:440–8.
[157] Yen ZC, Meyer IM, Karalic S, Brown CJ. A cross-species comparison of X-chromosome inactivation in Eutheria. Genomics 2007;90:453–63.
[158] Gronbaek K, Hother C, Jones PA. Epigenetic changes in cancer. APMIS 2007;115:1039–59.
[159] Boyes J, Bird A. DNA methylation inhibits transcription indirectly viaa methyl-CpG binding protein. Cell 1991;64:1123–34.
[160] Nan X, Ng H-H, Johnson CA, Laherty CD, Turner BM, Eisenman RN, et al. Transcriptional repression by the methyl-CpG-binding protein MeCP2 involves a histone deacetylase complex. Nature 1998;393:386–9.
[161] Ng HH, Zhang Y, Hendrich B, Johnson CA, Turner BM, Erdjument-Bromage H, et al. MBD2 is a transcriptional repressor belonging to the MeCP1 histone deacetylase complex. Nat Genet 1999;23:58–61.
[162] Klose RJ, Bird AP. Genomic DNA methylation: the mark and its mediators. Trends Biochem Sci 2006;31:89–97.
[163] Bestor TH. The DNA methyltransferases of mammals. Hum Mol Genet 2000;9:2395–402.
[164] Grandjean V, Yaman R, Cuzin F, Rassoulzadegan M. Inheritance of an epigenetic mark: The CpG DNA methyltransferase 1 is required for de novo establishment of a complex pattern of non-CpG methylation. PLoS One 2007;2:e1136.
[165] Feltus FA, Lee EK, Costello JF, Plass C, Vertino PM. Predicting aberrant CpG island methylation. Proc Natl Acad Sci USA 2003;100:12253–8.
[166] Jair KW, Bachman KE, Suzuki H, Ting AH, Rhee I, Yen RWC, et al. De novo CpG island methylation in human cancer cells. Cancer Res 2006;66:682–92.
[167] Ito S, D'Alessio AC, Taranova OV, Hong K, Sowers LC, Zhang Y. Role of Tet proteins in 5mC to 5hmC conversion, ES-cell self-renewal and inner cell mass specification. Nature 2010;466:1129–33.
[168] Tahiliani M, Koh KP, Shen Y, Pastor WA, Bandukwala H, Brudno Y, et al. Conversion of 5-methylcytosine to 5-hydroxymethylcytosine in mammalian DNA by MLL partner TET1. Science 2009;324:930–5.

[169] Sutcliffe JS, Nelson DL, Zhang F, Pieretti M, Caskey CT, Saxe D, et al. DNA methylation represses FMR-1 transcription in fragile X syndrome. Hum Mol Genet 1992;1:397–400.
[170] Amir RE, Van den Veyver IB, Wan M, Tran CQ, Francke U, Zoghbi HY. Rett syndrome is caused by mutations in X-linked MECP2, encoding methyl-CpG-binding protein 2. Nat Genet 1999;23:185–8.
[171] Chen WG, Chang Q, Lin Y, Meissner A, West AE, Griffith EC, et al. Derepression of BDNF transcription involves calcium-dependent phosphorylation ofMeCP2. Science 2003;302:885–9.
[172] Levenson JM, Roth TL, Lubin FD, Miller CA, Huang IC, Desai P, et al. Evidence that DNA (cytosine-5) methyltransferase regulates synaptic plasticity in the hippocampus. J Biol Chem 2006;281:15763–73.
[173] Miller CA, Sweatt JD. Covalent modification of DNA regulates memory formation. Neuron 2007;53: 857–69.
[174] Feng J, Zhou Y, Campbell SL, Le T, Li E, Sweatt JD, et al. Dnmt1 and Dnmt3a are required for the maintenance of DNA methylation and synaptic function in adult forebrain neurons. Nat Neurosci 2010;13:423–30.
[175] Grayson DR, Guidotti A. The dynamics of DNA methylation in schizophrenia and related psychiatric disorders. Neuropsychopharmacology 2013;38:138–66.
[176] Zhubi A, Cook EH, Guidotti A, Grayson DR. Epigenetic mechanisms in autism spectrum disorder. Int Rev Neurobiol 2014;115:203–44.
[177] Zhou Z, Hong EJ, Cohen S, Zhao WN, Ho HYH, et al. Brain-specific phosphorylation of MeCP2 regulates activity-dependent Bdnf transcription, dendritic growth, and spine maturation. Neuron 2006;52:255–69.
[178] Martinowich K, Hattori D, Wu H, Fouse S, He F, Hu Y, et al. DNA methylation-related chromatin remodeling in activity-dependent *Bdnf* gene regulation. Science 2003;302:890–3.
[179] Pastor V, Host L, Zwiller J, Bernabeu R. Histone deacetylase inhibition decreases preference without affecting aversion for nicotine. J Neurochem 2011;116:636–45.
[180] Ficz G, Branco MR, Seisenberger S, Santos F, Krueger F, Hore TA, et al. Dynamic regulation of 5-hydroxymethylcytosine in mouse ES cells and during differentiation. Nature 2011;473:398–402.
[181] Mellen M, Ayata P, Dewell S, Kriaucionis S, Heintz N. MeCP2 binds to 5hmc enriched within active genes and accessible chromatin in the nervous system. Cell 2012;151:1417–30.
[182] Song CX, Szulwach KE, Fu Y, Dai Q, Yi C, Li X, et al. Selective chemical labeling reveals the genome-wide distribution of 5-hydroxymethylcytosine. Nat Biotechnol 2011;29:68–72.
[183] Yu M, Hon GC, Szulwach KE, Song CX, Zhang L, Kim A, et al. Base-resolution analysis of 5-hydroxymethylcytosine in the mammalian genome. Cell 2012;149:1368–80.
[184] Kaas GA, Zhong C, Eason DE, Ross DL, Vachhani RV, Ming GL, et al. TET1 controls CNS 5-methylcytosine hydroxylation, active DNA demethylation, gene transcription, and memory formation. Neuron 2013;79:1086–93.
[185] Rudenko A, Dawlaty MM, Seo J, Cheng AW, Meng J, Le T, et al. Tet1 isocritical for neuronal activity-regulated gene expression and memory extinction. Neuron 2013;79:1109–22.
[186] Barreto G, Schafer A, Marhold J, Stach D, Swaminathan SK, Handa V, et al. Gadd45a promotes epigenetic gene activation by repair-mediated DNA demethylation. Nature 2007;445:671–5.
[187] Ma DK, Jang MH, Guo JU, Kitabatake Y, Chang ML, Pow-Anpongkul N, et al. Neuronal activity-induced Gadd45b promotes epigenetic DNA demethylation and adult neurogenesis. Science 2009;323:1074–7.
[188] Sultan FA, Wang J, Tront J, Liebermann DA, Sweatt JD. Genetic deletion of gadd45b, a regulator of active DNA demethylation, enhances long-term memory and synaptic plasticity. J Neurosci 2012;32:17059–66.
[189] Moonat S, Sakharkar AJ, Zhang H, Pandey SC. The role of amygdaloid brain-derived neurotrophic factor, activity-regulated cytoskeleton-associated protein and dendritic spines in anxiety and alcoholism. Addict Biol 2011;16:238–50.
[190] Williams JN Jr, Nichol CA, Elvehjem CA. Relation of dietary folic acid and vitamin B12 to enzyme activity in the chick. J Biol Chem 1949;180:689–94.
[191] Hamid A, Wani NA, Kaur J. New perspectives on folate transport in relation to alcoholism-induced folate malabsorption–association with epigenome stability and cancer development. FEBS J 2009;276:2175–91.

[192] Barbier E, Tapocik JD, Juergens N, Pitcairn C, Borich A, Schank JR, et al. DNA methylation in the medial prefrontal cortex regulates alcohol-induced behavior and plasticity. J Neurosci 2015;35:6153–64.
[193] Ravindran CR, Ticku MK. Changes in methylation pattern of NMDA receptor NR2B gene in cortical neurons after chronic ethanol treatment in mice. Brain Res Mol Brain Res 2004;121:19–27.
[194] Ravindran CR, Ticku MK. Effect of 5-azacytidine on the methylation aspects of NMDA receptor NR2B gene in the cultured cortical neurons of mice. Neurochem Res 2009;34:342–50.
[195] Qiang M, Denny AD, Ticku MK. Chronic intermittent ethanol treatment selectively alters *N*-methyl-D-aspartate receptor subunit surface expression in cultured cortical neurons. Mol Pharmacol 2007;72:95–102.
[196] Qiang M, Denny A, Chen J, Ticku MK, Yan B, Henderson G. The sites specific demethylation in the 5'-regulatory area of NMDA receptor 2B subunit gene associated with CIE-induced up-regulation of transcription. PLoS One 2010;5:e8798.
[197] Guidotti A, Dong E, Gavin DP, Veldic M, Zhao W, Bhaumik DK, et al. DNA methylation/demethylation network expression in psychotic patients with a history of alcohol abuse. Alcohol Clin Exp Res 2013;37:417–24.
[198] Autry AE, Monteggia LM. Brain-derived neurotrophic factor and neuropsychiatric disorders. Pharmacol Rev 2012;64:238–58.
[199] Joe KH, Kim YK, Kim TS, Roh SW, Choi SW, Kim YB, et al. Decreased plasma brain-derived neurotrophic factor levels in patients with alcohol dependence. Alcohol Clin Exp Res 2007;31:1833–8.
[200] Zanardini R, Fontana A, Pagano R, Mazzaro E, Bergamasco F, Romagnosi G, et al. Alterations of brain-derived neurotrophic factor serum levels in patients with alcohol dependence. Alcohol Clin Exp Res 2011;35:1529–33.
[201] Kerns RT, Ravindranathan A, Hassan S, Cage MP, York T, Sikela JM, et al. Ethanol-responsive brain region expression networks: implications for behavioral responses to acute ethanol in DBA/2J versus C57BL/6J mice. J Neurosci 2005;25:2255–66.
[202] Jeanblanc J, He DY, Carnicella S, Kharazia V, Janak PH, Ron D. Endogenous BDNF in the dorsolateral striatum gates alcohol drinking. J Neurosci 2009;29:13494–502.
[203] Wolstenholme JT, Warner JA, Capparuccini MI, Archer KJ, Shelton KL, Miles MF. Genomic analysis of individual differences in ethanol drinking: evidence for non-genetic factors in C57BL/6 mice. PloS One 2011;6:e21100.
[204] Gavin DP, Kusumo H, Zhang H, Guidotti A, Pandey SC. Role of growth arrest and DNA damage-inducible, beta in alcohol-drinking behaviors. Alcohol Clin Exp Res 2016;40(2):263–72.
[205] Murawski NJ, Moore EM, Thomas JD, Riley EP. Advances in diagnosis and treatment of fetal alcohol spectrum disorders. Alcohol Res 2015;37:97–108.
[206] Zhou D, Lebel C, Lepage C, Rasmussen C, Evans A, Wyper K, et al. Developmental cortical thinning in fetal alcohol spectrum disorders. Neuroimage 2011;58:16–25.
[207] Biermann T, Reulbach U, Lenz B, Frieling H, Muschler M, Hillemacher T, et al. *N*-methyl-D-aspartate 2b receptor subtype (NR2B) promoter methylation in patients during alcohol withdrawal. J Neural Transm (Vienna) 2009;116:615–22.
[208] Hillemacher T, Frieling H, Hartl T, Wilhelm J, Kornhuber J, Bleich S. Promoter specific methylation of the dopamine transporter gene is altered in alcohol dependence and associated with craving. J Psychiatr Res 2009;43:388–92.
[209] Zhang J, Xing B, Song J, Zhang F, Nie C, Jiao L, et al. Associated analysis of DNA methylation for cancer detection using CCP-based FRET technique. Anal Chem 2013;86:346–50.
[210] Zhao Y, Zhou H, Ma K, Sun J, Feng X, Geng J, et al. Abnormal methylation of seven genes and their associations with clinical characteristics in early stage non-small cell lung cancer. Oncol Lett 2013;5:1211–8.
[211] Bleich S, Degner D, Sperling W, Bonsch D, Thurauf N, Kornhuber J. Homocysteine as a neurotoxin in chronic alcoholism. Prog Neuropsychopharmacol Biol Psychiatry 2004;28:453–64.

[212] Bleich S, Lenz B, Ziegenbein M, Beutler S, Frieling H, Kornhuber J, et al. Epigenetic DNA hypermethylation of the HERP gene promoter induces downregulation of its mRNA expression in patients with alcohol dependence. Alcohol Clin Exp Res 2006;30:587–91.
[213] Bonsch D, Lenz B, Kornhuber J, Bleich S. Homocysteineassociated genomic DNA hypermethylation in patients with chronicalcoholism. J Neural Trans (Vienna) 2004;111:1611–6.
[214] Bonsch D, Lenz B, Kornhuber J, Bleich S. DNA hypermethylationof the alpha synuclein promoter in patients with alcoholism. Neuroreport 2005;16:167–70.
[215] Jasiewicz A, Rubiś B, Samochowiec J, Małecka I, Suchanecka A, Jabłoński M, et al. DAT1 methylation changes in alcohol-dependent individuals vs. controls. J Psychiatr Res 2015;64:130–3.
[216] Ruggeri B, Nymberg C, Vuoksimaa E, Lourdusamy A, Wong CP, Carvalho FM, et al. Association of protein phosphatase PPM1G with alcohol use disorder and brain activity during behavioral control in a genome-wide methylation analysis. Am J Psychiatry 2015;172:543–52.
[217] Philibert RA, Penaluna B, White T, Shires S, Gunter T, Liesveld J. A pilot examination of the genome-wide DNA methylation signatures of subjects entering and exiting short-term alcohol dependence treatment programs. Epigenetics 2014;9:1212–9.
[218] Lubin FD, Roth TL, Sweatt JD. Epigenetic regulation of BDNF gene transcriptionin the consolidation of fear memory. J Neurosci 2008;28:10576–86.
[219] Anier K, Malinovskaja K, Aonurm-Helm A, Zharkovsky A, Kalda A. DNA methylation regulates cocaine-induced behavioral sensitization in mice. Neuropsychopharmacology 2010;35(12):2450–61.
[220] Han J, Li Y, Wang D, Wei C, Yang X, Sui N. Effect of 5-aza-2-deoxycytidine microinjecting into hippocampus and prelimbic cortex on acquisition and retrieval of cocaine-induced place preference in C57BL/6 mice. Eur J Pharmacol 2010;642:93–8.
[221] Brueckner B, Boy RG, Siedlecki P, Musch T, Kliem HC, Zielenkiewicz P, et al. Epigenetic reactivation of tumor suppressor genes by a novel small-molecule inhibitor of human DNA methyltransferases. Cancer Res 2005;65:6305–11.
[222] LaPlant Q, Vialou V, Covington HE, Dumitriu D, Feng J, Warren BL, et al. Dnmt3a regulates emotional behavior and spine plasticity in the nucleus accumbens. Nat Neurosci 2010;13:1137–43.
[223] Pulipparacharuvil S, Renthal W, Hale CF, Taniguchi M, Xiao G, Kumar A, et al. Cocaine regulates MEF2 to control synaptic and behavioral plasticity. Neuron 2008;59:621–33.
[224] Im HI, Hollander JA, Bali P, Kenny PJ. MeCP2 controls BDNF expression and cocaine intake through homeostatic interactions with microRNA-212. Nat Neurosci 2010;13:1120–7.
[225] Spear LP. Adolescent neurobehavioral characteristics, alcohol sensitivities, and intake: Setting the stage for alcohol use disorders? Child Dev Perspect 2011;5:231–8.
[226] Pascual M, Boix J, Felipo V, Guerri C. Repeated alcohol administration during adolescence causes changes in the mesolimbic dopaminergic and glutamatergic systems and promotes alcohol intake in the adult rat. J Neurochem 2009;108:920–31.
[227] Guidotti A, Grayson DR. DNA methylation and demethylation as targets for antipsychotic therapy. Dialogues Clin Neurosci 2014;16:419–29.
[228] French SW. Epigenetic events in liver cancer resulting from alcoholic liver disease. Alcohol Res 2013;35: 57–67.
[229] Varela-Rey M, Woodhoo A, Martinez-Chantar ML, Mato JM, Lu SC. Alcohol, DNA methylation, and cancer. Alcohol Res 2013;35:25–35.
[230] Garro AJ, McBeth DL, Lima V, Lieber CS. Ethanol consumption inhibits fetal DNA methylation in mice: Implications for the fetal alcohol syndrome. Alcohol Clin Exp Res 1991;15:395–8.
[231] Kutay H, Klepper C, Wang B, Hsu SH, Datta J, Yu L, et al. Reduced susceptibility of DNA methyltransferase 1 hypomorphic (Dnmt1N/ +) mice to hepaticsteatosis upon feeding liquid alcohol diet. PLoS One 2012;7:e41949.
[232] Bartel DP. MicroRNAs: genomics, biogenesis, mechanism, and function. Cell 2004;116:281–97.

[233] Cheng LC, Pastrana E, Tavazoie M, Doetsch F. miR-124 regulates adult neurogenesis in the subventricular zone stem cell niche. Nat Neurosci 2009;12:399–408.
[234] Kosik KS. The neuronal microRNA system. Nat Rev Neurosci 2006;7:911–20.
[235] Miller BH, Wahlestedt C. MicroRNA dysregulation in psychiatric disease. Brain Res 2010;1338:89–99.
[236] Stark KL, Xu B, Bagchi A, Lai WS, Liu H, Hsu R, et al. Altered brain microRNA biogenesis contributes to phenotypic deficits in a 22q11-deletion mouse model. Nat Genet 2008;40:751–60.
[237] Charrier A, Chen R, Chen L, Kemper S, Hattori T, Takigawa M, et al. Connective tissue growth factor (CCN2) and microRNA-21 are components of a positive feedback loop in pancreatic stellate cells (PSC) during chronic pancreatitis and are exported in PSC-derived exosomes. J Cell Commun Signal 2014;8: 147–56.
[238] Chen YP, Jin X, Xiang Z, Chen SH, Li YM. Circulating microRNAs as potential biomarkers for alcoholic steatohepatitis. Liver Int 2013;33:1257–65.
[239] Francis H, McDaniel K, Han Y, Liu X, Kennedy L, Yang F, et al. Regulation of the extrinsic apoptotic pathway by microRNA-21 in alcoholic liver injury. J Biol Chem 2014;289(40):27526–39.
[240] Tsai PC, Bake S, Balaraman S, Rawlings J, Holgate RR, Dubois D, et al. MiR-153 targets the nuclear factor-1 family and protects against teratogenic effects of ethanol exposure in fetal neural stem cells. Biol Open 2014;3:741–58.
[241] Sathyan P, Golden HB, Miranda RC. Competing interactions between micro-RNAs determine neural progenitor survival and proliferation after ethanol exposure: evidence from an ex vivo model of the fetal cerebral cortical neuroepithelium. J Neurosci 2007;27:8546–57.
[242] Mantha K, Laufer BI, Singh SM. Molecular changes during neurodevelopment following second-trimester binge ethanol exposure in a mouse model of fetal alcohol spectrum disorder: from immediate effects to long-term adaptation. Dev Neurosci 2014;36:29–43.
[243] Pappalardo-Carter DL, Balaraman S, Sathyan P, Carter ES, Chen WJA, Miranda RC. Suppression and epigenetic regulation of MiR-9 contributes to ethanol teratology: evidence from zebrafish and murine fetal neural stem cell models. Alcohol Clin Exp Res 2013;37:1657–67.
[244] Qi X, Yang X, Chen S, He X, Dweep H, Guo M, et al. Ochratoxin A induced early hepatotoxicity: new mechanistic insights from microRNA, mRNA and proteomic profiling studies. Sci Rep 2014;4.
[245] Tal TL, Franzosa JA, Tilton SC, Philbrick KA, Iwaniec UT, Turner RT, et al. MicroRNAs control neurobehavioral development and function in zebrafish. FASEB J 2012;26:1452–61.
[246] Wang LL, Zhang Z, Li Q, Yang R, Pei X, Xu Y, et al. Ethanol exposure induces differential microRNA and target gene expression and teratogenic effects which can be suppressed by folic acid supplementation. Hum Reprod 2009;24:562–79.
[247] Pietrzykowski AZ, Friesen RM, Martin GE, Puig SI, Nowak CL, Wynne PM, et al. Posttranscriptional regulation of BK channel splice variant stability by miR-9 underlies neuroadaptation to alcohol. Neuron 2008;59:274–87.
[248] Guo Y, Chen Y, Carreon S, Qiang M. Chronic intermittent ethanol exposure and its removal induce a different miRNA expression pattern in primary cortical neuronal cultures. Alcohol Clin Exp Res 2012;36: 1058–66.
[249] Steenwyk G, Janeczek P, Lewohl JM. Differential effects of chronic and chronic-intermittent ethanol treatment and its withdrawal on the expression of miRNAs. Brain Sci 2013;3:744–56.
[250] Lewohl JM, Nunez YO, Dodd PR, Tiwari GR, Harris RA, Mayfield RD. Upregulation of microRNAs in brain of human alcoholics. Alcohol Clin Exp Res 2011;35:1928–37.
[251] Yadav S, Pandey A, Shukla A, Talwelkar SS, Kumar A, Pant AB, Parmar D. miR-497 and miR-302b regulate ethanol-induced neuronal cell death through BCL2protein and cyclin D2. J Biol Chem 2011;286:37347–57.
[252] Tapocik JD, Solomon M, Flanigan M, Meinhardt M, Barbier E, Schank JR, et al. Coordinated dysregulation of mRNAs and microRNAs in the rat medial prefrontal cortex following a history of alcohol dependence. Pharmacogenomics J 2013;13:286–96.

[253] Gorini G, Nunez YO, Mayfield RD. Integration of miRNA and protein profiling reveals coordinated neuroadaptations in the alcohol-dependent mouse brain. PLoS One 2013;8:e82565.
[254] Vlachos IS, Kostoulas N, Vergoulis T, Georgakilas G, Reczko M, Maragkakis M, et al. DIANA miR Path v.2. 0: Investigating the combinatorial effect of micro-RNAs in pathways. Nucleic Acids Res 2012;40: W498–504.
[255] Balaraman S, Winzer-Serhan UH, Miranda RC. Opposing actions of ethanol and nicotine on microRNAs are mediated by nicotinic acetylcholine receptors in fetal microRNAs and ethanol toxicity cerebral cortical-derived neural progenitor cells. Alcohol Clin Exp Res 2012;36:1669–77.
[256] Wei X, Li H, Miao J, Liu B, Zhan Y, Wu D, et al. miR-9*-and miR-124a-Mediated switching of chromatin remodelling complexes is altered in rat spina bifida aperta. Neurochem Res 2013;38:1605–15.
[257] Chiang HR, Schoenfeld LW, Ruby JG, Auyeung VC, Spies N, Baek D, et al. Mammalian microRNAs: experimental evaluation of novel and previously annotated genes. Genes Dev 2010;24:992–1009.
[258] Tapocik JD, Barbier E, Flanigan M, Solomon M, Pincus A, Pilling A, Sun H, Schank JR, King C, Heilig M. microRNA-206 in rat medial prefrontal cortex regulates BDNF expression and alcohol drinking. J Neurosci 2014;34:4581–8.
[259] Pan Z, Zhang M, Ma T, Xue ZY, Li GF, Hao LY, et al. Hydroxymethylation of microRNA-365-3p regulates nociceptive behaviors via Kcnh2. J Neurosci 2016;36:2769–81.
[260] Pando R, Even-Zohar N, Shtaif B, Edry L, Shomron N, Phillip M, et al. MicroRNAs in the growth plate are responsive to nutritional cues: association between miR-140 and SIRT1. J Nutr Biochem 2012;23:1474–81.
[261] Prins SA, Przybycien-Szymanska MM, Rao YS, Pak TR. Long-term effects of peripubertal binge EtOH exposure on hippocampal microRNA expression in the rat. PLoS One 2014;9:e83166.
[262] Zuo L, Lu L, Tan Y, Pan X, Cai Y, Wang X, et al. Genome-wide association discoveries of alcohol dependence. Am J Addict 2014;23:526–39.
[263] Zakhari S. Alcohol metabolism and epigenetics changes. Alcohol Res 2013;35:6–16.
[264] Ponomarev I, Wang S, Zhang L, Harris RA, Mayfield RD. Gene coexpression networks in human brain identify epigenetic modifications in alcohol dependence. J Neurosci 2012;32:1884–97.

SECTION

SUMMARY AND OUTLOOK

CHAPTER

NEUROPSYCHIATRIC DISORDERS AND EPIGENETICS: SUMMARY AND OUTLOOK

19

J. Peedicayil*, D.R. Grayson, D.H. Yasui†**

**Department of Pharmacology and Clinical Pharmacology, Christian Medical College, Vellore, Tamil Nadu, India; **Department of Psychiatry, College of Medicine, University of Illinois, Chicago, IL, United States; †University of California, Davis, CA, United States*

CHAPTER OUTLINE

19.1 INTRODUCTION

This book has discussed in detail the role of epigenetics in the pathogenesis of several neuropsychiatric disorders (Section II). In addition, the translational aspects of the role of epigenetics in some neuropsychiatric disorders, such as, cognitive disorders, pervasive developmental disorders, intellectual disability, and drug addiction have been discussed. In Section I, the role of environmental factors in the pathogenesis of neuropsychiatric disorders, and the potential value of epigenetic biomarkers in the management of some neuropsychiatric disorders were also discussed. In many of the disorders discussed, psychiatric symptoms are not the major symptoms of affected patients. Rather, the major symptoms are neurological. However, all the disorders covered in this book have an epigenetic component in their pathogenesis, and a psychiatric component in their clinical presentation. As is clear from Section II, we are presently in the early stages of investigating the role of epigenetics in the pathogenesis of most neuropsychiatric disorders. It is also apparent that translational aspects of epigenetics, such as epigenetic therapy, have great potential in the future in the clinical management of patients with neuropsychiatric disorders like cognitive disorders, multiple sclerosis, and brain tumors. This chapter

Neuropsychiatric Disorders and Epigenetics. http://dx.doi.org/10.1016/B978-0-12-800226-1.00019-8

discusses some potentially important aspects of the role of epigenetics in neuropsychiatric disorders, including possible future strategies in the study of the role of epigenetics in these disorders.

19.2 DIFFERING ROLES OF EPIGENETICS IN THE PATHOGENESIS OF DIFFERENT NEUROPSYCHIATRIC DISORDERS

The role of epigenetics in several neuropsychiatric disorders has been well characterized and elucidated. Such disorders have a major neurological component and include the following: (1) Fragile X syndrome (FXS), for which as far back as 1991, three studies identified, by positional cloning, that this disorder is due to the trinucleotide repeat CGG copy number instability in the *fragile X mental retardation-1* (*FMR1*) gene on the X chromosome [1–3]. Normally, the size of the repeat is from 7 to about 60. In most FXS patients, the repeat is greatly expanded to more than 230 repeats and becomes abnormally hypermethylated, leading to silencing of the *FMR1* gene [4]. More recently, it has been shown in peripheral blood mononuclear cells obtained from FXS patients that there are novel *FMR1* regions of chromatin and DNA methylation and hydroxymethylation associated with *FMR1* epigenetic silencing [5]. (2) Rett syndrome, where the majority of cases identified were due to mutations in the gene *MECP2* encoding X-linked methyl-CpG-binding protein 2 (MeCP2) in 1999 [6,7]. (3) Several types of brain tumors including gliomas and medulloblastomas for which epigenetic changes like those involving DNA methylation are being elucidated and are facilitating molecular diagnosis and classification of tumors [8–12]. For many neuropsychiatric disorders, like Rett syndrome, Huntington's disease, and FXS, there are already available standardized and established laboratory tests which are made use of in the clinical management of affected patients [13–15].

Unlike in the neuropsychiatric disorders discussed earlier, in other neuropsychiatric disorders which have a major psychiatric component, the epigenetic basis of the pathogenesis of disease has proved more elusive, although there appear to be many solid leads in the path to the elucidation of the underlying epigenetic abnormalities. Such disorders include the following: (1) schizophrenia, where the underlying epigenetic abnormalities are thought to include hypermethylation of the promoters of genes encoding reelin and GAD67 in GABAergic neurons in the prefrontal cortex [16]; hypomethylation of the gene encoding brain-derived neurotrophic factor (BDNF) in the frontal cortex [17]; and dysfunction of microRNAs (miRNAs) like miR-9-5p and hsa-miR-34a [18,19] in the brain. (2) Bipolar disorder, where hypermethylation of the promoters of genes encoding reelin and GAD67 in GABAergic neurons in the prefrontal cortex is thought to contribute to disease pathogenesis [16]; and where there is dysregulation of miRNAs like miR-499, miR-708, and miR-1908 in the brain [20]. (3) Major depressive disorder, where a decrease in DNA methylation of the BDNF gene [17], and abnormal levels of miRNAs like miR-185 and miR-491-3p, are thought to contribute to the disease [21]. To date, none of these or other findings on the role of epigenetics in the pathogenesis of schizophrenia, bipolar disorder, and major depressive disorder, have been translated into laboratory tests useful for the clinical management of patients with these disorders. Significantly, in these neuropsychiatric disorders, environmental factors, such as psychosocial factors have a major role in disease pathogenesis [22–25].

In the light of the above data, why have epigenetic studies on neuropsychiatric disorders with a greater neurological component like FXS and Rett syndrome been more successful than epigenetic studies on neuropsychiatric disorders with a greater psychiatric component like schizophrenia and major depressive disorder? The answer to this question probably lies in three interrelated facts: (1) the epigenetic studies of the neuropsychiatric disorders with a greater neurological component, as alluded to above, started earlier, that is, in the 1980s and 1990s whereas the epigenetic studies of neuropsychiatric disorders with a greater psychiatric component started in earnest later, in the first decade of the 21st century. (2) The neuropsychiatric disorders with a relatively great neurological component are biologically more simple. (3) There is a greater role played by environmental factors like psychosocial factors in the pathogenesis of neuropsychiatric disorders with a greater psychiatric component, as referred to earlier.

19.3 EPIGENETICS OF NONNEURONAL CELLS IN NEUROPSYCHIATRIC DISORDERS

To date, most studies investigating the role of epigenetics in the pathogenesis of neuropsychiatric disorders have included only neurons (or nerve cells). However, in addition to neurons, the brain also contains glia (or glial cells) which surround the cell bodies, axons, and dendrites of neurons [26,27]. Glia differ from neurons morphologically and do not form dendrites or axons. Glia also differ from neurons functionally in that they are not electrically excitable and are not involved in electrical signalling [27]. Glial cells are far more in number in the human brain: there are 2–10 times more glial cells than neurons [27]. Glial cells are of two major types: microglia and macroglia. Microglias are immune cells and are mobilized to present antigens and become phagocytes during injury, infection, or degenerative diseases. Regarding macroglia, there are three major types: oligodendrocytes, Schwann cells, and astrocytes. In the human brain, 80% of all glial cells are macroglia, about half of these being oligodendrocytes and the other half being astrocytes [27]. For the normal functioning of neurons in the brain, functional integration is required between neurons and the glial cells [28]. In addition to neurons, the development and functioning of glial cells are also subjected to epigenetic control [28–30]. In recent years, there is experimental evidence that abnormalities of epigenetic mechanisms of gene expression of glial cells can contribute to the pathogenesis of neuropsychiatric disorders like schizophrenia [31,32], major depressive disorder [33], autism and Rett syndrome [34], and multiple sclerosis [35]. In the light of these relatively recent developments, the role of epigenetic dysregulation of glial cells may have to be given more attention and importance in the study of the role of epigenetics in neuropsychiatric disorders.

19.4 LONG NONCODING RNAs AND NEUROPSYCHIATRIC DISORDERS

Noncoding RNAs (ncRNAs) are RNAs that are not translated into proteins. They have been classified recently into short ncRNAs that are less than 200 nucleotides long, and long ncRNAs (lncRNAs) that are greater than 200 nucleotides long [36]. To date, scores of studies have investigated the role of ncRNAs in the pathogenesis of neuropsychiatric disorders. However, these studies have mainly focused on one type of short ncRNA, miRNAs.

In recent years it has become clear that lncRNAs are the most common type of ncRNAs in mammals, and novel lncRNA genes are being identified at a rapid pace [37]. Mammalian genomes are now thought to encode tens of thousands of lncRNAs [38]. Moreover, as many as 40% of the lncRNAs are expressed specifically in the brain, where they show precisely regulated temporal and spatial expression patterns [38]. In addition to being involved in brain development, function and plasticity, lncRNAs are thought to contribute to diseases of the brain when they malfunction, and are potential targets for drug therapy [38–41]. Hence, in addition to ncRNAs like miRNAs, the role of lncRNAs in the pathogenesis of neuropsychiatric disorders may also need to be adequately investigated.

19.5 FUTURE STRATEGIES IN THE EPIGENETICS OF NEUROPSYCHIATRIC DISORDERS

The study of the role of epigenetics in the pathogenesis of neuropsychiatric disorders is exquisitely and particularly difficult, because not only is the human brain inordinately complex, but, it is also enclosed within the bony skull which is quiet impenetrable. However, human ingenuity may soon help get over the latter problem, thanks to a new field in science called optogenetics. Optogenetics is the combination of genetic and optical methods to achieve gain or loss of function of well-defined events in specific cells of a living tissue pioneered by the noted scientist Karl Deisseroth [42]. It is a technology that allows targeted, fast control of precisely defined events in biologic systems [42]. The general strategy of optogenetics involves introducing a light-sensitive protein to a specific cell type, illuminating the targeted cells with defined parameters, and obtaining a readout of the cellular behavior [43]. Optogenetics enables precise perturbation of distinct cell types of the brain based on molecular signatures, functional projections, and intracellular biochemical signalling pathways, and could help dissect the mechanisms underlying neuropsychiatric disorders [43], including epigenetic mechanisms [43,44].

Another new technology relevant to the epigenetics of neuropsychiatric disorders is epigenetic (and epigenomic) editing. This technology enables targeted epigenome manipulation at defined loci [45]. The technology is based on the fusion of a DNA recognition domain with a catalytic domain of a chromatin-modifying enzyme to generate targeted EpiEffectors which comprise designed DNA recognition domains and catalytic domains from the chromatin-modifying enzyme. The DNA recognition domain serves to bind a unique DNA sequence and deliver the annexed functional domain to defined target loci in the genome, where it can change the chromatin modification state. This way gene expression, cellular differentiation, and other biologic processes can be altered [45]. Epigenetic editing could help induce gene expression modulation and correct disease- associated epimutations [46]. It could also provide valuable insights into the issue of cause versus consequence of epigenetic changes like changes in DNA methylation and histone modifications in relation to disease pathogenesis [46].

ABBREVIATIONS

BDNF Brain-derived neurotrophic factor
FMR1 Fragile X mental retardation-1
FXS Fragile X syndrome
GAD67 Glutamic acid decarboxylase67
lncRNA Long non-coding RNA
miRNA MicroRNA
ncRNA Noncoding RNA

GLOSSARY

Glioma A malignant tumor arising from glial cells in the brain
Medulloblastoma A malignant tumor of the brain arising in the cerebellum, especially in children

REFERENCES

[1] Verkerk AJ, Pieretti M, Sutcliffe JS, Fu Y-H, Kuhl DPA, Pizzuti A, et al. Identification of a gene (*FMR-1*) containing a CGG repeat coincident with a breakpoint cluster region exhibiting length variation in fragile X syndrome. Cell 1991;65:905–14.

[2] Oberlé I, Rousseau F, Heitz D, Kretz C, Devys D, Hanauer A, et al. Instability of a 550-base pair DNA segment and abnormal methylation in fragile X syndrome. Science 1991;252:1097–102.

[3] Kremer EJ, Pritchard M, Lynch M, Yu S, Holman K, Baker E, et al. Mapping of DNA instability at the fragile X to a trinucleotide repeat sequence p(CCG)*n*. Science 1991;252:1711–4.

[4] Jin P, Warren ST. Understanding the molecular basis of fragile X syndrome. Hum Mol Genet 2000;9: 901–8.

[5] Brasa S, Mueller A, Jacquemont S, Hahne F, Rozenberg I, Peters T, et al. Reciprocal changes in DNA methylation and hydroxymethylation and a broad repressive epigenetic switch characterize *FMR1* transcriptional silencing in fragile X syndrome. Clin Epigenet 2016;8:15.

[6] Amir RE, Van den Veyver I, Wan M, Tran CQ, Francke U, Zoghbi HY. Rett syndrome is caused by mutations in X-linked *MECP2*, encoding methyl-CpG-binding protein 2. Nat Genet 1999;23:185–8.

[7] LaSalle JM, Yasui DH. Evolving role of MeCP2 in Rett syndrome and autism. Epigenomics 2009;1: 119–30.

[8] Huttner A. Overview of primary brain tumors: pathologic classification, epidemiology, molecular biology, and prognostic markers. Hematol Oncol Clin North Am 2012;26:715–32.

[9] Masui K, Cloughesy TF, Mischel PS. Molecular pathology in adult high-grade gliomas: from molecular diagnostics to target therapies. Neuropathol Appl Neurobiol 2012;38:271–91.

[10] Yong RL, Tsankova NM. Emerging interplay of genetics and epigenetics in gliomas: a new hope for targeted therapy. Semin Pediatr Neurol 2015;22:14–22.

[11] Samkari A, White JC, Packer RJ. Medulloblastoma: toward biologically based management. Semin Pediatr Neurol 2015;22:6–13.

[12] Sturm D, Orr BA, Toprak UH, Hovestadt V, Jones DT, Capper D, et al. New brain tumor entities emerge from molecular classification of CNS-PNETs. Cell 2016;164:1060–72.

[13] Kalman LV, Tarleton JC, Percy AK, Aradhya S, Bale S, Barker SD, et al. Development of a genomic DNA reference material panel for Rett syndrome (*MECP2*-related disorders) genetic testing. J Mol Diagn 2014;16:273–9.

[14] Saft C, Leavitt BR, Epplen JT. Clinical utility gene card for: Huntington's disease. Eur J Hum Genet 2014;22:e1–3.

[15] Amancio AP, de O, Melo CA, de M, Vieira A, Minasi LB, de M, e Silva D, da Silva CC, et al. Molecular analysis of patients suspected of fragile X syndrome. Genet Mol Res 2015;14:14660–9.

[16] Grayson DR, Guidotti A. The dynamics of DNA methylation in schizophrenia and related psychiatric disorders. Neuropsychopharmacology 2013;38:138–66.

[17] Mitchelmore C, Gede L. Brain derived neurotrophic factor: epigenetic regulation in psychiatric disorders. Brain Res 2014;1586:162–72.

[18] Hauberg ME, Roussos P, Grove J, Børglum AD, Mattheisen M. Schizophrenia Working Group of the Psychiatric Genomics Consortium. Analyzing the role of microRNAs in schizophrenia in the context of common genetic risk variants. JAMA Psychiatry 2016;73:369–77.

[19] Lai C-Y, Lee S-Y, Scarr E, Yu Y H, Lin Y T, Liu C-M, et al. Aberrant expression of microRNAs as biomarker for schizophrenia: from acute state to partial remission, and from peripheral blood to cortical tissue. Transl Psychiatry 2016;6:e717.
[20] Forstner AJ, Hofmann A, Maaser A, Sumer S, Khudayberdiev S, Mühleisen TW, et al. Genome- wide analysis implicates microRNAs and their target genes in the development of bipolar disorder. Transl Psychiatry 2015;5:e678.
[21] Serafini G, Pompili M, Hansen KF, Obrietan K, Dwivedi Y, Shomron N, Girardi P. The involvement of microRNAs in major depression, suicidal behavior, and related disorders: a focus on miR-185 and miR-491-3p. Cell Mol Neurobiol 2014;34:17–30.
[22] Peedicayil J. The importance of cultural inheritance in psychiatric genetics. Med Hypotheses 2002;58:164–6.
[23] Verghese A. The integration of psychiatry and neurology. Indian J Psychiatry 2016;58:104–5.
[24] Millan MJ. An epigenetic framework for neurodevelopmental disorders: from pathogenesis to potential therapy. Neuropharmacology 2013;68:2–82.
[25] Millan MJ, Andrieux A, Bartzokis G, Cadenhead K, Dazzan P, Fusar-Poli P, et al. Altering the course of schizophrenia: progress and perspectives. Nat Rev Drug Discov 2016;15:485–515.
[26] Götz M, Huttner WB. The cell biology of neurogenesis. Nat Rev Mol Cell Biol 2005;6:777–88.
[27] Kandel ER, Barres BA, Hudspeth AJ. Nerve cells, neural circuitry, and behavior. In: Kandel ER, Schwartz JH, Jessell TM, Siegelbaum SA, Hudspeth AJ, editors. Principles of neural science. 5th ed. New York: McGraw-Hill; 2013. p. 21–38.
[28] Yu Y, Casaccia P, Lu QR. Shaping the oligodendrocyte identity by epigenetic control. Epigenetics 2010;5: 124–8.
[29] Liu J, Casaccia P. Epigenetic regulation of oligodendrocyte identity. Trends Neurosci 2010;33:193–201.
[30] Mitchell A, Roussos P, Peter C, Tsankova N, Akbarian S. The future of neuroepigenetics in the human brain. Prog Mol Biol Transl Sci 2014;128:199–228.
[31] Goudriaan A, de Leeuw C, Ripke S, Hultman CM, Sklar P, Sullivan PF, et al. Specific glial functions contribute to schizophrenia susceptibility. Schizophr Bull 2014;40:925–35.
[32] Chen X-S, Huang N, Michael N, Xiao L. Advancements in the underlying pathogenesis of schizophrenia: implications of DNA methylation in glial cells. Front Neurosci 2015;9:1–8.
[33] Nagy C, Suderman M, Yang J, Szyf M, Mechawar N, Ernst C, Turecki G. Astrocyte abnormalities and global DNA methylation patterns in depression and suicide. Mol Psychiatry 2015;20:320–8.
[34] Maezawa I, Calafiore M, Wulff H, Jin LW. Does microglial dysfunction play a role in autism and Rett syndrome? Neuron Glia Biol 2011;7:85–97.
[35] Huynh JL, Garg P, Thin TH, Yoo S, Dutta R, Trapp BD, et al. Epigenome-wide differences in pathology-free regions of multiple sclerosis-affected brains. Nat Neurosci 2014;17:121–30.
[36] Allis CD, Caparros M-L, Jenuwein T, Lachner M, Reinberg D. Overview and concepts. In: Allis CD, Caparros M-L, Jenuwein T, Reinberg D, editors. Epigenetics. 2nd ed. New York: Cold Spring Harbor Laboratory Press; 2015. p. 47–115.
[37] Volders P-J, Verheggen K, Menschaert G, Vandepoele K, Martens L, Vandesompele J, et al. An update on LNCipedia: a database for annotated human lncRNA sequences. Nucleic Acids Res 2015;43:D174–80.
[38] Briggs JA, Wolvetang EJ, Mattick JS, Rinn JL, Barry G. Mechanisms of long non-coding RNAs in mammalian nervous system development, plasticity, disease, and evolution. Neuron 2015;88:861–77.
[39] Qureshi IA, Mehler MF. Emerging roles of non-coding RNAs in brain evolution, development, plasticity and disease. Nat Neurosci 2012;13:528–41.
[40] Qureshi IA, Mehler MF. Long non-coding RNAs: novel targets for nervous system disease diagnosis and therapy. Neurotherapeutics 2013;10:632–46.
[41] Roberts TC, Morris KV, Wood MJA. The role of long non-coding RNAs in neurodevelopment, brain function and neurological disease. Philos Trans R Soc B 2014;369:20130507.
[42] Deisseroth K. Optogenetics. Nat Methods 2011;8:26–9.

[43] Mei Y, Zhang F. Molecular tools and approaches for optogenetics. Biol Psychiatry 2012;71:1033–8.
[44] Konermann S, Brigham S, Trevino AE, Hsu PD, Heidenreich M, Cong L, et al. Optical control of mammalian endogenous transcription and epigenetic states. Nature 2013;500:472–6.
[45] Kungulovski G, Jeltsch A. Epigenome editing: state of the art, concepts, and perspectives. Trends Genet 2016;32:101–13.
[46] de Groote ML, Verschure PJ, Rots MG. Epigenetic editing: targeted rewriting of epigenetic marks to modulate expression of selected target genes. Nucleic Acids Res 2012;40:10596–613.

Index

A

D

E

I

J

K

L

M

N

O

R

S

T

Printed in the United States
By Bookmasters